Tree Rings and Climate

Trees and stones will teach you that which
You can never learn from masters.

St. Bernard of Clairvaux

Cross section from a drought-affected *Pseudotsuga menziesii* (Douglas-fir) which began growth in A.D. 1113 and was cut in A.D. 1240. From Lion House, Johnson Canyon, Colorado.

Tree Rings and Climate

H. C. FRITTS

Laboratory of Tree-Ring Research
University of Arizona, Tucson, Arizona, U.S.A.

1976

ACADEMIC PRESS
LONDON   NEW YORK   SAN FRANCISCO
A Subsidiary of Harcourt Brace Jovanovich, Publishers

ACADEMIC PRESS INC. (LONDON) LTD.
24/28 Oval Road,
London NW1

United States Edition published by
ACADEMIC PRESS INC.
111 Fifth Avenue
New York, New York 10003

Copyright © 1976 by
ACADEMIC PRESS INC. (LONDON) LTD.
Second printing 1978

All Rights Reserved
No part of this book may be reproduced in any form by photostat, microfilm, or any other means, without written permission from the publishers

Library of Congress Catalog Card Number: 76-1079
ISBN: 0 12 268450-8

Printed in Great Britain by
Willmer Brothers Limited, Birkenhead

Preface

The organization, information, and principles in this volume represent my views on the current status of the interdisciplinary field of dendroclimatology. This science is broadly defined to include tree-ring studies involving climate-related problems which touch on such diverse fields as archaeology, hydrology, oceanography, biology, forestry, and geophysics. Subject matter is presented in a set of graded chapters, starting with the basic biological facts and principles of tree growth, developing important quantitative methods, and ending with examples of reconstructing past climate.

Chapter 1 introduces and defines the most important terms, principles, and concepts of the science; chapters 2 through 4 cover the basic biology governing the response of ring width to variation in climate; and chapter 5 describes details of the climatic and environmental system and presents models of the growth–climatic relationships. Chapters 6 and 7 treat basic statistics and methods of analysis of these relationships; and chapters 8 and 9 describe various examples and applications using the methods described earlier in the text.

The reader is assumed to be at least a third-year college student with some experience and background in at least one science. However, most technical terms are described when they are first mentioned, and for purposes of review, definitions are provided in the Glossary at the end of the text. Because of the international interest, scientific names of tree species are used throughout, although they are listed with the common names in the Appendix, for which the following are among the references consulted: Bailey (1970); Kelsey and Dayton (1942); Little (1953); and Willis (1973).

The origins of this book can be traced to my graduate studies on forest tree growth. They began with the development of a dendrograph which was used to record reversible variations in stem size as well as the cumulative variations representing radial growth. At that time I was skeptical of tree-ring dating and of the possibility for climatic inference because I had read criticisms of the science by Waldo Glock (1955), Sampson (1940), and Sampson and Glock (1942).

Therefore, after joining the staff at the Laboratory of Tree-Ring Research in 1960, I embarked on an extensive program to examine and test the fundamental basis of such work. As it turned out, all major principles and concepts originally formulated by the founders of the discipline proved correct. Results of my tests and the botanical basis for them are treated in this book.

During the first years at the Laboratory of Tree-Ring Research, David G. Smith, Marvin A. Stokes, and a number of students worked with me on various aspects of these tests. Studies were conducted in the San Francisco Mountains of Arizona on the effects of moisture stress on ring width, and at Mesa Verde, Colorado, on the biological basis and climatological relationships governing the ring-width response. Subsequent investigations, which involved other students and associates, focused on *Pinus longaeva* (bristlecone pine) in the White Mountains of California and *Pinus ponderosa* (ponderosa pine) in the Santa Catalina Mountains north of Tucson, Arizona.

As the principles of dendrochronology became more firmly established, attention was increasingly directed to the utilization of climatic information in tree-ring widths. A resultant paper (Fritts, 1965) is the first major effort to analyze both the spatial and temporal variations in growth and to use them to infer variations in climate. Increased experience with statistics lead to the most important single discovery of my work which was made while on sabbatical leave at the University of Wisconsin in the spring of 1969. It involved the use of eigenvectors along with multiple regression and canonical analysis to calibrate climate and ring widths for objectively reconstructing past climate. At the same time Webb and Bryson (1972) were applying the technique to pollen analysis; Imbrie, several years earlier (Imbrie and Kipp, 1971), had applied similar methods of factor analysis to calibrate marine fossil data with sea-surface temperatures and to reconstruct past climate. T. J. Blasing, Bruce P. Hayden, and John E. Kutzbach joined me in the formal development and documentation of the method, and they are responsible for the precision and clarity of the mathematical descriptions in the paper that resulted from this work (see Fritts *et al.*, 1971).

There are physiological and ecological details of tree growth which, according to our present knowledge, appear to have little bearing on the year-to-year variations in ring width, and are therefore outside the scope of this book. These include some factors that *could* have an important influence on growth, but which have not as yet been proven to exert a significant effect. Other factors which are highly influential receive more attention than is usually given to them in standard texts. For example, genetic variation, soil factors, and many forest influences are considered only briefly because these particular factors are more or less constant for a particular tree and site, and they do not contribute markedly to the year-to-year variations in ring width. On the other hand, factors affecting diurnal, seasonal, and yearly variations in stem growth receive more than the usual attention because the net effect of these variations can obviously contribute markedly to the variations in the resultant ring width. In short, the emphasis in the biological discussion is upon those factors and conditions that appear most relevant to tree-ring research.

It is recognized, however, that future studies of the growth–climate system will undoubtedly lead to new possibilities and refinements in which factors originally thought to be irrelevant will turn out to have a measurable effect. When this occurs, the newly discovered factors must be added to the list of others already considered important.

In addition, I would like to emphasize that this particular text is essentially a progress report on a rapidly developing subject. I hope no reader will consider these pages as unquestionable statements of fact. Rather, they should be viewed as a treatment of best inferences of dendroclimatology as we know it, including applications, conclusions, and principles that are *reasonable deductions from the accumulated facts*.

The treatment of statistics is equally pragmatic in that only the most pertinent techniques are included in this text. The juxtaposition of the biological and statistical discussions is a result of my firm belief that biological insight is essential for obtaining the most meaningful statistics, and that statistical methods are similarly important for the accurate quantifying and testing of the biological results.

Since this volume does not attempt to give a comprehensive review of all tree-growth research, the bibliography is designed to direct the reader to the most recent data sources and to provide more in-depth reading for the interested scientist. Because a large part of the text represents my own experience, the reference choices are heavily weighted towards my own work and the work of my close colleagues and students. I am well aware of the fact that important work has been accomplished elsewhere, both inside and outside the United States, and I regret that the focus on my own views and experiences excluded valuable studies from being individually cited in the book. I hope the texts and review articles that are included will assist the reader in finding some of the omitted works.

I am indebted to colleagues, friends, and students who have played a very important role in the development of this discipline as described in the text. The present and past staff of the Laboratory of Tree-Ring Research should be mentioned at the outset, as well as the many colleagues in botany and dendrochronology scattered throughout the world who have participated, either directly or indirectly, in this work.

Some of the most important contributors appear as authors and co-authors of papers cited in the text.

The following people deserve special mention for their important contributions to the development, preparation, and critical review of this book. First and foremost are Karen Babcock McDougall and Judith A. Sherwood, who most ably carried out many editorial responsibilities during the three most important years of the preparation of this work. Others who shared this responsibility are Nelle W. Noble, Marna C. Ares, and Emily De Witt. Marilyn J. Huggins is responsible for most of the scientific drawings and figures in the text.

A number of scientists made specific contributions in addition to their publications listed in the references. The most important of these are T. J. Blasing, Charles W. Stockton, and Valmore C. LaMarche, Jr. Prodyot K. Bhattacharya and James R. Gebart advised me on various statistical problems; William D. Sellers made constructive suggestions regarding the energy and water balance, and Fiorenzo C. Ugolini assisted with the section on soils development. Others, among them Martha A. Wiseman, G. Robert Lofgren, C. Larrabee Winter, and Linda G. Drew, also contributed to the scientific content.

Many of the above, as well as others, have read at least portions of the manuscript and have made numerous helpful suggestions and comments. Special thanks are expressed to Linda G. Brubaker, David J. Shatz, T. J. Blasing, G. Robert Lofgren, Martha A. Wiseman, Thomas Nash III, and Miriam Colson, whose reviews were most helpful in the editing of the text.

Various unpublished materials represent work funded by NOAA Grants E-231-68(G) and E-41-70(N), NOAA contract 1-352441. National Science Foundation Grants GB-1025, GB-3658, GA-26581, and ATM75-17534, and AFOSR Grant 72-2406. The opportunity to undertake this particular volume was made possible by a sabbatical leave from the University of Arizona and by a John Simon Guggenheim Fellowship for 1968–1969. Facilities of the University of Arizona Computer Center were used both for statistical analyses and for preparation of the manuscript.

HAROLD C. FRITTS

September, 1975

Contents

Preface v
Dedication xiii

Chapter 1. Dendrochronology and Dendroclimatology

    I. Introduction 1
    II. Historical Background 4
    III. Scope of Dendrochronology 10
    IV. Some Basic Principles and Concepts of Dendrochronology 14
    V. Dendroclimatic Procedures and Analyses . . . 28
    VI. Examples of Analysis 38
    VII. Further Definitions and Concepts 46
    VIII. The Climate–Growth System and its Dendroclimatological Significance 52

Chapter 2. Growth and Structure

    I. Introduction 55
    II. Gross Structure 56
    III. The Vascular Cambium 63
    IV. Growth—A Variable Process 68
    V. Variations in Shoot Growth 70
    VI. Variations in Radial Growth 74
    VII. Phenology and its Relation to Ring Growth . . 103
    VIII. Systematic Variations in the Width and Cell Structure of Annual Rings 107
    IX. Growth of Roots 113
    X. Significance of Growth and Structure to Dendroclimatology 115

Chapter 3. Basic Physiological Processes: Movement of Materials and Water Relations

I.	Limiting Factors and Plant Processes	118
II.	Some General Terms and Basic Concepts	120
III.	Cell Water Status	123
IV.	Transpiration	127
V.	Soil Moisture	133
VI.	The Soil System and Factors Affecting its Development	136
VII.	Absorption of Water	140
VIII.	Internal Water Relations of Trees	142
IX.	Moisture Stress and Tree Form	147
X.	Soil Factors Affecting Root Growth, Ring-width Sensitivity, and Longevity	148
XI.	Uptake of Mineral Salts	151
XII.	Translocation	152

Chapter 4. Basic Physiological Processes: Food Synthesis and Assimilation of Cell Constituents

I.	Introduction	156
II.	Photosynthesis and Respiration	157
III.	Synthesis of Foods and Assimilation	159
IV.	Measurement of Photosynthesis and Respiration	162
V.	Factors Affecting Photosynthesis and Respiration	163
VI.	The Annual Net Photosynthetic Regime for a *Pinus Ponderosa* on a Semiarid site	178
VII.	Some Implications of these Physiological Measurements	182
VIII.	The Distribution of Foods and Interactions with Growth	184
IX.	Essential Mineral Salts	189
X.	Growth-regulating Substances	190
XI.	Physiological Preconditioning and Correlating Systems	193
XII.	Changes in the Physiological Seasons with Varying Elevation of the Tree Sites	202

Chapter 5. The Climate–Growth System

I.	Introduction	207
II.	The Energy and Water Balances	208
III.	Site Factors which can Modify the Energy Balance	213

IV. Biotic and Other Nonclimatic Factors: Dendrochronological Examples 219
V. Modeling Relationships in the Ring-width and Climatic System 223
VI. A Model for Factors Affecting Cambial Activity and hence Ring Width 226
VII. Modeling the Effects of Temperature and Precipitation on Ring Width 231
VIII. The Concept of the Climatic "Window" . . . 238
IX. The Concept of the Response Function 240
X. Suitability and Limitations of the Growth Model . . 242

Chapter 6. The Statistics of Ring-Width and Climatic Data

I. Reliability of Measurements 246
II. General Statistics 254
III. Standardization 261
IV. Filtering Techniques 268
V. Other Methods for Assessing the Growth Curves . . 277
VI. Analysis of Variance 282
VII. Analysis of Chronology Error. 290
VIII. Correlation Analysis 293
IX. Power Spectrum and Cross-Power Spectrum Analyses . 295
X. Variability in Statistical Characteristics of Ring Widths Among Sites 300
XI. Statistical Characteristics of Ring Widths Within a Tree. 304

Chapter 7. Calibration

I. Introduction 312
II. The Procedure of Calibration 313
III. The Role of Statistics and Sample Size 321
IV. Degrees of Freedom and the Effective Sample Size . . 323
V. Selecting the Statistical Model 325
VI. The Diversity in Variable Selection 327
VII. Testing the Association between Variables . . . 329
VIII. Multivariate Techniques 340

Chapter 8. Interpretation of Climatic Calibrations, Reconstruction, and Verification

    I. Introduction 376
    II. Response Functions 377
    III. Strengths and Weaknesses of Response Function Analysis 400
    IV. Significance of Response Function Capability. . . 401
    V. Assessing Effects on Growth of Varying Climate . . 402
    VI. Climatic Reconstruction and Verification . . . 405
    VII. Inferences from Chronologies with Different Growth Responses 407
    VIII. Reconstruction Using Multivariate Transfer Functions . 412

Chapter 9. Reconstructing Spatial Variations in Climate

    I. Introduction 434
    II. The General Nature of Dendroclimatographic Analysis . 437
    III. The Statistical Model 438
    IV. A Feasibility Study 439
    V. Recalibration 450
    VI. Climatological Studies 455
    VII. Summarization of Reconstructions for Winter Using the Pressure Types 470
    VIII. Verification of Reconstructions for Winter . . . 476
    IX. Summarization of Reconstructions for Summer using the Pressure Types 488
    X. Verification of Reconstructions for Summer Using Independent Tree-ring Data 491
    XI. Verification Using Journals, Historical Data, and Various Proxy Records of Climate 499
    XII. Applications to Climatological Problems . . . 500
    XIII. Present and Future Prospects of Dendroclimatology . 503

Appendix—Scientific and Common Names of Trees . . . 507

Bibliography 511

Glossary 530

Author Index 547

Subject Index 553

DEDICATED TO

Colleagues and students throughout the world
who share with me the excitement and challenges
of this new scientific field.

Chapter 1

Dendrochronology and Dendroclimatology

I	Introduction	1
II	Historical Background	4
III	The Scope of Dendrochronology	10
	A. Subfields of dendrochronology	10
	B. Value of dendroclimatology	10
	C. Biological constraints on dendroclimatology	11
IV	Some Basic Principles and Concepts of Dendrochronology	14
	A. The uniformitarian principle	14
	B. The principle of limiting factors	15
	C. The concept of ecological amplitude	16
	D. Site selection	17
	E. Sensitivity	19
	F. Crossdating	20
	G. Repetition	23
	H. Standardization	25
	I. Modeling growth–environmental relationships	25
	J. Calibration and verification	27
V	Dendroclimatic Procedures and Analyses	28
VI	Examples of Analysis	38
VII	Further Definitions and Concepts	46
	A. The domain of climate	46
	B. The domain of environment	48
	C. The domain of physiological processes	51
VIII	The Climate–Growth System and its Dendroclimatological Significance	52

I. Introduction

It is well known that the approximate age of a temperate forest tree can be established by counting the growth rings in the lower part of the stem. It is less well known that the patterns of wide and narrow rings can be compared among trees to establish the exact year in which the rings were formed. The same sort of comparison can be made among wood fragments of unknown age and the rings in living trees in order to establish the date when the fragment was part of a living, growing tree. Thus, tree rings can be used to establish the year in which an event took place, as long as the event involved the maiming or killing of a tree. Precise dates can be established for the building of a medieval cathedral or American Indian pueblo; the occurrence of an earthquake, landslide, volcanic eruption, or fire; and even the date when a panel of wood was cut for a Dutch painting.

Such precise dating is possible from ring widths for a very simple reason: tree growth is frequently affected by variations in climates, and the yearly sequence of favorable and unfavorable climate (wet and dry or warm and cold years) is faithfully recorded by the sequence of wide and narrow rings in large numbers of trees. These patterns of wide and narrow rings are observable not only in living trees in an area, but also in stumps and the wood of trees that have grown in nearby areas. The procedure of matching ring patterns among trees and wood fragments in a given area is referred to as *crossdating* (Fig. 1.1). The procedure is used to identify the year in which each ring was formed and to assign exact calendar dates to the rings. The matching is necessary to identify special cases where rings may be absent or where two or more apparent rings have been formed during one year. When the ring patterns are matched carefully enough to identify the problem rings, the date of the outermost ring of a tree indicates exactly when it died or was sampled and the innermost one when the tree was a seedling. In addition, an old tree used as a construction timber can be crossdated with the early record in a living tree of the same area, as long as their life spans overlap. The rings of the construction timber in turn can be crossdated with a beam from an older structure to extend the ring-width sequences back in time (Fig. 1.1).

In dating of past historical and natural events, the climatically induced sequences of wide and narrow rings are used only to identify the year in which each ring was formed. The climatic information in the rings can, however, be analyzed in its own right to give a view of what the climate has been in the past and what it is most likely to be in the future. The science of reconstructing past climate by use of tree rings is known as *dendroclimatology*, which is a branch of the more general discipline of *dendrochronology*. The prefix *dendro* is from the Greek word for tree, *dendron*, and the word *chronology* is the name of the science that deals with time and the assignment of dates to particular events.

Such reconstruction of past climate is accomplished by taking the following steps: (1) Comparing modern meteorological records with the widths of tree rings produced during the same period of time; (2) establishing a statistical equation for the relationship between the two; and (3) substituting the widths of the dated rings in the equation to obtain

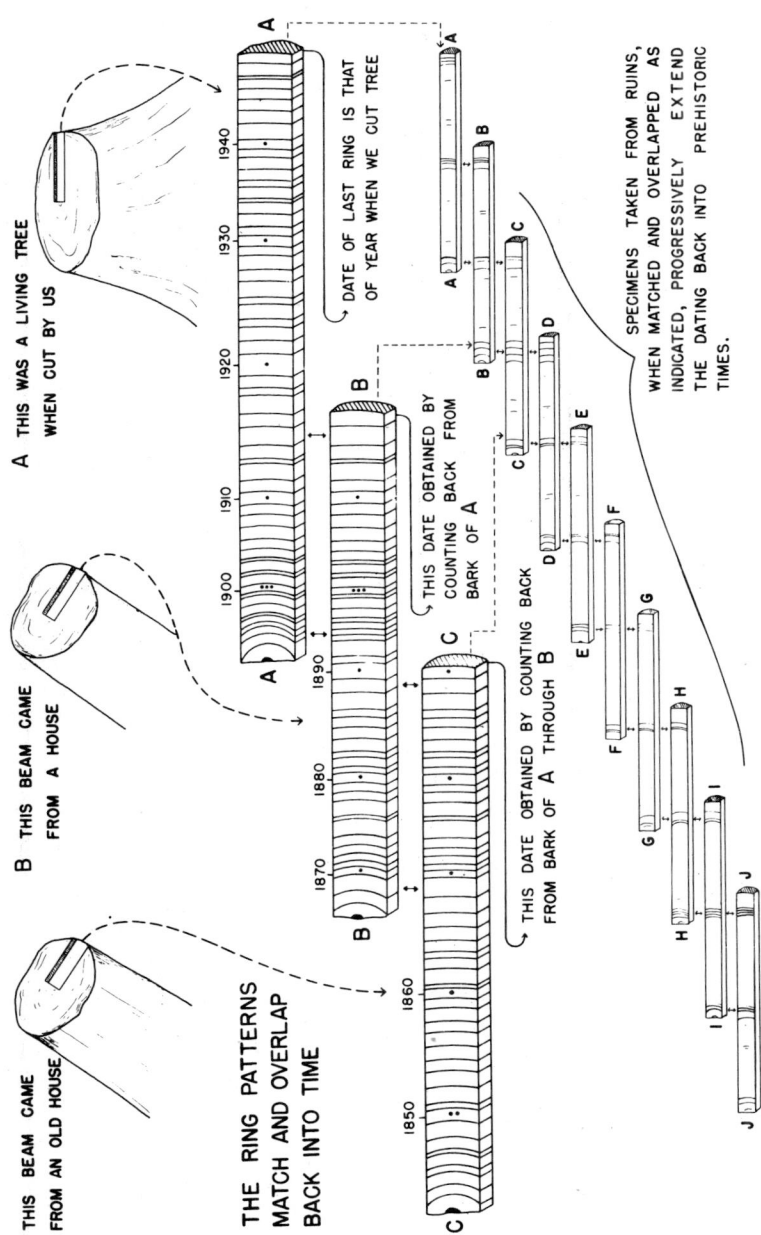

FIG. 1.1. A diagram illustrating crossdating and the extension of a dated ring-width chronology backward in time. (After Stallings, 1949)

a statistical estimate of the climate for previous years. Thus, the estimates of climate from tree rings can substitute for meteorological records and provide valuable information for periods and areas where no meteorological information exists.

However, as simple as the principle of reconstruction sounds, a variety of biological and statistical variables and many complicated relationships must be understood to make accurate reconstructions. When the principles and techniques are applied properly, dendroclimatic analysis can resolve a surprising variety of problems encountered in many fields of research. This volume will deal with both the basic knowledge and the specific techniques necessary for making many types of climatic reconstructions and for testing the validity of the results.

II. Historical Background

Andrew E. Douglass (Fig. 1.2) is the acknowledged father of dendrochronology. He started his career as an affiliate of Harvard College Observatory and in 1894 traveled to Flagstaff, Arizona, where he became the "First Assistant" to the director of the Lowell Astronomical Observatory.

One of Douglass's research interests was sunspots. He thought sunspot activity might influence the weather on earth and was looking for a relationship between the cyclic activity of sunspots and the behavior of climate, particularly precipitation (Dean, in press). The records of weather available at that time were too scanty and discontinuous to establish a clear relationship between the two.

In 1901, while Douglass was on a long buckboard trip through the forest of northern Arizona, he hit upon an idea which allowed him to test the weather–sunspot relationship. He noticed that the rings exposed on a log cut from a pine at Flagstaff exhibited variations in width. He had already been struck by the differences between the forests in Arizona and those he had known in New England (Dean, in press). The trees in Arizona forests were widely spaced, while in New England the forests were dense with much undergrowth. He knew that the growth of New England trees was influenced mostly by shading and competition within the forest. Douglass wondered if trees in Arizona were influenced more by available moisture than by competition within the stand and if moisture stress had a corresponding effect on the width of the rings. He reasoned further that if this were so, the dry years might be recorded as narrow rings, and he could use the tree-ring widths as a proxy for a long record of weather and climate.

FIG. 1.2. A. E. Douglass, the founder of dendrochronology, holds a corer (borer) which is used to extract a sample of wood 4 mm in diameter. (Courtesy of Charles W. Hubert, Western Ways Photo, Tucson)

Douglass began to examine stumps in the vicinity of Flagstaff. In 1904 he acquired six log sections from a local sawmill and noticed that the 1st, 3rd, 6th, 9th, 11th, and 14th rings in from the bark were considerably smaller than their neighbors. He was particularly impressed by a group of small rings that occurred 21 rings inside the bark. Later he examined an old stump in a rancher's field and observed the same pattern, but there were 11 rather than 21 rings between the group of small rings and the bark (Douglass, 1914). Douglass reasoned that the tree must have been cut in 1894. He surprised the owner as he correctly identified the year the tree had been felled.

It was not until 1911 that Douglass recognized the real significance of his observations. He identified a pattern of wide and narrow rings in trees from Prescott, Arizona, some 50 miles (81 km) southwest of Flagstaff which was similar to those he had noted in the Flagstaff trees. In this manner Douglass established crossdating, a procedure which he recognized could be applied in areas where ring growth is frequently limited by climate.

Dean (in press) in describing Douglass's contribution points out

> Douglass was not the first to discover crossdating (Heizer, 1956; Studhalter, 1955). That honor goes to the French naturalists Duhamel and Buffon, who in 1737 discovered that a conspicuous frost-damaged ring occurred 29 rings in from the bark on each of several newly felled trees. Other investigators confirmed their observation, and this feature was subsequently used as a marker for the 1709 ring. Crossdating based on relative ring widths was independently recognized in 1827 by A. C. Twining in Connecticut, in 1838 by the mathematician Charles Babbage in England, in 1859 by Jacob Kuechler in Texas, and in 1904 by Douglass. Although Twining was the first to grasp the full significance and potential of crossdating, it was Douglass who utilized the principle as the basis for the development of the science of dendrochronology.

Douglass noted two major implications of his discovery (Dean, in press). The first was that crossdating could be used as a chronological tool to identify the exact calendar year in which rings were produced by studying the pattern of wide and narrow rings. All that was required was that the year for the outermost ring of the stem be known, and that the same relative variations in ring structure be observable in many trees. Wood samples of unknown age could be dated to the year by matching their ring-width patterns with that of a previously dated ring-width chronology.

The second implication of Douglass's finding was that the ring-width patterns themselves represent a record of environmental conditions over a region. While some of the ring-width variation was attributable to local conditions within and surrounding the habitats of the trees, a large portion of the variation was observable in all trees and thus could only reflect factors occurring over the entire region, such as the yearly variations in climate. Douglass hoped that this surrogate of climate would allow him to test for a causal link between sunspot variations and climatic conditions, as well as to extend the record of sunspots several centuries back into the past.

He proceeded to derive this record of climate by comparing and combining the ring records of many trees, a procedure now called *chronology building*. After studying hundreds of trees from the Flagstaff–Prescott area, Douglass noted that the most reliable patterns for crossdating were found in the narrow rings. He also observed that the trees at the lower elevational boundary of the species exhibited the most variability in ring width.

By 1914, Douglass had succeeded in building a composite chronology of nearly 500 years from the rings of *Pinus ponderosa*.* He demonstrated from the available weather records that ring width was directly related to precipitation of the preceding winter. From these observations he hypothesized that precipitation replenished soil moisture which in turn affected the amount of growth in the spring. He also noted that trees exhibited "a reserve power or vitality which may run low or be built up by varying environment". He called this the conservation term and adjusted a chronology for it by obtaining a running mean or by applying a numerical coefficient which was thought to adjust the ring widths for the effects of prior conditions (Douglass, 1919). Modern statistical techniques have now replaced many of Douglass's early quantitative procedures.

Douglass traveled to the eastern United States in 1914, where he presented a talk on his research at the Carnegie Institution of Washington, D.C. Among the dignitaries in the audience was Clark Wissler, an anthropologist at the American Museum of Natural History, New York, who was attempting to date the prehistoric and historic pueblo sites of arid southwestern North America and was intrigued by the chronological implications of Douglass's method. Would it not be possible to date the rings from logs in the pueblos as Douglass had dated the old stumps? Wissler discussed the possibility with Douglass and sent him specimens of the prehistoric wood (Douglass, 1921, 1929, 1935).

After studying several specimens, Douglass was able to crossdate six that had been collected from a huge ruin at Aztec, New Mexico, and he used them to establish a composite chronological sequence of 139 years (Dean, in press). He could not match this Aztec chronology with the living trees from Flagstaff so calendar dates could not be applied to it, and it remained for several years as a chronology "floating" in time. In 1920 Douglass received nine sections from the large ruin of Pueblo Bonito located in Chaco Canyon 50 miles (81 km) south of the Aztec ruin. Douglass found that seven of these specimens crossdated, yielding a 100-year-long chronology, and the series crossdated with the Aztec chronology. The Pueblo Bonito series ended 39 years earlier than the Aztec rings from which he concluded that the Pueblo Bonito was under construction 40 years before parts of the Aztec ruin were built. Thus, the first precise time-relationship among the prehistoric pueblos was established, although the absolute dates of the chronologies were still "floating" in time.

According to Dean (in press), Neil M. Judd, who was excavating Pueblo Bonito, was one of the first archaeologists to realize the dating

---
* A full list of species referred to in this text is given in the Appendix, together with the common names of each.

potential of tree rings. He envisioned extending the prehistoric floating chronology forward in time toward the present by crossdating ring series from sites known to postdate Aztec. At the same time the Flagstaff chronology could be pushed backward in time by selecting materials from early historic structures, such as Hopi Indian pueblos, and from late prehistoric sites. He realized that when the Flagstaff and Aztec chronologies overlapped sufficiently in time to establish the dating, it would be possible to assign calendar dates to all prehistoric sites linked to the floating chronology. In 1922 Judd proposed his scheme to Douglass and helped him obtain funding from the National Geographic Society, with the result that three expeditions to archaeological sites were arranged. The first and second, in 1923 and 1928, resulted in a 585-year floating chronology for the prehistoric sites, and an extension of the Flagstaff chronology back to A.D. 1260. No overlap appeared to exist between the two sets, however.

The third expedition focused on sites that appeared to occupy the temporal gap between the two tree-ring chronologies. On June 22, 1929, a charred beam marked as HH-39 was excavated which extended the Flagstaff record back to A.D. 1237. Work that evening revealed that the inner rings of HH-39 crossdated with the outer part of the floating sequence. Absolute dates could now be assigned to all the prehistoric sites that had been crossdated with the floating chronology (Douglass, 1929; Haury, 1962). The spectacular cliff dwellings of Mesa Verde, Canyon de Chelly, and Tsegi Canyon were dated in the A.D. 1200's, Aztec was built between 1111 and 1120, and Pueblo Bonito was constructed in the late 11th century. The ring chronology through the former gap period was confirmed by many additional materials. Douglass had made a significant contribution to archaeology and he had obtained for himself a tree-ring record of climatic variations that extended from A.D. 700 to 1929.

Today the southwestern archaeological tree-ring chronology has been extended back to 322 B.C. and more than 20,000 dates have been derived from nearly 1000 sites (Dean, in press). Tree-ring dating has also been extended to many other regions of the world, including nonarid sites, and it has been applied to many different kinds of phenomena.

In 1906 Douglass moved to the University of Arizona in Tucson and in 1937 he established the Laboratory of Tree-Ring Research, which became the first institution devoted exclusively to tree-ring studies. Two students were especially helpful to Douglass in developing and extending tree-ring analysis to other problems, new species, and a wide range of sites. Waldo S. Glock, a geologist, was instrumental in developing qualitative analytical techniques while Edmund Schulman (Fig. 1.3), an astronomer and climatologist, leaned more toward statistical analysis and dendroclimatic techniques.

FIG. 1.3. Edmund Schulman was a student of Douglass who pioneered dendroclimatic studies and, in a search for old trees, discovered the ancient high-elevation *Pinus longaeva* of California. Schulman is surfacing a mounted tree core with a razor blade. (Courtesy of Laboratory of Tree-Ring Research, Tucson)

Glock left the Laboratory of Tree-Ring Research and wrote several extensive reviews of tree growth and climatic relationships (Glock, 1955; Agerter and Glock, 1965). Maintaining contact with the developing science in Europe, he championed detailed anatomical and other botanical investigations on the subject.

Schulman continued his work in southwestern North America at the University of Arizona, where he obtained the title of dendrochronologist in 1941. His two best known contributions are a monographic study on dendroclimatology (Schulman, 1956) and his discovery of 4500-year-old *Pinus longaeva* (Bailey, 1970), the ancient bristlecone pine (Schulman, 1958).

Bruno Huber at the Forest-Botany Institute in Munich, Germany, was the first to initiate tree-ring dating studies in Europe (Huber, 1941). He was followed by Karol Ermich in Poland, O. A. Hoeg in Norway, Bo

Eklund in Sweden, Peitsa Mikola and Ilmani Hustich in Finland, and Erik Holmsgaard in Denmark, to mention only a few.

III. The Scope of Dendrochronology

A. Subfields of Dendrochronology

Some authors have restricted the term *dendrochronology* to the use of tree rings to date events (de Martin, 1970; Bitvinskas, 1974). However, the techniques of dendrochronology are applied to a variety of problems of environment and climate. In this volume the term is defined broadly to include all tree-ring studies where the annual growth layers have been assigned to or are assumed to be associated with specific calendar years.

Dendrochronology may be divided into a number of subfields, several of which focus on its application to problems of environment and climate. The prefix *dendro* is used in conjunction with the name of the particular scientific discipline, so that just as the term *dendroclimatology* refers to dendrochronological investigations of past and present climates, the term *dendroclimatography* refers to the application of tree-ring analysis to the mapping of past and present climates. Similarly, *dendroecology*, *dendrohydrology*, and *dendrogeomorphology* (Alestalo, 1971) refer to the application of dendrochronology to the study of, respectively, the ecology of past biotic communities, river flow and flooding history, and geomorphic processes. The choice of the indentifying term is arbitary and many dendrochronological studies may fall legitimately into more than one of these subfields. In this volume examples are drawn from many subfields.

B. Value of Dendroclimatology

Tree rings have characteristics which make them an exceptionally valuable source of paleoclimatic information. Among such attributes are the facts that (1) the width of rings is easily measured for a continuous sequence of years, and these measurements may be calibrated with climatic data, and (2) the rings can be dated to the specific years in which they were formed, so that the climatic information is precisely placed in time. Few other sources of paleoclimatic information can provide both continuity and precise datability (Gates and Mintz, 1975), and few can be replicated and quantified as easily as tree rings.

Tree-ring data can be used to reconstruct the yearly variations in climate that occurred prior to the interval covered by direct climatic measurements. These reconstructions can extend the climatic record backwards in time and increase its length sufficiently to improve the existing statistics on climatic variability. Such improvements could help man to better anticipate possible future climatic changes as well as to better understand those of the past (Gates and Mintz, 1975). For example, the works of Lamb (1963, 1969a, 1972), Lamb *et al.* (1966), and Ladurie (1971) have served to establish estimates of the characteristics of natural climatic variability for the past thousand years. These data, however, are largely restricted to the North American and North Atlantic–European sectors of the world, and climatic variability elsewhere in the Western Hemisphere is not well known.

It will be shown in Chapters 8 and 9 that tree rings of North America do provide significant information on past variations in climate over eastern Asia and the North Pacific Ocean as well as over the North American continent. In addition, as tree-ring analysis is applied to new areas and species, it is reasonable to expect that information will become available on past variations in climate for additional large sectors of both the Northern and Southern Hemispheres (Fritts *et al.*, 1971; LaMarche, 1974a).

Although the actual prediction of future climatic change is a task for the atmospheric scientist, the detailed reconstructions of past climatic variability obtained from tree rings should give these scientists an improved base from which to estimate the possible climatic modes, their frequencies, and geographic extent. Such knowledge also can help to improve the modeling of atmospheric circulation (Sheppard, 1966), allow man to discriminate between natural climatic variability and unnatural changes accentuated by him, and help to estimate probable future occurrences of devastating climates, such as prolonged drought.

C. *Biological Constraints on Dendroclimatology*

The link between past climate and ring width occurs because plant growth is affected by certain conditions in the forest environment. A large number of these environmental conditions vary throughout the life of a plant, and at times they may limit growth and affect the form of many plant structures.

The specific limiting conditions which can affect plant growth may be classed as either external or internal factors. Some of the most important external limiting factors are water, temperature, light, carbon dioxide, oxygen, and soil minerals. Some of the most common internal limiting conditions are the amounts of available food, minerals, growth regulators, enzymes, and water. In reality the level of the internal factors is often a result of external factors that were limiting at some prior time in the life of the plant. Complex interactions can occur between external factors, internal conditions, physiological processes, and growth, but few of these interactions have been studied adequately.

There is a considerable lack of understanding, especially of the relationship between large trees and their environment. Most of our knowledge has been obtained by laboratory experimentation on seedlings or relatively small plants grown in controlled environmental chambers, green houses, tree nurseries, or plantations, and measurements may not accurately reflect the same processes as they occur in mature trees in the complex natural environment. In spite of the limits of our knowledge and the obvious complexities in biological systems, however, there are many valid inferences that can be made using existing information about environmental factors, internal conditions, and structures in plants.

In recent years the growth layers, or rings, in woody plants have been studied in this light and used to make many inferences about conditions within plants and their surrounding environments. These growth layers are, in reality, sheaths of cells that appear as concentric rings in a cross section of the stem. Each ring is usually the result of a single yearly flush of growth which begins in the spring and ceases in the summer or in early autumn, so that one layer is produced every year. A sharp boundary occurs between rings (Fig. 1.4) because cells along the inner boundaries of rings are larger and have thinner walls than those along the outer boundaries.

Not all rings are distinct annual increments of growth, however. Sometimes when factors are highly limiting, growth cannot begin and no ring is produced. At other times, a stress period occurring in the middle of the growing season may cause two or more growth layers to form within a particular year. When there is a possibility that no ring has been formed in some years and more than one ring may have been produced in other years, a simple count from the outside to the center of the stem obviously cannot be used to determine the year in which each ring was formed, so crossdating with other specimens is necessary.

There are many species, geographic areas, and instances where dendrochronology and its related subdisciplines cannot and should not be applied because tree-ring patterns cannot be or have not been dated. For instance, many woody species, such as those growing in the tropics or semitropics, can produce several growth layers per year, and the number and features of the growth layers are often not coincident from tree to tree or on opposite sides of the stem from the same tree (Alvim, 1964). Similarly, *Cupressus arizonica*, several species of *Juniperus*, and other troublesome species have traditionally been avoided by dendrochronologists, because several rings can be produced during a single growing season and crossdating cannot be obtained (Glock, 1951; Schulman, 1951; Glock et al., 1960).

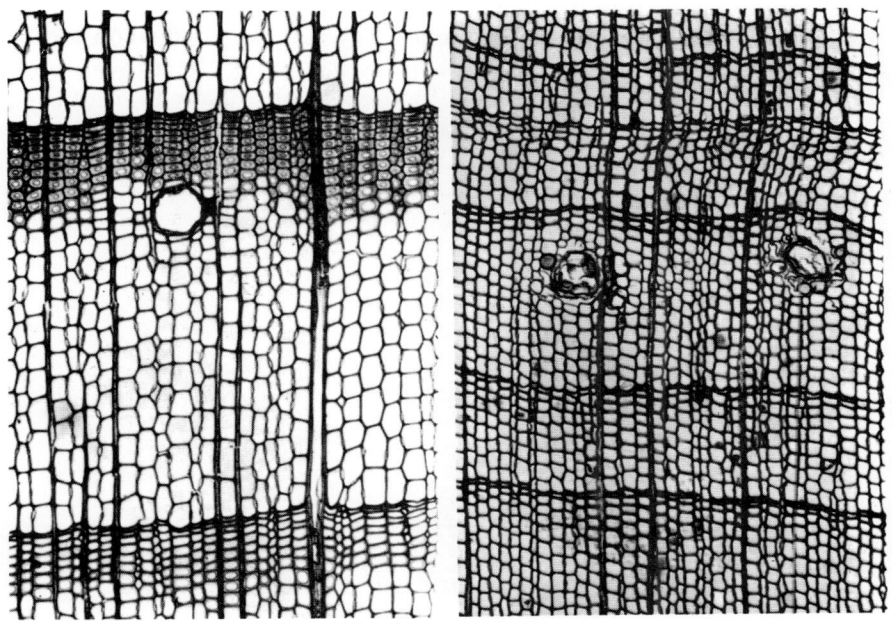

FIG. 1.4. Annual rings result from varying physiological conditions that affect the cells formed throughout the growing season. At the beginning of growth in the spring, large thin-walled cells are produced. As the end of the season nears, smaller cells with thicker walls are formed. The ring boundary is the abrupt change in cell size between the small cells formed at the end of one growing season and the large cells formed at the beginning of the next. The large openings are resin ducts. *Pinus sylvestris* is on the left and *Pinus cembra* on the right. (Photo courtesy of Fritz H. Schweingruber, Eidgenössiche Anstalt für das forstliche Versuchswesen.)

Other species may exhibit anatomical features that provide little contrast in the cell structure throughout the ring, so that the boundary of an annual growth layer is not visible except when thin cross sections of the stem are made (Fahn, et al., 1963).

In still other cases the woody plants may produce an annual growth layer, but there may be little difference in structure or appearance from one ring to the next. This can arise because climatic factors do not vary sufficiently to limit processes affecting growth. In such cases, factors arising from local site conditions and internal conditions of the plant override the climatic factors of the forest environment. Effects of climate on ring growth are not apparent when the rings are wide, when their structure is the same from year to year, or when the structural variations are unique for each individual tree and therefore cannot be associated with any particular condition of climate. A wood specimen is considered undated if there is *any* doubt about the synchrony of the growth rings. Table 1.I lists those genera known by the author to have been utilized successfully for dendrochronological work.

TABLE 1.I Genera with at least one species that is known to crossdate

Abies	*Cupressus*	*Populus*
Agathis	*Ephedra*	*Pseudotsuga*
Alnus	*Fagus*	*Purshia*
Amelanchier	*Fitzroya*	*Quercus*
Araucaria	*Fraxinus*	*Sequoia*
Artemisia	*Juniperus*	*Sequoiadendron*
Austrocedrus	*Larix*	*Taxodium*
Betula	*Libocedrus*	*Thuja*
Carya	*Picea*	*Tsuga*
Cedrus	*Pinus*	*Ulmus*
Celtis	*Pistacia*	*Zygophyllum*
Cercocarpus	*Podocarpus*	

IV. Some Basic Principles and Concepts of Dendrochronology

Every science accumulates a body of knowledge from which generalizations can be made. These generalizations, based upon repeated observations and experience, may be described as *principles* or concepts which embody the fundamental truths of the science. The following pages will describe the most widely used and basic principles and concepts in the field of dendrochronology, with primary focus on the subfield dendroclimatology. Some principles and concepts may be familiar, as they often represent modifications borrowed from other sciences. The actual biological and physical bases for these principles and concepts will be described in subsequent chapters.

A. The Uniformitarian Principle

A principle basic to any study of the past is the principle of "uniformity in the order of nature," originally proposed by James Hutton in 1785. It is commonly stated as:

> The present is the key to the past.

Applied to dendrochronology, the uniformitarian principle implies that the physical and biological processes which link today's environment with today's variations in tree growth must have been in operation in the past. Likewise, the types of weather variations and climatic patterns observed today also must have occurred in the past. This does not imply that the paleoclimate was the same as the climate at the present. However, it does imply that the same *kinds* of limiting conditions affected the same *kinds* of processes in the same ways in the past as in the present; only the frequencies, intensities, and localities of the limiting conditions affecting growth may have changed. Therefore, one can establish the relationship between variations of tree growth and variations in present-day climate and infer from past rings the nature of past climate.

In order to make this kind of inference, however, it is important that the entire range of variability in climate that occurred in the past is included in the present-day sampling of environment. For example, at least one year's climate should be as dry and one as wet in the modern period of calibration as was the extreme climate of the past. If this is not the case and the past conditions lie outside the domain of modern climate, then the inferences from tree rings may be inappropriate and inaccurate. The uniformitarian principle is assumed in all dendrochronological inferences, and, as in all sciences of the past, if this principle does not hold, no conclusions regarding the past can be made. Checks of actual past climatic conditions, using data and evidence independent of tree rings (examples given in Chapters 8 and 9), support the validity of the uniformitarian principle to dendroclimatological research. Thus, it is reasonable to conclude that inferences may safely be made about climates of the recent past by applying dendroclimatological techniques.

B. *The Principle of Limiting Factors*

The well-known biological law or principle of limiting factors may be simply stated: a biological process, such as growth, cannot proceed faster than is allowed by the most limiting factor. The same factors may be limiting to some extent in all years, but the degree and duration of their limiting effects vary from one year to the next. If a factor changes so that it is no longer limiting, the rate of plant processes will increase until some other factor becomes limiting. For example, during periods of ample but not excessive moisture, growth will increase until it is limited by some other condition. This principle is important to dendrochronology because ring widths can be crossdated only if one or more environmental factor becomes critically limiting, persists sufficiently long, and acts over a wide enough geographic area to cause ring widths or other features to vary the same way in many trees.

The principle implies that the narrower rings provide more precise information on limiting climatic conditions than do the wider rings. During years when rings are generally wide, factors may become limiting to different degrees in each tree, depending on its locale, ecological position in the site, and a great variety of nonclimatic factors. As a result considerable variation in growth patterns may occur from tree to tree. If the growth of a tree is never limited by some climatic or environmental condition, there will be no information on climate in the widths of rings and they will not crossdate. No amount of study of ring widths from such trees can be expected to reveal information on past climates. The principle of limiting factors including nonlinear complications is described more fully later in Chapter 3 and in other sections of this book.

C. The Concept of Ecological Amplitude

Each species, depending upon hereditary factors which determine its phenotype, may grow and reproduce over a certain range of habitats. This range is referred to as its *ecological amplitude*. Some species grow on a wide range of habitats because their heredity allows a large ecological amplitude. Other species are limited to a small number of specific sites because their heredity allows a small ecological amplitude, and still other species such as *Pinus radiata* and *Sequoiadendron giganteum*, which appear to have large ecological amplitudes, have become restricted to certain natural localities because of isolation resulting from geographic factors.

Near the center of its geographical distribution, a species is often found on the widest varieties of sites, and climatic factors may rarely be limiting to growth except in years of most extreme climate and in the most limiting local situations. Near the margin of its natural range, a species may occur on a relatively small variety of sites, and climate frequently becomes highly limiting to physiological processes, including growth.

Often the growth of trees near arid forest limits is most affected by drought, while the growth of trees near the upper elevational or high latitudinal forest limits is most affected by low temperatures. Species differing in ecological amplitude may be limited by the same climatic factor if the individual trees are growing in comparable habitats within their own ecological amplitudes. For example, a high elevation or northern species and a low elevation or southern species which are both near their arid limits can respond similarly to a year of drought, even though they may be responding to different moisture amounts.

In many situations such factors as length of the daylight period, shade, and low amounts of soil minerals, which do not vary significantly from one year to the next and do not involve variations in climate, can limit plant distribution but have little influence on the variability in ring width. Other nonclimatic factors such as fire, insect attack, or disease can affect both plant distribution and ring width, and the latter cannot be used to infer climatic variation except in cases where these factors may themselves be affected by variations in climate.

D. *Site Selection*

Critics in the past have questioned the validity of selecting the most drought-sensitive trees for sampling rather than selecting trees randomly (Glock, 1955). Such judgment fails to recognize that the

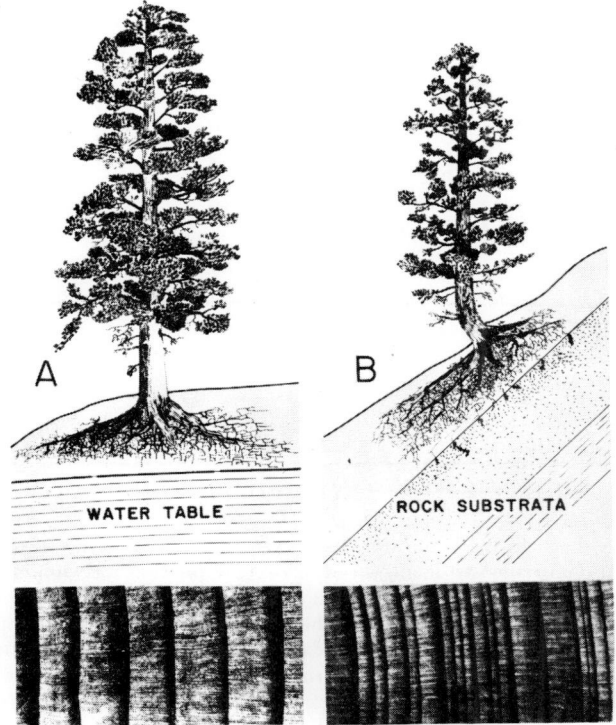

FIG. 1.5. Trees growing on sites where climate seldom limits growth processes produce rings that are uniformly wide (A). The ring widths provide little or no record of variations in climate and are termed *complacent*. Trees growing on sites where climatic factors are frequently limiting produce rings that vary in width from year to year depending upon how severely limiting climate has been to growth (B). These are termed *sensitive*.

dendrochronologist has a particular strategy in mind which requires that his samples be affected similarly by a given set of growth-limiting factors. The sampling is deliberately stratified to draw observations from that population of ring widths which contains the desired information, and it is further restricted to a particular species to keep genetic response more or less constant. Within these restrictions the sampling may be drawn by

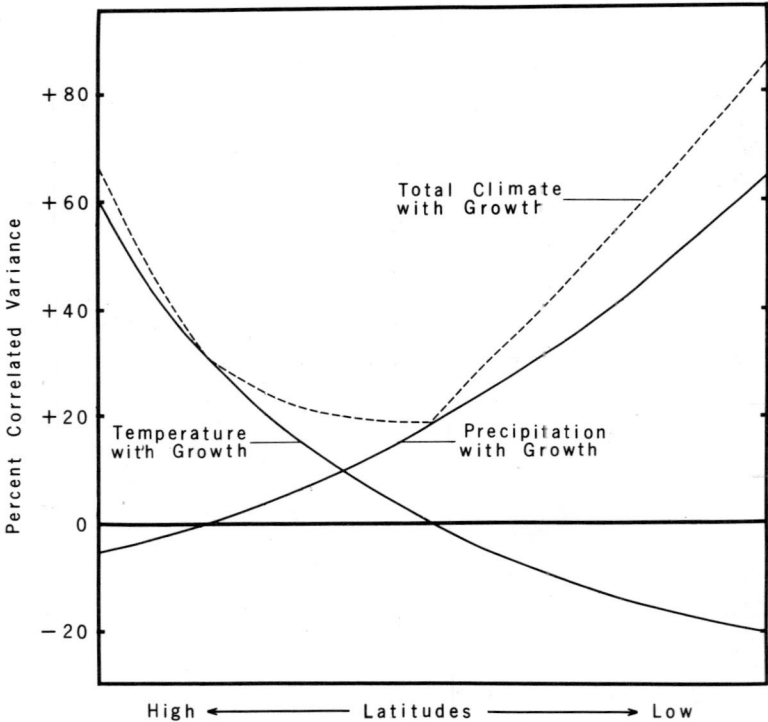

FIG. 1.6. A diagram modeling the changes in the amount of correlation expected between variations in ring width and variations in precipitation and temperature plotted as a function of latitude of the tree site. Solid lines represent the percentage correlation and the sign of the relationships, and the dashed lines express the total amount of variation accounted for by both climatic variables neglecting the sign of the relationship. At low latitudes, as well as at low elevations, variations in ring width often are highly and positively correlated with variations in precipitation and negatively correlated with variations in temperature. These values diminish in magnitude toward midlatitudes because the two climatic factors are less often limiting to growth. At high latitudes, as well as at high elevations, the correlation with temperature is likely to reverse in sign and to increase as temperature becomes directly correlated with growth. There is little correlation of growth with precipitation in many high latitude and high elevation sites, though sometimes the correlation becomes inverse and has a negative sign.

some random design. The selection of drought-sensitive arid-site trees is no different than a forester's selection of a homogeneous stand to test a particular management practice. Selection involves limiting the sampling space to a small number of variables, thus eliminating a variety of possible circumstances not relevant to the question at hand.

Dendrochronologists must apply the law of limiting factors and the concept of ecological amplitude when they obtain their research materials in order to assure selection of trees which will give them the information they desire. This selction is referred to as *site selection*. Thus, in studies of ring widths and drought, it is important to rely upon trees growing in the driest sites, for those are the individuals in which ring width is most likely to have been limited by drought (Fig. 1.5b). In studies of temperature, the ring widths of trees near the species' upper elevational or latitudinal limits will yield the most reliable information on temperature. In regions of temperate and moist climates, only the ring widths of trees in the most extreme environments are sufficiently limited by climatic factors to allow easy crossdating and dendroclimatic analysis of the tree rings (Fig. 1.6). Certain techniques that are described elsewhere in the book can be applied to facilitate analysis of materials from temperate and moist climates.

E. Sensitivity

An inspection of rings under a magnifying glass can provide a clue as to how often climate becomes limiting to growth. The more the tree has been limited by environmental factors, the more the tree will exhibit variation in width from ring to ring (Fig. 1.5a). The dendrochronologist refers to this variability in ring width as *sensitivity* and to the lack of width variability as *complacency*. Such fluctuations in ring width can be estimated qualitatively by visual inspection of the rings; or it can be calculated from measurements of width and expressed as a statistic called *mean sensitivity*, which is a measure of the relative differences in width between adjacent rings. The details of this statistic are described in Chapter 6.

Douglass (1937) likened the contrast between sensitive and complacent ring series to the difference between the Morse telegraph code of dots and dashes and an undifferentiated continuum of sound. By breaking the sound continuum into discrete contrasting units, the dots and dashes, we can produce coherent messages. In much the same way, the sequence of narrow (dots) and wide (dashes) rings in a sensitive ring series conveys messages about the life of the tree (Dean, in press).

F. Crossdating

Crossdating is the most important principle of dendrochronology. Its application provides a type of "experimental" control because it assures the proper placement in time of each growth layer. The yearly ring widths must be crossdated among all radii within a stem and among different trees in a given stand, as well as among ring-width patterns of neighboring stands. The variations in ring characteristics, especially ring widths, are examined and synchronously matched among all samples from a given region. If there is sufficient covariation among rings in

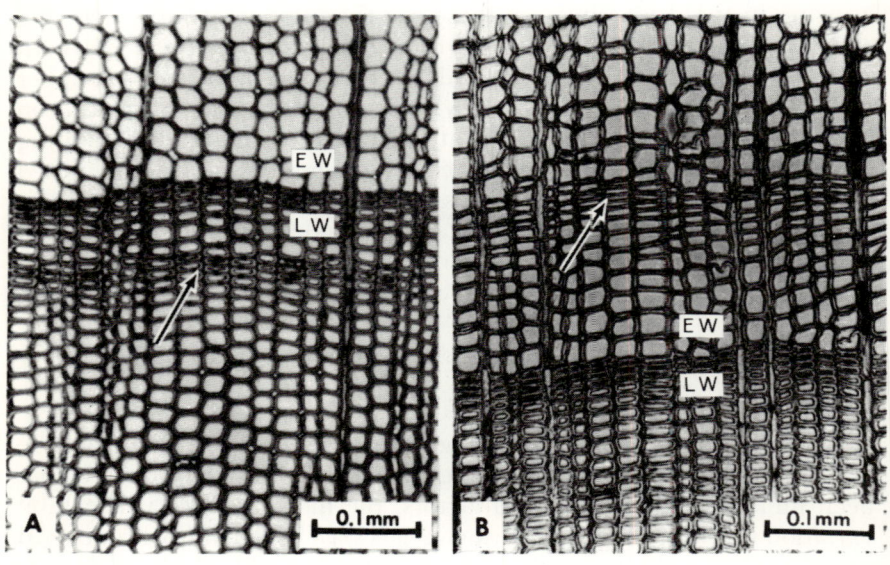

FIG. 1.7. Annual growth layers or rings are formed because the wood cells produced early in the growing season (EW) are large, thin-walled, and less dense, while the cells formed at the end of the season (LW) are smaller, thick-walled, and more dense. An abrupt change in cell size between the last-formed cells of one ring (LW) and the first-formed cells of the next (EW) marks the boundary between annual rings. Sometimes growing conditions temporarily become severe before the end of the season and cause subsequently formed cells to be smaller with thicker walls (arrows). When more favorable conditions return, the subsequently formed cells are larger and have thinner walls. The resulting dark bands within the growth layer are called *intra-annual growth bands* or *false rings* and are usually easily identified by the gradual transition in cell size on both margins of the band. Occasionally, these intra-annual bands are indistinguishable from the true annual ring and the problem must be resolved by crossdating. In A the false ring is within the latewood formed near the end of the growing season. In B it is within the earlywood formed near the beginning of the growing season. Growth is in the upward direction. (Adapted from Kuo and McGinnes, Jr., 1973)

different trees and the sample is large enough, the year in which each ring was formed can be correctly ascertained.

Crossdating is possible because the same or similar environmental conditions have limited the ring widths in large numbers of trees, and the year-to-year fluctuations in limiting environmental factors that are similar throughout a region produce synchronous variations in ring structure. The fact that crossdating can be obtained itself is evidence that there is some climatic or environmental information common to the sampled trees.

To continue Douglass's Morse code analogy, crossdating involves the recognition of identical "messages" in the "dot–dash" sequence of ring-width variability in the sensitive ring series of different trees (Dean, in press).

Sometimes during a year of extreme climate a tree may not form a ring on all portions of the stem. The ring is then said to be *partial, locally absent*, or *missing* along certain stem segments or radii. At other times a change in cell structure will occur within an annual growth increment so that the layer resembles the boundary of a true annual ring. Such features are called *intra-annual growth bands* or *false rings* (Fig. 1.7).

When rings are absent from a sampled radius or if an intra-annual growth band appears to be an annual layer, the variations in ring widths for that particular specimen do not crossdate exactly with variations in widths seen in samples from other trees or from other portions of the stem (Fig. 1.8). The problem rings can be spotted because, if the rings are properly matched on one side of the problem area, they will appear not to match on the other side. When the discrepancy is recognized (see arrows in Fig. 1.8) and the ring-width sequence is corrected for the absence of certain annual growth layers or for the presence of false layers, the patterns of widths throughout the entire length of a specimens are observed to be consistent with respect to variations in other specimens from nearby sites. If, however, there is little sensitivity and low correlation of ring-width variation among trees, or if there are large numbers of locally absent or false rings, the dating may be uncertain, and the samples must be laid aside as unsuitable for dendrochronological work.

Thus, crossdating includes matching of ring-width patterns among specimens; examining the synchrony; recognizing any lack of coincidences; inferring where rings may be absent, false, or improperly observed; testing the inference by examining carefully the ring structure in other specimens; and finally arriving at the correct regional chronology with agreement among the growth sequences of trees in neighboring stands.

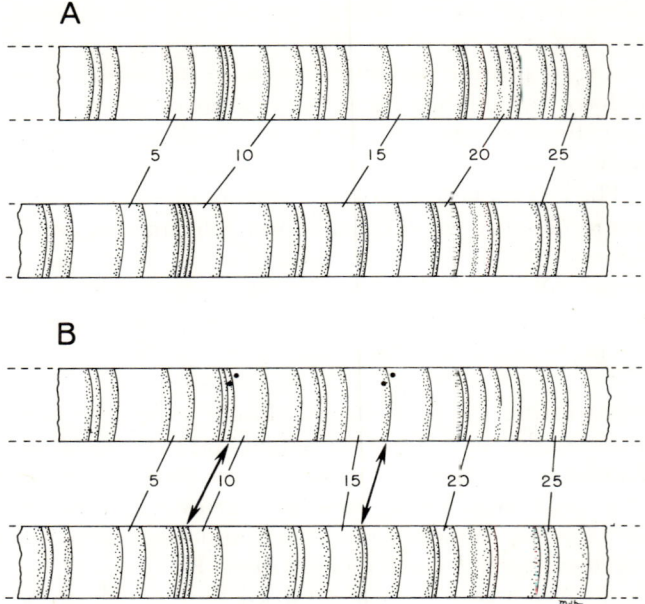

FIG. 1.8. Crossdating makes it possible to recognize areas where rings are locally absent or where an intra-annual growth band appears like a true annual ring. The patterns of wide and narrow rings are compared among specimens. Every fifth ring is numbered in the diagram and in A the patterns of wide and narrow rings match until ring number nine, after which a lack of synchrony in pattern occurs. In the lower specimen of A, rings 9 and 16 can be seen as very narrow, and they do not appear at all in the upper specimen; while rings 21 (in the lower) and 20 (in the upper) show intra-annual growth bands. In the upper specimen of B, the positions of inferred absence are designated by two dots, the intra-annual band in ring 20 is recognized, and the patterns in all ring widths are synchronously matched. (Drawing by M. Huggins).

It was shown in Figure 1.1 how ring-width chronologies are constructed and lengthened beyond the age covered by living trees by crossdating their ring-width patterns with those of older pieces of wood which have inner portions that extend the chronology back in time (Stokes and Smiley, 1968; Bannister, 1969; Dean, in press).

Remnants of dead bristlecone pine *Pinus longaeva* (recently *P. aristata*, renamed by Bailey, 1970) from southern California have been crossdated with living trees and with each other to develop an especially long chronology. The arid-site chronology developed from the ancient bristlecone pines is now more than 8,200 years in length, which is considerably beyond the maximum age of 4,600 years for the oldest living tree (Ferguson, 1970b). Also, a 5405-year-long chronology has been developed for the same species at the upper tree line (LaMarche and Harlan, 1973). In other areas wood and charcoal from archaeological and historical sites have been used to extend the chronologies backward in time (Dean, in press).

G. *Repetition*

A number of specimens must be examined and crossdated from any given site to avoid the possibility that all collected specimens could be missing a ring for any one year or could have an intra-annual growth band appearing like a true annual ring. Further verification is obtained when several independently dated samples are compared and no inconsistency appears.

In addition, the *average* of replicated measurements from a large number of trees provides the best estimate of climate, because the growth variation that is associated with climatic variation, which is common to all trees, is retained when such averages are made. A large portion of the effects of nonclimatic factors which differ among individuals and from site to site is minimized by the averaging process.

Repetition or replication in sampling of more than one stem radius per tree, as well as from more than one tree, allows statistical comparisons of variability within the same tree as compared with variability between trees and between groups of trees. Measurements of this variability provide valuable information on how factors of the site and climate control tree growth. If climate is highly limiting to growth, all replicated samples within and among trees will show approximately the same ring-width variation and the rings will be easy to crossdate. In such cases, a reliable ring-width chronology may be developed from a relatively small number of radii and trees. If climate is not highly limiting, variations in factors of the site may cause marked differences in ring sizes between trees, and differences in growth on opposite sides of the same trees may result from variations in the structure of the forest stand, lean of the stem, and competition from neighboring trees. A greater amount of replication may be necessary to achieve a reliable chronology from such a site.

Statistical techniques described in Chapter 6 can be used to ascertain the amount of replication needed to obtain the desired reliability. These techniques utilize the amounts of ring-width variability in a particular sample to estimate what is referred to as *variance*, including not only the average variability in the chronology of a site but also the differences within trees, among trees, and among sites.

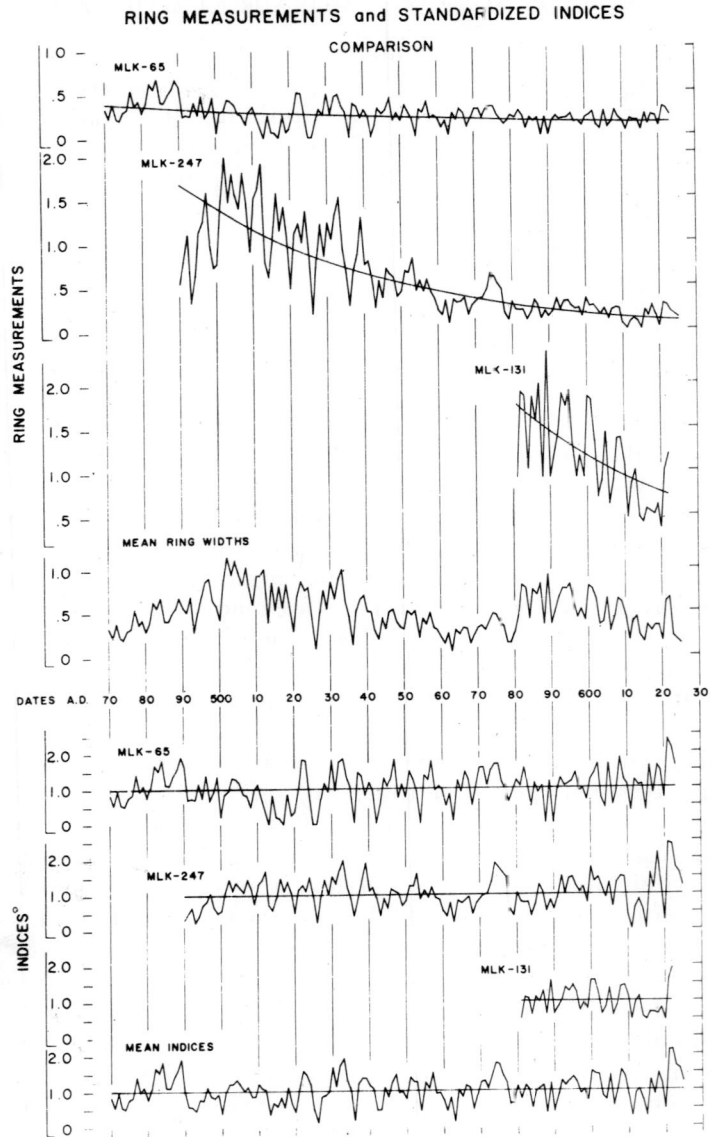

FIG. 1.9. Standardization of ring-width measurements is necessary to remove the decrease in size associated with increasing age of the tree. If the ring widths for the three specimens shown in the figure are averaged by year without removing the effect of the tree's age, the mean ring-width chronology shown immediately below them exhibits intervals of high and low growth associated with the varying age of the samples. This age variability can be removed by fitting a curve to each measured series, and dividing each ring width by the corresponding value of the curve. The resulting values shown in the lower half of the figure are referred to as indices and may be averaged among specimens differing in age to produce a mean chronology for a site.

H. Standardization

Standardization is such a basic procedure in dendrochronology that it is considered by some to be a principle. It will be shown in Chapter 2 that ring widths can vary not only with fluctuations in environmental conditions, but also with systematic changes in tree age, height within the stem, and conditions and productivity of the site. In studies of ring-width variability associated with climate, it is usually most convenient to estimate the systematic changes in ring width associated with age and to remove them from the measurements (Fig. 1.9). This correction of ring width for the changing age and geometry of the tree is known as *standardization* and the transformed values are called *ring-width indices*. The indices generally have no linear trend; their mean value is one; and the larger variability in ring width of the younger, fast-growing portions of a tree is made comparable to the lesser variability in the ring width of the older, slower-growing portions of the tree. The standardized indices of individual trees are averaged to obtain the mean chronology (*mean standardized indices*) for a sampled site (see lowest plot, Fig. 1.9). The terms *chronology* or *ring-width chronology* as used in this text, refer to the averaged yearly standardized indices for a number of trees sampled from a particular site. Details of the standardizing procedure are described more fully in Chapter 6.

I. Modeling Growth—Environmental Relationships

Inferences about past environments and climates are based upon some idea or model of how the environment affects growth. A *model* may be a statement, equation, or diagram which represents a basic set of facts and their interrelationships. They can range from extremely simple preconceived notions to highly complex equations derived from literature reviews, field observations, and extensive experimental research. Models may be used in many different ways.

Models presented as diagrams or equations may be used to illustrate particular phenomena, such as the model in Fig. 1.6 which illustrates the association of ring-width variations with variations in two climatic factors.

Models of various types may serve as *hypotheses* which are checked by comparison with information derived from observation. Such models are often revised repeatedly when inconsistent situations appear or as contradicting data dictate. When models closely resemble actual relationships, the model may be accepted as the best available inference, though it is subject to further testing and revision as more knowledge is obtained or as new ideas suggest improvements.

A model in the form of a mathematical or statistically derived equation can serve as an objective and precise description of the linkages between the inputs and outputs of a particular system. In such models numbers can be assigned to the variables and the equations can be solved to obtain statistical or mathematical estimates of the effect of changes in inputs on the system outputs. Observations of the modeled phenomena can then be compared with the estimates, and equations adjusted to improve the estimates.

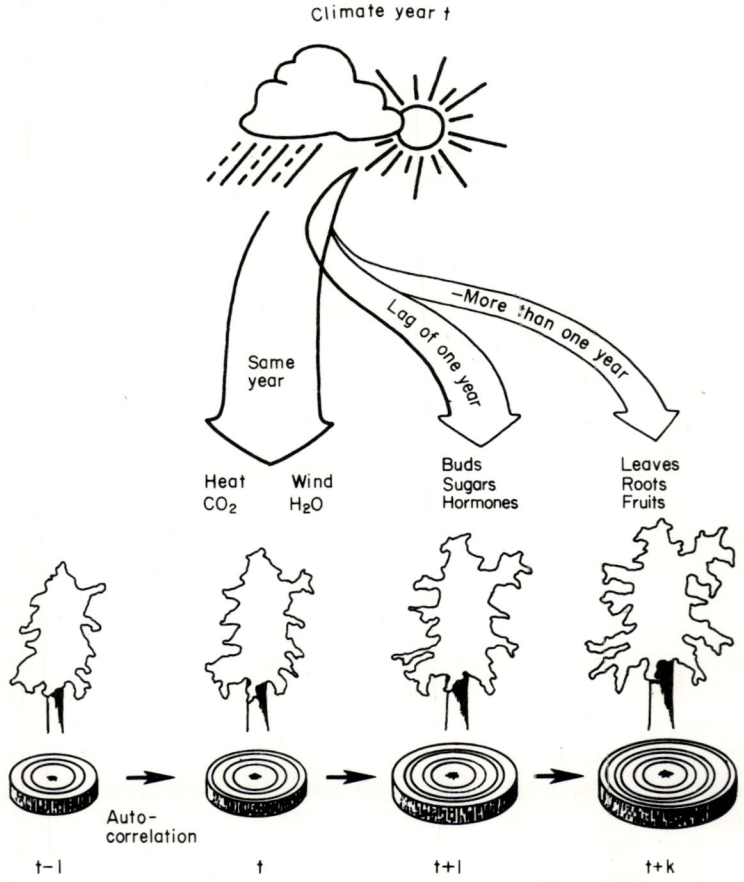

FIG. 1.10. Climate of a given year t is modeled to have a large effect on ring width for the same year through the heat, wind, carbon dioxide, and water that impinges on the tree. However, it can also affect ring width in the following years, $t+1$ up to $t+k$, through effects on buds, sugar, hormones, and the growth of leaves, roots, and fruits. Because of these linkages the width of the ring in year $t-1$ is statistically related to the ring width in year t. This effect is modeled as autocorrelation in ring width. (Drawing by M. Huggins)

The diagram in Fig. 1.10 is a pictorial representation which models the way climate of a given year (t) affects ring widths in a tree not only for the same year t but in subsequent years so that the response lags 1 to k years behind the occurrence of climate. The diagram also portrays the lagging effect of autocorrelation, an association between ring width of year $t-1$ and the widths of the subsequently formed rings for years t, $t+1$ up to $t+k$, which can perturb the causal relationships.

Some early workers thought that ring growth of trees on dry sites was influenced only by soil moisture available during the period of growth. Others working on trees near their polar limits or at high elevations modeled growth as a simple and direct function of temperatures during the period of growth (see Fig. 1.6). Both models, especially the one for arid sites, have been shown to be overly simplified. They were good first approximations for the relationships known at that time, but as knowledge has increased and better statistical techniques have become available, new and more complex models have been developed. The most recent ones, which describe the important biological relationships between climate and ring widths are presented in Chapters 5 and 8, and details on development of statistical models are described in Chapters 6 through 9.

J. Calibration and Verification

The units of ring widths, like those of a thermometer, can be calibrated with units of environmental variables. This can be accomplished by constructing a statistical model (Fig. 1.11) resembling the actual relationships (Fig. 1.10), establishing the values for the statistical coefficients of the model, and then applying the coefficients to tree-ring indices to reconstruct climate for early time periods where ring-width indices are available but where no record of past environments exists. The statistical association giving rise to the calibration may result from a cause-and-effect relationship between tree growth and climate, or it may represent purely correlated effects. Such correlated relationships are useful for reconstructing past climatic variations even though it may be impossible to determine the actual chain of cause and effect.

It is desirable, especially when dealing with correlated relationships, to hold back some of the environmental data for use as an independent check. The *reconstructions* derived from tree rings are compared to actual environmental data to verify the accuracy of the estimate. Such verification is the proof necessary to demonstrate that the reconstructed variations did in fact exist. Historical records, qualitative paleoclimatic information, and other data about past environments can also serve as

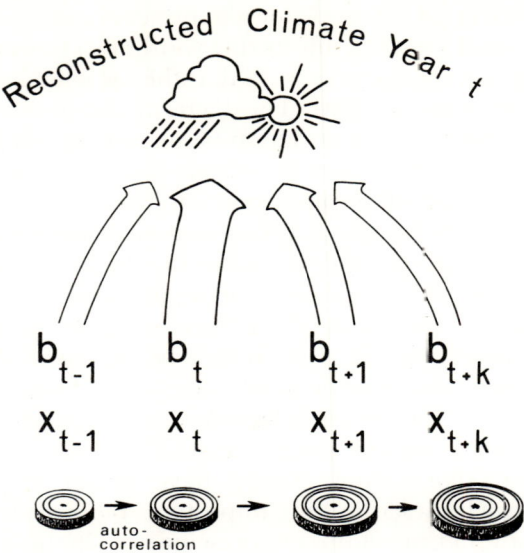

FIG. 1.11. A model for the statistical calibration of the system portrayed in Fig. 1.10, which is achieved by obtaining coefficients or scaling factors, b_{t-1}, b_t, b_{t+1}, and b_{t+k}, which can be multiplied by the corresponding ring widths and indices, x_{t-1}, x_t, x_{t+1}, and x_{t+k}, and the results added to obtain an estimate or reconstruction of the climate in year t. The scaling factors are then applied to ring-width indices for early time periods to reconstruct past climate. The scaling factor b_{t-1} is included in the model to correct the causal linkages for any statistical autocorrelation effects. (Drawing by M. Huggins)

additional checks. A variety of approaches to calibration and reconstruction have been used. Those which have proven most useful are described in Chapters 7, 8, and 9.

V. Dendroclimatic Procedures and Analysis

Now that the terms *dendrochronology* and *dendroclimatology* have been defined and the major principles of the field have been described, one might ask a number of questions: Just what is dendroclimatic analysis? How are the principles to be applied? What kinds of problems can it solve? How are the appropriate trees found and sampled? What are the actual laboratory procedures that are employed? How are the data processed and analyzed? What kinds of analyses can be made regarding past climates? This section includes a step-by-step description of how a hypothetical problem might be approached and how the principles are employed. The details and the scientific basis for these procedures are described throughout the remaining pages of this book.

For purposes of illustration, let us follow a present-day study through the basic steps of dendroclimatic analysis. Let us assume that a climatologist is working in a semiarid area for which there exists only a 20- or 30-year climatic record. By using the extended record provided by tree rings, he hopes to answer several questions: Are the mean and the variations expressed in the climatic records representative of the long-term mean and variations as reflected by the rings of 200- to 300-year-old trees? Is the recorded climate wetter or drier than that of the past? What is the time structure of recurring climatic extremes, such as drought? He may wish also to model the atmospheric circulation in hopes of evaluating possible climatic changes in the future and in the more distant past. For such an endeavor he could use available meterological records, plus the extended record from tree rings as a *proxy* of climate, to ascertain variability of past climate and to guide him in deciding whether his estimates of future changes are within the ranges of past climatic variations.

How could this scientist proceed to collect materials, process data, and exploit information in tree rings? He would first collect data from a selected site, perhaps a relatively undisturbed and open forest where the climatic factor of interest (in this example, drought) is most limiting to tree growth. If he wishes to calibrate his collections with a particular weather record, he would choose an area where a weather station has been maintained for a number of years. He need not be overly concerned about the precise location, as trees may be sampled at distances of 20 or more miles (32 km) from a weather station without substantially reducing the correlation between ring width and climate at the recording station (Julian and Fritts, 1968). It will be shown in Chapter 9 that not every tree-ring sample need be associated with a nearby weather station when large-scale variations in climates are to be reconstructed from tree growth.

The sampled trees would need to be of the sensitive type and located at the lower elevational or drier limits for the particular species. The form and structure of the tree often provide clues as to its age and the quality of the ring-width record (Fig. 1.12). Within these restrictions of site, the researchers could select trees randomly, though most workers search for the oldest trees with the most ring-width variability to maximize the amount of information on climate that can be obtained.

The samples are usually collected with an *increment corer* (or borer) (see Fig. 1.2), which is used to extract a thin cylinder of wood from the stem. The corer usually does not harm the tree and exuded resin from conifers soon seals the wound. The core holes have, however, been shown to be penetration points for certain diseases (Hart and Wargo, 1965), and in

FIG. 1.12. Tree form provides valuable clues as to the tree age and degree of stress on the site. Characteristics of old age exhibited by this *Austrocedrus chilensis* from Aconcagua Province, Chile, are (1) gnarled snag top from which all small branches have disappeared, (2) the narrow strip of living bark, shown on the left, which formed an asymmetrical buttressed trunk, and (3) the small cluster of dense foliage on the living branch, intermixed with dead and dying twigs. Some or all of these characteristics are exhibited by many different species of old-age conifers (Schulman, 1956). The tree is on a typical barren, rocky slope, is over 1000 years old, and has a sensitive and datable ring-width pattern reflecting moisture stress. (Courtesy of V. C. LaMarche, Jr.)

such cases simple precautions such as sealing the hole, using a disinfectant, or not sampling during times of infection may be advisable.

It is generally more expedient to locate the best sites by first coring a few trees and immediately examining their rings to check for width variability and crossdating. If the rings are generally wide and exhibit little width variation, then the researcher can conclude that climatic factors have not been very limiting to tree growth in the site, and he can move to drier, more exposed, or rockier sites where climate is likely to be more limiting. If the second choice of sites is too extreme, the worker will find that the samples exhibit many *partial* (locally absent) rings, so that the accuracy of crossdating is questionable. He would then move to slightly less extreme sites or sample a different species. A knowledge of nearby chronologies and a brief visual examination of the specimens for synchronization during the critical growth years can be a great help in selecting the optimum sample.

Once the general ecological niche and appropriate species for good crossdating and high sensitivity are located, at least two radii from 20 or 30 trees of the same general age in a given site are sampled. Replication by sampling two radii per tree is sufficient to allow analysis of the variations within and among trees. If one is interested in evaluating the chronologies and the site factors of specific trees, he may choose to replicate with a large number of cores per tree, although collection of more than two radii per tree is usually not worth the effort for many dendroclimatic analyses.

After returning to the laboratory, the scientist will dry, mount, and surface the cores for purposes of examination. Crossdating, as practiced in North America, is a time-consuming process (Table 1.II) in that it

TABLE 1.II Average work time required to collect and process a 200-400 year ring-width chronology

Task	Time (in person-hours)[a]			
	Mean minimum	Mean	Mean maximum	Mean %
1. Collection[b]	11	15	23	7
2. Specimen preparation	7	12	17	6
3. Dating	51	72	120	33
4. Dating check	10	17	28	8
5. Measuring	30	39	53	18
6. Measuring check	20	22	27	10
7. Keypunching	6	6	7	3
8. Computer set-up	4	6	8	3
9. Output check	5	15	19	7
10. Project supervision	8	11	15	5
TOTALS	152	215	317	100%

[a] Based upon estimates for each category made by 8 persons, figures rounded to nearest hour. [b] Assuming collection of 50 cores from 25 trees by 2 persons. Figures include field documentation but not travel time.

requires both tedious examination of the specimens by eye, an operation that can be facilitated by certain graphical techniques (Stokes and Smiley, 1968; Ferguson, 1970a; Bitvinskas, 1974), and computer analysis. It may be helpful to start with a small number of the most easily dated or oldest specimens and to gradually incorporate the more difficult specimens in this group.

In areas of mild and moist climate, such as western Europe, the variations in ring width may be too small to permit easy dating by eye. In such cases all specimens may be measured before dating, and certain computations can be made to help identify the correct yearly sequence (Eckstein and Bauch, 1969; Baillie and Pilcher, 1973; Bitvinskas, 1974). However, if some rings are missing from any portion of the samples, it is difficult to use these particular computer methods and dating may be best accomplished by eye first. Even when a computer program is used, the worker must reexamine the various matches made by the computer and make a final judgement based upon his knowledge of the sample and visual inspection of the wood. Only after exact dating is obtained for each ring can the data from many trees be pooled and the attempt be made to reconstruct variations in past climate from the mean ring-width variations in the trees.

At the Laboratory of Tree-Ring Research, Tucson, as each core is dated the specimen is marked to facilitate the process of measurement. The rings identified with the end of each decade, each half century, and each century are marked differently with a needle prick, as are difficult areas where rings may be partial, absent, or hard to see (Stokes and Smiley, 1968). In the case of extremely arid sites, 10 or more trees are selected which have the longest, most sensitive, and most complete ring-width records, as shown by the two most complete radii sampled from each tree. If the sites are less extreme, if the rings do not vary greatly in width, or if crossdating between trees is poor, a sample of 20, 30, 40, or more trees may be an appropriate sample size. Sometimes samples of 60 or more trees have been obtained.

After the width of each dated ring is measured, the values are coded on computer cards or written on magnetic computer tape. All of the coded data are computer-processed through a preliminary cleanup routine; the data are listed to assist checking for tabulation and coding errors; and certain calculations are made to facilitate selection of materials for subsequent analysis. The mean ring widths for 20-year intervals, overlapping by 10 years, are plotted to assist in choosing the appropriate standardizing curve. After the output from this cleanup routine has been checked, errors corrected, and obvious anomalies of growth deleted, the data are resubmitted for final processing and analysis. Table 1.II includes estimates of the time involved in collecting and processing a typical dendroclimatic sample.

With field work and laboratory work done, the researcher is ready to analyze his data. Analysis usually proceeds in four basic phases:

(1) A growth function of some form is fitted to each measured radius by means of a curve-fitting computer technique (see Chapter 6). Different curves for different specimens may be specified by using appropriate computer control cards. Standardization is accomplished by dividing the ring width by the value of the fitted curve for the particular year, which removes the systematic changes in ring-width values associated with increasing tree age. The resulting indices from all rings formed in each year are averaged to obtain a mean index chronology for individual trees, subgroups, and sites, depending on the sampling design (Fig. 1.9).

(2) Certain statistical characteristics are calculated from the indices of each measured series. Additional statistics are obtained from all chronologies representing the merger of two or more sets of ring-width measurements. Finally, if a worker desires more extensive statistical measurements, the necessary computations are made. A variety of statistics described in Chapter 6 can be used to characterize a sample, which in turn can guide further collection or subsequent analysis.

(3) The relationships between ring-width indices and climate are modeled, and the degree and character of the relationships are assessed. A variety of statistical tools is described in this text which can assist in this part of the analysis.

(4) The relationships between ring width and climatic variation established in phase 3 are then used to reconstruct past climate, and these reconstructions are verified by independent data. After verification is obtained, the reconstructions may be compared directly with meteorological records to test the proposed climatological question.

The most rapid dendroclimatological advances in recent years have been made because new statistical techniques have greatly increased our capabilities in the last two phases of analysis. Models in the past were developed in a rather rigid *a priori* fashion, often using preconceived notions about growth processes. Meteorological data were transformed by the model to a single synthesized variable which supposedly resembled the way the trees responded to climate, and the growth data were used only for comparison with the synthesized variable to see whether a relationship existed. If the results were judged to be unsatisfactory, the model was abandoned and a new *a priori* model developed.

For example, precipitation sometimes was summed over an interval of several months to a year, because the sum was thought to represent the moisture available for tree growth. In more sophisticated studies precipitation and temperature were combined in a way to simulate conditions of water stress (Palmer, 1965; Zahner and Stage, 1966). In studies at high latitudes temperature was averaged for several-month intervals or transformed in a fashion to resemble the energy available to growth (Dahl and Mork, 1959).

However, the growth itself provides valuable information which can be used in model development. This is accomplished by starting with a general and flexible *a priori* model and applying statistics in *a posteriori* fashion to shape the general model to the specific yearly tree-ring responses.

For example, variables of climate derived from monthly meteorological records, such as precipitation, temperature, or evaporation, can be compiled for a number of intervals corresponding to different seasons (Fritts *et al.*, 1965c; Fritts 1969). Several of these variables are used to estimate ring-width indices statistically, using multiple regression techniques to obtain the optimum relationship. The resulting statistical equation is an *a posteriori* model in which coefficients are the statistical expression of the way the trees are responding to variations in climate. A number of statistical equations may be developed using different types of measurements and applying more and more complex models in a stepwise fashion, adding one variable at a time, and testing for improvements. The gradual increase in complexity of the general model allows assessment of the relative importance of each variable at each stage of complexity. The final models are derived in *a posteriori* fashion by selecting those that appeared physiologically most reasonable, as well as those that provided the most significant statistics.

The final models may become so complex that the individual coefficients of the equations cannot be interpreted directly. In such cases the statistical equations can be solved by substituting values for one climatic variable at a time. Calculated changes in growth associated with specified changes in each climatic variable are used to estimate the relative importance to growth of each climatic variable. Now, however, through the use of principal component analysis it is possible to utilize the characteristics of the meteorological data in compiling the variables for regression analysis. Fewer *a priori* decisions are required, the statistical problems are simplified, and more precision is obtained in the final analysis. Details on this new approach to statistical modeling are described in Chapters 7 and 8.

Once the tree-growth responses to climatic variables have been suitably tested, past variations in tree growth can be used to deduce climate. The simplest approach is to plot the variations in past growth, observe the intervals of unusually high and low growth, and to use the best tree-growth response model to deduce past climate.

A more elaborate approach is to select ring-width chronologies which respond to the same climatic factors and to use the spatial variations in growth to infer spatial variations in climate. The growth values can be scaled for purposes of comparison; the data are then averaged by decade, these values plotted on maps and isograms drawn to facilitate analysis (Fig. 1.13) (Fritts, 1965). The areas of higher than average and lower than average growth are inferred to be areas dominated by the appropriate anomalies in climate. Hence, the extremely high growth at the beginning of the 20th century followed by low growth in the 1930's (Fig. 1.13b) is one of many lines of evidence suggesting that the climate changed around the turn of the century.

The anomalies in growth for a particular period of time may be compared to the anomalies in another period of time to deduce the extent of a change. For example, the mean ring width and its variability as measured by the standard deviation for western North American chronologies from 1901 to 1960 were compared to the same variables for the longer period from 1631 to 1932 (Fig. 1.14). Large positive anomalies in growth and, by inference, high moisture supply can be inferred for many areas in the West. High variability in moisture may also be inferred especially for the Pacific Northwest. As in the previous example, the particular climatic anomaly is inferred from a knowledge of the growth response to climate.

However, the same multiple regression and principal component techniques used for *a posteriori* modeling of the tree growth response may be applied to the calibration of the tree-ring data and the statistical reconstruction of past climate. For example, it will be shown in Chapter 7 that spatial patterns in tree growth throughout western North America can be calibrated with spatial patterns of surface air pressure during 1900-1962 for half of the Northern Hemisphere. A series of equations is derived which transfers variations of ring-width indices into estimates of variations in pressure. Ring-width indices are then applied to these transfer functions to reconstruct the associated patterns of surface pressure as well as other variables of climate. Maps are constructed of the estimated anomalies in pressure for past periods and areas where few measurements of pressure are available.

Such statistical methods are significantly more powerful than the simple mapping and inference techniques, because a large number of tree-ring chronologies can be handled simultaneously and their

36 TREE RINGS AND CLIMATE

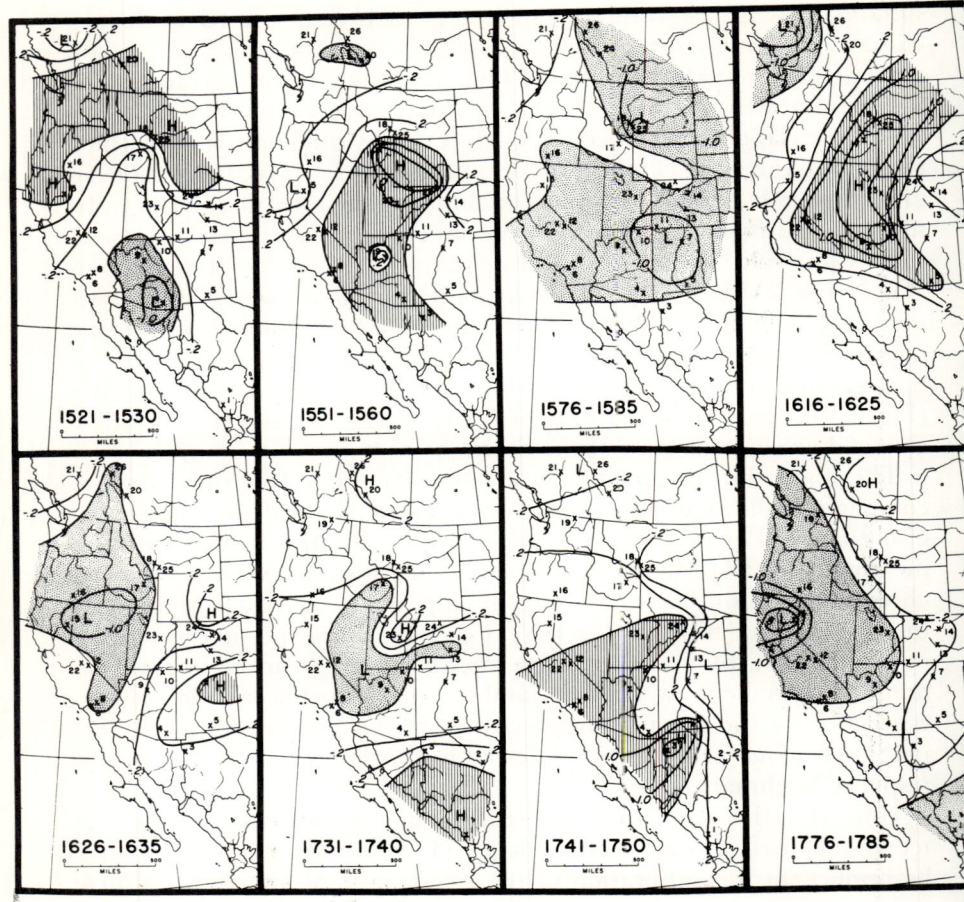

FIG. 1.13a

differences as well as their similarities utilized to estimate anomalies in a much larger spatial grid of climate. Other statistical techniques of importance to dendroclimatic reconstruction include analysis of variance, conditional probability, power and cross-power spectral analysis, and digital filtering. The details of these and other techniques are described in subsequent chapters.

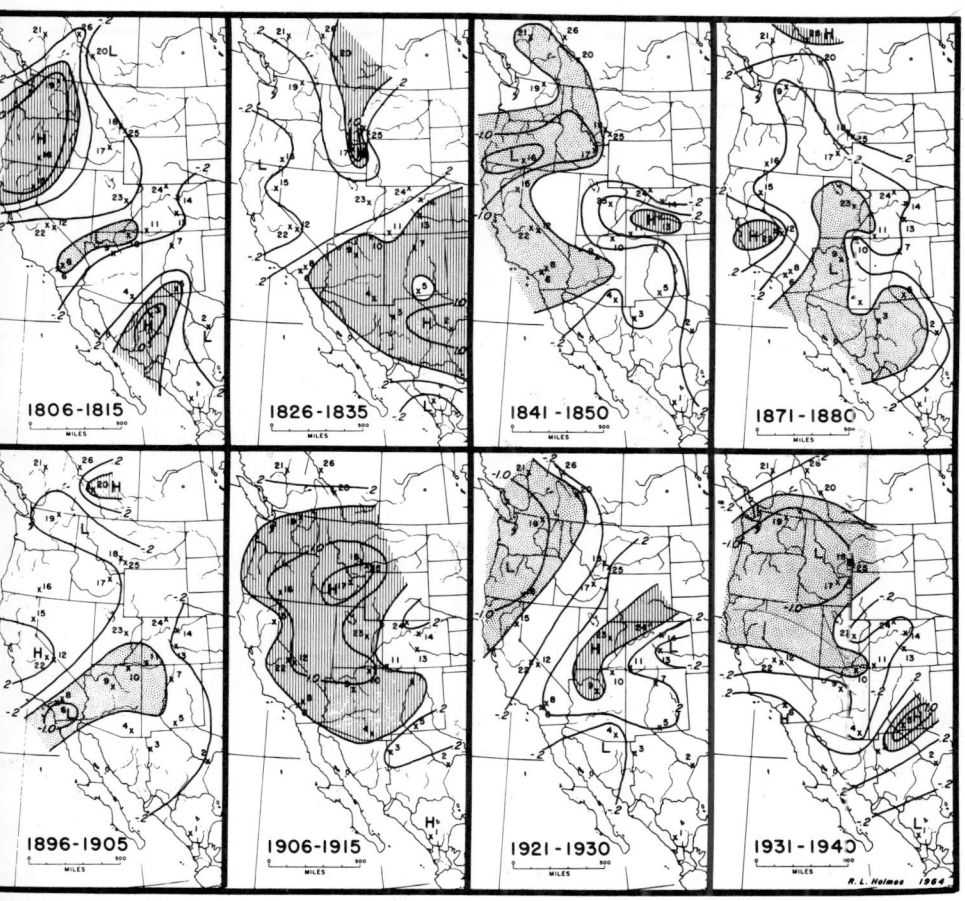

Fig. 1.13b

FIG. 1.13a and b. Decades since A.D. 1500 for which there are large departures from average in ring width over western North America. Data are from arid-site trees and are scaled to standard units by subtracting the means and dividing by the standard deviations for the period A.D. 1651–1920. Isograms join values of equal growth departure and the shaded areas of high growth (H) and low growth (L) designate those departure values equal to or exceeding values of 0·6. There are some similarities between the growth patterns prior to 1900 and after 1900, but marked differences occur. These differences indicate that the climate of the present is somewhat different from the climate of the past. (From Fritts, 1965)

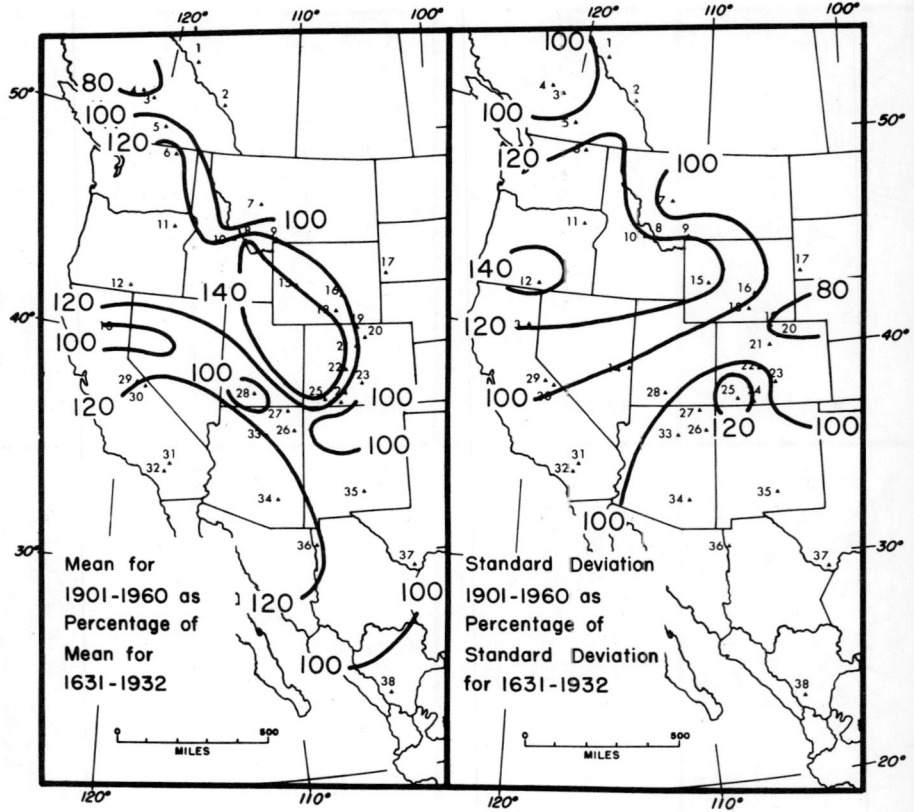

FIG. 1.14. Ring-width data for different time periods can be compared to reveal particular anomalies in climate. The means and standard deviations of 38 tree-ring chronologies for A.D. 1901–1960 are expressed as a percentage of the means and standard deviations for the longer period, A.D. 1631–1932. The mean growth and, by inference, the moisture supply have been anomalously high in the extreme Southwest, along the western slope of the central Rocky Mountains, and in the Pacific Northwest during recent times. The variability of moisture as revealed by the standard deviation of ring width has been high in the Northwest and locally in the Southwest. (From Fritts, 1971)

VI. Examples of Analysis

While the sequence of steps and the data used are similar for many kinds of dendroclimatological analyses, there are alternative procedures and innovations that have been applied. The following paragraphs describe some of the most interesting examples.

LaMarche and Harlan (1973) sampled and built a ring-width chronology for *Pinus longaeva* at the upper elevation limits for the species. The chronology in these trees was statistically modeled with temperature and precipitation, and narrow rings were found to be correlated with low temperatures (LaMarche, 1974a). Also it was noted that the tree-growth response lagged behind and averaged the effects of climate for periods as long as 15 or more years. LaMarche recognized that if he used the standard indexing procedures, a large portion of the averaged and lagged response would be removed. There were portions in the early part of the trees' ring records, however, that reflect the vigorous growth of youth. Therefore, he discarded those portions of his ring-width record and then averaged the actual ring-width measurements occurring in the remaining record. In order to bring out the long-term changes in growth, he averaged the ring widths for 100-year intervals (Fig. 1.15).

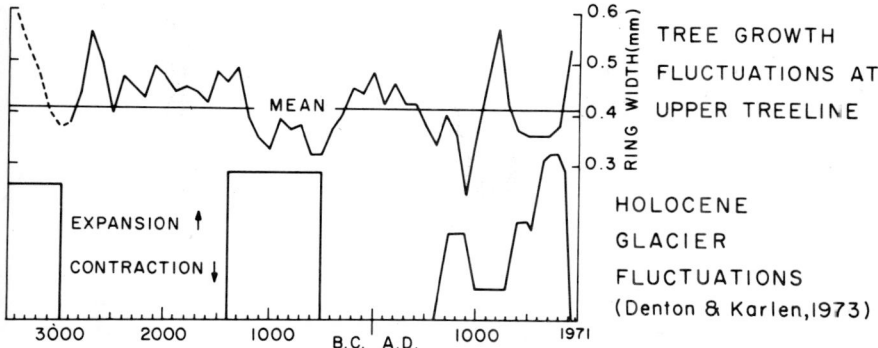

FIG. 1.15. Upper plot shows the average ring-width in temperature-sensitive *Pinus longaeva* in the White Mountains of California taken for successive 100-year periods centered on the first year of each century during the past 5300 years. Lower plot shows the periods of expanding and contracting mountain glaciers. Good comparison is seen between periods of inferred low summer temperature and periods of growth and advance of glaciers. (Courtesy of V.C. LaMarche, Jr.)

Since LaMarche lacked sufficient climatic data to verify his results by 100-year intervals, he examined the glacial record in the nearby Sierra Nevada of California and in other regions of the Northern Hemisphere and found good agreement between periods of inferred low temperature in summer and periods of growth and advance of mountain glaciers (Fig. 1.15).

Another study by LaMarche and Fritts (1971b) involved two

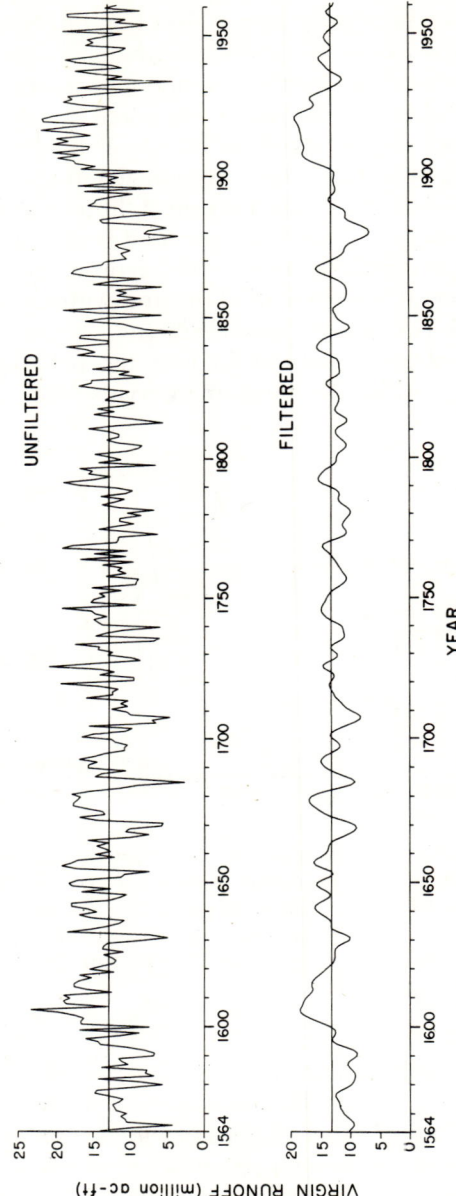

FIG. 1.16. The annual virgin streamflow of the Colorado River at Lee Ferry as reconstructed by using ring-width index variation in trees on 17 sites in the Colorado River Basin. The calibration period was 1896–1961, and growth for each year and the three following years was used to statistically estimate streamflow. The smoothed curve shown below the reconstruction represents essentially a running 10-year mean (see Chapter 8). (Reprinted by permission from "Long-term Streamflow Records Reconstructed from Tree Rings", Paper of the Laboratory of Tree-ring Research at the University of Arizona, No. 5, by C. W. Stockton, Tucson: University of Arizona Press, © 1975.)

replicated samples from high elevation *Pinus cembra* in the Alps for which ring-width variations were associated with the percentage of glaciers that were advancing in the Alps. The relationship between the two variables turned out to be linear, so that plots of the tree-ring chronologies were used to identify times in the past marked by glacial advance and retreat.

LaMarche and Fritts (1972) utilized a variety of tree-ring chronologies from arid southwestern North America in making statistical tests for possible linkages between ring width and sunspot numbers. They were unable to establish the existence of any significant relationship and concluded that further search for empirical associations between tree-ring indices and the record of sunspot numbers is likely to prove unrewarding. However, they noted significant periodicities in many tree-ring records at frequencies of approximately 2 years and 22 through 29 years.

Stockton (1971, 1975), and Stockton and Fritts (1971a) describe a study which utilized ring-width chronologies from sites throughout river basins to extend the relatively short streamflow records backward in time. Replicated tree-ring samples were obtained from 3 different river basins of varying size and the several chronologies calibrated with the sequence of streamflow, precipitation, and temperature data for the area. Equations were derived which were applied to the variations in past ring widths to reconstruct streamflow in past years for which no hydrologic record exists (Fig. 1.16).

Stockton and Fritts (1973) studied an environmental problem in Canada created by the construction of a dam. Closure of the dam gates had reduced the normal high flow in the Peace River during the spring and early summer. Before closure of the gates, the high water impounded the water in Lake Athabasca and caused flooding in the waterways forming the Peace–Athabasca delta system. The record of water levels began in 1935, and it was thought the characteristics of flooding for the 1935-1966 period might not be a proper representation of the long-term features of the ecological system in which the plant and animal communities of the area developed. An estimate of the character of flooding for a longer period of time was needed to make decisions for the proper management of the delta area and the releasing of water from the dam.

The nearest trees with a record of climate were at the Arctic tree line some 300 miles (483 km) to the north. However, stands of *Picea glauca* growing along the channels, sloughs, and minor depressions within the Peace–Athabasca delta were found to crossdate and certain differences in the ring-width chronologies were apparent from site to site. It was inferred that soil moisture regimes associated with the water levels in the

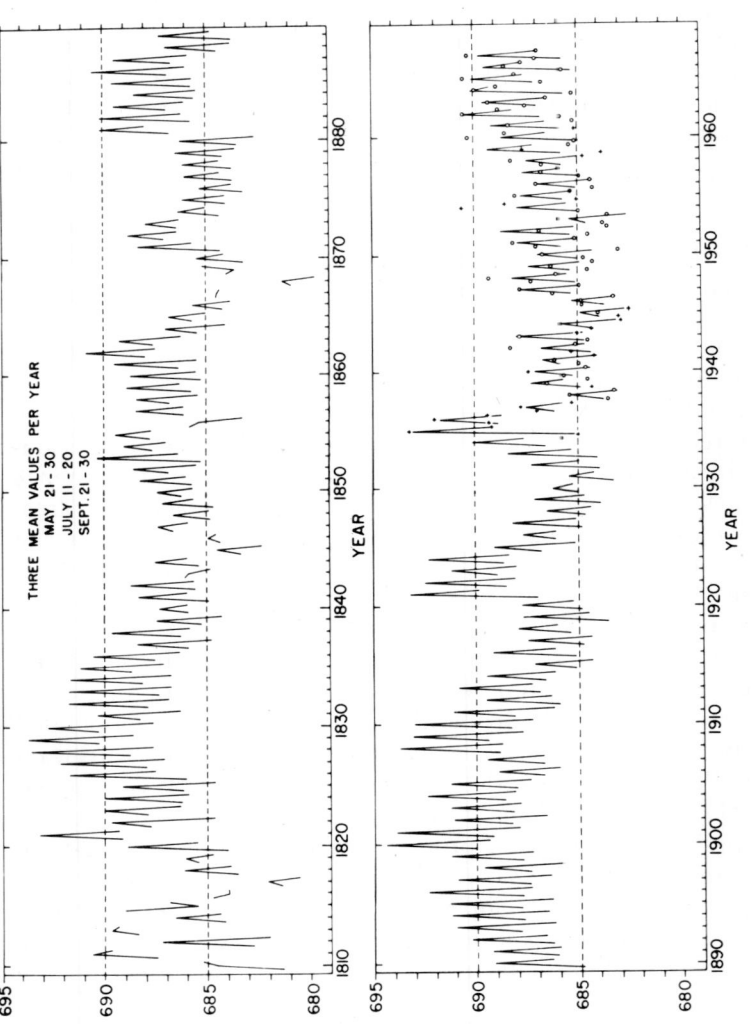

FIG. 1.17. Levels for Lake Athabasca, Alberta, Canada, as reconstructed from tree-ring data. Tree rings indicate that prior to 1935 there was greater variability in lake levels during May and July, but there was less variability in lake levels for September than during the recent calibration period. Points indicate actual lake levels used for calibration and the crosses indicate estimated data. The lines connect the three estimates from tree rings representing mean lake level for May 21–30, July 11–20, and September 21–30. Points are not

channels had influenced tree growth. These relationships were statistically modeled, and the water level records for the lake observed for May 21–30, July 11–20, and September 21–30 were calibrated with the ring-width variation for each year as well as for the following year. Calibration accounted for 57% to 80% of the lake level variation. The calibration equation was utilized to reconstruct the water levels from 1810 through 1934 (Fig. 1.17). The variations in water levels, especially those early in the season, were found to be considerably greater in the longer reconstructed record than in the 33-year calibration period. That is, if the variations in water levels had been estimated from the short, 33-year period, those for the early spring would have been only one-third the amplitude observed for the longer period of 1810–1967. The extended record of lake levels obtained from the trees helped the Canadians to decide on an appropriate management system for the delta area to counter the deleterious effects of the dam.

Dendrochronological studies may be used to solve many environmental problems. There is considerable interest in air pollution damage and its effect upon the growth of trees (Vinš and Tesař, 1969). Trees in the San Bernadino Forest in Southern California that were supposedly killed by the Los Angeles smog (Taylor, 1973) show reduction in the width of the outermost growth rings (Fig. 1.18).

Another study (Ashby and Fritts, 1972) represents an attempt to use tree rings to measure possible inadvertent climatic changes at LaPorte, Indiana. An increase of rainfall had been attributed to increased industrialization and subsequent air pollution. Growth at first increased with the supposed increase of rainfall; but, as the amounts of pollution and rainfall reached their apparent maximum extent, growth became markedly reduced. It was thought that deleterious effects on the trees due to the increase in pollution may have countered the more favorable effects of increased precipitation, and it was concluded that the ring record in oak neither supported nor refuted the anomaly in precipitation. The studies of tree rings and air pollution suggest that tree-ring analysis can be used in a variety of ways to diagnose pollution effects in their early stages before death begins to take its toll on the trees.

New techniques have been pioneered by H. Polge (1966, 1970) in France and developed by M. L. Parker (M. Parker, 1971, Parker and Meleskie, 1970; Jones and Parker, 1970; Parker and Kennedy, 1973) in the use of X-rays and densitometry to obtain measurements of wood density within the ring (Fig. 1.19). The X-ray film negative is scanned with a beam of light, and a photocell measures the intensity transmitted by the film. The resulting electrical signal can be plotted or converted to units of wood density by comparison with a standard. Measurements at selected distances across the ring can be fed directly into a computer and a program used to calculate ring width, density, and cell structure variation making up the earlywood and the latewood. Densitometry could provide a means of gathering much more information on past environmental conditions than is available from only ring width. Parker and Henoch (1971), Heger *et al.* (1974), and Cleaveland (1975) present evidence that X-ray analysis will be especially meaningful for studies in

FIG. 1.18. A section from a Jeffrey pine (*Pinus jeffreyi*) in the San Bernardino National Forest which showed apparent damage from the Los Angeles smog. Note that near the bark (dark area at the top) there is a reduction in the amount of latewood and in ring width that occurred about nine years before damaged foliage was noted and the tree was cut. (Photograph by James Harsha.)

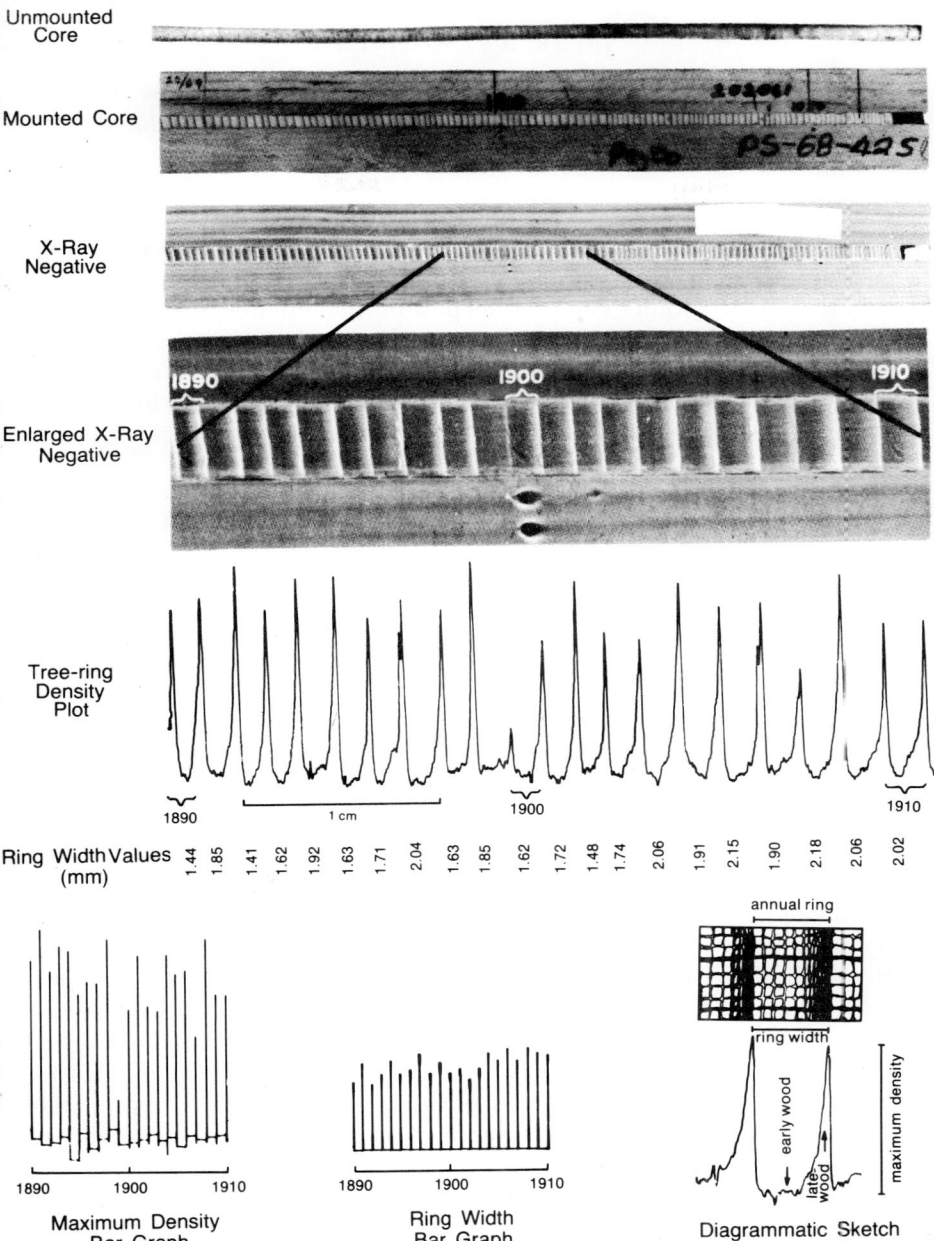

Fig. 1.19. A core extracted from a tree can be mounted and X-rayed to measure changes in wood density. The X-ray negative is then scanned with a beam of light and photocell. The electrical output can be calibrated in terms of wood density. Ring widths can also be measured as the distances between peaks in maximum density. This allows measurement of several features in ring structure which can result from different environmental factors and can provide a variety of information on past environments (From Jones and Parker, 1970)

temperate regions and on mesic sites, where variations in the environment of the growing season are often more highly correlated with changes in cell structure and wood density than with variations in ring width.

The X-ray method is now included as an important part of certain new dendroclimatic and dendroecological work. The applied aspects of the field are changing very rapidly as new improvements are made. This volume calls attention to some of the possibilities, but does not attempt to cover systematically all the important developments, nor does it do justice to the new innovations and applications. The interested reader should watch the forestry, wood technology, and dendrochronology literature for current developments.

VII. Further Definitions and Concepts

This chapter has reviewed some of the historical roots of dendrochronology as practiced in western North America. The field and its various subfields have been defined, and some of the basic principles and concepts have been described. A few selected applications have been mentioned, and the value of dendroclimatology has been stressed. This chapter will conclude with discussion of a few additional terms and concepts which apply to the phenomena which actually link variations in ring width with variations in climate.

A. *The Domain of Climate*

Climate may be defined simply as an expression of meteorological phenomena representing weather occurring over a long period of time. It includes processes of exchange of energy and mass between the earth and the atmosphere. Energy may be represented by temperature, by radiant energy, and by the movement and condition of the earth's fluids, including both the liquids and gases. The aggregate of climatic conditions expressed as means, variance, and extremes for a region over a period of many years is referred to as the *macroclimate*. The data used to describe the macroclimate are often based either upon 30 or more years of observation from a single station or upon the average of several measurements from stations at diverse locations within a particular region.

According to Gates and Mintz (1975) the climate may be regarded as a system with certain properties and processes that produce the climate including its variations:

> The properties of the *climate system* may be broadly classified as thermal properties, which include the temperature of the air, water, ice, and land; kinetic

properties, which include the wind and ocean currents, together with the associated vertical motions, and the motion of ice masses; aqueous properties, which include the air's moisture or humidity, the cloudiness and cloud water content, groundwater, lake levels, and the water content of snow and of land and sea ice; and static properties, which include the pressure and density of the atmosphere and ocean, the composition of the (dry) air, the oceanic salinity, and the geometric boundaries and physical constants of the system. These variables are interconnected by the various physical processes occurring in the system, such as precipitation and evaporation, radiation, and the transfer of heat and momentum by advection, convection, and turbulence.

The term *macroclimate* implies that climate is the expression of a steadystate or a more or less constant system when, in fact, macroclimate may be more appropriately portrayed as an average frequency of occurrence of different *climatic states* which may persist for periods of a month to many years (Table 1.III). Changes in the frequency of these climatic states may be attributed to slowly varying characteristics of the climate system, such as variations in temperatures of the ocean and its circulation, changes in the coverage by permanent ice and snow, changes in the transparency of the atmosphere, or possibly changes in radiation output from the sun. Sometimes climatic states in the past have persisted long enough to be considered climatic change (Gates and Mintz, 1975). There is no general agreement, however, among scientists as to how long a climatic state must persist or how large the variation must be to be designated a climatic change.

TABLE 1.III Components of the climate–tree system involved in dendroclimatology and their spatial and temporal domains

Level	Component	Spatial Domain	Temporal Domain
1	Macroclimate	Regional	Many years
2	Climatic state	Regional	One month to a number of years
3	Weather	Regional and local	Minutes to days
4	Microclimate	Local	Minutes to years
5	Operational environment	Plant surface	Instantaneous to years
6	Plant process	Molecular	Instantaneous to hours

Estimates of the earth's actual macroclimates from station records can be in error if the number of meteorological stations for a given region is small, if station locations are not representative of the area, if observations are poorly made, if the instruments are unreliable, or if the records are short. For example, the averages of climate for the first 30 years of the current century as reported by many stations in western North America can be shown to be biased by anomalously high precipitation and high tree growth (Figs. 1.13, 1.14).

The term *climate* includes a seasonal component representing variations throughout the annual cycle. The term *weather* refers to short-term variations in the climate system, including all observable meteorological phenomena which can change abruptly within a few minutes and vary from day to day. Such phenomena include the precipitation, heat, light, wind, and related factors that affect the day-to-day conditions of the biosphere.

A particular organism experiences only that portion of the weather that occurs within its particular site. A tree, for example, is affected by the moisture conditions of the soil around its roots, the radiant energy on its leaves, and the temperature and movement of gases surrounding its leaves, stems, and roots. The aggregate of meteorological elements occurring within a localized site may be defined as a *microclimate*.

The term *microclimate*, like macroclimate, may refer to the aggregate of local weather for many years. However, in practice, the word *microclimate* often includes relatively short-term measurements in a specific locale which just as easily could be considered the weather or *micrometeorology* of the local site. In this text the term *microclimate* will be used for both short-lived micrometeorological conditions and the longer-term aggregates of measurements for a local site (Table 1.III).

The microclimate varies markedly for different organisms and different parts of the same organism. For example, the top of a forest canopy on a clear day may be exposed to relatively intense light, heating of the leaf surface, and wind, which can dissipate energy and remove moist air that surrounds the leaves. Simultaneously, near the forest floor there may be little light and air movement, high humidity, and leaf temperatures near those of the surrounding air. The microclimate of a hill crest may differ from that of a valley because of marked differences in exposure to wind or because of cold air drainage at night.

B. *The Domain of Environment*

Not all elements of the microclimate influence plants; only those that limit some physiological process will exert an effect on growth. Mason and Langenheim (1957) propose that the term *operational environment* be used for all relationships occurring between micrometeorological elements and the physiological processes which produce such an effect. They propose further that only the specific phenomena which enter such operational relationships at a given time should be considered to be *environmental factors* (Fig. 1.20). Thus the specific factors of the operational environment may change markedly throughout the day, season, and year as well as throughout the life history of the organism (Table 1.III). For example, temperature and moisture may be the most important factors of the operational environment to the germinating seed, while light and minerals in the soil become more important as the seedling grows older, and the operational environment may become increasingly complex as a seedling grows into a mature forest tree.

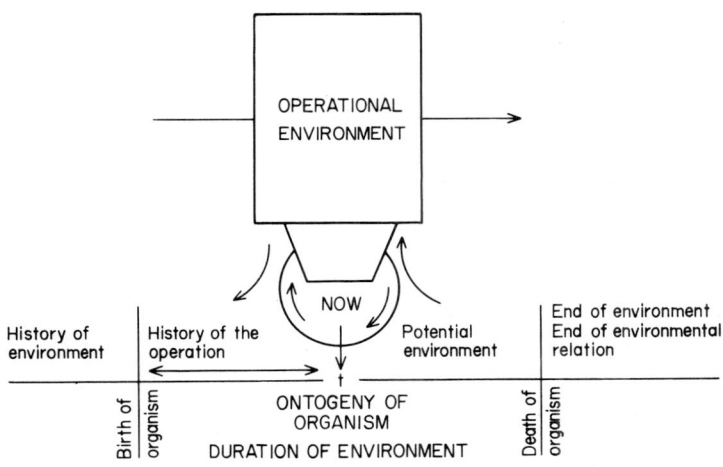

FIG. 1.20. A diagrammatic representation of the concept environment as described by Mason and Langenheim (1957). The environment is an organism-related concept and exists only when it has an operational relationship with the organism at time t, shown in the diagram as NOW. The components of the operational environment vary with the life history of the organism. The environment begins with the birth of the organism and ends with its death. Portions of the potential environment become used in the operation. Old relationships become history of the operation. The operational environment may be influenced by the history of the environment as it is evolved through geological time and successional changes in the biotic community on the site. (Redrawn from Mason and Langenheim, 1957)

Each environmental factor can enter into numerous interacting relationships in different parts of the plant, at different times throughout the year, and at different times throughout the life of the tree. For purposes of growth analysis it may be practical to consider an environmental factor as detrimental or beneficial only if its net effect is to decrease or increase growth. For example, an increase in solar radiation may increase light intensity and raise leaf temperatures. With more light there is an increase in photosynthesis and the making of foods, but with the increase in temperature there is also increased respiration and increased consumption of food. The higher temperatures can enhance the potential for loss of water from the tree, and the increase in solar radiation may influence processes in other parts of the plant. It may heat the soil surface, increase the evaporation of water from the soil, and reduce soil moisture. If the reduction of moisture in the soil impinges upon water absorption, growth can be severely curtailed. The net effect of increased solar radiation, then, is favorable only if the beneficial effects of increased photosynthesis are not offset by the detrimental effects of increased respiration and reduction in available water.

There are many phenomena that we have traditionally regarded as environmental factors which are not a part of the operational environment as defined above. Some of these factors such as the presence of a disease or insect pest in the area, the potential for fire, or the absence of an important insect pollinator are capable of entering into a reaction with an organism but may not yet have done so. Mason and Langenheim (1957) class such factors as the *potential environment* (Fig. 1.20). As the environmental relationship between the microclimate and plant processes proceeds through time, the potential environment may be portrayed as being used in the operation, and a history of the operation is generated. The variations in ring width and cell structure are in one sense a recorded history of the past variations in the tree's operational environment.

Any phenomena not directly affecting the organism should be excluded from the concept of *environment* (Mason and Langenheim, 1957). Factors excluded are indirect factors and interactions that ultimately result in a phenomenon to which an organism later reacts directly, and factors to which the organism is indifferent. Such phenomena are a part of the history of the environment as depicted in Fig. 1.20, but they are not a part of the operational environment. The history itself can be important, however, because of its place in a sequence of events which leads to the final conditions that become part of the operational environment.

The difference between the history of the environment and history of the operation is also illustrated in Fig. 1.20. The latter, which is most important to tree-ring analysis, starts with the birth of the organism and extends through the development of the organism which can be viewed from any given point in the plant's development shown in the figure as time, t, or NOW. The operational environment ends at death of the organism because death marks the termination of the environmental relationship.

C. The Domain of Physiological Processes

The physiological processes are the physical and biochemical changes (such as diffusion phenomena and chemical reactions) which occur along plant surfaces, throughout all living structures of the organism, and sometimes in structures that are dead. All reactions, developments, and structures of an organism are the cumulative effects of the environment acting on growth processes. However, all processes also depend upon the genetic constitution of the individual organism. The relationships between heredity and environment as they affect processes and plant structures, such as ring width, may be diagrammed as follows:

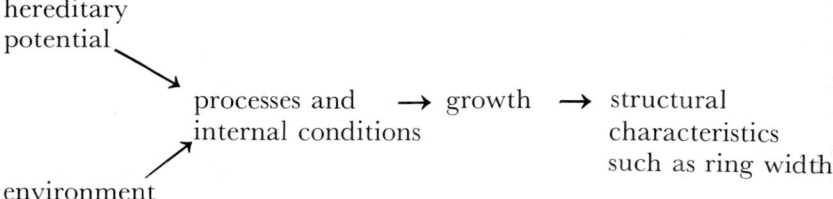

Heredity determines what processes can occur, what environmental factors can affect them, and what range of each environmental factor will have a limiting effect. As mentioned earlier, internal conditions can also occur which modify the effects of the external environment.

Some factors of the operational environment, such as frost or extreme summer drought, may have immediate effects on ring growth and structure. They influence plant processes and can produce a chain of reactions in the tree which, over a period of minutes or hours, affects processes that modify or limit growth. Other factors of the operational environment, such as wintertime drought, may have no immediate effect on growth, since the growing tissue is dormant. They may, however, influence plant processes or change conditions within the plant so that at some later time they become limiting to growth. Factors such as competition for moisture, shading, and soil development also produce systematic changes in the operational environment. The effect of these factors is seen as variation in ring structure and size, which change systematically or vary slowly throughout the life history of the plant. Other factors such as death of a neighbor, fire, pollution, and defoliation by insects can result in sudden changes in the environment and in the structure of the tree. An abrupt increase or decrease in growth is usually followed by a systematic readjustment while the tree and its neighbors gradually attain a new equilibrium with the imposed change.

VIII. The Climate-Growth System and Its Dendroclimatological Significance

For purposes of this book the components listed in Table 1.III are considered parts of a system in which the elements of weather and climate affect the tree microclimates, the operational environments, and the physiological processes which in turn control growth. Chapters 2 through 5 deal with the variety of physical and biological processes that are involved in the environment–growth linkages, and models described in Chapter 5 illustrate the many interactions and interlocking relationships. However, in spite of many interactions and complicating feedback relationships the primary flow of the system is unidirectional, starting with the climatic input and resulting in the ring-width output.

The degree of cause-and-effect linkage, that is, the influence which one phenomenon has upon another, is referred to as *coupling*; while the association between variations in two phenomena without regard to cause and effect is referred to as *correlation*. Phenomena that are directly coupled are generally correlated, but there are many phenomena which are correlated but not directly coupled. Usually, however, correlation between two phenomena is due to at least indirect coupling. Long chains of interacting cause-and-effect linkages can be involved.

For example, variations from day to day in the microclimates of a forest are more or less coupled with variations in the regional weather, because the weather systems control the inflow and outflow of energy and moisture available in the microclimates. The temperature of a single leaf is tightly coupled to the temperature of the surrounding air, the energy that is absorbed and lost by the leaf, and the disposition of the energy within the leaf. A number of physiological processes in the leaf can be affected by temperatures, which in turn can be coupled with conditions affecting growth. However, if growth is limited by other conditions or processes at the particular time, there can be no immediate and direct coupling between leaf temperatures and growth.

Since there are many leaves in the crown of a tree, the activities of growing tissues at the base of the stem are loosely coupled with the conditions in any particular leaf, but they can be more tightly coupled to the mean condition of all the leaves. In addition, the width of the ring is the average of growth processes throughout the entire growing season; so the width is very loosely coupled to conditions within the tree at any particular time. Thus, ring width is not correlated with specific conditions within the tree but with conditions averaged through both space (the tree tissues) and time (a number of months throughout the year).

In the discussion that follows concerning the many processes occurring throughout a tree, there are so many alternatives for cause-and-effect relationships and there is such a variety of microclimates from the top of the crown to the deeply buried roots, that possibilities for significant correlations between growth and climate appear exceedingly remote. However, the practice of site selection assures that only those trees are sampled in which growth-controlling processes are correlated with variables of climate; and the procedure of replication assures that a large number of rings are sampled so that much of the variation within and between trees is averaged for each site.

Thus the variations from one year to the next in the ring-width chronology are correlated with the variations in macroclimate not only because plant processes are coupled with environment, but also because the diversity in the microclimates and the physiological relationships have been averaged by the growing tree as well as averaged in the procedures of chronology development. The correlation between ring widths of many trees and regional climate makes it possible to statistically calibrate ring widths with certain variations in regional climate and thereby to reconstruct past climatic variations from past variations in ring widths.

If trees are in environments altered by nonclimatic factors such as fire, disease, or logging, changes in growth may occur which are not correlated with variations in the macroclimate. The ring record from such trees will show anomalies in growth which are not the same in other trees of the region. In such instances ring-width variation should not be used to represent variations in macroclimate. However, an investigator who is more interested in nonclimatic factors may seek out such trees, because the anomalies can be used to date and measure the magnitude of a nonclimatic change. Intensive sampling of rings from trees throughout a disturbed site can document the extent of the disturbance. Affected trees may be limited to some extent by both climatic and nonclimatic factors, however, and when this is the case it is possible to estimate the variations in past climatic factors while documenting site disturbance. Climatically related patterns are subtracted from those of the affected trees, and the remaining variance is attributed to the nonclimatic effects (Jonsson, 1969; Alestalo, 1971; Nash *et al.*, 1975).

The sampling, evaluating, and ultimate use of ring-width data can vary depending upon the problem under investigation. In most cases, the success of a study depends upon the investigator's ability to decipher the anatomical evidence in the tree ring, to relate this evidence to the processes of the tree that affect growth, to model appropriately the relationships between variations in environmental factors and growth, and to infer or estimate statistically the reconstructed environment or associated anomaly in climate.

The following four chapters describe the features of growth and discuss the physiological processes that contribute most to yearly variations in ring width. The important environmental variables that may limit each process are described, and these are related to the meteorological factors and energy and water budgets in the site. Biological models are then presented which relate climatic factors of precipitation and air temperature to variations in ring width. Chapters 5 through 9 deal with statistical characteristics and statistical analyses of tree rings and describe ways of calibrating variations in ring width with variations in climate. Various approaches to reconstructing climate using tree-ring data are presented, along with illustrations drawn from published as well as some unpublished works including current efforts.

Chapter 2

Growth and Structure

I	Introduction	55
II	Gross Structure	56
III	The Vascular Cambium	63
IV	Growth—A Variable Process	68
V	Variations in Shoot Growth	70
VI	Variations in Radial Growth	74
	A. Measuring radial growth	75
	B. Diurnal periodicities in radial growth	80
	C. Seasonal periodicities in radial growth	84
VII	Phenology and its Relation to Ring Growth	103
VIII	Systematic Variations in the Width and Cell Structure of Annual Rings	107
IX	Growth of Roots	113
X	Significance of Growth and Structure to Dendroclimatology	115

I. Introduction

A knowledge of tree growth and wood structure is necessary for constructing realistic models relating environmental factors to ring widths or other anatomical features of the ring. For example, it would be difficult to specify in what ways variations in climate affect the width of rings without knowing where and when a tree grows, how growth can vary during each season and throughout the life of a tree, and how climatic factors affect the various chemical and physical processes that influence growth. An understanding of the manner in which seasonal growth begins and ends is also essential to explaining how two rings can form in one growing season or how a ring can be absent in certain parts of a stem. This chapter will describe the variety of ways in which the growing processes and woody structures can vary throughout the stems and roots of a tree. It also will set forth other relevant facts about growth and structure necessary to the application of dendroclimatic techniques dealt with in subsequent chapters of the book.

II. Gross Structure

In order to establish a working vocabulary for dendroclimatology, it is necessary to review some basic botany and to define several of the important terms and processes related to tree growth, beginning with tree structure. There are six principal organs of a tree: the roots, stems, and leaves, which are *vegetative organs*, and the flowers, fruits, and seeds, which are *reproductive organs*. Each of these organs is made of cells organized into various tissues, which are identified on the basis of appearance, structure, location, or function.

For example, a tissue named on the basis of appearance, its green color, is *chlorenchyma*, which includes all cells colored by the green pigment, *chlorophyll*. A tissue named on the basis of appearance and structure is *parenchyma*, relatively large cells with thin walls. Both *epidermis* and *mesophyll* are tissues identified on the basis of location, the former being

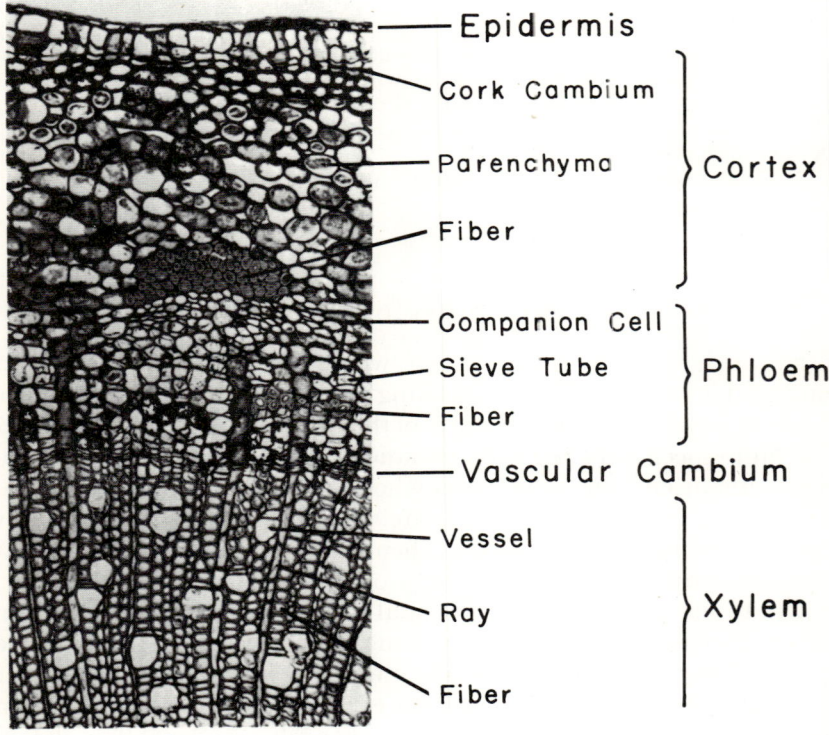

FIG. 2.1. A cross section of a young stem of *Populus deltoides* showing various tissues. (Courtesy of Tillman Johnson, Ohio State University)

one cell in thickness and located on the outside of leaves, young stems, and young roots, and the latter including all cells located between the upper and lower epidermis of a leaf. Two tissues named on the basis of their function in the vascular system are *xylem*, the water-conducting tissue which makes up the woody cylinder of a tree, and *phloem*, the food-conducting tissue in the bark (Fig. 2.1).

All tissues which contain cells that are capable of dividing and producing new cells make up a functional class of tissues called *meristems*. The thin layer of meristematic tissue joining the wood and the bark produces most of the vascular system and is referred to as the *vascular cambium*, while the cork-producing meristem (if present) in the outer bark is referred to as *cork cambium* (Fig. 2.1). Whenever the word *cambium* is used without its modifier, it refers to the vascular cambium. The *apical meristem* is the tissue capable of dividing at the tips of roots and stems (Fig.

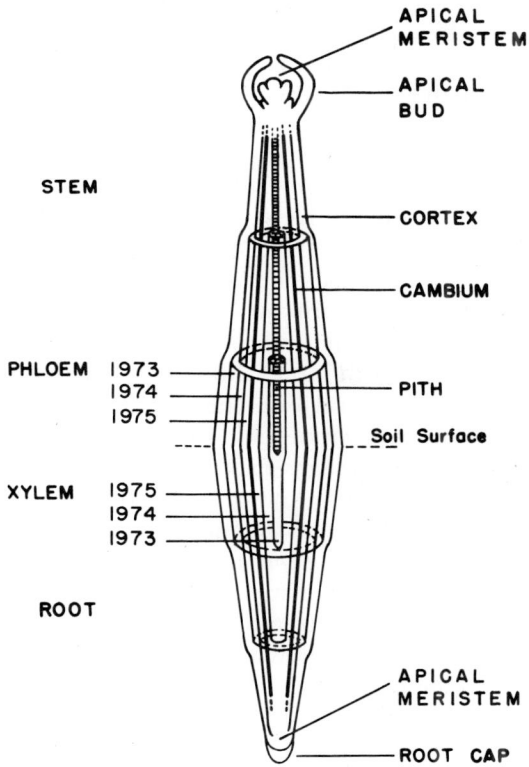

FIG. 2.2. A diagrammatic sketch of a three-year-old seedling showing the relationships of the various tissues to the growth layers in the phloem and xylem.

2.2). Meristematic tissue, which is found in all buds, is capable of producing new stems and leaves (*vegetative buds*) or flowers, fruits, and seeds (*reproductive buds*).

All organs and structures, except for certain parts of seedlings, originate either directly or indirectly from the meristems. A longitudinal section through a vegetative apical bud (Fig. 2.2) would show a meristem at the stem tip to be made up of tightly-packed, small cells with thin cell walls and dense protoplasm. The short appendages surrounding the stem tip in the drawing (Fig. 2.2) are embryonic leaves. Cells produced in the lower and central portion of the meristem enlarge and become the soft parenchyma tissue at the center of the stem called the *pith* (Figs. 2.2 and 2.3). A sheath of cells surrounding the pith becomes the first xylem and phloem. Other cells between the xylem and phloem continue to divide and this tissue becomes the vascular cambium. Parenchyma tissue, which lies outside the phloem in young stems, becomes *cortex* and epidermis (Fig. 2.1). Eventually, the cortex and epidermis are ruptured and are replaced by the corky tissue which develops from the cork cambium formed first within the cortex and later within the phloem (Fig 2.1). Although the first-formed xylem and phloem originate from the apical meristem, all subsequent xylem and phloem cells formed in the stem are generated from vascular cambium. Cells that are produced on the outside of the cambial layer become phloem, and cells that are produced on the inside become xylem (Figs. 2.1 and 2.2). The tissues outside the vascular cambium are referred to collectively as the bark.

Apical buds at the tips of dormant branches contain a meristem that can grow into a shoot if dormancy is broken. There are also lateral buds at the base of leaves and below the apical bud that contain similar meristems which can develop into lateral branches or shoots. As the bud grows throughout the season, new embryonic leaves or flowers are produced. The lower and older appendages expand and grow into well-defined leaves. The entire bud expands and grows into the one-year-old stem.

The root differs from the stem in that no buds are formed, the meristem is under a structure called a *root cap*, no pith is present, and the lateral roots originate from internal meristems within young roots, rather than from buds. For a more detailed treatment of meristems in woody plants, see Romberger (1963) and Kozlowski (1971a and b).

Figure 2.2 is a diagrammatic sketch of a three-year-old seedling showing the relative positions of the layers of xylem and phloem in the successive segments of growth. At the end of the third year, the stem segment formed during the first year includes three layers of xylem surrounding the pith which are in turn surrounded by the vascular cambium, three layers of phloem, cortex, and an epidermis. The most recently formed xylem and phloem lie next to the cambium. These layers are continuous with the outermost xylem and innermost phloem layers of the two-year-old stems as well as the xylem and phloem of the one-year-old twig, leaves, branches, and secondary roots.

The cambium of the stem and root may be visualized as two paraboloids, joined at their widest circumferences at the soil surface and terminating in the apical meristems in the bud and under the root cap. As the stem grows in length and width, new layers of xylem and phloem are

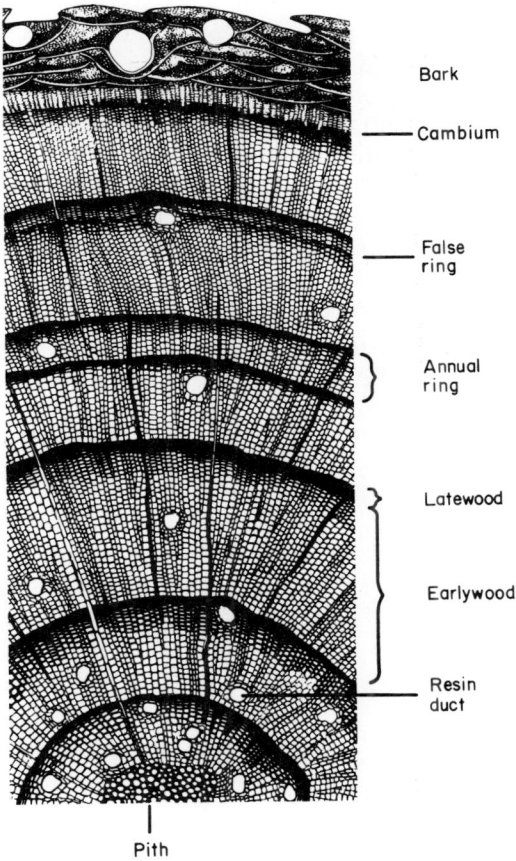

FIG. 2.3. Drawing of cell structure along a cross section of a young stem of a conifer. Tracheids are the predominant cells in the xylem. The earlywood is made up of large and relatively thin-walled tracheids, and the latewood is composed of small, thick-walled tracheids. Variations in tracheid thickness produce false rings in either earlywood or latewood. The large circular openings are resin ducts, which form a complex system of tubes throughout a coniferous tree.

produced along the paraboloid surface. The oldest phloem which lies nearest the epidermis or bark is crushed and becomes part of the bark, but the xylem, which is more rigid, maintains its shape.

A transverse or cross section through the stem of a tree (Fig. 2.4) will expose the growth layers as concentric rings. *Tracheids* are the predominant elements in the rings of *gymnosperms,* the nonflowering seed plants (Fig. 2.5a). These are vertically oriented cells with relatively thick lignified walls which die when they become functional as compared to parenchyma which have thin walls and can function as living cells. The tracheids may range from 0·5 to 15·0 mm in length (Kramer and Kozlowski, 1960), and their breadth (in a circumferential direction, see Fig. 2.4) can vary from 15 to 80 microns (Brown and Panshin, 1940). However, the tracheid width and wall thickness in a radial direction, that is, from the center to the outside of the stem (Fig. 2.4), can vary markedly within any one annual ring. Tracheids along the inner portions of a ring are wide and have the thinnest walls. These wide cells form the *earlywood*, sometimes called *springwood*, which is a porous tissue, low in density, and light in color (Fig. 2.3). Somewhere in the midportion of the ring the tracheids become markedly flattened along the radial dimension, and toward the outside of the ring the walls of the tracheids appear noticeably thicker. These narrow and flattened cells with thick walls form the *latewood* or *summerwood*, which is less porous, more dense, and darker in color than the earlywood portion of the ring (Figs. 2.5a and b).

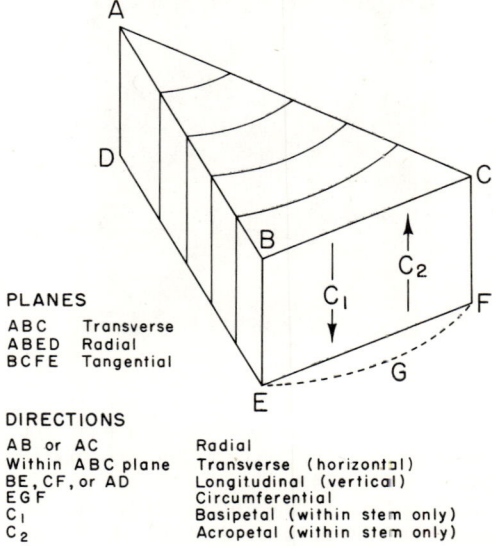

FIG. 2.4. Terminology used to describe directions and surfaces in a stem.

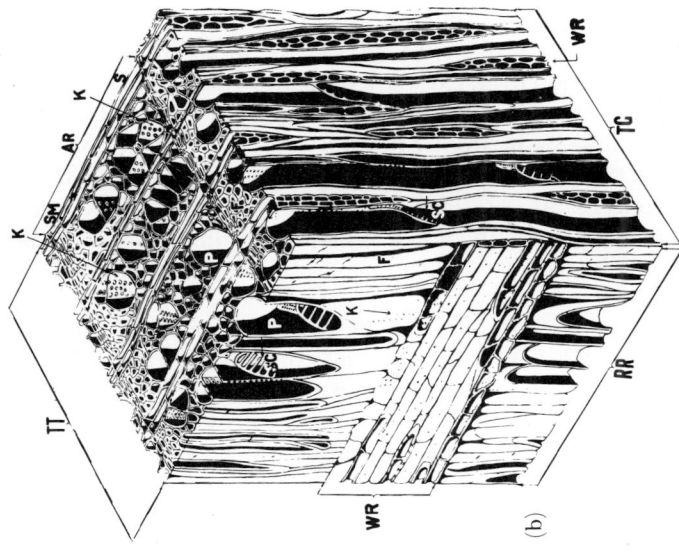

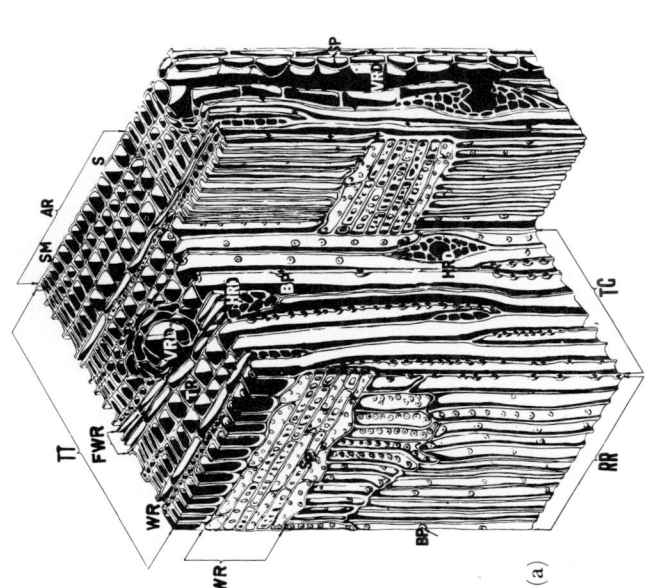

FIG. 2.5a and b. Drawing showing the detailed structure in wood blocks of gymnosperms (nonflowering seed plants) and of angiosperms (flowering seed plants). TT is the transverse plane, RR the radial plane, and TG the tangential plane (See Fig. 2.4). (a) Gymnosperm wood showing an annual ring, AR, including earlywood, S, and latewood, SM, which are made up mostly of tracheids, TR, with walls perforated by simple pits, SP, and by more complex structures called bordered pits, BP. Horizontally oriented parenchyma cells are wood rays, WR and FWR. HRD designates horizontal resin duct, and VRD designates vertical resin duct. (b). Angiosperm wood including earlywood, S, and latewood, SM, which include (1) large water-conducting vessels or pores, P, with end walls open or ending with a grate, SC, and with walls perforated by pits, K, (2) small vertically oriented thick-walled fibers, F, and (3) horizontally oriented parenchyma cells which form the wood rays, WR. (Courtesy of U.S. Forest Service, Forest Products Laboratory, Madison, Wisconsin)

The transition in cell size and wall thickness from earlywood to latewood is gradual in many species. In the so-called hard pines, such as *Pinus ponderosa*, the transition from earlywood to latewood is more abrupt. Some workers describe and have tried to distinguish as many as five different layers within growth rings (Studhalter *et al.*, 1963). However, such minute distinctions, when discernible at all, are not needed for dendroclimatic analysis at present, although the future for such distinctions looks good through the use of X-ray densitometry. Two basic types of cells as represented in the earlywood and latewood will be the focus of this discussion.

A widely used quantitative definition of latewood in *Picea* was put forward by Mork (1928), which includes as latewood all tracheids in which the width of the common wall between two cell cavities is equal to or greater than the width of one cell cavity. All measurements are made in a radial direction. Mork's formula is simple and objective, and has been applied to a variety of other gymnosperm genera, but Larson (1957) and others have found that the marked color change from earlywood to latewood in the rings of certain gymnosperms is equally satisfactory and much simpler to use. The formula is not applicable to *angiosperms* (the flowering plants) because they form vessels and fibers rather than tracheids (Fig. 2.5b).

The sharp transition between the last-formed latewood cells of one growth layer and the first-formed earlywood cells of the next is usually distinct from the more gradual transition between the earlywood and latewood within the same ring. Usually a trained worker with an adequately surfaced specimen will have little difficulty using low (10–$20 \times$) magnification to distinguish the sharp boundaries between most rings from the more diffuse variations of structure within the annual ring.

Figures 2.3 and 2.5 show most of the cell types which make up the wood of gymnosperms. Thin-walled cells forming lines or ribbons radiating from the pith are called *wood rays* and often remain alive for many years in the *sapwood*, the outer light-colored portion of the stem. Metabolic wastes move inward through the living rays and are dumped at the point where these cells are no longer living so that the wood inside the sheath of living rays turns darker and is referred to as *heartwood*. The sapwood–heartwood boundary is influenced by the vigor of the tree, the retention of leaves, and the area of the crown (Grier and Waring, 1974).

In gymnosperms the vertically oriented tubular spaces sheathed by living parenchyma are *resin ducts*, which form an interconnected intercellular system both horizontally and vertically within the tree (Figs. 2.3 and 2.5a). Resin ducts may be scattered throughout the annual growth ring, clustered, restricted to the outer ring boundary, or totally absent. The exudation of resin from the ducts into wounded tissue helps to protect the living tissue from insect attacks or disease. As the vigor of a tree decreases with increasing age, or with increasing stress of the environment, the pressures of the resin in the ducts usually declines and the tree thereby becomes more vulnerable to disease.

The wood of angiosperms, or so-called hardwoods, is made up of *vessels*, various types of *fibers*, and longitudinally oriented parenchyma called wood rays or *xylem rays* (Fig. 2.5b). The most prominent feature is the vessels, which consist of joined vertically oriented tubular cells from which the end walls have disappeared. The widths of vessels range from 20 to 400 or more microns, and their lengths vary from a few centimeters to many meters (Kramer and Kozlowski, 1960).

The wood of angiosperms is designated as either *ring-porous* or *diffuse-porous*. In ring-porous wood, such as *Ulmus, Fraxinus,* and *Quercus*, the vessels in the earlywood are markedly wider than those in the latewood, and therefore yearly growth rings are relatively simple to identify. In diffuse-porous wood, such as *Fagus, Acer,* and *Liriodendron*, the vessels are approximately the same diameter throughout the ring. The greater part of the xylem in angiosperms consists of thick-walled and narrow diameter wood fibers which are smallest or most dense in the latewood portion of the ring. In some genera, such as *Populus*, a distinct layer of parenchyma is formed at the boundary of the ring.

III. The Vascular Cambium

The growth of a cell involves many physical and biochemical processes. There are three basic growth phases that can be recognized, though the phases overlap and involve more or less continuous changes. (1) Cell division involves the duplication of hereditary materials and formation of a new wall which partitions the cell into two smaller daughter cells; in the case of *Pinus strobus* the actual divisions are reported to take approximately one day (Wilson, 1964). Skene (1972) reports that the time from the beginning of one division to the next for a given cambial cell in *Tsuga canadensis* decreases during the course of the growing season from 28 to 10 days for vigorously growing trees and from 35 to 20 days for slowly growing trees. (2) Cell enlargement or expansion occurs by means of pressures developed within the cell which stretch the wall as new materials are added to the cell. Skene (1972) found that in *Tsuga canadensis* the enlargement of newly formed tracheids continues for 18 days in the early portion of the growing season but later in the season for only 9 days. (3) Cell differentiation and maturation follows cell enlargement or accompanies the late stages of the growing period as further materials are added to the wall and protoplasm. Early in the growing season the time required for deposition in the walls of *Tsuga canadensis* tracheids is 10 days. Near the end of the growing season wall deposition continues for up to 50 days (Skene, 1972). Wodzicki (1971) has shown that in *Pinus sylvestris* the rate of cell wall thickening increases from a low rate in late May to a maximum rate in early July, and for the rest of the season a steadily greater number of days is required before wall thickening is complete.

Differentiation and *maturation* as the names imply are the processes by which the growing cells take on the structural features of the tissues of which they will become a part. When cell enlargement is complete and the structures are fully formed, no more growth occurs. In the case of vessels and tracheids, death of the cells must occur before they can become functional in water transport. According to Skene (1972) this transition of function occurs over a period of about four days. When a cell is dead, no further change in structure is possible, though it may expand and contract slightly due to changing hydrostatic pressures and tensions within the tree.

Diameter growth of a tree occurs as the cambium divides and derived cells undergo enlargement and differentiation, producing new xylem and new phloem. Simultaneously, the cambial sheath expands as new cells are formed on the inside of the cambium. The term *vascular cambium* is sometimes used to designate only a single layer of cells which is capable of dividing and differentiating new cells on either side of the layer, but the term is more commonly applied to the entire meristematic zone, including portions of the phloem and xylem where the cells are enlarging but are still able to divide (Wilson *et al.*, 1966; Fig. 2.6).

During those months of the year when the vascular cambium is inactive, the cambium is usually two to four cell layers thick. These cells are rectangular in cross section, and the walls are sharply and distinctly outlined (Ladefoged, 1952). In the spring, the walls become semitransparent and more plastic. The protoplasm changes from a viscous semisolid state to a more fluid or liquid state, the cells become extended in a radial direction, and the cambial zone may double in thickness. At this time, bark slippage can occur and the young cells are easily damaged by severe freezing or other adverse conditions. Injury may interrupt growth, causing some cells to remain in an immature state and causing other cells to become anomalous in size or structure. These abrupt changes in structure can be used to identify both the cells which were growing when the tree was injured and the year and season of injury.

At the onset of the growing season, the active cambium may be discontinuous with cell division beginning beneath the buds (Kozlowski, 1971b), but reactivation progresses *basipetally* (Fig. 2.4) that is, from the stem tip down to the base of the tree. Basipetal movement of cambial initiation appears to be most rapid in ring-porous trees, and sometimes the lapse in time between initiation in the top and bottom of the tree is too short to be detected. There is some evidence (Phipps, 1967) which suggests that the earlywood cells of some ring-porous species are produced by division of the cambium during autumn prior to the beginning of the current growing season. When environmental conditions are favorable, the cells start to enlarge and differentiate almost

MATURE PHLOEM

DIFFERENTIATING PHLOEM	MATURING PHLOEM	
	RADIALLY ENLARGING PHLOEM	
	DIVIDING PHLOEM (Phloem mother cells)	CAMBIAL ZONE
CAMBIUM	CAMBIAL INITIAL (dividing)	
	DIVIDING XYLEM (Xylem mother cells)	
DIFFERENTIATING XYLEM	RADIALLY ENLARGING XYLEM	
	MATURING XYLEM	

MATURE XYLEM

FIG. 2.6. The relative positions of tissues in the cambial zone. (From Wilson *et al.*, 1966)

simultaneously throughout the tree. Actual reactivation of the cambium may not occur until much of the earlywood is fully developed.

In large conifers, the basipetal migration of cambial initiation takes about a week before it reaches the stem base (Kozlowski, 1971b), and in large, diffuse-porous trees, intervals from three weeks to more than a month can elapse before basipetal migration of cambial initiation is complete.

Approximately 90% of the cambium is made up of elongated, spindle-shaped cells called *fusiform initials*. They give rise to the longitudinally arranged elements in the xylem and phloem (Fig. 2.5a and b). Growth of the stem in a radial direction occurs by division of the fusiform initials throughout the entire length of the cell and by formation of the new wall along the tangential plane (Fig. 2.4) (Kozlowski, 1971b). There are also cube-shaped cells in the cambium which divide along the tangential plane and give rise to the horizontally oriented rays of the xylem and phloem.

Division of the cambial cells forming a wall along a radial rather than a tangential plane occurs about 1% to 2% of the time. Sometimes a wall along the radial plane is formed by a division oblique to the radius and the newly formed short cells elongate as well as enlarge in a circumferential direction (Kozlowski, 1971b). Such division increases the number of cambial initials and allows the circumference of the cambial sheath to increase.

The active cambial initials may include 12 to 40 cells in the radial direction in fast-growing trees and six to eight cells in slow-growing trees (Figs. 2.1, 2.3). Once cambial activity is initiated and the size of the active zone is established, it usually remains more or less the same size until latewood begins to be differentiated. As the rate of cell division gradually declines, the width of the cambial zone decreases to that of the dormant state, although some of the last-formed cells can continue to enlarge and differentiate after cell division ceases.

A model proposed by Wilson (1964) and Wilson and Howard (1968) is useful for describing how cells are produced from the cambium of *conifers*, the conebearing gymnosperms. The cambium is treated as a factory in which building material and energy sources are continuously fed in from the active xylem and phloem. Complex structures are manufactured from these building materials, and they are transferred to the xylem and phloem as new cells. It is assumed for the model that the number of cells in the cambial zone is constant during the period of rapid growth, that all cells in the zone divide at the same rate, and that the division cycle for all cells takes the same length of time. The cambial cell is at a minimum size immediately after division. The cell manufactures new structural components, doubles in size, and then divides. This division increases the cell number in the cambial zone. Assuming that the number of dividing cells must remain at a constant number, the division results in the transfer of a cell on the periphery of the cambial zone to the zone of radial enlargement, where it can no longer divide (Fig. 2.7). If a cell in the center of the cambial zone divides, then both daughter cells reenter the division cycle, and another cell from the edge of the cambial zone differentiates. If a cell on the edge of the zone divides, then the outer daughter cell differentiates and the inner reenters the division cycle. Hari and Sirén (1972) describe a more general model dealing with varying physiological conditions of the tree and the changing ecological factors of the site.

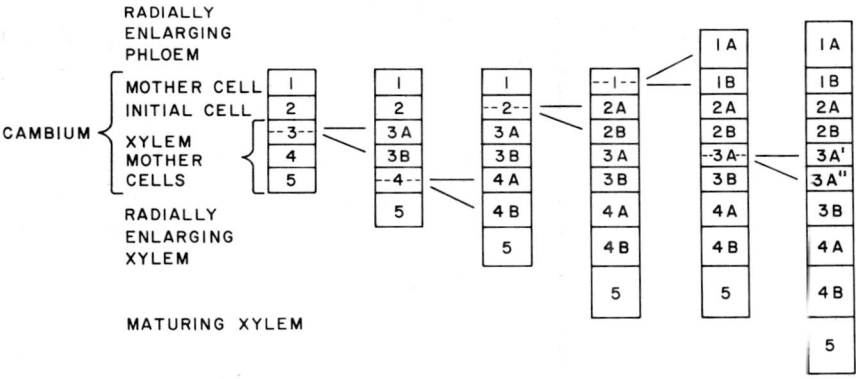

FIG. 2.7. A hypothetical radial section of a population of cambial cells developing from five cells in the cambial zone. The initial cells give rise to mother cells which are capable of further division. Five cambial divisions give rise to four xylem cells (3B, 4A, 4B, and 5) and one phloem cell (1A). (Redrawn from Wilson, 1964)

Cell enlargement is readily followed by differentiation and maturation, which are most apparent in the structure of the cell wall. A variety of materials makes up the cell wall constituents (Wardrop, 1964a); some contribute to the matrix and framework of the first-formed cell wall, while others are added at later stages. The primary substance in the framework is *cellulose*, and the first substances produced in the cell wall matrix are *polysaccharides* such as *hemicellulose*. At later stages in differentiation a hard material, *lignin*, is added. After the cell wall is fully impregnated with lignin, the protoplasm disappears and differentiation of the tracheid is complete. The walls of ray cells are usually less lignified, and they can remain alive as long as they are a part of the sapwood, as has been mentioned. The walls of xylem cells often contain thin areas or pits, and in conifers the wall near the pit is raised so it is referred to as a *bordered pit* (Fig. 2.5a).

The bark in a several-year-old stem (Fig. 2.3) consists of an outer layer of corky tissue and an inner layer of living phloem. The cortex and epidermis formed on the one-year-old twig (Fig. 2.1) is ruptured and sloughed off. The phloem includes cells that are large, nonwoody, and thin-walled called *sieve tubes* and associated cells that are smaller, called *companion cells* (Fig. 2.1). The sieve tubes are involved in translocation of organic substances throughout the tree. In many species, the sieve tubes become nonfunctional after one season, and they are crushed by the newly developing phloem. Sometimes, however, the sieve tubes become reactivated during the first part of a second season, and in a few species they remain functional for a longer time. The phloem region also includes parenchyma and thick-walled fibers (Fig. 2.1), the latter of which sometimes occur in vertical bundles scattered throughout the bark.

Divisions of elements in the outer phloem cells give rise to arcs of cork cambium (Fig. 2.3). These cells divide toward the outside of the stem and form cork cells technically called *phellem*. While the phellem cells differentiate, the cellulose walls become impregnated with a waxy water repellant called *suberin*. As the stem grows in width and the bark is pushed outward, parenchyma cells in the inner bark divide and expand, thus accommodating the strains of circumferential expansion. The cork cambium and the layers of parenchyma cells create zones of weakness which are eventually ruptured, causing the bark to exfoliate in patterns which differ among species (Kozlowski, 1971b). The thickness of bark is a function of the tree's heredity, the vigor of the tree, and its age. Trees with thick bark, such as old or rapidly growing individuals, have additional protection which insulates the enclosed cambium from extreme temperatures due to fire or frost and protects the fragile tissue from mechanical injury.

IV. Growth—A Variable Process

It has been stated in the previous section that growth includes the basic processes of cell division, cell enlargement, and cell differentiation as the cells take on the attributes of the various plant tissues. It will be shown later that there are many other processes that can affect the three basic growth processes, and that their rates can be limited directly or indirectly by a wide variety of external environmental factors and internal conditions within the tree. As a result the rates of growth may vary throughout the day, throughout the season, and throughout the lifetime of the tree. For example, many growing tissues exhibit diurnal expansion and contraction which is traceable in large part to varying environmental conditions around the tree (Kramer and Kozlowski, 1960). Likewise, the activity of apical meristems and the vascular cambium varies throughout the year. These growing tissues are dormant during the winter; dormancy is broken in the spring; and they become most active during the early part of the frost-free season. This activity slows down as the season progresses, until it ceases totally and dormancy again sets in. Usually growth is most vigorous and the active season longest for young trees and less vigorous and the season shorter for old trees.

The period of growth also varies for the many organs and tissues of the tree. For example, height and radial growth at the stem apex begin at about the same time, but height growth can cease earlier in the season than radial growth. Root growth may begin somewhat later in the spring than stem growth, and it can continue later in the season. Under a warm winter climate, there is evidence that roots can grow almost any month of the year (Fig. 2.8; Krueger and Trappe, 1967). Apical growth in roots may occur at a different time than diameter growth (Kozlowski, 1971b).

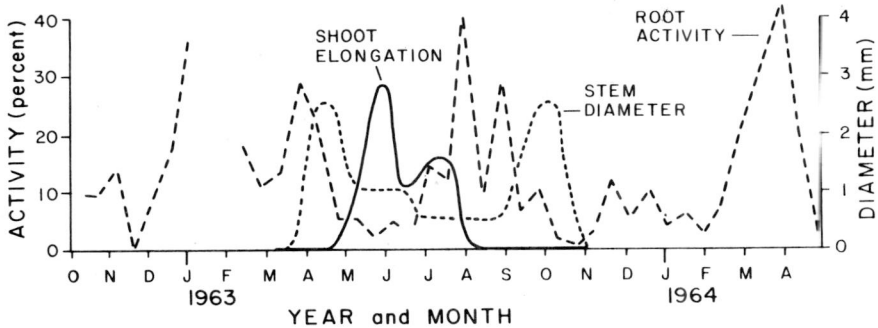

FIG. 2.8. Periodicities in growth throughout the year in shoots, stems, and roots of *Pseudotsuga menziesii* seedlings. (Redrawn from Krueger and Trappe, 1967)

Complicated interactions can occur among the growth processes in various parts of the tree. The interactions often involve the movement of growth regulators, the distribution of foods, and the changing rates of processes in different parts of the tree. In the following sections, growth of the shoot (height growth) and growth in width (radial growth) will be discussed as they vary (1) throughout the 24-hour day, (2) throughout the year, and (3) throughout the life of a tree. Variations among the different parts of the tree, among different species, and among different habitats will also be described.

V. Variations in Shoot Growth

The cambial activity in both conifers and hardwoods is initiated and maintained by growth hormones or regulators (see Chapter 4). The size and structure of the xylem cells, as well as the rate of cell division, also may be controlled by these growth regulators, which are produced by the actively growing stem apex or by the leaves. Therefore, the width and structure of the annual ring in trees are intimately related to the amount and duration of shoot growth (Kozlowski, 1971b).

A daily rhythm in shoot growth can be observed, most often with more rapid elongation occurring at night than during the day. At other times, when nights may be unusually cold and limiting to growth processes or when daytime conditions are especially favorable, growth may be more rapid during the day. Kienholz (1934) reported a high degree of correlation between the amount of night growth and minimum air temperature. Growth of shoots for certain species was observed to be negligible when temperatures were 42°F (5.6°C) or lower. Thus, during the beginning of the growing season when nights are unusually cold, the rate of shoot growth may be markedly reduced.

The general cumulative pattern of height growth in temperate species follows a *sigmoid* (*s*-shaped) *curve* throughout each season. Growth starts, accelerates rapidly, reaches a relatively high but constant rate during what is referred to as the "*grand*" period of growth, and finally decreases as the cumulative curve levels off sometime during the summer (Fig. 2.9). Hari et al. (1970) present an interesting dynamic model describing the variations in the daily height increment of plants as they change throughout the year.

As soon as temperatures are sufficiently high and days are sufficiently long, shoot growth is initiated. Most species in a given locality begin growth within a short time span in the spring. However, cessation of shoot growth may vary over a wide range of time (Fig. 2.9). Cessation often occurs as a result of internal limiting factors long before air temperatures or day lengths have changed sufficiently to become directly limiting to the growth processes. However, the internal limiting factors may sometimes be traced to physiological factors that in turn are limited by unfavorable conditions of climate. Severe water deficits late in the spring or in early summer can inhibit cell enlargement and affect shoot growth, especially in those species which have a long growth period and exhibit intermittent flushes of shoot growth (Kozlowski, 1964). However, the climate of the prior growing season (the time when buds are formed) often exerts a greater influence on the amount of height growth than the climate for the period in which the stem is elongating.

During a particular season, growth of individual shoots may deviate from a smooth, sigmoid curve. Such variability is especially common in the tropics where internal conditions are usually the most limiting and growth can occur theoretically at any time. The growing point of some tropical trees may be permanently active, while in other species it is intermittently active, with growth temporarily slowing down or ceasing after a certain number of leaves are formed (Alvim, 1964). For *deciduous* trees in areas where one season is unfavorable to growth, the period and rate of growth will be more highly correlated among individuals and among species. However, the synchronization of growth with a particular

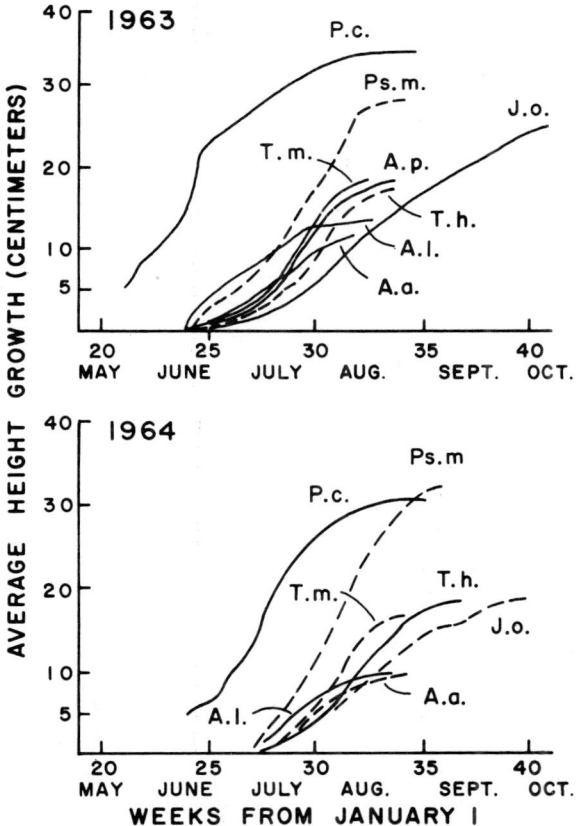

FIG. 2.9. Cumulative height growth for different coniferous species from western North America. P.c.—*Pinus contorta*, lodgepole pine; Ps.m.—*Pseudotsuga menziesii*, Douglas-fir; T.m.—*Tsuga mertensiana*, mountain hemlock; A.p.—*Abies procera*, Noble fir; J.o.—*Juniperus occidentalis*; T.h.—*Tsuga heterophylla*, western hemlock; A.l.—*Abies lasiocarpa*, subalpine fir; and A.a.—*Abies amabilis*, Pacific silver fir. (Redrawn from Williams, 1968).

season does not assure that the amount of growth will be correlated with the climate of that season.

Certain angiosperms and gymnosperms in temperate regions also can reinitiate apical growth several times during a given season (Kozlowski, 1964). Growth may resume, after apparent dormancy has set in, with the bursting and elongation of the apical bud or by development of lateral buds at the base of the apical bud. Many *Pinus* species appear to be *uninodal* in that only one series of needles is produced each year by the apical bud. Others, such as *P. echinata* and *P. taeda*, are *multinodal*. Buds in these species may produce several series of needles during one season's growth, and the xylem may exhibit multiple growth rings (Kozlowski, 1971b).

Vigorous branches of *Pinus edulis* are reported to produce summer shoots from the growth of buds that are formed after initiation of apical growth but have not passed through a period of rest (Lanner, 1970). Several cycles of growth occur in stands of *Pinus halepensis* in southern France under conditions of the Mediterranean climate (Serre, 1973). Despite recurrent growth flushes, the location of the primary whorl of numerous thick branches is usually distinguishable from the secondary height increments that succeed the initial growth flush. Recurring growth flushes in shoots are not necessarily accompanied by similar growth variations in the main stem, though multiple growth rings are often attributed to multiple flushes in the shoot apex.

There are certain variations in the seasonal pattern of shoot growth associated with tree age. In the spring, shoot growth for all age classes begins at approximately the same time, although the younger individuals close to the warm ground layer may slightly precede others breaking dormancy. The period of rapid shoot growth is shorter in mature trees than in seedlings, however. Likewise, multiple flushes are more likely to occur in young trees than in older trees. The flattened crowns so typical of old trees result in part from the decreased vigor of shoot growth that accompanies increasing age.

Growth of most lateral buds is inhibited by active growth of the stem apex. This inhibitory action controls crown form and is referred to as *apical dominance*. For example, the young apical buds in several species of *Pinus* and *Pseudotsuga menziesii* are strongly dominant, so that a single main stem with few lateral branches is almost always formed. On the other hand, trees of *Pinus longaeva* and many species of *Juniperus* and *Ulmus* have apical buds which exert less dominance, and as a result multistemmed trees are common. Growth rings formed in trees having multiple stems and several actively growing branches can exhibit variable widths as the branches feeding different segments of the cambium change in mass and change in their relative importance to the total foliage of the tree. For example, a period of rapid growth in one of several branches may temporarily inhibit the growth of the lesser branches, which in turn may cause a reduction in the widths of rings receiving nourishment from them. Those trees in which growth varies among several branches throughout the crown may exhibit associated variations of ring widths along the main stem depending upon the proximity and location of the branches.

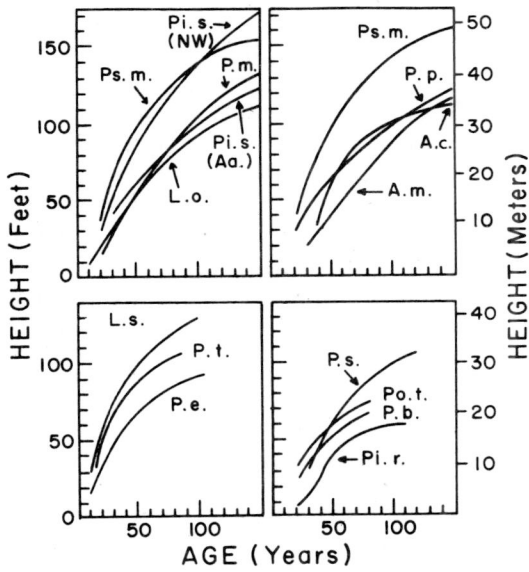

FIG. 2.10. Cumulative height growth for different species with increasing tree age. A.c.—*Abies concolor*, white fir; A.m.—*Abies magnifica*, red fir; L.o.—*Larix occidentalis*, western larch; L.s.—*Liquidambar styraciflua*, red gum; P.b.—*Pinus banksiana*, jack pine; P.e.—*Pinus echinata*, shortleaf pine; P.m.—*Pinus monticola*, western white pine; P.p.—*Pinus ponderosa*, Ponderosa pine; P.s.—*Pinus strobus*, eastern white pine; P.t.—*Pinus taeda*, loblolly pine; Pi.r.—*Picea rubens*, red spruce; Pi.s.—*Picea sitchensis*, Sitka spruce, (Aa. = Alaska, NW = Northwest); Po.t.—*Populus tremuloides*, aspen; Ps.m.—*Pseudotsuga menziesii*, Douglas-fir. (Redrawn from Baker, 1950)

The rate of growth can be expressed as a function of increasing age. When cumulative shoot growth is plotted as a function of tree age, it approximates the sigmoid curve. A brief period of increasing growth rate occurs during the seedling stage, a period of high growth rate occurs during the sapling stage, and a decreasing growth rate occurs as the tree matures and approaches old age (Fig. 2.10). The sigmoid curve is stretched over a longer period of time for trees with long life spans. For trees on stress sites, the period of rapid juvenile growth may be shorter than for those on more optimum sites, but the period of declining growth rate can be longer. Since the juvenile stage for trees on adverse sites may be shorter than for trees on optimum sites, the trees are not as tall and the crown becomes more spherical and flattened unlike the conical shaped crowns of young trees.

The rates of shoot growth as shown in Fig. 2.10 may vary for different species as well as for *ecotypes* representing variants of the same species. For example, when seeds from populations differing in elevation or latitude such as *Picea sitchensis* from Alaska and the Pacific Northwest (see Fig. 2.10) are grown together on the same site, differences may be observed in the response of shoot growth to the temperature and the length of the daylight period. These differences occur because the local environments have favored the survival of certain heredities which are most in tune with the natural climatic regime (Billings, 1957). Squillace and Silen (1962) found inherent differences in the rate of height growth of *Pinus ponderosa* related to differences in seed source over the species range. Such hereditary variations may influence the way a tree is affected by climate, but these factors are constants throughout the life of an individual tree and therefore are not contributors to year-to-year variations in ring width.

VI. Variations in Radial Growth

Usually climatic information derived from tree rings is based upon differences in width and stem girth that occur from one year to the next; but climatic information may also be found in wood density, a parameter which varies as a function of cell size and cell wall thickness. Both sources of climatic information reflect daily and seasonal variations in growth rates as they are affected by environmental conditions. However, these same growth rates are also influenced by other factors such as conditions of the plant in prior years, the age and structure of the tree, limitations of the site, and the hereditary potential of the particular tree. All variations in growth can be traced to variations in limiting and interacting plant processes which can vary through time and throughout the particular tree. It is, therefore, impossible when dealing with growth to compartmentalize completely a discussion of factors, for no one factor can operate independently of prior conditions or of other factors occurring at the same time. The remaining portion of this chapter will describe the ways radial growth can vary throughout the stem and root, from short to long periods of time, from young to old trees, and from site to site. Detailed description of the causal factors and the plant processes linking them to variations in ring width and wood density will be dealt with in subsequent chapters of the book.

A. Measuring Radial Growth

Although the annual increment of radial growth is easily measured by examining the ring width, it is often helpful to know when and how the ring was formed by measuring the amount of growth that occurred over a number of time intervals throughout the growing season. This can be accomplished by sampling a piece of the wood and bark at the beginning and end of the interval and examining any differences in the outermost ring. Samples can be obtained by means of a simple tubular leather punch. The collected materials are then fixed to preserve the cells, and sections are made with a microtome or razor.

A major disadvantage in using wood samples is that wounding is severe, and growth rates of the cambium may be altered as a result of the sampling technique. Wolter (1968) circumvents this difficulty by periodically inserting thin insect mounting pins horizontally through the bark into the wood. The insertion injures a few growing cells and marks those differentiating into xylem. The pin is left in the tree for future reference. The injured cells are examined after growth is complete, and the cell rows forming at each insertion date can be identified microscopically. Serre (1973) successfully uses a microsampler to obtain cambial material every two weeks without undue injury to the tree. However, the various means of sampling here described involve materials from different locations as well as for different times during the growing season so that large errors of measurement often are present and prohibit detailed analysis of growth rates. For the more precise analyses it is better to measure the same location on the stem throughout one or more growing seasons.

A variety of instruments and techniques has been used to measure and record the changes in stem girth or in a single radius. They range from devices such as forks, calipers, and range finders, all of which are used to measure gross size, to extremely precise mechanical or electronic sensors which record minute changes in size occurring over minutes and hours.

Dendrometers are widely used instruments which are installed on stems and are read to obtain a measurement of gross stem size. One type of dendrometer consists of a thin metal band with a vernier scale which allows precise measurement of the stem circumference (Hall, 1944; Liming, 1957). Other types of dendrometers measure the size of a single radius from a fixed point anchored into the wood of the tree (Reineke, 1948; Daubenmire, 1945; Bormann and Kozlowski, 1962). A measurement is made by placing a dial indicator on a permanently installed mount which guides the instrument to the same measuring point. Differences in the measurements between successive readings supposedly represent the amount of growth, but sometimes there can be substantial variations due to changes in *hydrostatic* forces, temperature changes, or ruptures in the bark.

Among the limitations of these dendrometers are the following. (1) They all require visual reading at some designated time. (2) Diurnal variations in stem size which may not necessarily represent growth may be large, so that the reading must be made early in the morning when the stem is most fully distended. (3) Since two measurements are involved to obtain a size change, a reading error causing an overmeasurement for one period will be accompanied by an undermeasurement for the following period. A permanently attached dendrometer constructed by Kuroiwa (1957) uses mirrors to obtain an exceptionally precise measurement.

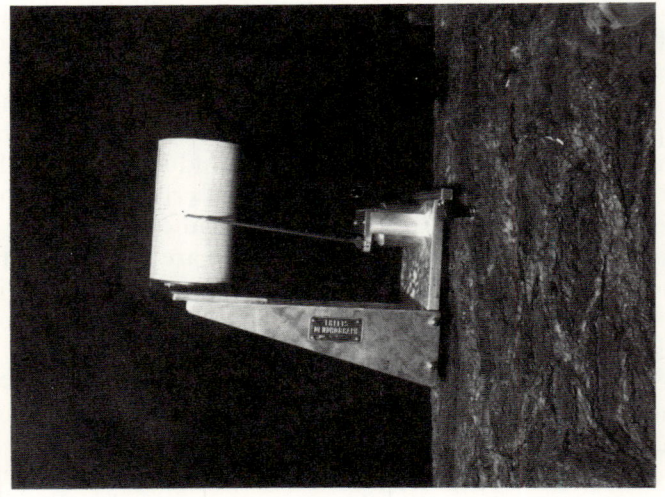

FIG. 2.11. A dendrograph used to continuously record the size of single radii in the stems of trees. The instrument is mounted on three lag screws inserted into the wood. A rod resting on a prepared bark surface bears on a lever connected to the arm and pen. Any change in the distance between the instrument and the bark surface is magnified 100 times and recorded by the pen on the clock-driven drum which rotates once every eight days. (Fritts and Fritts, 1955)

Dendrographs involve the same kind of measuring systems as dendrometers, but the stem size is continually recorded. Mitscherlich uses a dendrograph designed by Friedrich that continually recorded changes in size of a band stretched around a tree stem (Mitscherlich et al., 1966). A different kind of dendrograph was developed by Fritts and Fritts (1955) which mechanically magnifies and continuously records the size along a single stem radius (Fig. 2.11). A number of electronic devices have been developed to monitor stem size changes on electrical data loggers and recorders (Cooper and Herrington, 1959; Kuroiwa, 1959; Phipps and Gilbert, 1960; Impens and Schalck, 1965; Mitscherlich et al., 1966; Dobbs, 1969; Odin and Openshaw, 1971; LaPoint and VanCleve, 1971; Phipps and Yater, 1974).

In order to avoid error due to disturbance, dendrographs must be mounted carefully. The contact points of measurement should be prepared with hardened and waterproof surfaces to eliminate bark swelling and abrasion. Charts must be changed without disturbing the record or the mounting of the instrument. However, the continuity of measurement from dendrographs is an important advantage over the records from dendrometers because it allows assessment of diurnal variations in stem size which, along with seasonal variations, can be related meaningfully to growth, to the plant processes affecting growth and cell hydration, and to variable environmental conditions around the tree.

Some workers, especially those who have used dendrometers, have been impressed with the differences that they have observed at various locations throughout the same stem. For example, some have reported that one side of a stem can grow more during some weeks while the opposite side can grow more during other weeks (Young, 1952; Small and Monk, 1959; Bormann and Kozlowski, 1962; Kozlowski and Winget, 1964). Some of these variations result from difficulties inherent in the particular measuring device. Dendrometers are especially susceptible to errors due to differences in the placement of the reading instrument, while dendrographs, the mirror dendrometer, and electronic devices which obtain relatively continuous records are subject to these errors only when disturbed by changing of charts and other adjustments. Differences are also traceable to real features of the stem, including differences in internal water stress, wood structure, flexibility of the radius which is measured, the effective size of the radius measured, hygroscopic changes from wetting and drying of wood and bark, variations in instrument pressure, and variations in patterns of upward water transport.

Numerous records of stem size have been made by this writer using dendrographs mounted on a number of different species and on different sides and heights of the same trees (Fig. 2.11). Some dissimilarities are often apparent within and among trees, but they are rarely as large and marked as the similarities that occur. For example, the same patterns of diurnal and seasonal variations can be discerned easily at different locations in the same as well as in different trees, and even in those of different species.

Worrall (1966) found good correlation between short-term changes in stem diameter and internal water stress. Xylem sap tension, as well as diameter changes, were measured, and a linear regression analysis was applied to adjust the observed diameter to a standard tension. In this manner the growth measurements were separated from those reversible changes resulting from variations in water stress.

Dendrograph records show radial growth most clearly when diurnal reversible variations in stem size due to hydration changes are small and the annual net increase in size which is due to cell growth is large (Fritts, 1958, 1962c). However, the diurnal and seasonal reversible changes in stem size for trees in arid sites may be greater than the annual increment of growth (Fritts *et al.*, 1965c; Fritts, 1969), and stem diameters may

---

FIG. 2.12. Selected records of stem size for 1966 as obtained from a dendrograph (Fig. 2.11). The height of the pen was usually adjusted whenever charts were changed. An adjustment factor representing the difference in the trace before and after the change must be added to the reading. The instrument is dendrograph B, which was mounted at middle stem position on an arid-site *Pinus ponderosa* (see Figs 2.15, 2.19, 2.20, 2.21, and 2.23). The June record shows the characteristic stem size fluctuation at a time of low soil moisture and high potential evaporation. Diurnal variation in size of the stem is small because little soil moisture is available, little water is moving through the stem, and growth is slow. Deviations from the smoothly varying expansion of the stem at night and shrinkage during the day are irregularities related to varying amounts of cloud cover and variations in temperature, both of which influence loss of water from the tree. During the afternoons of 6/25 through 6/28 the small irregularities are the result of cloud cover in the afternoons and of light rains which fell in the evening of 6/26 (0·10 inches or 2·54 mm) and in the afternoon of 6/28 (0·73 inches or 18·54 mm). A 1·10 inch (27·94 mm) rain fell on 6/29. There was consequently more moisture available for both absorption and loss by the tree, so the dendrograph trace for the following week shows more variation in stem size. Periods of cloud cover in the morning and in the afternoon of 7/4 are reflected in the irregularities of the dendrograph trace. Dense clouds during midday on 7/8, 7/9, and 7/12 cause marked irregularities in the record of stem size. Light rains fell on 7/16 and 7/17. A 0·70 inch (17·78 mm) rain on 7/18 and a 0·36 inch (9·14 mm) rain on 7/19 accelerated stem rehydration and caused considerable expansion in stem size. Light rains and dense clouds during midday for the week of 7/20 to 7/26 produced the most marked midday irregularities in stem size. During 10/6 to 10/8 there were scattered clouds producing slight variations in stem size during midday. Other days for October were clear. The daily maximum stem size declined at this time due to an inactive cambium and increasing water stress. Fig. 2.21 provides more information on the climatic regime for the tree.

2. GROWTH AND STRUCTURE

INCHES X 10^{-2}

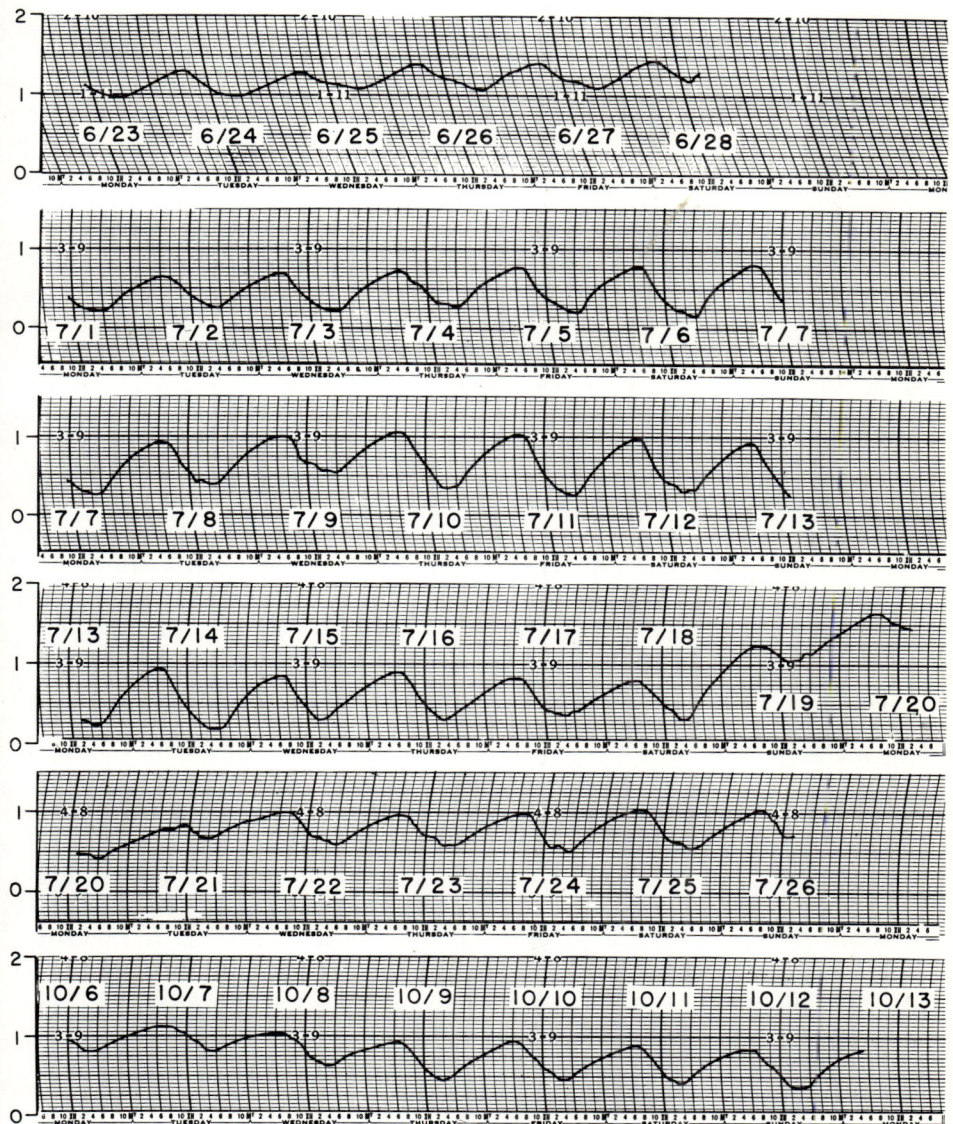

decrease even though the cambium remains active. In these cases, the dendrograph record is a valuable measurement of water status in the stem and must be supplemented by some more direct measurement of cambial growth if annual increment of cells is to be determined. Shrinkage and expansion similar to that in stems has been reported for leaves, roots, branches, and reproductive structures (Kozlowski, 1965, 1971b).

B. Diurnal Periodicities in Radial Growth

It might appear that diurnal variations in the rates of various processes occur too rapidly and are too transient to significantly affect ring width. Apparently, transient factors may affect the structures within the tree, however, if they occur repeatedly. It is therefore important to consider the minute-by-minute and hourly changes occurring in processes, because they are the basic ingredients accumulated over days, months, and years which produce the variations in ring width.

Figure 2.12 shows a portion of the dendrograph record from a single tree for June 13 through October 13, 1966. The variations in stem size reflect changes in cell volume resulting from variation in their water content. During the day, loss of water from the leaves, called *transpiration*, removes moisture faster than it is absorbed by the roots and thus produces a net decrease in cell water content. During the night transpiration from the leaves occurs more slowly than water absorption in the roots and there is a net increase in cell water content. During June and July cell division in the example tree was occurring in the cambium and expansion of the growing xylem and phloem resulted in a net increase of these tissues. The expansion each night was greater than the shrinkage during the day and there was a net increase proportional to the amount of growth.

The rate of change in size of tissues in the stem can be shown to be a function of both the difference between water uptake and water loss by the plant, and the ability of water to move, to be absorbed, and to be held by the tissues of the plant. For example, the growing tissues can often compete with the surrounding nongrowing tissues for water, so they are not as likely to shrink during the day and may even continue to expand while other tissues shrink. Whenever a plant is full of water and both water loss and water gain are small, the cells remain fully distended and little diurnal variation in size of the stem can occur except for expansion resulting from growth. Likewise, under conditions of severe water deficit, as may occur during an extended drought, little variation in stem size can occur. Cell water content is depleted, little water enters the roots because the soil is dry, and the loss of water by transpiration is low because there is little water to evaporate. Neither expansion nor contraction of the stem is possible. However, if rain replenishes soil moisture, water moves rapidly into the roots, and there is rapid expansion of the tissues within the tree (see day 7/19 in Fig. 2.12).

Under environmental conditions favorable to both rapid water absorption and rapid water loss, changes in stem size can be used to infer hydrostatic conditions within the tree. For example, the relative rate of transpiration can be deduced from the changes in stem size during the day (see Chapter 3, Figs. 3.14 and 3.15). During the morning the dendrograph traces in Fig. 2.12 record declining stem sizes when the rate of water loss from leaves is accelerating rapidly. However, continued transpiration during the day leads to increasing water deficits which cause increased absorption of water by roots, so that a unit change in stem volume in the afternoon represents a somewhat larger amount of transpiration than it does in the morning. Midday cloud cover and showers may markedly reduce the rate of water loss, shrinkage may diminish, and expansion of the stem may begin (see Fig. 2.12, days 7/8, 7/9, and 7/24). Conversely subsequent clearing weather may enhance evaporation, increase the rate of water loss, and cause the rate of shrinkage to increase again. As evening approaches, the rate of water loss diminishes until it equals water gain, at which time shrinkage ceases. Subsequent reduction in the rate of water loss or increase in water gain causes cell volumes to increase, and swelling begins. This swelling of the stem is rapid early in the evening when water content of tissue in the stem is low and moisture gradients throughout the plant are steep. As water uptake continues during the night and cells become more distended, moisture gradients within the plant diminish and the rate of swelling declines (Fig. 2.12). Usually shrinkage begins on the next day at sunrise before cells are fully distended and before the stem reaches potential maximum size.

The combination of temporary reduction of water loss and rehydration of cells during the afternoon cloudy weather so common in summer may in itself favor continuation of growth, which in turn could lead to the formation of a wider ring.

Since the stem radius measured by a dendrograph includes the bark, the cambium, and a few centimeters into the wood, the record of changes in stem size (Fig. 2.13) includes the net effect of moisture changes in the outer bark, mature phloem, and mature xylem, as well as growth resulting from cell division in the cambial region. If the mature tissues are shrinking faster than the growing tissues are expanding, it is possible that a net decrease in stem size can be obtained at a time when the cambium is active. However, if such progressive shrinkage has not set in, the changes in the maximum stem size from one day to the next can be used as an

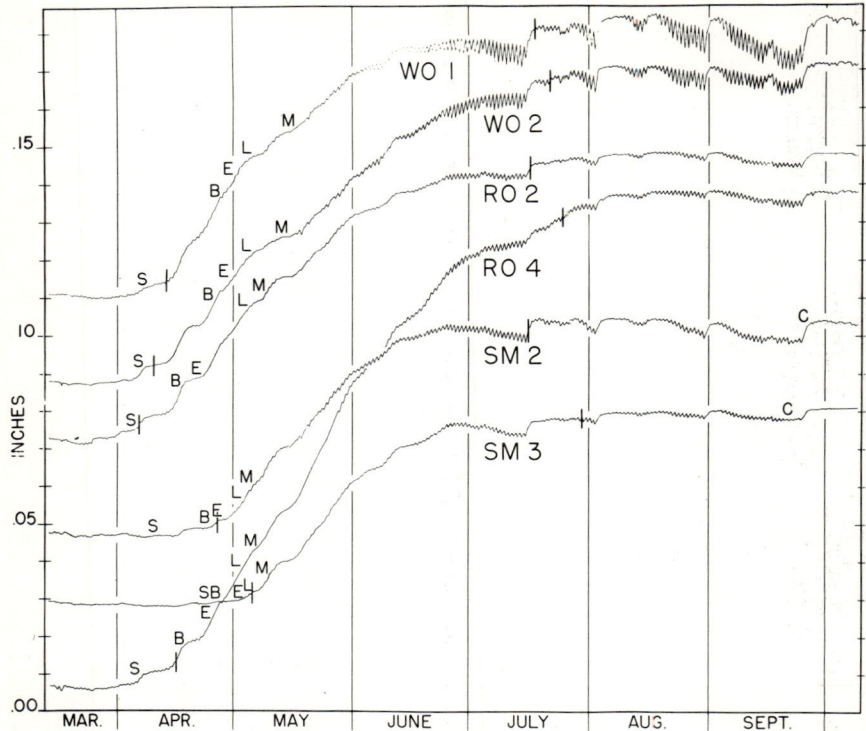

FIG. 2.13. Daily maximum and minimum size of tree stems recorded by dendrographs show cumulative growth. Both *Quercus alba*, white oak (WO 1 and 2), and *Q. rubra*, red oak (RO2 and 4), began growth during early or mid-April following swelling of the buds (S). The vertical bar on the left delineates 5% of the total season's growth. Several periods of rapid swelling, especially marked in mid-April, may be noted as buds opened (B) and as the leaves began to expand (E). Leaves expanded to $\frac{1}{2}$ their full size (L) and matured (M) more rapidly in *Q. rubra* than in *Q. alba*. *Acer saccharum*, sugar maple (SM2 and 3), started growth at the end of April and reached peak rates of stem expansion after leaves reached maturity early in May. Coincident changes in the rate of stem expansion in the six trees during May are associated with variations in temperature. As soil moisture declined and air temperatures rose in June, daytime water loss exceeded water uptake, and diurnal shrinkage and swelling occurred. The rate of stem expansion in June was proportional to the amount of moisture in the soil of the site (RO 4>RO 2; WO 2>WO 1; SM 3>SM 2). Although 95% of the growth was not completed until July (vertical bar in middle), a pronounced reduction in growth occurred earliest in *Acer* and latest in the moist site *Q. rubra* (RO 4). Several drying cycles during July through September enhanced diurnal variations in stem size and initiated shrinkage in the stems. Each drying cycle was terminated by cloudy weather and replenishment of soil moisture, which resulted in expansion of all measured radii. At the end of September, leaf coloration commenced in *Acer* (C), abscission of leaves occurred, and diurnal variations in stem size were reduced. The leaves of the *Quercus* began to change color at about the same time, but they did not abscise as rapidly from the tree.

index of daily growth (Fig. 2.13; Fritts, 1962b), though the measurements may include some increases in size by tissues other than those of the ring. In spite of this limitation, these indices of daily growth provide useful data for statistical analysis of factors affecting growth (Fritts 1958, 1959, 1960, 1962c, 1969). Plots of such data (Figs. 2.13 and 2.14) can also be used to infer when water relations and other factors are most favorable to cell expansion, when and how fast cell enlargement proceeds, and, indirectly, how rapidly cell division occurs.

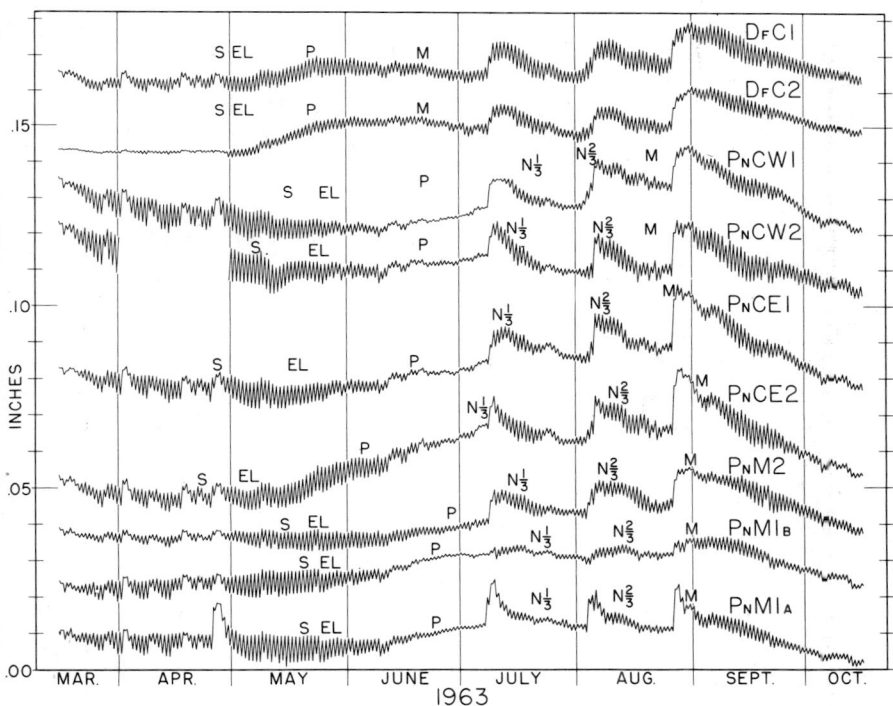

FIG. 2.14. Daily maximum and minimum size in the radii of mature tree stems which show diurnal and seasonal variations for 1963. Trees are on semiarid sites in Mesa Verde National Park, Colorado. Stems swell during moist periods and shrink during periods of drought. Shrinkage in dry years may exceed the total annual increment of growth. Df—Douglass fir, *Pseudotsuga menziesii*; Pn—pinyon, *Pinus edulis*; S—bud swelling; EL—bud elongation; P—pollen shed; N 1/3 and 2/3—needles 1/3 and 2/3 of mature size; M—needles mature size. The letters on the right designate different sites: C—canyon bottom, W—west-facing slope, E—east-facing slope, M—mesa-top. For details and/or similar plots during 1962, see Fritts *et al.*, 1965c.

D

Some trees, especially those which are slow-growing and subject to frequent and severe drought, exhibit diurnal variations in stem size induced by hydrostatic conditions within the tree which can mask the changes in stem size due to cambial activity (Fig. 2.14). Dendrograph or dendrometer measurements from such trees must be supplemented by other growth measurements if the exact period and rates of cambial activity are to be ascertained (Fritts *et al.*, 1965c).

C. Seasonal Periodicities in Radial Growth

It is apparent from many observations, including dendrograph measurements, that significant variations in stem size occur throughout the 24-hour day. When the daily stem sizes are plotted throughout the season, longer-term variations also appear (Figs. 2.13, 2.14, and 2.15). Such seasonal periodicities in radial growth may involve (1) differences in time between the initiation of cambial activity and its cessation, that is, the length of the growing season; (2) differences in the rates of cell division and cell enlargement leading to various rates of increase in stem girth, and (3) differences in size and structure of cells that become a part of the annual ring.

1. Length of the Growing Season

In general, cambial growth begins about the same time as in other meristems, but it may remain active later in the season than shoot growth. The period of cambial activity varies from one year to the next, as a function of differences in the yearly climates, differences in the environment from one site to the next, and differences in species (Ladefoged, 1952). Radial growth in trees of temperate climates may start in April, May, or June and cease in August or September. The growing season for broad-leaved trees usually lasts from two to more than three months. For some trees at high latitudes or high elevation sites it can be as short as four to six weeks. For other trees on warm, moist, and low latitude sites it can continue for a period as long as six months and in some cases for longer periods of time.

Conifers in temperate and moist environments generally have a longer period of cambial activity than broad-leaved trees in the same environments. The period of cambial activity in both types of trees tends to be longer for young than for old trees, and longer in tropical climates than in temperate climates.

The duration of cambial activity for conifers on arid or cold sites is generally less than for conifers in more optimum habitats. Fritts (1969) reports that ring growth in high elevation drought-subjected *Pinus*

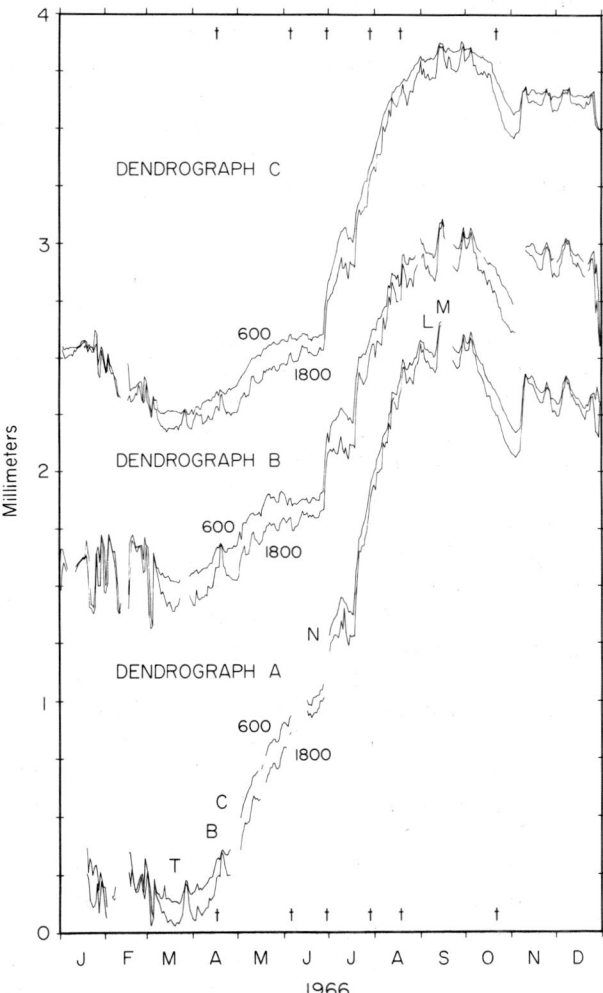

FIG. 2.15. The stem size measured by dendrographs each day at 6.00 a.m. (0600) and 6.00 p.m. (1800), 1966, at the base (A), middle (B), and upper (C) positions in the main stem of an arid-site *Pinus ponderosa* in southern Arizona. T—apical buds swelling; B—buds opening; C—cambial activity initiated; N—needles emerge; L—latewood cells lignified; M—needles reach mature size. The symbol *t* marks the days when the tree was enclosed in a plastic tent to obtain physiological measurements. See Chapters 3 and 4, and Figs. 2.19, 2.20, and 2.23.

longaeva continues for approximately 45 days. Growth cessation occurs even though environmental factors appear favorable. Giddings (1943) and Rees (1929) report that radial growth of tree line *Picea* at high latitudes continues from four to six weeks, while Fraser (1962) and Žumer (1969) report a two- to three-month growing period for the same genus in southern Canada and southeastern Norway near its low-latitude limits.

As stated earlier, growth initiation begins at the shoot apices and proceeds down the stem. However, in suppressed or old trees, especially during years of severe drought, the wave of cambial initiation may not always reach the ground line. In severe cases the initiating stimulus from the shoot apex is so weak that small patches of cambial cells become active beneath only the most vigorous branches. In less extreme cases, growth may be initiated throughout the stem except in localized cambia where the amount of growth is usually the least.

Some authors have reported that cessation of growth occurs first in or near the buds, and progresses basipetally with varying rapidity (Studhalter *et al.*, 1963). A larger number of workers report that cessation occurs first near the stem base, often in patches of cambia that are far removed from vigorous branches and roots. Acropetal growth cessation appears to occur in environmentally limited or forest-grown conifers such as *Pinus ponderosa* (Fritts *et al.*, 1965a; Budelsky, 1969). The basipetal migration of growth initiation and the acropetal migration of growth cessation helps to account for the observation that rings are generally narrowest in the lower *bole* (main stem) of the tree, where the growing season would be shortest. However, differences in growth rates do occur between the bole and top, and these differences are probably more important in causing differences in ring width than the growing season length (Skene, 1972).

Kozlowski and Peterson (1962) report that young trees continue growth throughout the stem, but many large dominant trees stop growing at the stem base during midseason while continuing to grow higher up in the stem. They also report a greater difference in duration of growth between upper and lower stem heights of dominant trees than for trees not included in the dominant class. Radial growth may cease sooner in suppressed trees with small crowns entirely below the forest canopy than in dominant trees with large exposed crowns (Winget and Kozlowski, 1965; Mitscherlich *et al.*, 1966). Trees in unthinned stands frequently have a shorter growing season than trees in thinned stands. The lower branches along the main stem of old trees grow for a shorter time than the young twigs and upper portions of stems.

Pinus edulis, a species widely distributed throughout Mesa Verde, Colorado, and well adapted to the arid environments on the mesa, begins to increase in stem size early in June. By mid-July the stems, which have been monitored by dendrographs, reach their maximum sizes and then progressive shrinkage of the stems begins (see Fig. 2.14). Samples from the cambium of *Pinus edulis* for 1962 (Fritts *et al.*, 1965c) show that initial swelling and differentiation of the phloem occurs in mid-May, with the first xylem cells being formed during the last week of May. Latewood cells are not evident until early August, several weeks after the stems have begun to decrease in size. By August 9, 1962, all cells had been differentiated and all, except the last-formed row of latewood cells, were lignified.

2. Rate of Increase in Stem Girth

The cumulative curve of radial growth throughout a season is similar in form to the cumulative curve for shoot growth over the life of a plant; it approximates a modified sigmoid curve (Figs. 2.13 and 2.15). The rate of radial growth increases rapidly to a peak near the beginning of the growth period and declines more slowly. The early, short period of increasing growth rate may last from one to several weeks; it is followed by a period of rapid growth which can continue for several weeks. Then the growth rate gradually declines, though the activity of some cambia can continue for a period of several weeks to several months after the period of maximum growth rate (Studhalter *et al.*, 1963).

The increasing size of a stem is rarely a smoothly varying function, however. Growth rates change and there are many variations in the cumulative curves. Short-term variations in growth rates are more apparent when plotted as growth-per-unit-of-time, rather than as a cumulative curve. For example, the cumulative curves of growth reported by Young and Kramer (1952) (Fig. 2.16) appear to be very similar at the stem base and at the top, though at the end of May there is a change that is obvious only in the growth of the stem base. The plots of weekly growth bring out variations from one week to the next. Except for the last of May and the third week of June, the periods of high growth rate at the top of the stem coincide with periods of high growth rate at the stem base. Measurements were also obtained at two intermediate stem heights and on trees with reduced amounts of foliage. Young and Kramer (1952) noted that when growth at one stem location either increased or decreased from that of the previous week, the growth usually changed in like manner at other stem locations and in trees in different crown classes. Many of these coincident changes were attributed to variations in environmental factors which influenced the water content of the trees.

As shown by the plots in Fig. 2.16, growth late in the season is often more rapid in the upper portions of the stem than in the lower portions.

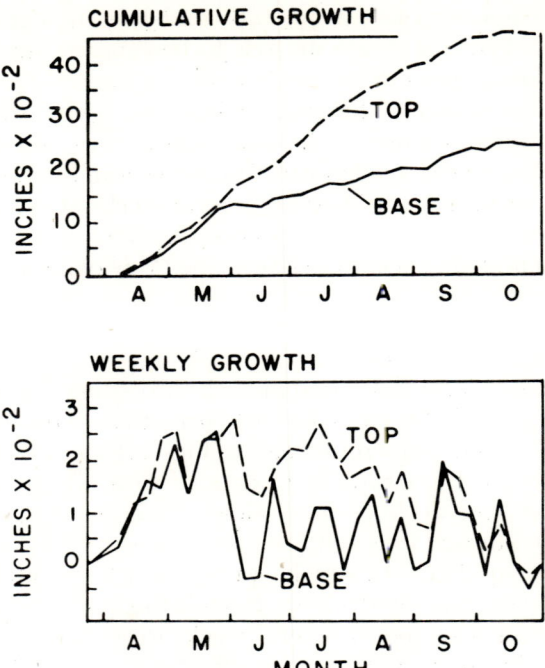

FIG. 2.16. Cumulative growth and weekly rates of growth as shown by stem diameter changes during the 1947 growing season at the base (4·5 ft = 1·37 m above ground) and at the top (80% total height) as measured by growth bands on *Pinus taeda*. The crowns of the trees occupied approximately 50% of the stems' height. (Redrawn from Young and Kramer, 1952)

The growth of young trees or of the younger tissues of the main stem frequently exhibits higher growth rates than older tissues and trees. Resumption of growth after a midsummer cessation is more likely to occur in young twigs than in the bole of the tree. Some variability in radial growth rate arises from strictly genetic differences among the trees.

Certain ring-porous trees, such as some species in the genus *Quercus*, exhibit two maxima in the rate of stem enlargement, one maximum occurring at the time of bud break and the other occurring after the leaves have enlarged to their full size (Fig. 2.17). Many trees representing diffuse-porous species growing under the same environmental regime show only one peak in growth rate coincident with the second maximum in *Quercus* (Fig. 2.18). Phipps (1961) attributes the first surge in stem growth in ring-porous trees to rapid enlargement of vessels formed before cambial activity begins.

Fritts (1960) could not attribute the reduction in rate of stem enlargement between the first and second growth maxima in *Quercus alba* and *Q. rubra* to any change in the external environment. He noted that the first peak in growth was correlated with other phenological changes such as bud break and leaf growth, and that the total increase in size during the first growth peak was comparable to the full width of earlywood in *Quercus alba* and to two-thirds of the earlywood in *Q. rubra*. This suggests that the period of reduced growth rate may be governed by internal conditions, perhaps associated with the transition from expansion of preformed earlywood to cambial activity and differentiation of latewood.

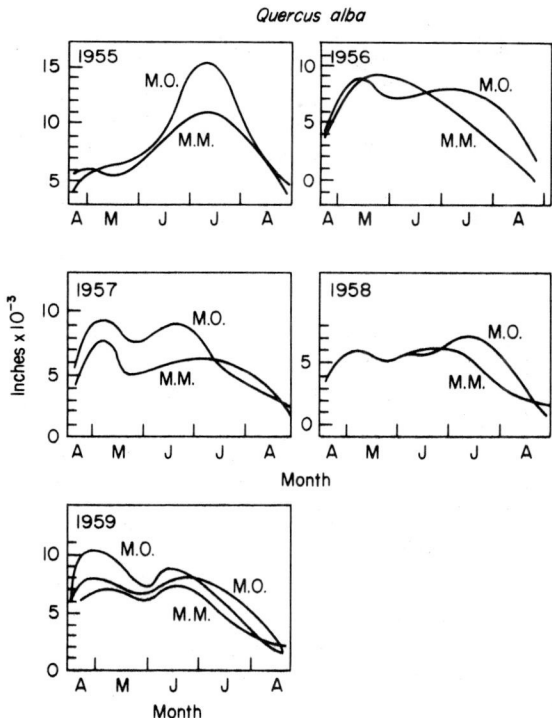

FIG. 2.17. Seasonal variability in the amount of stem enlargement per week, as measured by dendrometers on *Quercus alba* (a ring-porous species) growing in different plant communities in southern Ohio. MO—mixed oak; MM—mixed mesophytic. Five-week moving averages were calculated and the resulting smoothed curves were fitted to data points. (Redrawn from Phipps, 1961)

Data published by Kozlowski et al. (1962) show two periods of rapid growth in *Quercus ellipsoidalis* which correlated with two periods of precipitation. They did not recognize any indigenous rhythm in their trees. Zasada and Zahner (1969) report a continuous rate of cell enlargement in the xylem layer of *Quercus* throughout the period in which the reduced growth rate is supposed to occur. Their work suggests that the apparent bimodal peaks in growth cannot be attributed to xylem differentiation and suggest the possibility that the first peak in growth arises from rapid differentiation of phloem at the beginning of the growing season and not from xylem growth.

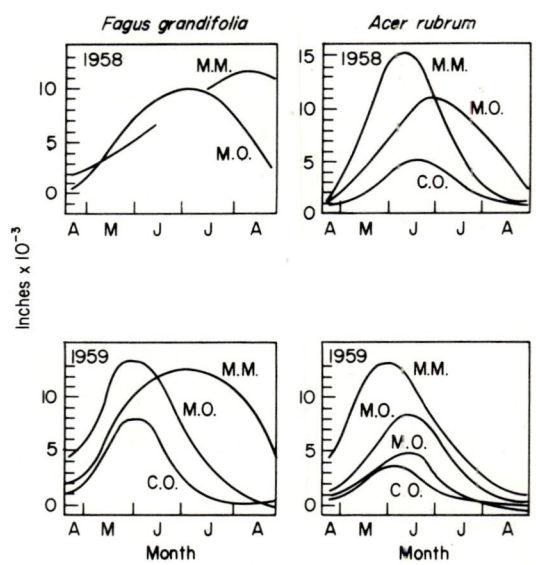

FIG. 2.18. Seasonal variability in the amount of stem enlargement per week, as measured by dendrometers on two diffuse-porous species in three plant communities in southern Ohio. MO—mixed oak, MM—mixed mesophytic, CO—chestnut oak. Five-week moving averages were calculated and the smoothed curves were fitted to data points. (Redrawn from Phipps, 1961)

Many of the fluctuations in growth rate during the year can be explained by variations in limiting factors of the environment. For example, low temperatures may become limiting to cell division or may affect other growth-controlling processes. Cold soils may restrict water uptake by the roots, thus retarding the hydration phase of growth. When soil moisture is low or when it is extremely high (causing low soil oxygen), the absorption of water by the roots may be low and growth may be reduced. High air temperature, high wind, high solar radiation, and low humidity are among the many factors which can enhance water loss, cause water deficits to occur, and thereby limit growth. And whereas clear weather followed by rain and cloudy conditions can favor increased hydration and induce swelling of tree stems, prolonged cloudy weather can cause the growth rate to decline by reducing photosynthesis and influencing the production of growth regulators in leaves. When clearing weather follows cloudy weather, there can be an increase in water loss and reduction in water content of the stem. This may reduce the rate of stem enlargement (Fritts, 1958) and can cause shrinkage to occur in the stem until high photosynthetic activity is resumed and increases the substances available for growth. A thorough discussion of environmental factors, particularly those affecting water relations and food-making processes, is presented in Chapters 3, 4, and 5.

Variations in rates of growth may arise from a variety of interactions within the tree. Reinitiation of apical growth or an increase in activity of the apical meristems may induce resurgence of radial growth in the twig and occasionally some growth resumption in the older portions of the stem. Defoliation, fire, pruning, or other mutilation of the crown can cause growth to increase, to decrease, or to stop entirely. Any injury to living tissue due to factors such as frost, lightning, or extracting a core from the stem, will produce wound tissue which, in turn, may induce anomalously high or low growth in the area surrounding the wound.

Phipps (1961, 1967) reports that the data shown in Fig. 2.17 were collected from dendrometers mounted on the tree stem in the spring of 1955. Anomalously high growth was noted in the second growth peak for 1955 but not the first, while in 1956 the first growth peak was higher than the second. These observations were offered as evidence that the earlywood cells were cut off from the cambium prior to insertion of the dendrometers. Since the cells of the earlywood were produced in the autumn, the wounding due to insertion of the dendrometer screws affected only the latewood of the same year and then the earlywood of the next.

Localized injury can result from the feeding activity of forest animals, the impact from falling limbs and rocks, and a variety of miscellaneous damaging events. A certain amount of the statistical error in tree-ring analysis is due to localized damage or effects of this sort. For more detailed reviews of the voluminous literature on radial growth rates, the reader may refer to Glock (1955), Bannan (1955), Studhalter et al. (1963), Agerter and Glock (1965), and Kozlowski (1971a and b).

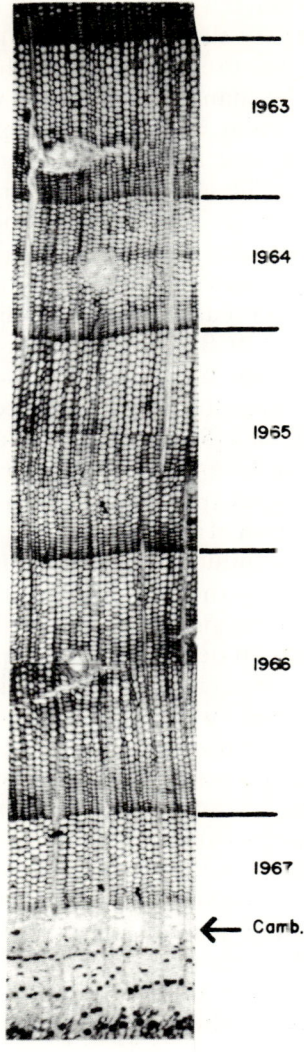

FIG. 2.19. The variability in cell size as shown in five annual rings from the base of a 110-year-old *Pinus ponderosa* growing in southern Arizona on a semiarid site at an elevation of 8,500 ft (2591 m). The photograph shows rings along a cross section of the stem sampled on August 8, 1967. A number of enlarged, but as yet unlignified, tracheids lie inside the cambium and outside the small diameter lignified cells that have already matured in the 1967 ring.

3. Variations in Cell Enlargement and Cell Differentiation

Variations in cell enlargement and differentiation occur throughout the growing season and give rise to variability in the cell structure and the general appearance of annual rings. At the beginning of the growing season, conifers form large-diameter tracheids with thin walls. As physiological conditions change within the plant and the surrounding environment becomes generally warmer and drier, the diameters of tracheids begin to diminish and the walls thicken. Usually, the smallest tracheids with the thickest walls are those formed at the end of the growing season along the outer margin of the ring (Figs. 2.3, 2.19, and 2.20).

The diameter and the location of the largest earlywood cells may vary from ring to ring (Fritts, 1969). Frequently, there is an increase in diameter along the *radial file* (or tier) of the first few rows of earlywood cells. Then a gradual reduction in tracheid diameters occurs throughout the remaining portion of the annual ring (Fig. 2.20). In ring-porous angiosperms the large earlywood vessels are differentiated early in the season and mature quickly, often prior to full expansion of the new leaves. Small vessels and thick-walled fibers are formed throughout the remaining portions of the growing season (Fig. 2.5b).

The large-diameter tracheids found in the rings of coniferous species are associated with periods of optimum soil moisture, favorable temperatures, rapid shoot and root growth, and a high rate of physiological activity within the tree (Zahner, 1968). Larson (1960, 1962) emphasizes the importance of rapid stem elongation and needle growth to the formation of large earlywood-type cells. As the activity of the apical bud and elongating needles declines and the amount of soil moisture diminishes, the size of mature, differentiated tracheids declines. This decrease in size of cells may be noted first in the lowest levels of the stem and last in the one-year-old twig. In older stems an increase in wall thickness normally accompanies the decrease in cell diameter, while in young stems which are close to the stem apex, the walls of tracheids in the latewood frequently remain thin. The initiation of wall thickening commonly occurs subsequent to cessation of stem elongation, sometimes preceding and sometimes following the decrease in cell diameter (Larson, 1964).

Replenishment of soil moisture or the reduction in water stress may be accompanied by a resurgence of shoot and radial growth, causing the unlignified tracheids to expand to the larger earlywood size. This increase in size of maturing cells can produce a band of earlywood between layers of latewood, forming a false ring as has been shown in Figs. 1.7 and 2.3. The resumption in growth usually occurs first in the one-year-old twigs or

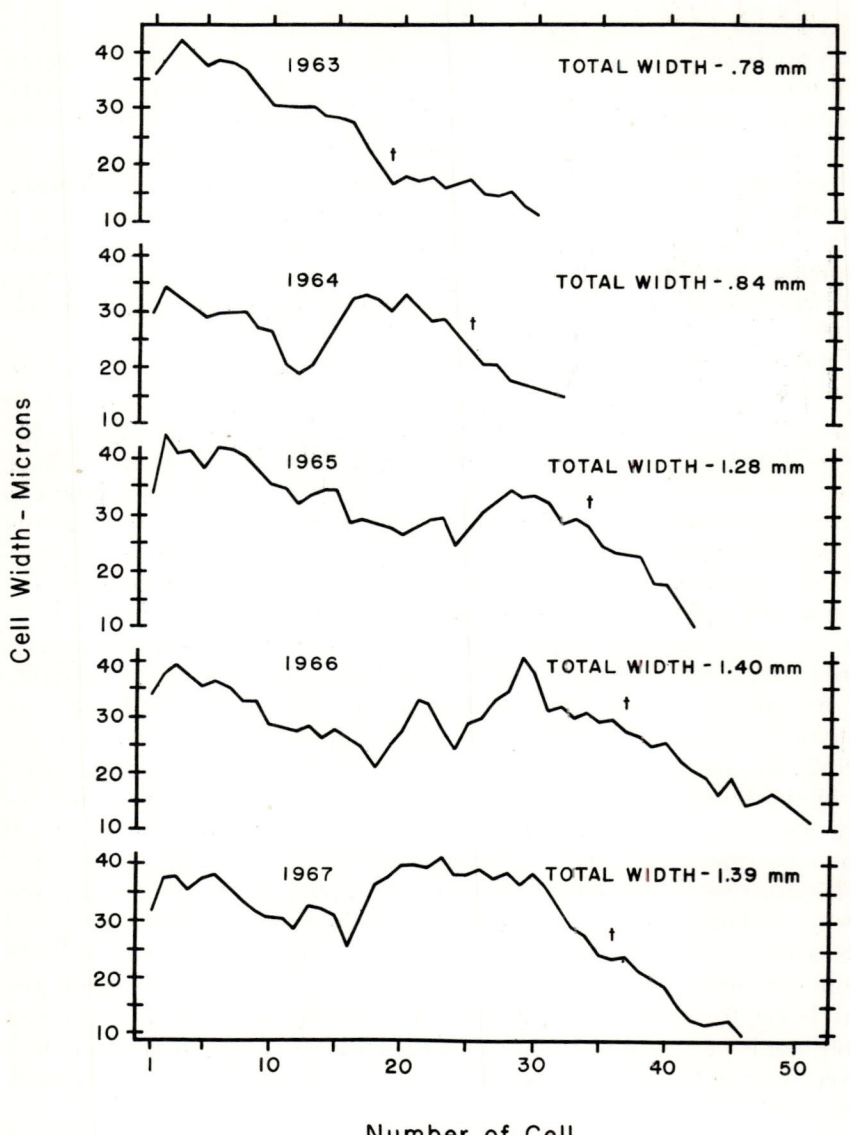

FIG. 2.20. A plot of average diameters of cells from five radial files in each of five annual rings shown in Fig. 2.19. Tracheid widths are plotted against cell number in the radial file. The letter t indicates the first cell with a thickened wall marking the late-season change from the thin-walled earlywood to the thick-walled latewood. Small cells occurring before t form intra-annual bands of latewood associated with drought in June.

in the uppermost portion of the stem, and a wave of increased expansion of tracheid diameter progresses down the stem of the tree. The wave may travel to the base of the stem or stop part way down if less favorable environmental conditions return and activity of the shoots declines. Since the resumption in formation of earlywood-like cells within the zone of latewood is closely associated with flushes in shoot growth, false rings are most typically found in small branches and twigs, in the stem portions near large, vigorous branches, and in young fast-growing trees. As the end of the season nears and activity of the shoots diminishes, the mature size of tracheids diminishes, cell walls thicken throughout the stem, and growth cessation progresses acropetally up the stem.

Budelsky (1969) noted that the sizes of mature tracheids in *Pinus ponderosa* were reduced and walls were thickened in the upper bole during a dry period in June, while no marked changes in the tracheid size and wall thickness occurred in the lower bole. Similarly, Fritts *et al.* (1965a) found that in the outer portions of a stem from an old tree, there is a greater tendency for intra-annual bands to be formed in the upper bole than in the lower bole, and they occur most commonly in the latewood portions of the ring. In a young stem, however, intra-annual latewood bands occur more frequently in the earlywood than in the latewood.

The annual rings of conifers which grow in an area with a long frost-free season and in an environment which may alternate between favorable and unfavorable conditions, such as moist winter, dry spring, and wet summer, are particularly likely to exhibit one to several intra-annual bands of latewood (Glock *et al.*, 1960; Glock and Agerter, 1962). In extreme cases, complete growth cessation may occur before the onset of the second period favorable to growth. For arid sites summer rains usually initiate the second growth flush, and if they are ample and early enough, a resurgence of growth may occur throughout the stem with the formation of wide tracheids next to very narrow ones, and thus more than one distinct ring is formed in a single growing season. The cambium, however, usually is not dormant in all radii of the same tree or in all trees, so that some growth layers formed in the same year have a recognizable growth band within a larger obvious annual ring. Crossdating among trees allows recognition of such circumstances so that the growth layers are placed in the appropriate growing year.

There is an increasing number of studies of tropical woods in which a variety of techniques has been used to ascertain the seasonal behavior of the cambium and the nature of the growth layers (Chowdhury, 1940; Mariaux, 1967). Some species are reported to produce distinct annual layers so that the rings can be counted to determine tree age. As yet there has been no definitive study demonstrating crossdating in tropical species. Much work is needed on species which do have distinct annual rings as to possible relationships between ring characteristics and variations in climate. It is hoped that botanical surveys in the tropics will utilize dendrochronological techniques as well as densitometry, dendrometry, and anatomical analysis to help find species which offer potential for dendroclimatic work.

4. Variations in Cells within Rings of Pinus ponderosa

Figures 2.19 and 2.20 show the variability in width of tracheids in five annual rings of an arid-site *Pinus ponderosa*, growing under the biseasonal precipitation regime in southern Arizona. The growth of several trees was monitored from 1963 through 1967, and the climatic regimes for the particular growing seasons are shown in Fig. 2.21. The tree which was sampled bore three recording dendrographs mounted at different levels on the stem (see Figs. 2.11 and 2.12). Cambial activity was assessed by periodic sampling of the cambial tissue, and measurements or observations on the development of buds, needles, and fruiting structures were obtained and used to characterize the *phenology* (the changing appearance of the tree during its yearly cycle including bud growth, needle development, flowering, fruiting, and cambial activity).

The environmental regime and factors of climate were monitored during the five years by means of a thermograph, a rain gauge, and soil moisture sensors which were buried in the soil around the roots of the tree (Fig. 2.21). Information on environment and growth was collected most systematically and frequently during the 1966 and 1967 seasons (Fig. 2.15) and, as a result, the changes in growth and the tree's environment are defined most precisely at that time. The climate and environmental occurrences, the general phenology, and growth of the tree and its effect on cell structure are described in the following pages.

(a) Environment and Growth of 1963. The environmental regime of the spring of 1963 (not shown in Fig. 2.21) was dry, and only 0·04 inches (1 mm) of rain fell during May and June. However, little rain falls in Arizona during this time. Dry conditions persisted throughout most of July (Fig. 2.21), as rains were sparse throughout the desert region and amounted to only 2·86 inches (72·6 mm) of moisture on the particular tree site. Heavy rains in August added 8·70 inches (221·0 mm) of water and replenished soil moisture, which was then plentiful through the remainder of the growing season.

Shoot growth was rapid in May but decreased with the greater water stress in June, and little growth occurred in July because rains were light. Some resurgence of growth was measured in August associated with the onset of heavier rains.

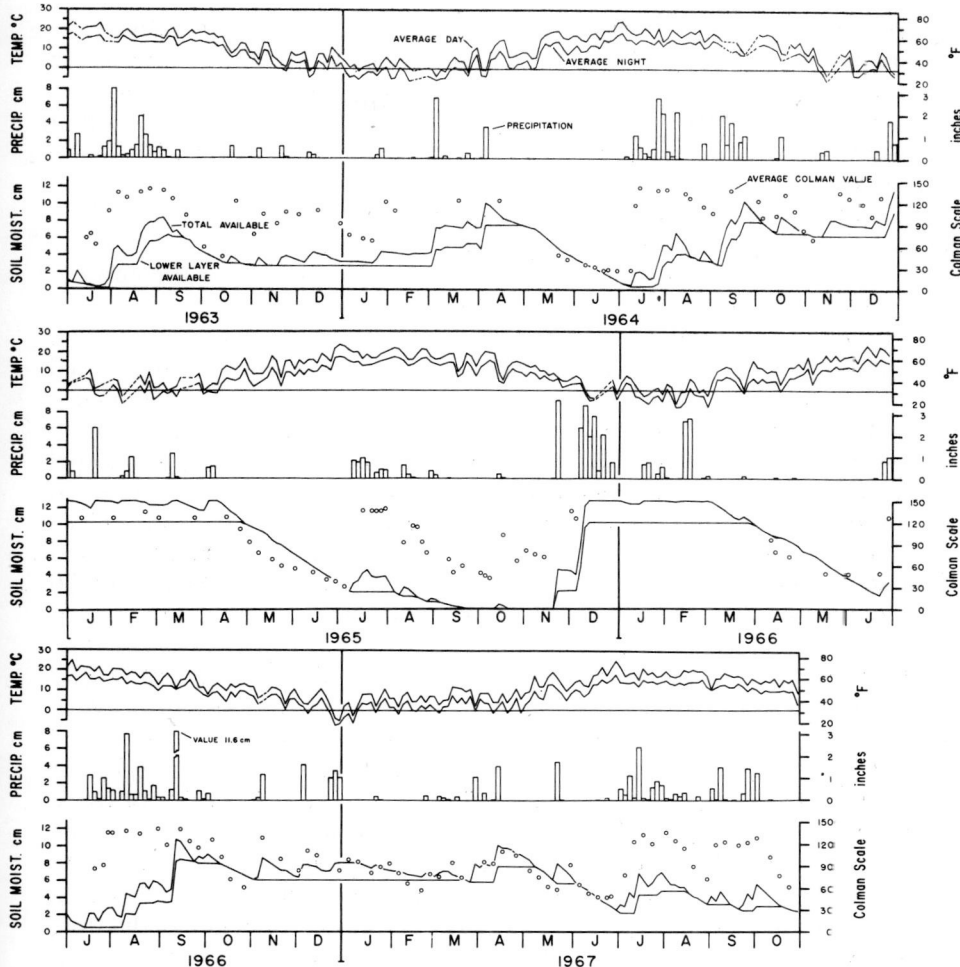

FIG. 2.21. Environmental regimes during the five years which included the seasons of growth for the rings shown in Fig. 2.20. Air temperature, precipitation, and soil moisture were measured by means of a thermograph, rain gauge, and Colman soil moisture meter. Total available and lower layer soil moisture were calculated using a soil moisture accounting scheme.

The first rows of tracheids formed in the ring averaged 40 microns along their radial dimension, but the size of successively formed cells declined (Fig. 2.20). This decrease in cell size correlates with the diminishing water supply of June. Cells in row 19 of Fig. 2.20 mark the beginning of latewood, and their thicknesses in the radial dimension average 17 microns. The period of latewood formation appears to correspond roughly with the onset of heavier summer rains. The moderate resurgence in shoot growth did not result in any increase in tracheid size at the base of the stem, but the period of cambial activity appeared to be extended so the percentage of latewood in the particular ring was greater than for rings formed during the other years. However, in the upper portions of the stem and in a nearby young tree resurgence of radial growth occurred, and enlarged thin-walled cells were formed in the outer portion of the ring.

(b) Environment and Growth of 1964. Soil moisture during the autumn and winter of 1963 and the spring of 1964 was the lowest of the five years studied. Above-average temperatures and a total lack of rain in May and June resulted in extreme water stress within the trees. Heavy rains started in mid-July and totaled 7·36 inches (186·9 mm) by the end of the month.

The radial diameter of the first 10 rows of tracheids (which would reflect early season growth when shoots were elongating rapidly) ranged from 25 to 35 microns. Their average sizes were less than those formed in the other four years reported here. The sizes of cells at rows 11 and 13 were only 19 to 20 microns wide and formed an intra-annual latewood band which correlates with the severe June drought. Rows 14 through 16 included larger cells associated with the period of improved hydration due to the July rains. The radial dimensions of all cells formed at that time increased to a maximum diameter of 33 microns. After row 20, the radial cell size began to decline again, and the first true latewood cells with heavy walls occurred in row 25 and are 24 microns wide.

It appears that the large earlywood cells formed after row 15 represent the response of the tree to the summer rains in mid-July. In this case more than half the ring width (the first 15 cells) was formed under the influence of moisture stored in the soil from the prior winter and spring rains. As would be expected, the change in cell size in the upper portions of the tree stem was more marked than the changes observed at the base of the tree, and very small cells produced during the June drought formed an even more distinct intra-annual latewood band.

(c) Environment and Growth of 1965. The precipitation of the winter of 1964 was moderate, and by January the soil moisture was completely recharged. However, soils again began to dry during a rainless May and June. July brought a moderate 3·92 inches (99·6 mm) of precipitation, while only 1·61 inches (40·9 mm) fell in August, making this the driest August of the five years.

Both radial growth and shoot growth were rapid from May through August, and even though the tracheids produced in June were somewhat smaller than those formed earlier, the activity of the cambium remained relatively high.

As a result, cells are wide, exceeding 40 microns at first, and then they are narrower with row 24 exhibiting the narrowest earlywood cells averaging 24 microns wide. These appear to have been formed early in July before the rains. Cells formed after row 24 are associated with the mid-July rains and they increase in size up to row 28. Latewood cells averaging 28 microns wide appear in row 34. Low moisture in August limited the growth rate, and cell diameters declined more rapidly across the latewood than in any other ring. The average diameter of the last-formed latewood cells in row 42 is less than 10 microns. The ring produced in one of the upper branches of a tree contained a distinct and marked cell size change representing a false ring. This suggests that more stressful conditions were present in the upper portions of the stem.

(d) Environment and Growth of 1966. The autumn of 1965 was dry, the winter was wet and cold, and the spring of 1966 was warm, with less than average precipitation. However, 1·98 inches (50·3 mm) of rainfall replenished soil moisture for a short period near the end of June. Drier conditions returned briefly in July, and heavy rains started on July 19; by the end of September over 16 inches (406·4 mm) of moisture had fallen since the beginning of summer rains. As would be expected, intra-annual latewood bands were formed in some of the trees during this growing season (Figs. 2.20 and 2.22)

The first radial growth on the monitored tree was measured in April, but the increase in stem size was slow (Figs. 2.12 and 2.15). In May the most rapid growth occurred, followed by drier conditions in June which caused a slight reduction in the rate of both shoot and radial growth. A temporary and limited resurgence of growth was measured in this tree during the brief moist interval of late June and early July (Figs. 2.12 and 2.15). The rainy season began on July 19 and the growth rate increased. Needle elongation ceased early in September, but some radial growth occurred until late September or early October.

A decrease in cell diameter is apparent up to cells in row 18 (Fig. 2.20), at which point diameters averaged 21 microns. Cell rows 19 through 21 are associated with temporary rehydration in late June and exhibit an increase in maximum size, while the smaller cells which follow in rows 22 through 24 reach a minimum diameter of 24 microns and reflect the declining soil moisture. Cells in rows 25 through 29 exhibit a second increase in maximum size which is associated with the onset of the summer rains on July 19. Rings observed in the branches of the tree show the same two distinct alternating layers of small and large diameter cells that were observed at the base of the tree. The remaining cells in rows 30 through 51, shown in Figs. 2.19 and 2.20, exhibit a gradual reduction in mature tracheid size, although soil moisture in August and September remained high. Latewood began to form with cells in row 37 with diameters averaging 28 microns. Another tree measured by Budelsky (1969) exhibited only one band of narrow latewood cells (Fig. 2.22), and it was best developed in the upper portions of the stem.

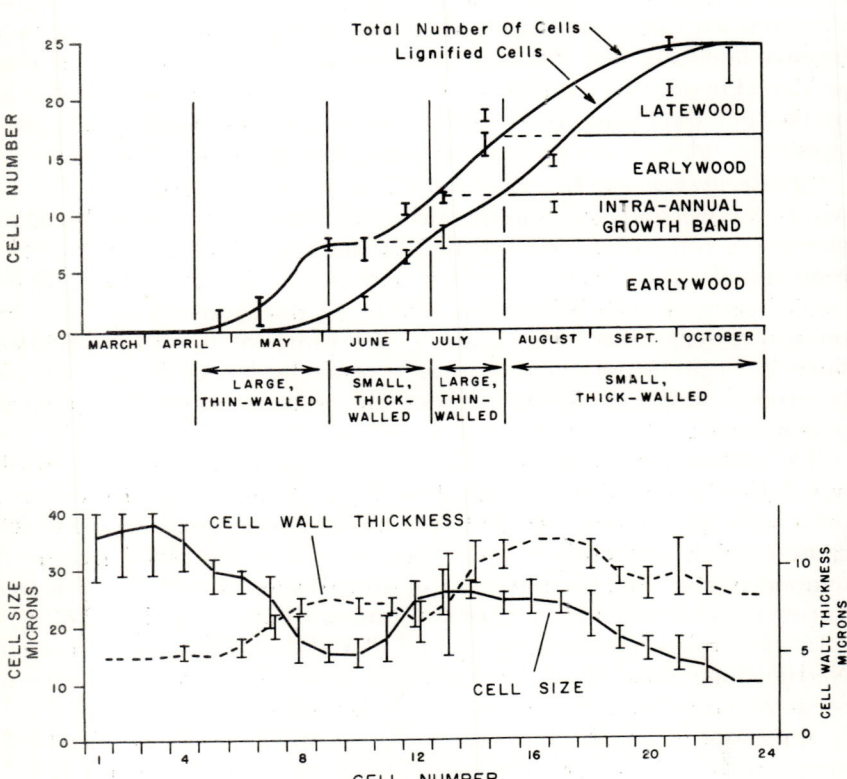

FIG. 2.22. Numbers and wall thickness of cells that were differentiated during the growing season of 1966 in the upper stem of a *Pinus ponderosa* in southern Arizona. A dry period in June (Fig. 2.21) reduced cell size and caused wall thickening which formed a latewood band within the earlywood portion of the ring. The number of tracheids was estimated from cambial samples harvested throughout the growing season. The size of the cells in a radial direction and wall thickness were measured along four radii of the stem sampled during the following spring. Brackets show the range of individual measurements. (Redrawn from Budelsky, 1969)

(e) Environment and Growth of 1967. Ample soil moisture was available during the winter of 1966 and was further sustained by precipitation during March, April, and May in the amounts of 1·57, 1·97, and 1·78 inches (39·9, 50·0, 45·2 mm), respectively. April, May, and June were unusually cool, and light rains fell in late June. Heavy rains started early in July, and by the end of the month 6·76 inches (172 mm) of moisture had fallen.

Shoot and radial growth were initiated during April, but early growth was slow during the cool weather, and mature xylem cells were not evident until June. With somewhat drier conditions in June, the rate of shoot growth declined, although cambial activity appeared unchanged. The rate of shoot growth increased in July but declined again in late August and September.

The first 15 cells of the growth ring for 1967, as shown in Fig. 2.20, reveal less change in cell size than any of the earlywood cells of the four previous rings. Cool and moist conditions maintained slow rates of shoot growth and a favorable water balance in the tree. Xylem cells produced throughout May and June were relatively large, with minimum diameters of 29 and 26 microns in cells of rows 12 and 16. Rows 13 through 15 represent larger diameter cells which were apparently produced during the first light rains of summer. The diameters of cells in rows 17 to 20 exhibit an increase in size, which coincides with the onset of the heavy summer rains. The cells in rows 20 through 30 were formed in late July and early August when moisture was not particularly limiting, and cell diameters averaged 35 to 40 microns. Cells in rows 31 through 46 are associated with declining late summer growth rates and exhibit a gradual reduction in size. Latewood formation with cells averaging 24 microns in diameter starts in row 36. The five outer rows of latewood cells are unusually narrow and produced an especially distinct annual ring boundary.

5. Variations in Cell Sizes and Ring Widths throughout the Length of the Main Stem

It may be noted from the results described in the prior sections that certain features of the latewood in the form of intra-annual growth bands were marked in the upper portions of the tree stem. In order to document this variation in more detail, certain growth rings were examined along 18 cross sections through the main stem of the tree. Four particular rings were selected which portray the kinds of variations that were observed. Fig. 2.23 presents the results as a plot of the ring width at each cross section and as a plot of earlywood and latewood thickness wherever they were observed. Much variability in ring width was observed to occur throughout the stem. The 1951 ring is the thickest of the four rings, indicating a season of very rapid growth. The intra-annual band of latewood within the earlywood undoubtedly marks those cells forming in the dry period of late June prior to the summer rain. A band of latewood within the earlywood was also formed in 1954, but the rate of cell growth throughout the tree was less, as indicated by the short distance from the beginning of the ring to the first-formed latewood. In 1955 there appeared to be no resumption of rapid growth and no band of earlywood associated with the beginning of summer rains. During 1956 the resumption of growth occurred only in certain stem segments within the lower crown and at the stem base. The data in the figure also support the inference that rings formed in the stem within the lower crown are the

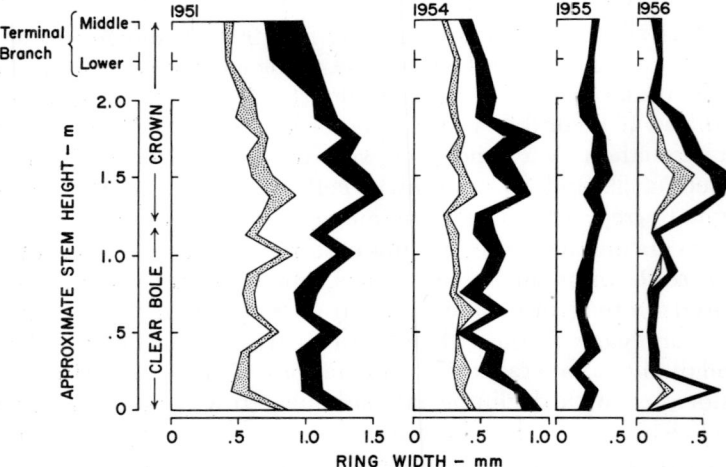

FIG. 2.23. The ring widths and location of latewood within selected growth layers in an arid-site *Pinus ponderosa* that was frequently subjected to water stress prior to the summer rains. Measurements were made along four radii on 18 cross sections cut from the main stem in 1969. The positions of each section are shown in relation to the terminal branch, crown, and clear bole. Heavy shading indicates latewood formed in the normal position along the outer margin of the ring. The light shading represents thick-walled cells forming intra-annual growth bands. The first thick-walled cells are usually formed at about the same time each year and represent the cells that were enlarging and differentiating during the hot, dry weather of June. Earlywood outside these first formed thick-walled cells indicates resumption of rapid growth in July, usually attributed to increased moisture from rains. If the summer rains are late or the growth rate low, the cambium may become dormant and a second layer of earlywood will not form (see 1955). The cells comparable to those formed as intra-annual rings in more moist years become the normal latewood. In other years, the resumption of earlywood formation may occur only in the crown or at the base of the tree so only in these portions of the stem will an intra-annual ring be formed (see 1956). Additional information about the tree and its environment is presented in Figs 2.15, 2.19, and 2.20.

widest and are the ones most likely to show a second earlywood layer resulting from resumption of rapid growth in midsummer. However, the abruptness between the intra-annual latewood and the subsequently formed earlywood is often more marked in the upper portions of the stem.

The details included in the diagram and the discussion of cell size and cell wall thickness show that considerable environmental information may be locked in the cell structure of the annual ring. It is not always information that is easily disentangled from the many complexities of cell growth, but, as new techniques are developed, cell structure producing wood density variations should be more easily and more objectively measured as well as better understood.

VII. Phenology and Its Relation to Ring Growth

It should be noted from the foregoing discussion that certain phenological phenomena such as the swelling, elongation, and opening of buds, the elongation of the stem and needles, maturation of the needles, flowering, and fruiting are often associated with specific stages in cambial activity and certain structural variations in the ring. In fact, observations solely on phenological development can provide much useful information on the timing and behavior of growth processes.

In the case of *Quercus* (Fig. 2.13), a ring-porous species, the first growth is observed at the base of the stem when the buds first begin to swell, and the initial surge of early growth has passed at the time the first-formed leaves are fully expanded. In the case of *Acer saccharum* radial growth does not commence until just before the unfolding of buds and the cambium becomes most active after the leaves attain full size. In *Pseudotsuga menziesii* (Fig. 2.14), the flush of needle growth occurs when the cambium becomes active, pollen is shed several weeks later, and needles reach mature size well before cambial activity ceases. Various arid-site *Pinus* species have slightly different phenological sequences (Figs. 2.14, 2.15). Cambial activity begins at about the time of bud swelling, but a month or more may elapse before the needles emerge from the bud and the pollen cones open. Cell division in the cambium ceases at about the time needles become fully elongated.

Phenological observations can be used to ascertain what limiting conditions are possible. For example, observation of the early rapid growth of *Pseudotsuga menziesii* needles suggests that full photosynthetic capacity of the newly formed foliage is attained sufficiently early in the growing season to influence the food-making potential for that summer and to affect the ring that is forming. In the case of arid-site *Pinus*, however, the needles appear much later and they are not fully formed until near the end of the growing season. Because of this delay, the effect

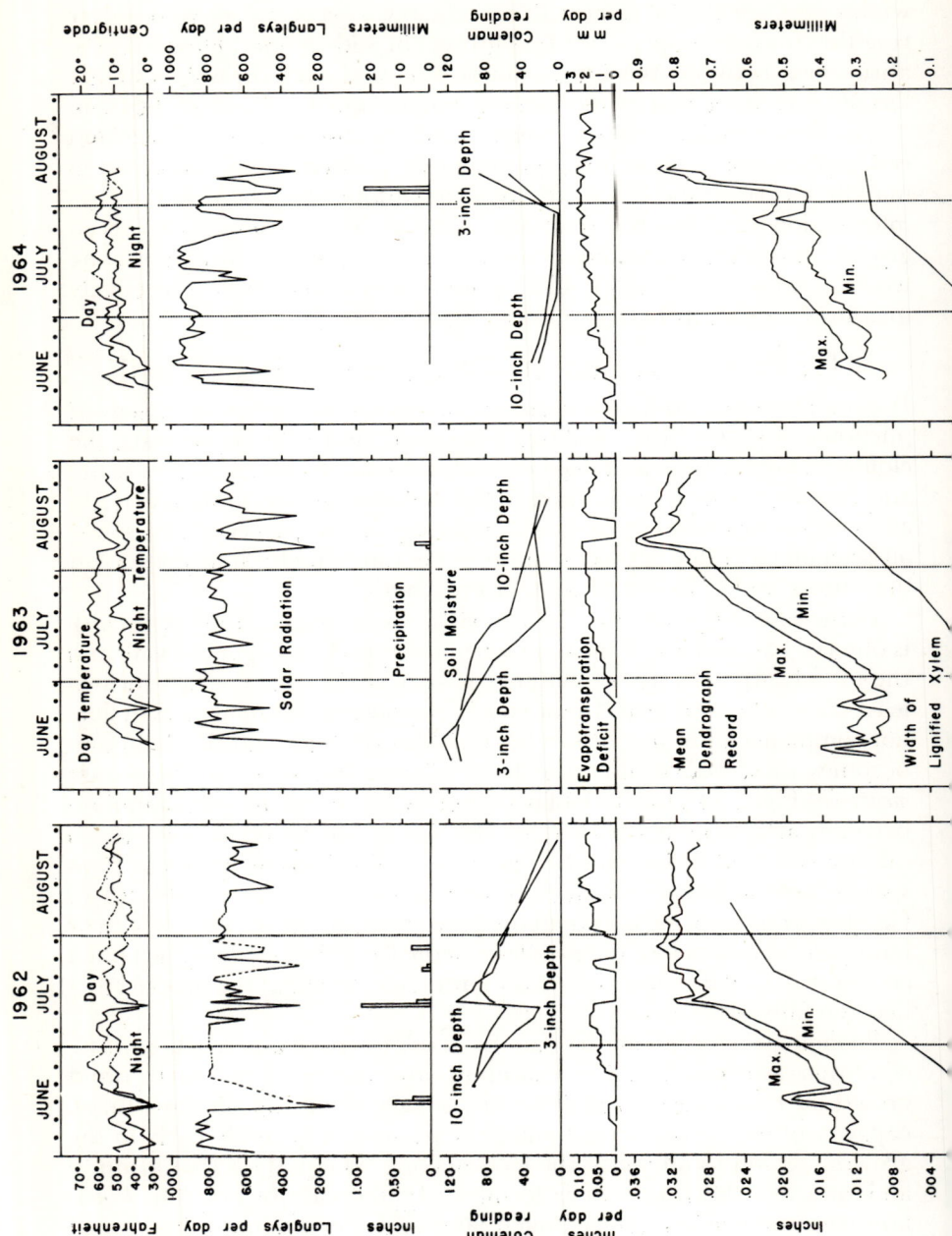

2. GROWTH AND STRUCTURE

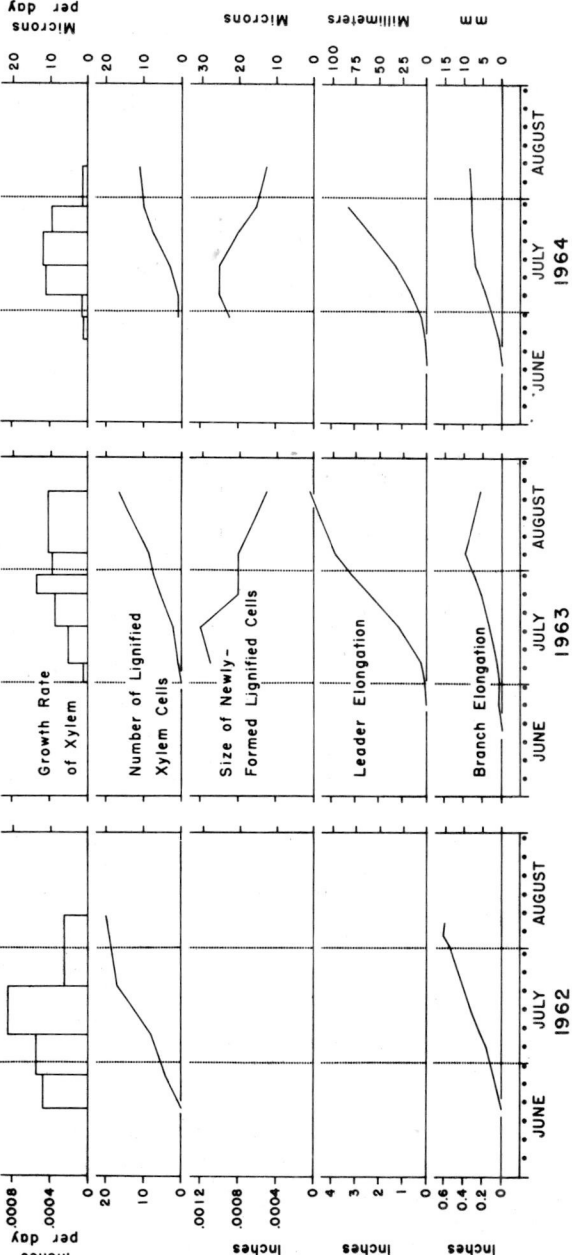

FIG. 2.24. The seasonal march of selected environmental parameters and various measurements of growth in *Pinus longaeva* in the White Mountains of California at an elevation of 10,400 ft (3171 m) for three growing seasons. Environmental parameters are mean day and night temperatures, daily solar radiation, precipitation, soil moisture in Colman units (where 0 is near the wilting point) and calculated evapotranspiration deficit. Growth measurements include stem size as measured by dendrographs, the width of the lignified xylem, the growth rate of xylem, the number of lignified cells in the xylem, the size of the newly formed lignified cells, leader elongation, and branch elongation. (Reprinted by permission from "Bristlecone Pine in the White Mountains of California, Growth and Ring-width Characteristics", Paper of the Laboratory of Tree-Ring Research at the University of Arizona, No. 4, by H. C. Fritts, Tucson: University of Arizona Press, © 1969.)

of climate on ring-width growth is most likely to be transmitted to the next year's ring through the effect on needle length. However, growth regulators produced by the expanding needles can modify cambial activity and cell structure and in this way exert some indirect influence on the character of the currently forming ring.

Fig. 2.24 is an example of the correlation between growth measurements and certain environmental parameters for *Pinus longaeva* (Fritts, 1969) during three growing seasons, while Fig. 2.25 shows the dates of certain phenological stages along with dendrograph measurements. The 1962 season was relatively moist with two major storms, which caused marked swelling in the tree stems. Temperatures in

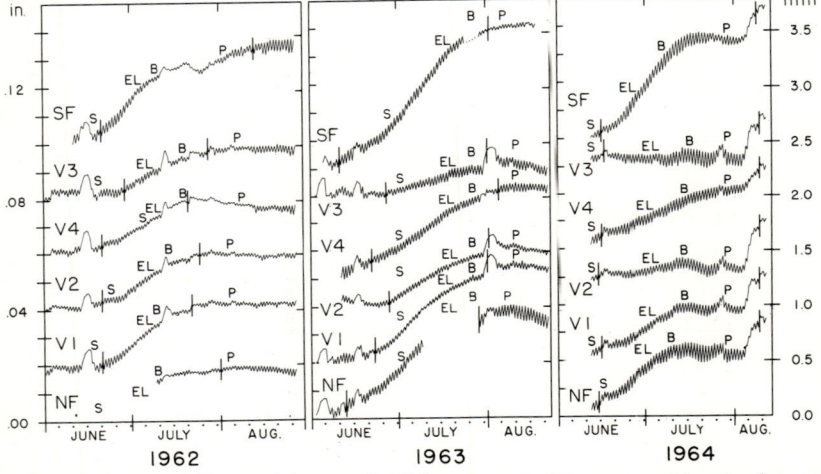

FIG. 2.25. The daily maximum and minimum stem size measured by dendrographs mounted on six *Pinus longaeva* trees during 1962, 1963, and 1964. Phenological changes in each tree are indicated as follows: S—bud swelling; EL—buds elongating; B—buds opening, needles beginning to emerge; and P—pollen shedding. Vertical lines designate time of 5% and 95% of total stem enlargement. Trees V1 through V4 are in the valley floor. The tree SF is on south-facing slope, and the tree NF is on a north-facing slope. The growth for the valley trees is summarized in Fig. 2.24. (Reprinted by permission from "Bristlecone Pine in the White Mountains of California, Growth and Ring-width Characteristics", Paper of the Laboratory of Tree-Ring Research at the University of Arizona, No. 4, by H. C. Fritts, Tucson: University of Arizona Press, © 1969.)

June were warmer than in the following two years; buds began to swell after the June storm, and the first cells of the ring were differentiated. The period of most rapid bud elongation, meristematic activity, and radial growth coincided early in July. Rain and snow in mid-July enhanced cell enlargement, though temperatures were near 0°C (32°F). The buds opened fully near the end of the growing period, and pollen was shed early in August after most of the cambial activity had ceased.

Bud and stem swelling began around the first day of July in 1963. Cell differentiation was observed several days later, but xylem growth reached its maximum rate late in July, while the buds were elongating. Pollination occurred and most growth was complete five or more days after a storm in early August.

The season of 1964 was the driest because of the lack of winter and spring precipitation. Bud swelling and the beginning of growth occurred in mid-June, approximately five days earlier than in 1962, but temperatures dropped shortly thereafter. The most rapid apical growth and cambial activity occurred near the end of June while the buds were elongating. However, stem swelling was slow, apparently as a result of low soil moisture and low temperatures. Pollen was shed two to three days before the end of July, as swelling in the stem and cambial activity diminished. Rain occurred during the first week in August and was accompanied by stem swelling, but not by any additional growth activity. The growing season averaged 45 days for all three seasons, and it ranged from 47 to 56 days for the younger trees and from 35 to 43 days for the older trees (Fritts, 1969).

Phenological development was observed on a variety of *Pinus longaeva* trees at different elevations, exposures, and slopes including those shown in Figs. 2.24 and 2.25 (see Fritts, 1969). Initial bud swelling and cambial activity began at low elevation sites at 9,400 ft (2,866 m) on June 20; at high elevations of 10,900 ft (3,323 m) and on north-facing slopes bud swelling and cambial activity began about 10 or 12 days later. Elongation of the buds began at low elevations on July 8–10, needles emerged from the bud scales on July 28–30, pollen was shed on August 6, and cell division ceased by August 17, but lignification was still in progress. At the high elevations the same phenological stages occurred about eight days later. For additional examples of phenological and tree-growth analysis see Douglas and Erdman (1967), Fritts (1958, 1960), Stokes (1965), and Serre (1973).

VIII. Systematic Variations in the Width and Cell Structure of Annual Rings

The annual growth layers of a tree exhibit systematic changes in width and cell anatomy which arise from the changing anatomy, physiological conditions, and environment of the aging tree. As was shown in Fig. 2.10, a young seedling may grow slowly at first, but as roots penetrate and spread throughout the soil and the crown enlarges, annual shoot growth may increase significantly. The seedling soon becomes a flourishing sapling in which both the apical meristems and cambia are growing at maximum rates. Eventually, the sapling becomes a maturing tree in which certain physiological and environmental conditions become limiting and the rates of shoot and ring growth decline. As the tree

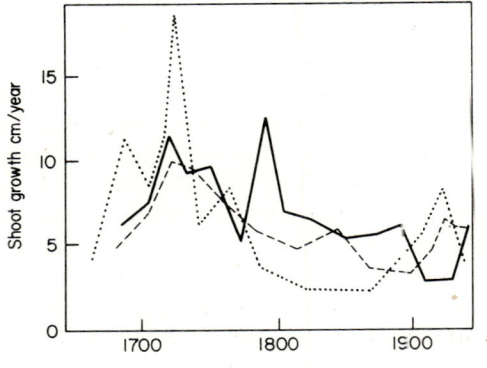

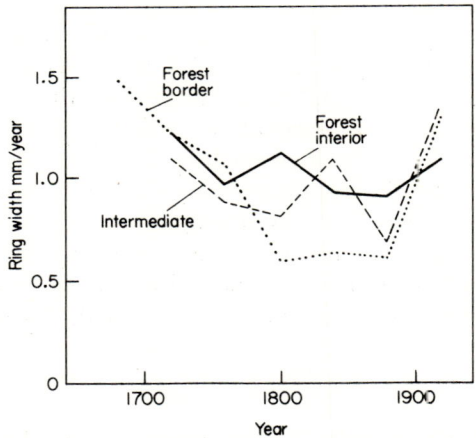

FIG. 2.26. Shoot growth and average ring widths observed on a number of transverse sections from the stems of *Pinus ponderosa* on three different sites near Flagstaff, Arizona. Length of shoot growth is reconstructed by comparing the distance along the stem between sections and the differences in dates of the innermost rings. Ring-width measurements are averages for 40-year intervals of time on all available sections. The forest border tree grew more rapidly than the other trees during the sapling stage, while the forest interior tree grew more rapidly than the other trees when they matured. The forest border and intermediate trees exhibited greater increase in growth rate with change in climate after 1900 than the tree on the forest interior site. (Data from Glock *et al.*, 1963, recompiled and described by Fritts *et al.*, 1965a)

approaches old age, the level of growth stabilizes to about the same rate throughout the main stem of the tree.

The changes in growth associated with increasing age may vary from site to site. Fig. 2.26 shows the changes in rate of shoot growth and average ring width throughout the entire stem of three trees of *Pinus ponderosa* growing in northern Arizona. The data on shoot growth are based upon distances between and ages of cross sections obtained from the main stems of the trees, while those on ring width are 40-year means plotted as a function of time. One tree was on a dry, forest border site (Glock *et al.*, 1963), a second on an intermediate site, and the third on a moist, forest interior site. The rate of shoot growth during the first 20 years was lowest for the tree on the driest site. However, at an approximate age of 80 years, the rate of shoot growth in the arid-site tree exceeded that of the trees on more moist sites. However, all three trees attained high rates of growth before their one hundredth year, and in the case of the forest interior tree, a maximum in shoot growth occurred at an age somewhat greater than 100 years. Rings at the base of the main stems attained maximum widths at an earlier age. In general, after the first 100 years, the rates of both shoot and ring width declined. This decline in growth rate was most marked for the tree on the dry site and least marked for the tree on the moist site. The growth rates approached a state of apparent equilibrium, where the mean growth rate remained about the same, first in the arid-site tree and last in the moist-site tree. A period of unusually high precipitation occurred in the early 1900's (Fritts, 1965; Fritts *et al.*, 1965a) and caused shoot growth and ring widths to increase, especially in the arid-site tree.

Certain systematic variations in the annual growth layer are associated with the circumference of the ring, the development of the crown, and the amount of competition with neighboring trees. Farrar (1961) illustrates how the width of a ring in a young forest-grown tree may vary from the stem apex to the stem base (Fig. 2.27A). The ring is comparatively narrow at the stem apex and reaches its first maximum width in the stem nearest the branches with the most foliage. The ring is often narrower in the lower part of the crown and along most of the bole. Ring width increases slightly to a second maximum at the ground line where there is root flare. However, if ring-width data are converted to cross-sectional area (dashed line in Fig. 2.27) the plot of growth as a function of bole height takes a different form. There is a marked increase in area from the stem apex to the portion of the stem just below the crown. Below this point there is only a slight increase in area with decreasing height until near the base where there is an increase in area associated with the root flare (also see Phipps, 1967).

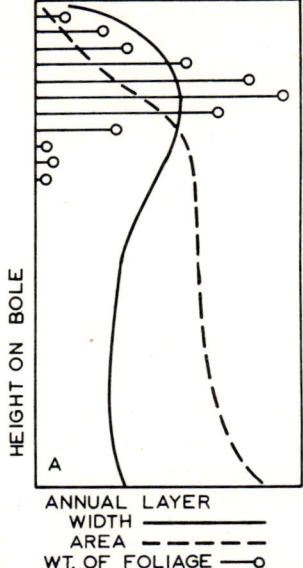 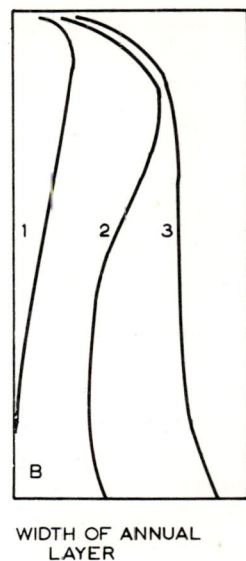

FIG. 2.27a and b. (a) The size of an annual growth layer as measured by its width and cross-sectional area and plotted as a function of height in the bole. The weights of foliage are indicated by length of the horizontal line at the indicated height. (b) The size of the annual growth layer as a function of height in the bole for (1) a crowded or pruned tree with a small crown, (2) a dominant tree in a young stand, and (3) an open-grown tree with live branches extending to the stem base. (From Farrar, 1961)

Figure 2.27B illustrates how the width of the annual ring can vary with differences in the density of the forest stand and tree crown. The rings are narrower in trees that are crowded or pruned to have a small crown (B1). In such a tree the rings become narrower in the bole the greater the distance from the tree crown. If the tree is severely crowded and the crown sufficiently small, the sheath of xylem may be discontinuous or absent in certain areas of the bole, especially near the stem base. Figure 2.28 is an extreme and obvious example where the xylem sheath produced during 18 continuous years was absent along a portion of the stem. In trees growing in more open stands (Fig. 2.27, B2 and B3), the rings are more nearly the same size throughout the lower portions of the stem.

Farrar (1961) further illustrates in Fig. 2.29 some of the changes in width that may occur in the stem of a young plantation-grown tree. During the first few years when the tree is free from competition, the xylem sheath may increase in thickness from stem apex to the stem base. A few years later the lower limbs become shaded by the upper limbs and

FIG. 2.28. A cross section from a coniferous stem with 18 growth rings absent on the right side of the photograph but present on the left. (Photo by James Harsha)

by neighboring trees, and the sheath is about the same width throughout the lower stem. As the tree ages some of the lower limbs may die, others become more shaded, and a zone of maximum ring width occurs near the crown base especially near those branches bearing the most foliage.

If the plantation is thinned and neighboring trees are removed, the rings will thicken, especially in the lower part of the stem, while in the crown the thickness may remain relatively unchanged. The ring width may increase from the stem apex to the stem base. As the canopy closes during subsequent years, shading of the lower branches increases and the xylem sheath resumes its normal form with maximum thickness near the area of maximum foliage and at the stem base.

As many conifers increase in age, there is an increase in specific gravity of the wood, an increase in the percentage of latewood, and a decrease or no change in the density of earlywood. The slow-growing, more suppressed trees exhibit a greater change in percentage of latewood from the center of the stem to the outermost ring, than fast-growing and dominant trees (Kramer and Kozlowski, 1960). However, Larson (1957) reports that there is apparently a certain intermediate ring width in the development of a tree at which point the specific gravity of the xylem is maximum. In the wide rings of young, fast-growing trees, such as six to eight year-old *Pinus* saplings, and in the very narrow rings of over-aged trees, a smaller percentage of latewood is formed and the wood is less dense.

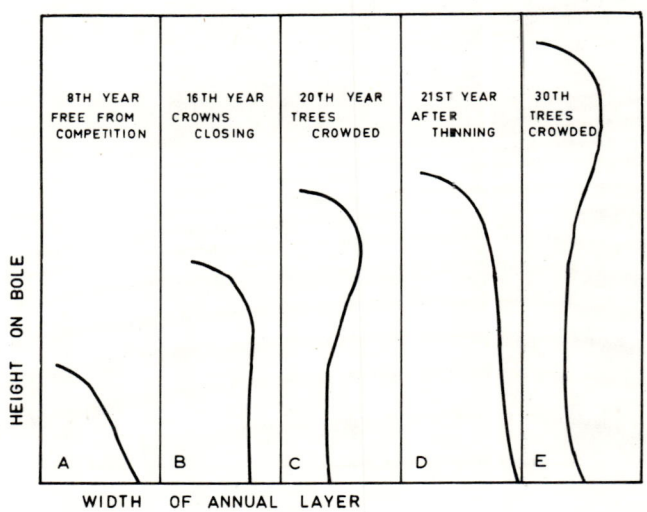

FIG. 2.29. The width of the annual growth layer throughout the tree bole for selected years in the life of a plantation-grown *Pinus resinosa*. Foliage covers the entire stem base in A, but by the 15th year lower branches become shaded (B), and by the 20th year (C) the lower branches have died. Thinning exposes the crown to full light in the 21st year (D), but by the 30th year (E) the crowns have closed again and the trees are crowded. The horizontal scale of width is approximately 1,000 times the vertical scale of height. (From Farrar, 1961)

The widths of both the earlywood and latewood in ring-porous angiosperms usually decrease with the increasing age of the tree, but the earlywood occupies an increasing proportion of the total width (Brown and Panshin, 1940) and there is decreasing year-to-year variation in width of earlywood (Phipps, 1967). Also, the percentage of latewood may vary systematically with varying site characteristics (Larson, 1957) and with varying seasonal environmental regimes (Larson, 1957; Smith and Wilsie, 1961; also see Fig. 2.22). Since a variety of factors can contribute to systematic variations in ring structure, there is no simple equation or general function that can describe all variations in growth that occur with increasing tree age. However, certain statistical methods are described in

Chapter 6 which can be used to fit a curve to some types of systematic variations, such as diminishing ring widths that are associated with increasing tree age. The curve represents a statistical assessment of all systematic changes related to tree age regardless of their causes.

The rings in leaning stems and lateral branches exhibit another kind of systematic variation leading to eccentricity in the width and structure of the ring. Such eccentricity is usually associated with particular characteristics in cell anatomy and is designated *reaction wood*. In coniferous species the reaction wood occurs on the lower side of the tree and is therefore referred to as *compression wood*. The rings with compression wood are wider and they contain a larger proportion of latewood than the rings on the upper side of the lean. Compression wood is also denser and more brittle than normal wood; the tracheids are heavily lignified and tend to be circular in transverse section so that more intercellular spaces are visible than in normal wood (Westing, 1968).

Reaction wood in angiosperm species occurs on the upper side of the lean, and is therefore referred to as *tension wood*. There are fewer and smaller vessels in tension wood than in normal wood; and there is a corresponding increase in the proportion of fibers which are conspicuously thick-walled. However, there is a reduced amount of lignification in tension wood (Hughes, 1965) and some fibers develop a gelatinous layer (Wardrop, 1964b). The presence of reaction wood in rings can be interpreted as evidence for a nonvertical orientation of the stem (Westing and Schulz, 1965) and changes from normal wood to reaction wood are often used to establish the date of events affecting the orientation of a tree (see Chapter 5; Alestalo, 1971).

IX. Growth of Roots

The apical growth of roots, unlike growth of stems, may exhibit no inherent dormant period, and it is not certain whether there is any inherent periodicity in root elongation (Kramer and Kozlowski, 1960). However, many workers have reported rapid root elongation in the spring and autumn, and less active growth in the dry period of summer. Radial growth in roots, unlike apical growth, appears to be influenced by what is occurring in the stem and neighboring roots, as well as by the surrounding environment. Thus the distribution, timing, and anatomy of radial growth in any part of the root system appear to be affected by growth conditions of the stem, the location of the root with respect to the stem, the activity of the root tips, and environmental conditions (Fayle, 1968).

Roots, like stems, produce annual growth layers, but in roots the distinction between layers is more difficult to discern with increasing distance from the stem base. The width of the growth ring generally tapers along with the diameter of the roots so that the widest part is nearest the stem and the narrowest at the root tip. Near the stem, the anatomy of the growth layer in roots appears much like the annual ring in the stem. Cell walls are thinner in the latewood of roots than in stems, the vessels in the earlywood of ring-porous species are smaller, and the cells are somewhat larger toward the distal portions of roots so that the tissues resemble diffuse-porous wood. For hardwood species, the farther the root is from the stem base, the greater the proportion of vessels and the less the proportion of fibers (Fayle, 1968). Near the root apices, however, some of these changes in structure may reverse. Fayle reports that the specific gravity of root wood in his trees was 20% lower than the specific gravity of stem wood.

Initiation of cambial activity in roots appears to be a continuation of that in the stem. The wave of growth initiation in the cambium is triggered by activity in the shoots and moves down the stem to the ground line where it may be delayed before it proceeds toward the root tip. The rates of cambial growth in roots are more variable than in the stem (Fayle, 1968). Growth cessation follows a more irregular pattern than growth initiation, but it generally proceeds from the root apices toward the stem-root base (Fayle, 1968).

Whereas the patterns of ring width along a cross section of the main root may show a marked similarity to the ring-width patterns in the lower portions of the stem (Schulman, 1945; LaMarche, 1968), there may be less or no correlation with the stem ring-width patterns in the finer subdivisions of root systems, at greater distances from the stem base, or with increasing age of the root. The growth layer is frequently distributed around the circumference of roots in a more eccentric manner than in the stems. Mechanical stresses transmitted into the root system, soil pressure, burial or exposure of the root, the development of adventitious shoots, and changes in the availability of growth regulators and food all affect root growth and contribute to the irregularity and multiplicity which are characteristic of the annual xylem sheaths, especially in the distal portions of roots (Fayle, 1968).

If the stems of stream-side trees are buried by flood deposits and they survive, new xylem tissue produced in the buried stems will appear more like root wood than like stem wood (Sigafoos, 1964). The incline of trees, the age of wounds, and the age of sprouts, along with changes in anatomy of stems buried by sediments, can be used to date and reconstruct past history of floods (Sigafoos, 1964).

Also, trees on *talus slopes* with exposed roots yield information on extant geological processes that are taking place in the sites (Eardley and Viavant, 1967; LaMarche, 1968). *Buttress roots* (which are elongated vertically) are formed by *Pinus longaeva* as a direct result of exposure (LaMarche, 1968). The cambia along the exposed surfaces die, leaving only the lower surfaces of living bark, which continue to grow and eventually produce an asymmetric vertical slab form. A well-preserved buttress root will show concentric growth rings around the root axis corresponding to the time when the root was fully buried in the soil. Reduction of the cambial area in old exposed roots may continue until only a narrow strip of living bark remains along the protected underside. The ages of trees and measurements of maximum depth of the highest (what was initially the shallowest) exposed roots, plus an estimate of the depth of initial root development, allowed Eardley and Viavant (1967) and La Marche (1968) to estimate the rates of slope degradation occurring around the trees.

X. Significance of Growth and Structure to Dendroclimatology

It has been shown in this chapter that growth of a tree may vary in many ways and over a wide range of time scales representing different frequencies. For example, there are hourly variations in tissue size that occur primarily as a response to changes in temperature and moisture or to changes in the diurnal variations between water uptake and water loss. There are seasonal changes that occur at frequencies of a few days to a few months, which may result from varying activity of the apical meristems, varying vigor of the cambial tissues, or varying environmental factors that become limiting to one or more growth-controlling processes. Many factors vary over the annual cycle and there are changes at frequencies of one or more years which often, but not always, result from year-to-year variations in one or more aspects of climate. Many of these variations can be complicated by internal preconditioning, structural changes in the roots, stems, and leaves, and by changes in the general physiological condition of the tree. Still longer-term variations in growth can occur at frequencies of several decades to several centuries, depending on the longevity of the tree. These often arise from systematic changes within the tree due to increasing age and alterations of the local environment. Sometimes long-term macroclimatic changes may be involved, but if this is so, similar growth changes will be found in many trees regardless of their age and position within the forest.

Ring widths can be influenced by environmental variations occurring at all of these frequencies. The annual variations in width are averages of the day-to-day and season-to-season variations in processes that have limited growth. They also reflect both the year-to-year variations and the longer-term changes which may or may not reflect variations in climate.

Certain dendrochronological techniques, including standardization, have been developed to emphasize variations at particular frequencies and to minimize others. For example, the standardization curve of some prescribed form may be fitted to the change in ring widths expressed as a function of tree age (Fig. 1.9). Such curves are designed to be flexible enough to take into account many of the systematic variations that can occur over the life of the tree, including those variations due to the location of the cross section throughout the stem or root and those variations that are a function of the changes in the surrounding competition, stand history, and productivity of the site.

The techniques of curve fitting do not distinguish whether the long-term variations are due to climatic or nonclimatic factors, though there is circumstantial evidence that most of these changes in both arid- and mesic-site trees is nonclimatic. However, long-lasting drought may produce changes in the forest density, causing death of certain trees and expansion of others into the vacated environment. The resulting gradual adjustment in the plant community causes ring widths to return to the average state even though the drought persists. Thus, curves fitted to century-long changes in growth in arid- and mesic-site trees are necessary in order to remove important nonclimatic variations; and fortunately there is so little climatic variation left in the ring widths at these time scales that little climatic information is lost.

Trees that are long lived and in cold sites such as *Pinus longaeva* at high elevations or *P. sylvestris* at high latitudes, do not appear to lose as much climatic information at time scales of centuries and millenia as do short-lived trees on warmer sites. This appears to occur because the changes in community structure and forest density resulting from long-term variations in temperature accentuate and reinforce the year-to-year climatic effects rather than leading to adjustment and a return to the average state as occurs in warmer sites. If the trees are of this type, the variations due to tree age and site factors may be considerably smaller than the variations due to changes in climate. In such cases curve-fitting procedures are applied with caution or not at all. It is still necessary, however, to identify and to try to separate any changes due to tree age and site conditions from those due to climate. This can sometimes be accomplished if the ring records are of sufficient length by discarding the first few decades of vigorous growth that are obviously not the result of climate. After these are removed, the remaining ring widths can be compared for intervals in common, adjusted, averaged, and the mean record used to infer variations in climate.

Considerable attention has been given in this chapter to the variations that occur in cell structure within the annual ring. It has been stated that these variations can be attributed to particular characteristics of the growing season and the associated climates. While such small-scale variations in cell size have little net effect on ring widths, they are an important source of information that can be measured by densitometric analyses. The close association between cell size, wood density, and factors of the environment point to a tremendous amount of information that must be present in the rings and which should be extractable by measuring density variations along with ring widths. If these new data on density can be successfully related to and calibrated with environmental data, the simultaneous analysis of density and ring width may provide significantly more information on past climate than is possible using only the widths of rings.

Chapter 3

Basic Physiological Processes: Movement of Materials and Water Relations

I	Limiting Factors and Plant Processes	118
II	Some General Terms and Basic Concepts.	120
III	Cell Water Status	123
IV	Transpiration .	127
V	Soil Moisture .	133
VI	The Soil System and Factors Affecting its Development	136
VII	Absorption of Water	140
VIII	Internal Water Relations of Trees	142
IX	Moisture Stress and Tree Form	147
X	Soil Factors Affecting Root Growth, Ring-width Sensitivity, and Longevity	148
XI	Uptake of Mineral Salts .	151
XII	Translocation .	152

I. Limiting Factors and Plant Processes

It was pointed out in Chapter 1 that a climatic factor can affect structural characteristics such as ring width only as it influences the operational environment of the plant and limits the rate of physiological processes that in turn influence growth. The actual relationships that are involved may vary, depending upon the condition of the growing tissues, the relative activities of the controlling processes, and the operational history of the plant's environment that has preconditioned it. Also, the most limiting conditions to plant processes can change markedly throughout the year, so that one particular climatic factor may be directly correlated with ring width at one time, inversely correlated with it at another time, and totally uncorrelated at still other times. For example, during the spring, growth may begin earlier in a warm year than in a cool year, because low temperature is most limiting to the growth-initiating process.

In such a case, temperature would be directly correlated with ring width. Later in the growing season, when temperatures are higher, the hottest weather may be most limiting because it can limit the production of enzymes and hormones which are necessary for certain processes to occur and because it accentuates the loss of water from tissues in which water is already deficient. Thus, temperature at this time of year becomes inversely correlated with ring width. During the dormant winter period, freezing temperatures may prevail so that variations in temperature do not affect any particular process, and therefore temperature variations are uncorrelated with growth. On the other hand, if warm weather in winter is associated with thawing of tissues and increased physiological activity, certain biochemical changes may occur which affect subsequent growth and lead to correlation of winter temperatures and ring width.

This example is only one of several possible explanations for the correlation of temperature with variations in ring width. In the following discussion it will become apparent that there are a number of physiological processes involved.

As implied by the law of limiting factors mentioned in Chapter 1, only a small number of environmental factors can limit a process at a given instant in time. This was first recognized in 1843 by Liebig who proposed what he called the Law of the Minimum: "When a process is conditioned as to its rapidity by a number of separate factors, the rate of the process is limited by the pace of the 'slowest' factor" (Meyer, *et al.*, 1973). Liebig's law has been shown to be an oversimplification because at times more than one factor can be limiting to a process. Thus, in 1909 Mitscherlich suggested that "plant yield" is sometimes limited by deficiencies in several factors. He modified Liebig's law to include interactions between several factors but he accepted the essential feature of Liebig's law, that usually the greatest change in yield will be produced by the one factor which is most deficient (Meyer *et al.*, 1973).

Lundegårdh (1931) stated the same principle in a different form, which he calls the Law of Relative Effects:

> The more nearly a factor is in minimum in relation to the other factors acting upon the plant, the greater is the relative influence of a change of that factor upon the growth of the plant. As a factor increases in intensity, its relative effect upon the plant decreases; and when the factor is in the region of its maximum, the effect of a change upon the plant is nil.

It was also noted that some environmental factors appear limiting to a plant when they are in extreme superabundance as well as when they are deficient (Lundegårdh, 1931; Daubenmire, 1959; Glock and Agerter, 1962). However, a careful analysis of the situation usually reveals that the so-called "superabundance" of one factor is often the result of another factor which becomes sufficiently minimal to limit the process. For example, an excess of soil moisture appears to limit certain processes because air is driven out of the soil and oxygen diminishes to sufficiently low levels that metabolism is limited in the roots. Thus, it was soil oxygen that was the operationally limiting factor, not high soil moisture. Both low and high temperatures appear to limit the rates of certain chemical reactions. However, the high temperatures may not limit the rates of a reaction directly, but rather alter enzyme systems that in turn become the operationally limiting conditions producing the effect.

Liebig's Law of the Minimum and Lundegårdh's Law of Relative Effects are important in the following discussion of physiological processes. They imply that it is not necessary to consider all factors at all times as limiting. Actual conditions and circumstances can usually be ascribed to a small number of critical factors. The following pages describe the most important growth-controlling processes and the factors most likely to limit them. The treatment will emphasize those relationships important to the growth processes in order that they may be properly handled in the models to be developed later in the book. The remainder of this chapter includes a discussion of the processes affecting the movement of materials, and the next chapter deals with the food-making processes and various types of physiological interactions that occur among the organs and tissues of a tree.

II. Some General Terms and Basic Concepts

The movement of substances through the soil, into the plant, from one cell to the next, and out of the plant is accomplished by two basic processes: *mass movement* and *diffusion*. Mass movement is the transfer of molecules or particles from one place to another through the application of some external force. Such external forces include gravity, hydraulic pressure, or surface phenomena affecting cohesion and adhesion of molecules. Mass movement is a flow of particles which are swept or moved all in one direction. Examples of mass movement are the percolation of water into the soil after a rain, the rise of moisture in the soil due to capillary action, the movement of water and minerals up through the xylem from the roots to the leaves, the movement of foods and minerals through the sieve-tube elements of the phloem, and the discharging of water droplets (*guttation*) from turgid leaves.

The movement of materials into and out of living cells in the plant occurs mainly by the process of diffusion. Unlike mass movement, diffusion results from the energy of molecules and ions, which produces ceaseless, random motion of the particles. Diffusion is the kinetic dispersion of molecules or ions of a particular kind from a region where there is a large number of them to a region where there are fewer numbers of them. The relative number of molecules for a substance is expressed as *concentration*, while the combined effects of concentration and molecular activity are referred to as *free energy*. In conceptual terms, free energy is a measure of the relative movement (kinetic energy) of molecules of a given substance. This movement will increase with an increase in the number of molecules (concentration). Diffusion may therefore be defined as a net movement of molecules (or ions) of a given substance from a region of high free energy (concentration) of that substance to a region of low free energy (concentration) of that substance (Meyer et al., 1973). The difference in molecular activity between two areas is called the *free energy gradient*. Each substance diffuses relatively independently of the diffusion of another substance until a state of equilibrium is reached. At equilibrium all molecules in an open system are randomly dispersed and the free energy of each substance is nearly uniform. Molecular motion continues, but no further change in concentration or net movement of molecules occurs.

The kinetic energy of molecular motion is essentially the heat within the substance, and the motion is directly proportional to the absolute temperature. Theoretically, then, at absolute zero ($-273°C$) there would be no molecular motion and no diffusion. Diffusion rates are highly dependent upon temperatures and, therefore, the radiant energy from the sun and the resulting heat energy of the environment may be regarded as the source of energy causing molecular movement and diffusion to occur throughout the plant. As temperatures increase in an open system, molecular movement is greater, and diffusion occurs more rapidly. Ordinarily materials move into and out of a plant more rapidly during the warm days of summer than during the cold days of winter. In general the rate of diffusion can be shown to increase by a factor of 1·2 to 1·3 for each 10°C rise in temperature (Meyer et al., 1973).

The size of the moving particles also affects the rate at which they can diffuse. Thus small, light molecules move more rapidly than large, dense ones. The steepness of the energy gradient, that is, the difference in free energy over a given distance, affects the rate of diffusion. The greater the free energy difference, the steeper the gradient and the faster the diffusion rate.

The diffusion of molecules and ions can also be affected if there is resistance to their movement by substances surrounding them. Thin layers or membranes in the cytoplasm, which lie inside the cell walls (Fig. 3.1), may vary in their permeability to different molecules diffusing throughout them. Such layers, which are referred to as *differentially*

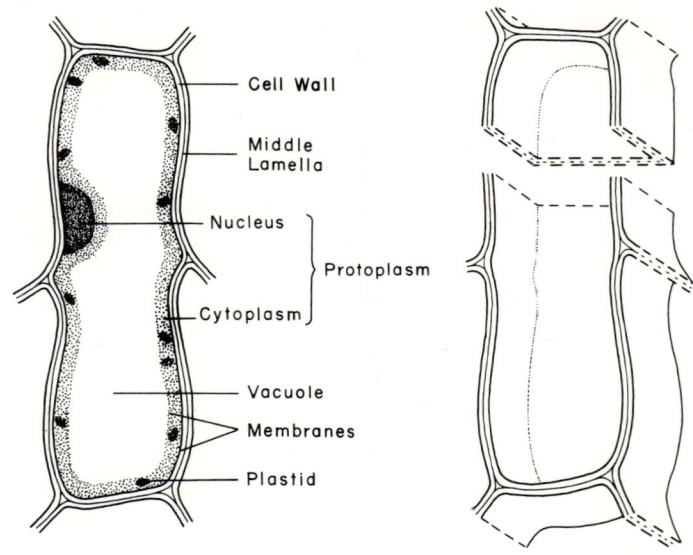

FIG. 3.1. Parts of a plant cell as seen through a microscope. The drawing on the right shows the cell wall in three-dimensional perspective. (Drawing by M. Huggins)

permeable membranes, may change in permeability and in this manner may regulate the movement of substances within the plant.

The cell wall is not alive and is generally permeable to most molecules. Exceptions occur when the cell wall is impregnated with waxy or fatty substances, such as *suberin* or *cutin*, which are usually found in the epidermis or bark. Since water is insoluble in these substances, it can move only with considerable difficulty through cell walls impregnated with them.

Diffusion is a very important process to the growth–environmental relationships, because it is involved in the transport of all materials into and out of living cells and it is responsible for many of the size changes that occur in the growing tissues of plants. Foods, minerals, water, and gases utilized in growth processes generally move through the membranes and into the growing cell by diffusion (Fig. 3.1); all substances made within cells must diffuse out of the protoplasm of the cell before they can be translocated (by either diffusion or mass flow) to other locations in the plant; and gases move between the leaf and atmosphere by diffusion.

The diffusion of water is so important to the plant that special terms are used for it. The loss of water by evaporation from the leaves and stems is referred to as *transpiration*. For the purposes of this text the diffusion of water through the differentially permeable membranes in the cytoplasm of cells is called *osmosis*.

There is one type of movement which is not strictly diffusion nor is it a mass flow phenomenon. It involves minerals or ions which are moved across cell membranes against a free energy gradient. Such movement may occur in well aerated roots and is referred to as *active transport*. It requires an expenditure of energy which is released by the metabolic activity of the plant.

III. Cell Water Status

The growth of trees is undoubtedly controlled more by the movement of water than by the movement of any other single substance. Growth cannot occur unless water is plentiful enough to maintain erect developing structures and expanded cell walls. The normal functioning of the physiological processes which control growth must occur in a water solution, raw materials move into and out of the growing cells in a water solution, and all variations in the sizes of growing cells are due to the movement of water into and out of the tree. This movement is regulated by the osmotic properties of the cells in the roots and leaves, and by pressures and resistances to water movement exerted by the various tissues within the tree.

Living cells consist of a viscous *cytoplasm* and *nucleus*, together called the *protoplasm*, which is surrounded by a differentially permeable membrane and a more or less rigid cell wall (Fig. 3.1). The cell wall is flexible but places some limit on the maximum cell volume. Thin-walled cells generally are more flexible than thick-walled ones. The cytoplasm is commonly distributed as a thin layer along the inner surface of the cell wall within the membrane, and it encompasses a nucleus containing the genetic material, and sometimes includes specialized protoplasm called *plastids*, and other bodies. A vacuole commonly occupies the central portion of the cell and is filled with a solution of water and dissolved substances, sometimes referred to as the *cell sap*. A second differentially permeable membrane separates the cytoplasm from the vacuole.

All water movement in plants may be viewed as occurring along gradients of decreasing free energy of water (Fig. 3.2). A measure of this free energy, called the *water potential*, is expressed in units of pressure, called *bars*, as the difference in free energy of pure water at atmospheric pressure and that in the system under consideration (Kozlowski, 1968; Kramer, 1969). In Fig. 3.2 water potential of the soil is shown to be near zero, the water potential of the plant ranges from -1 or -2 bars to -30 bars, and the water potential becomes markedly lower along the external cell surface of the leaf. The curves in Fig. 3.2 portray the differences that occur in profiles of the energy gradient under (1) normal moist conditions, (2) temporary wilting, and (3) permanent wilting after a prolonged drought.

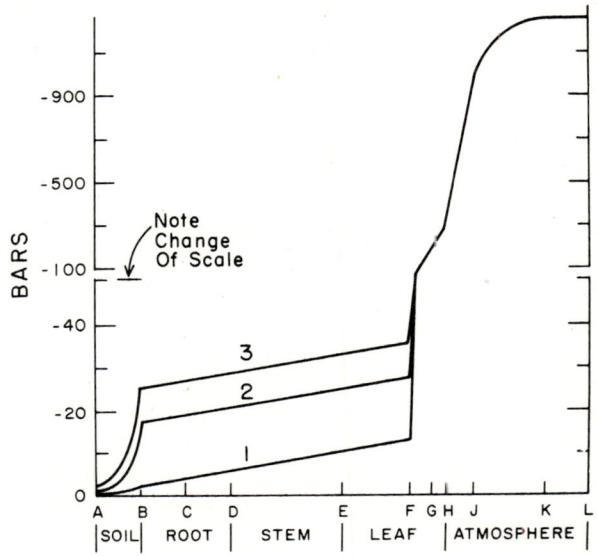

FIG. 3.2. Energy profiles in the stream of water passing through the soil/plant/atmosphere continuum under conditions of normal transpiration (curve 1), temporary wilting (curve 2), and permanent wilting (curve 3). The ordinate represents water potential as shown in units of pressure called *bars*. The abscissa represents locations along the transpiration path (A) soil, (B) root surface, (C) root cortex, (D) root endodermis, (DE) xylem in root, stem, and leaves, (F) mesophyll cells, (FG) intercellular space, (GH) stomate, (HJK) adhering boundary layer of air, and (KL) free atmosphere. (Redrawn from Philip, 1957)

The water potential (designated as Ψ) of an unconfined solution is simply a function of the amounts of dissolved solutes and hydrostatic forces. The component of water potential due only to the dissolved materials is referred to as *osmotic potential* (Ψ_s). However, if a solution is confined, as in the protoplasm and vacuole of a cell, the water potential is higher due to the positive pressure that develops inside the cell. This component of water potential is the *pressure potential* (Ψ_p) or the *turgor* pressure, and generally it has a positive value (Kramer, 1969). The water potential of a nonturgid cell (where the pressure potential is zero) will be the same as the water potential of an unconfined solution, except in the case where *colloids* may hold water or surface phenomena may exert an additional force. These forces due to surface phenomena are expressed as the *matrix potential* (Ψ_m) and like the osmostic potential have negative values.

Using the above symbols, the water status of a plant cell can be expressed as the sum of the various potentials by the following equation:

$$\Psi_{cell} = \Psi_s + \Psi_p + \Psi_m \tag{3.1}$$

When a cell is placed in a solution of water, it will not change shape if the water potential inside the cell is the same as the water potential outside (Fig. 3.3). However, diffusion (in this case, osmosis) will occur if there are differences in water potential between the inside and outside of

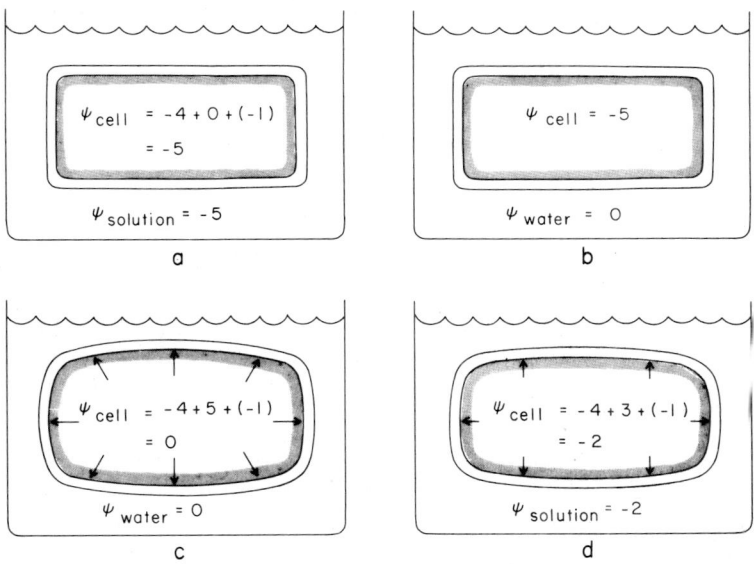

FIG. 3.3. (a) A hypothetical cell in which osmotic potential equals -4 bars, pressure potential equals 0, matrix potential equals -1 bar, and the total water potential for the cell is -5 (see equation 3.1). If the cell is placed in a solution with the same water potential as the cell, as in (a), there is no water potential gradient and the water is in a state of equilibrium. However, if the cell is placed in pure water (b), a difference amounting to 5 bars exists between the water potential inside and outside the cell. Water diffuses into the cell (c) until the pressure potential (arrows) exerts an outward force equal to the difference in water potentials (assuming changes in cell volume are insignificant) and a gradient no longer exists. The cell swells and is said to be turgid. If the water potential of the bathing solution is then changed from 0 to -2 bars (d), water will diffuse out of the cell until the pressure potential (arrows) is reduced from 5 bars to 3 bars, at which time the water in the cell reaches a new equilibrium and the cell is less swollen. Under actual conditions, water diffusing into the cell dilutes the solute, increasing the osmotic potential and decreasing the pressure potential to some value slightly less than indicated in the example.

the cell. For example, water diffuses into a cell if the water potential of the cell is lower than that of its surroundings. As water enters a cell, the water potential increases and the pressure potential exerted by the cell wall increases until the water potential inside the cell equals the water potential outside the cell. The water potential gradient no longer exists and osmosis ceases.

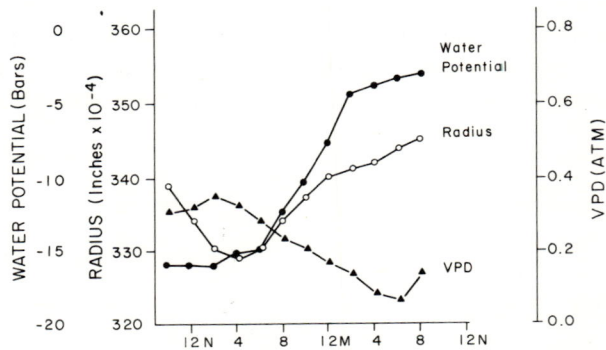

FIG. 3.4. Water potential of needles, measurement of the stem radius, and vapor pressure deficit in the air (VPD) surrounding a semiarid *Pinus ponderosa* at various times during May 22–23, 1965. The water potential was measured for needles periodically removed from a branch, the radius was measured by a dendrograph, and the vapor pressure deficit was measured as the difference between the actual vapor pressure of the air and the saturation vapor pressure at the same temperature. (Redrawn from Cunningham and Fritts, 1970)

Changes in water potentials and cell size reflect the movement of water through the plant. The changes in stem size and needle water potential that can be observed throughout the day in arid-site *Pinus ponderosa* are shown in Fig. 3.4 (Cunningham and Fritts, 1970). Early in the afternoon (before 4:00 p.m.), the water potential in the needles reaches its lowest value at a time when the stem is shrinking as water is drawn from it faster than it is replaced by the roots, and the *vapor pressure deficit* (the difference between the actual vapor pressure of the air and the saturated vapor pressure at the same temperature) is near its maximum for the day. After 4:00 p.m. the vapor pressure deficits of the atmosphere begin to decline in absolute value because air temperature and the resulting saturation pressure decrease. The water potential gradient decreases and the amount of water lost from the tree becomes smaller than the amount absorbed from the soil, more water moves into the stem than is drawn from it, and the sizes of cells in the stem and water potentials of the needles begin to increase. At 8:00 a.m., when the needles are turgid and exhibit the highest water potential, the tree attains its greatest water content. However, the vapor pressure deficit between the leaf and the air begins to decline by this hour and this causes the water loss by transpiration to increase. When water loss exceeds water uptake by the roots, the stem begins to shrink again and the water potential in the needles begins to decline.

Such diurnal variations in the water status of a tree can have substantial effects on a variety of processes involved in cell division, cell enlargement, the making and utilization of food, and the production of cell parts. Therefore, environmental conditions which alter the water status can also affect growth and are undoubtedly responsible for many of the short term variations in growth described in Chapter 2. Such effects may appear small in magnitude, but if they occur day after day their cumulative effects can substantially affect ring width.

IV. Transpiration

Transpiration includes the loss of water by evaporation from all parts of the exposed plant including stems, flowers, and fruits, as well as the leaves. That portion of transpiration which is attributed to the leaves occurs in two stages. Water evaporates from the moist cell walls into the substomatal cavities and other intercellular spaces within the leaf (Fig. 3.5), and the water vapor then diffuses from the intercellular spaces through the leaf epidermis to the outside air. Passage through the epidermis occurs through microscopic holes in the epidermis called *stomates* or through the epidermal cells. As mentioned earlier in the chapter, the outer walls of the epidermal cells are impregnated with a waxy substance called *cutin* (Fig. 3.5). This impedes diffusion so that the amount of water lost through the epidermal cells is usually considerably less than the water lost through the open stomates. Small amounts of

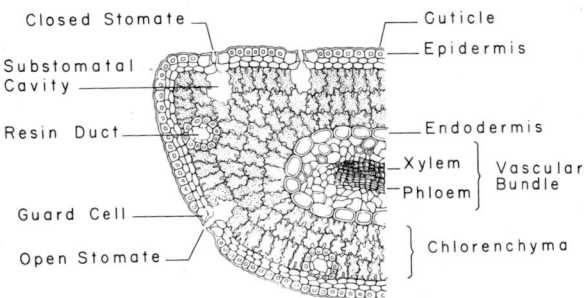

FIG. 3.5. A diagram of a cross section of half a pine needle showing various tissues and examples of open and closed stomates. (Drawing by M. Huggins)

water are also lost from the stem and twigs through *lenticels*, which are ruptures in the bark.

The stomates are generally numerous, but varied in structure, shape, and number from one species to another. Two cells called *guard cells* surround each stomate and govern the size of the opening. Stomates open whenever water diffuses into the guard cells and their volume increases. They close whenever water diffuses out of the guard cells and the volume decreases (Meyer *et al.*, 1973). The opening and closing occur because walls of the guard cells are thicker near the pore opening than on the side away from the pore and, as a result, the cell stretches more easily away from the pore, causing the two cells to arch away from each other as the volume of the guard cells increases. In general the stomates open in the light of morning and close with darkness in the evening. The opening and closing has a marked effect on transpiration rates.

In addition to the size and number of stomates on a leaf, the amount of energy absorbed by the leaf and the amount radiated from it can substantially affect leaf temperatures and influence the transpiration rates. During the night and in shade, the leaf may radiate more energy than it absorbs, and its temperature may drop well below the temperature of the air. Under conditions of high solar radiation during the day, the leaf may absorb more energy than it dissipates, and its temperature can rise above the temperature of the air (Fig. 3.6). Wind may cool the leaf and lower the leaf temperature during the day, but it can also transfer energy to a cooled leaf at night and raise its temperature closer to that of the air. Transpiration itself consumes energy because of the cooling power of evaporation.

The driving force of transpiration, according to Gates (1968b), is essentially the difference between the water potential or the vapor within the leaf and the water potential of the free air outside the leaf surface. Both moisture and temperature govern the free energy gradients, but several resistances may be encountered as the water diffuses from the inside of the leaf out to the free atmosphere. The cuticle layer on the walls of the epidermis offers resistance, the changing sizes and shapes of stomates can cause large variations in resistance, and any moist layer of air adhering to the external surface of the leaf can add further to the resistance.

Gates (1968b) has constructed a physical model of the various resistances and pathways affecting transpiration from the leaf and uses the model to calculate transpiration as a function of varying wind velocities, temperature, and humidities of the air, assuming that the stomatal pores remain unchanged. His calculations (Fig. 3.7) illustrate that transpiration can markedly increase or decrease with slight increases

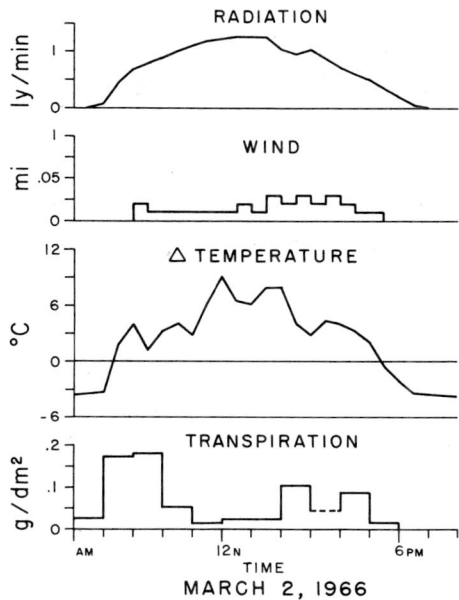

FIG. 3.6. Incident radiation on a horizontal surface, wind velocity, difference between temperature of leaf and surrounding air, and transpiration from a seedling of *Pinus ponderosa*. Early in the morning (8:00 to 10:00 a.m.) transpiration is rapid because stomates are fully open and water potential of the needle is high. Cooling by transpiration prevents needle temperatures from rising more than 3° or 4° above air temperature. By late morning water deficits and high needle temperatures cause reduction in the size of the stomates, increased stomatal resistance, and decreased water potential gradient between the needle and air. Less of the absorbed radiation is dissipated by transpirational cooling, and the needle temperatures rise considerably above those of the air. At times in the afternoon when wind velocities are greater than in the morning, they dissipate a portion of the absorbed energy and the temperatures of needles exhibit less departure from temperatures of the air. Later in the afternoon conditions favor more rapid transpiration, differences between needle and air temperatures decline until stomatal closure occurs with declining light levels after 5:00 p.m. At night the needles radiate energy and their temperatures fall below those of the surrounding air. (Redrawn from Brown, 1968)

in wind speed from conditions of still air. For example, a gentle breeze of less than 50 cm per second (Fig. 3.7) can increase transpiration substantially by removing the boundary layer or by increasing the temperature of the leaves. On the other hand, a 50 cm per second breeze can decrease transpiration if it dissipates the heat of radiation absorbed by the leaf and reduces the temperature of the leaf. Wind will increase or decrease transpiration depending upon which conditions are most limiting. The figure indicates that some changes in transpiration rates

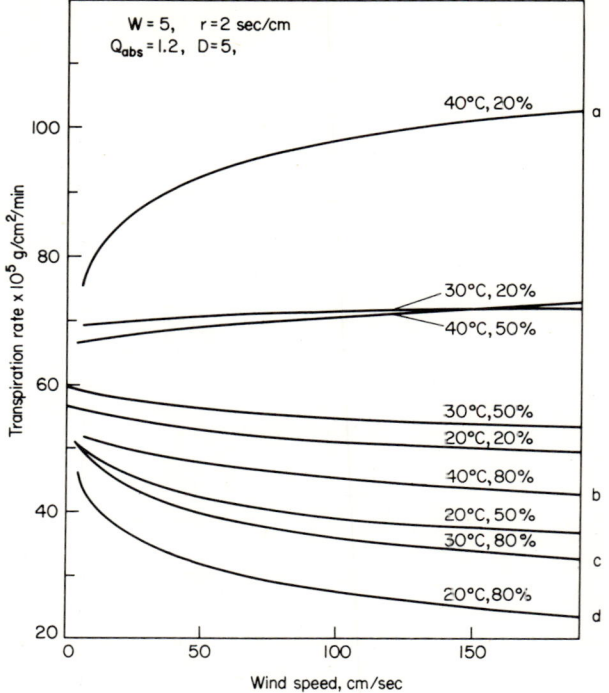

FIG. 3.7. Transpiration rate expressed as a function of wind speed as calculated from a model which takes into account differences in air temperatures and relative humidity percentages. Computations are for the same leaf area, leaf diameter, internal resistance, and radiant energy absorbed by the leaf. Under certain temperatures and humidities (curve a) transpiration rates increase with increasing wind speed, but under other conditions (curves b, c, d) transpiration rates decline. The greatest changes in transpiration rates occur between conditions of still air and wind speeds 50 cm per sec or less. See text for further discussion. (Modified from Gates, 1968b)

occur when wind velocities rise to 100 and 150 cm per second, but they are not as large as the difference between still air and velocities of 50 cm per second.

In the case of drought-sensitive trees which are growing exposed on windy and sunny sites, the leaves are highly ventilated. This simplifies the relationships, since leaf temperatures during the day are likely to remain close to the temperature of surrounding air and the stomates remain open during clear days unless large water deficits occur. In this case there is no boundary layer of high humidity near the leaf surface as occurs in still air or within a closed forest canopy, and the free energy gradient of water vapor near the leaf surfaces is steep. When humidity in the air is low for such ventilated trees, air temperature can be the most limiting factor to water loss and it can become highly correlated with changes in water status leading ultimately to changes in growth. However, the transpiration rates from such exposed trees can be limited by high air humidity, which in turn is directly correlated with moisture conditions of the area and antecedent precipitation falling on the site.

As mentioned earlier, stomatal sizes may change with variations in light intensity which can produce marked changes in leaf resistance and transpiration rate. For example, the opening of stomates at dawn and their closure at night has a very pronounced effect upon transpiration rate. Low light levels during cloudy weather may induce stomatal closure and reduce transpiration. The transpiration from shaded trees within a forest throughout the day is often affected more by variations in light which affect stomatal opening and leaf resistance than by variations in temperature and wind which affect the water potential gradient. Water status for such trees would be less well correlated with air temperature and more correlated with variation in light than for trees on more exposed and arid sites.

One method of measuring the size of stomatal pores in conifers is to pluck needles from trees, enclose them in a pressure chamber filled with an alcohol-water solution, and gradually increase the pressure while observing the needles under a microscope. At a given pressure the ethanol is forced through the stomatal opening, and the air in the leaf is displaced by the solution, so that the color of the needles becomes darker. The size of the stomatal opening is inversely proportional to the pressure required to induce infiltration (Fry and Walker, 1967). The infiltration pressure for needles of *Pinus ponderosa* can be expressed as an index where a value of 20 represents low infiltration pressures when the stomates are fully open, and an index of 60 or higher represents high pressures when the stomates are fully closed.

The infiltration indices throughout a 24-hour day are shown in Fig. 3.8. The tree used in this example was enclosed in a transparent, air-conditioned tent, so that the transpiration and net carbon dioxide exchange could be measured as well (Drew *et al.*, 1972). As light levels increased during the early morning, photosynthesis commenced and the stomates opened, causing the infiltration index values to decline. The stomatal opening reduced the internal resistance of the needles and transpiration increased. From 8:00 a.m. until noon, transpiration varied as a function of the difference between the water potential of the air and the leaf, because the fully opened stomates offered little resistance. Clouds at noon reduced light intensities, and by 1:00 p.m. the stomates began to close, causing leaf resistance to increase and transpiration to decline.

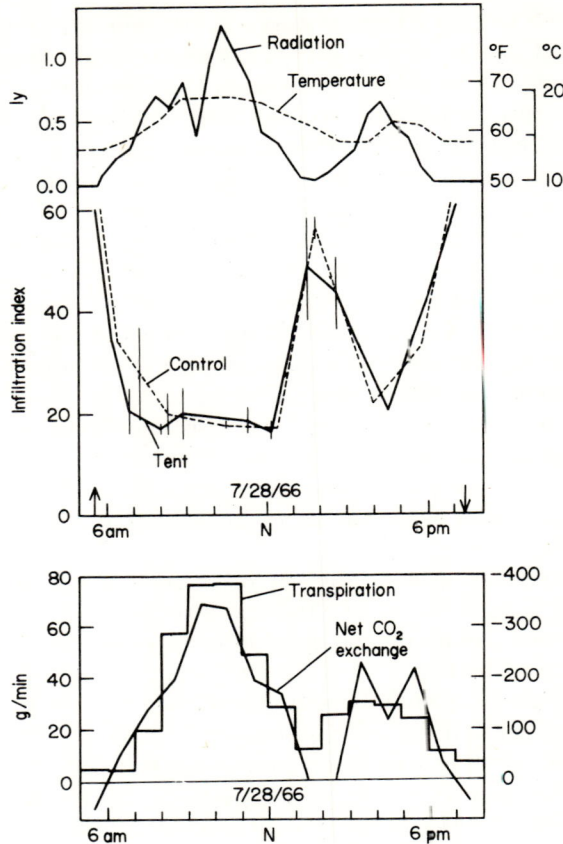

FIG. 3.8. Radiation and temperature of air in the crown of an arid-site *Pinus ponderosa*, the infiltration indices showing stomatal opening (low index) and closing (index of 60), and measured transpiration and exchange of carbon dioxide. The tree was enclosed in a polyethylene air-conditioned tent, and a nearby unenclosed tree was used as a control. Arrows indicate sunrise and sunset. (Drew *et al.*, 1972)

Later in the afternoon the clouds dissipated, light intensities increased, stomates opened, and transpiration increased. By 5:00 p.m., light levels had declined sufficiently to cause stomates to close again and for transpiration to decline.

Figure 3.6 shows a different situation where declining rates of transpiration from a drought-subjected pine were measured while in full sunlight from 10:00 a.m. to 12:00 noon. It was inferred by Brown (1968) and substantiated by later measurements (Drew *et al.*, 1972) that the stomates closed during the day in response to internal water deficits and high leaf temperatures, which in turn caused transpiration rates to decline. Other examples of diminished rates in transpiration during the day, apparently due to water deficits and stomatal closure, are illustrated later in the book (see Figs 3.14, 3.15, and 4.7).

V. Soil Moisture

Due to certain forces among water molecules and between water and soil particles, the water potential of the soil varies depending upon the soil moisture content (Figs 3.9 and 3.10). When all pore spaces within the soil are filled with water, the soil is said to be saturated and the water potential is near zero (high) (Fig. 3.9). A certain amount of this water, the *gravitational water*, is not available to plants since it drains from the soil in one or two days. Drainage ceases at potentials of -0.33 bars, and the soil water retained at this potential is said to be at *field capacity*. A large portion

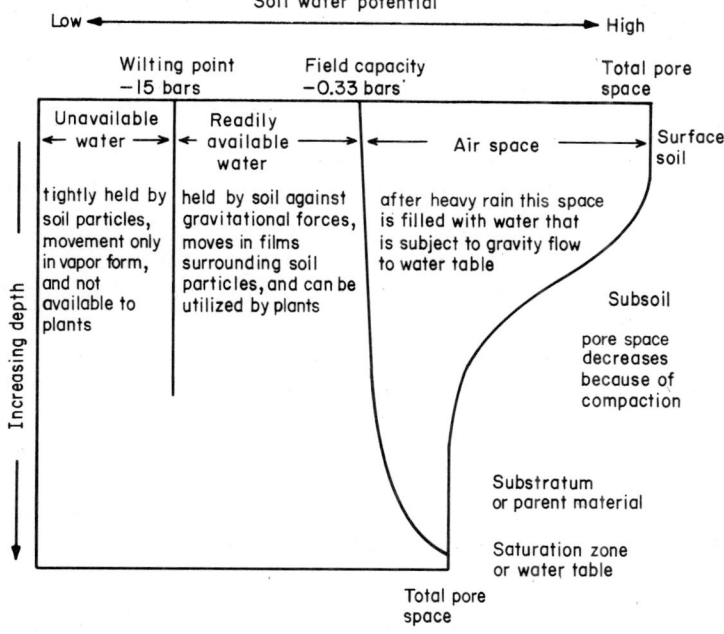

FIG. 3.9. Diagram of the types of water in a hypothetical soil of uniform texture. Most of the water that is held at extremely low water potentials (less than the wilting point) will not diffuse into the root and is unavailable to the plant. Similarly, water held at potentials greater than field capacity is subject to gravitational pull, drains from the soil, and so is also largely unavailable to the plant. Pore space and aeration capacity decrease with increasing depth because of soil compaction.

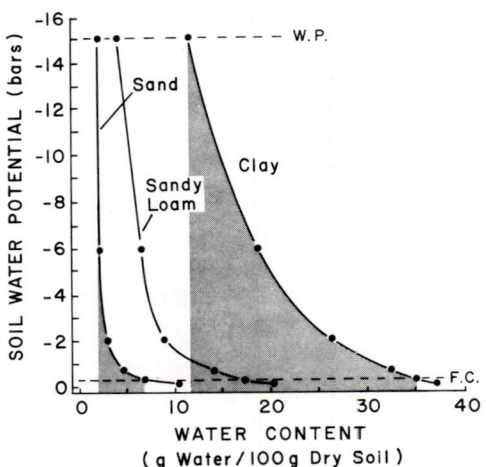

FIG. 3.10. Varying amounts of water are held in different soils by water potentials ranging from 0 bars (water saturation) to −15 bars (wilting point). The water available to plants is the difference between field capacity (F.C.) and wilting point (W.P.), which for sand and clay is represented by the shaded areas under each curve. Since more water is available in clay than in sand, water potentials decline more slowly per gram of water lost, and it often takes longer for such soils to reach wilting point. Though less water is available in sand than in clay, a larger proportion of this water is removed before water potentials are low enough to greatly reduce diffusion gradients in the plant. (Redrawn from data presented by Kramer, 1969)

of the water held in the soil at field capacity is readily available to plants and it can move in the soil as thin films around the soil particles or by molecular diffusion.

Soil moisture is lost by evaporation into the air and diffusion into the roots. At first relatively large amounts of water can be lost from the soils with little change in water potential (Fig. 3.10), but as the soil dries, smaller amounts of moisture are lost per unit change in water potential (Fig. 3.10). Plants begin to wilt during the day, and when water potentials approximate −15 bars many plants become permanently wilted and the soil water is said to be at the *permanent wilting point*. Though small amounts of water can diffuse into roots after the wilting point is reached, this water is generally considered to be unavailable (Figs 3.9, 3.10). For purposes of this discussion we can assume that saturation, field capacity, and wilting point are soil constants, though they actually vary over a range of values depending upon the species, soil temperatures, and other conditions of the site.

A moderate rainfall of 1 inch (25 mm) on a dry soil may wet only the upper 10 inches (254 mm) of soil raising its water content to field capacity, while the lower layers may remain dry. Unless more moisture is added, water will not move to deeper levels by gravity flow, although water can move slowly by capillary action and by diffusion. If more rain falls, the moisture at the soil surface temporarily rises above field capacity, moisture drains to lower levels wetting the soil until the entire moist layer reaches field capacity. At this point the downward movement of water due to gravity flow ceases.

Soils may differ greatly in the quantity of water they hold at field capacity (-0.33 bars) and at the wilting point (-15 bars). The maximum amount of soil water that is considered to be available to a plant is the difference between the water held at field capacity and that held at wilting point (see shaded areas in Fig. 3.10). Since soils of different *texture*, such as sands, loams, and clays can differ greatly in the water content at both field capacity and wilting point, they also can vary substantially in the amount of water available to the plant. For example, clay may hold five times as much available water as sand, whereas loam holds intermediate amounts (Fig. 3.10). Also, the grams of water remaining in clay at wilting point can be greater than that held by sand at field capacity. The sharper bend of the curve for sand than for clay in Fig. 3.10 indicates that there is a greater change in water potential per amount of water lost from sand than from clay. Thus, in sandy or gravelly soils, practically all the available water is loosely held at -2 bars or more, but in heavy clays, less than 40% of the available water is held that loosely. Thus, water potential gradients may be steeper for sands than for clay, even though the sand has less water content.

According to Kramer (1969), water in capillary films, pores, and cracks in the soil moves primarily as a liquid. The rate of water movement is materially affected by soil texture and conditions of moisture. Capillary movement in a moist soil is most rapid in the coarser sands and slowest in the finer clays, because there are more large pore spaces in the former. Under drier conditions, however, when the films of water are thinner, the effects of texture are reversed. The movement is most rapid in the latter, because there is greater surface area and more contact among the films of water.

A small amount of water may rise by capillary movement from the water table for short distances. The height of the capillary rise is greatest in clays and least in sands. However, in both kinds of soil, capillary movement of water from moist to dry regions is a very slow process, especially when the soil moisture is less than field capacity. During periods of rapid transpiration, the available water on soil particles in contact with roots moves into the roots faster than it is replaced by capillary action or molecular movement. This produces a cylinder of soil surrounding each absorbing root that is depleted of water, although a few millimeters away the moisture content of the soil may be near field capacity (Kramer, 1967). Over a period of time, capillary movement may replace some of this water. When soil moisture is low, long distance transport in the soil, including capillary rise from the water table, cannot provide enough moisture to meet the demands of a vigorously growing tree. However, capillary movement may provide sufficient moisture to prevent death of the tree in times of drought.

VI. The Soil System and Factors Affecting its Development

The previous section dealt with the soil as if it were a simple porous medium important only because it affects the water supplied to the roots of a plant. Actually, the soil is a highly complex physical–biological system that develops in the upper weathered layer of the earth's crust and which, according to Jenny (1941, 1958, and 1961) is a function of five soil-forming factors: (1) climate, (2) organisms growing in or on the soil, (3) topography of the site, (4) parent material from which the soil is derived, and (5) time. This section of Chapter 3 includes a brief description of how the soil profile—the assemblage of generic horizons—can vary with small changes in topography and drainage, and it illustrates how the distribution of the horizons and their characteristics can influence plant processes, which can in turn affect root distribution, water absorption, water balance, and ring width.

As soil development progresses, the original *parent material* becomes differentiated into distinct layers called *soil horizons*. Soil horizons are distinguished by their morphological, physical, chemical, and

FIG. 3.11. Soil texture percentages, the soil moisture constants, percent air space and numbers of roots vary with increasing depth and from one soil type to the next. Three soils are shown which form in the gently rolling glacial till in central Ohio. The Cardington soil is a moderately well-drained silt loam which supports a beech–maple forest, the Bennington soil is an imperfectly drained silt loam which supports an almost pure stand of beech, and the Marengo soil is a poorly drained silty clay loam which supports a swamp forest. A, B, and C designate the relative positions of the three master soil horizons. The sand, silt, and clay are expressed in relative amounts, the percent moisture is shown for wilting point (W.P.), field capacity (F.C.), and saturation (Sat). The difference between wilting point and field capacity represents the potential volume of moisture available to plants, while the difference between field capacity and saturation represents the water that can fill much of the air space immediately after a rain but drains from permeable soils in one or two days. The roots were counted along the face of a soil pit and their numbers vary with the amount of air space, textural features, and permeability of the respective soils. (Redrawn from Fritts and Holowaychuck, 1959)

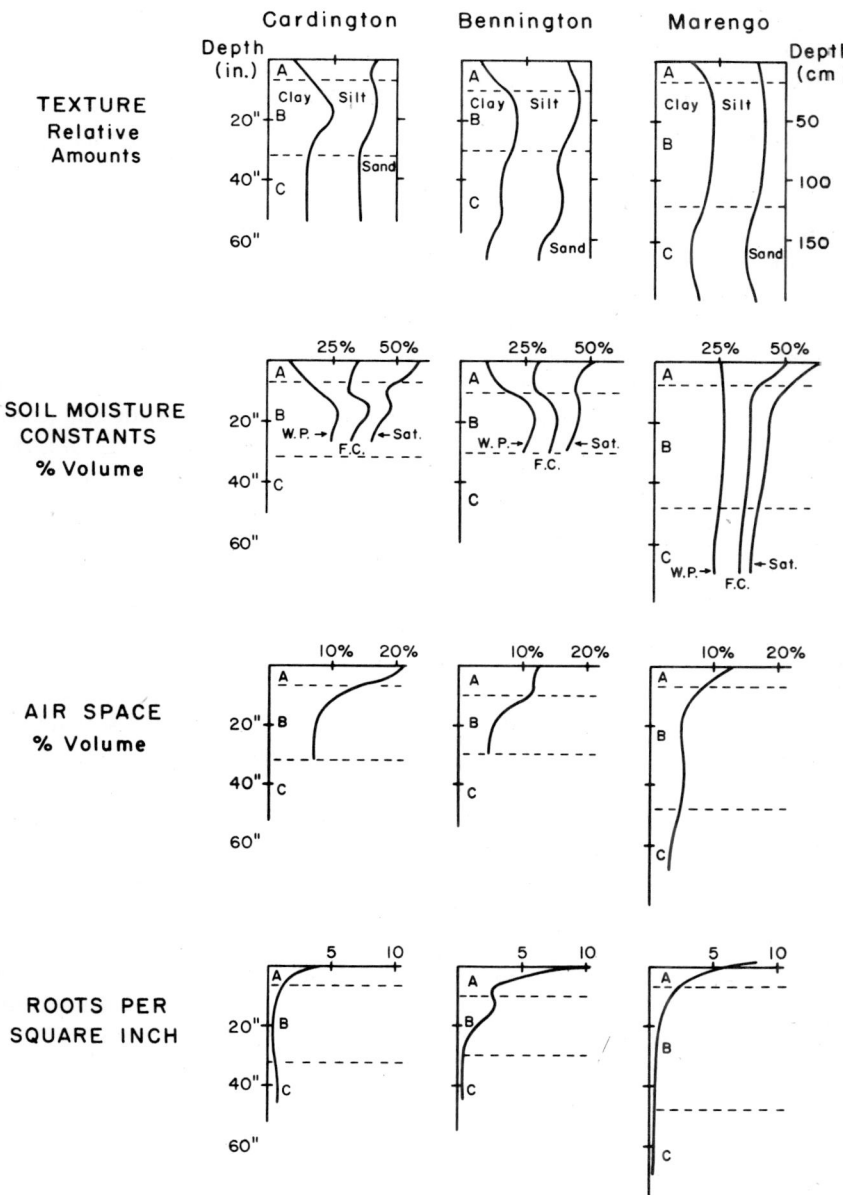

mineralogical properties. The vertical assemblage of soil horizons from the soil surface down to the lowermost layer of undifferentiated unweathered material is referred to as the *soil profile* (Fig. 3.11). It may be divided into four master horizons, one organic and three mineral. The organic horizon is often on the mineral soil surface and includes leaf litter and decomposed organic matter. The uppermost mineral horizon, the *A horizon* in Fig. 3.11, represents the portions of the soil from which certain materials have been removed and other materials added by the soil-forming processes; this is called the horizon of *eluviation*. The second mineral horizon, B, is the level into which many of the materials have moved from the A horizon above. The *B horizon* is called the horizon of *illuviation*. The third and lowermost layer is the *C horizon,* which includes relatively undifferentiated parent materials from which the two upper horizons were derived (Fig. 3.11). Horizons A and B constitute the *solum*. The A, B, and C horizons can be further subdivided into *A1, A2, A3, B1, B2, B3, C1, C2*, etc. Details on the morphogenesis, designations, and definitions of soil horizons are available from the publications of the Soil Survey Staff (1951, 1960, 1962, and 1967).

The relative importance of the soil-forming processes and chemical reactions differ among these three horizons and vary as a function of the five soil-forming factors. As a result, soil moisture constants, air space, different soil chemicals, root distribution, and amounts of sand, silt, and clay vary among the major horizons and from one soil to the next. For example, the Cardington soil in Fig. 3.11 is a well-drained soil which supports a beech–maple forest of *Fagus grandifolia* and *Acer saccharum*. The sites of the Bennington and Marengo soils are more gently-sloping leading to imperfect and poor drainage. The Bennington soil occurs at the margin of swampy depressions where the forest is dominated by *Fagus grandifolia*, while the Marengo soil represents the most poorly drained situations and supports a swamp forest, a mixture of tree species that can tolerate poor drainage.

Textural differences among the three soil profiles are a result of movement of materials such as clay from A to B, and other soil-forming processes. Clay reaches its maximum percent in the Cardington soil at a depth of 20 inches (51 cm), while the maximum accumulation of clay in the Marengo soil is less. Furthermore, the zone of clay accumulation in the B horizon is deeper. As in Fig. 3.10 the soil moisture constants in Fig. 3.11 generally increase with an increase in clay content. The highest porosity is found in the well-drained Cardington soil and the least in the poorly drained Marengo. Large numbers of roots occur in all A horizons, but they extend to the greatest depth in the well-drained Cardington. Most of the roots in the Cardington and Bennington soils are those of

Fagus grandifolia, whereas in the Marengo they are those of swamp forest species which are more tolerant than *Fagus* of high levels of moisture and poor aeration.

In general there is a decline in porosity and numbers of roots in a soil with increasing depth and with decreasing drainage. Kochenderfer (1973) reports that in north-central West Virginia, 77-89% of all observed root endings were found in the upper 0·6 m of soil. The number of roots was greatest in coarse soil with good aeration and decreased as the soil texture became finer. Roots are distributed more superficially in poorly-aerated soils, in heavy clays, and where a dense subsoil excludes soil air and causes the water table to be high (Kramer, 1969). If air is displaced from the larger noncapillary pores by an excess of moisture for an extended period of time, root growth and water absorption may be hindered by a deficiency of oxygen. However, if the soil is porous, and the excess moisture percolates to lower soil levels, air can move into the vacated noncapillary pores behind the moving water. Frequent wetting and drying and temperature fluctuation ventilates the air spaces and facilitates the exchange of oxygen and carbon dioxide between the soil and the atmosphere so that soil oxygen does not become limiting to the growth of roots.

Shallow-rooted trees on poorly drained soils may exhaust soil moisture more readily than trees on well-drained soils which are deeply rooted (Fritts, 1958; Huntington, 1914). This occurs because the total root surface and the volume of water available to a shallow-rooted tree is less than that available to a deeply rooted tree, even though the site may be classified as "wet" (Fritts, 1958; Fritts and Holowaychuck, 1959). Also, evaporation at the soil surface can remove a larger proportion of the water available to a shallow-rooted tree than to a deeply rooted one. Thus, if other factors are equal, trees with shallow root systems are more likely to have variable ring widths than trees with deep root systems.

On steep slopes erosion can remove large volumes of soil, so that the development of horizons cannot keep pace with the removal of soil. Such soils show less pronounced profile differentiation, and often the soils on steep slopes are dominated more by the texture of the original parent material than by characteristics arising from soil-forming factors. Other soil characteristics that can affect available soil moisture, distribution of roots, and soil productivity could be enumerated; but they are less dynamic, vary little from one year to the next, and have little influence on year-to-year ring-width variability.

VII. Absorption of Water

Most of the water absorbed by a tree enters through the roots. The rates for absorption are partly dependent upon the water potential within the plant which is generated by the transpiration of water from the leaves (Fig. 3.12). However, the uptake of moisture is also influenced by (1) the absorbing surface area of the roots, (2) the resistance to movement of water from the soil to the roots, and (3) the water potential gradient which decreases with decreasing available moisture in the soil surrounding the roots.

Figure 3.13 is a drawing of the tissues through which water must pass in moving from the soil into the xylem of the young root. The area of the

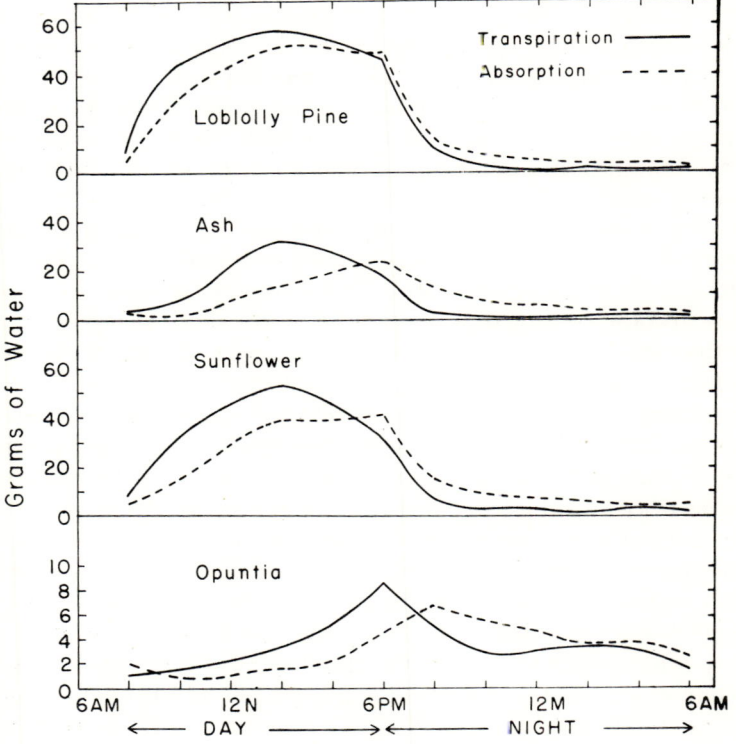

FIG. 3.12. Average rates of absorption and transpiration of four plant taxa throughout the day. Rates of absorption change in proportion to rates of transpiration but lag behind transpiration during the day. The tensions increase and shrinkage occurs in the xylem until the two curves cross, at which time stem swelling will begin. Plants were mounted on a rotating table in a greenhouse. (Redrawn from Kramer, 1937)

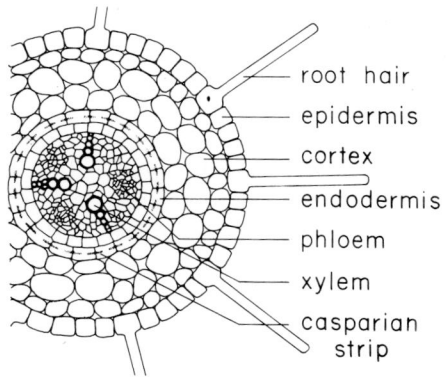

FIG. 3.13. Cross section of a young root showing tissues through which water must pass as it moves from the soil to the xylem in the root. (Drawing by M. Huggins)

absorbing surface of a root is enhanced by projections from the epidermal cells called root hairs or by fungi that may be associated with the root. Water may diffuse into root cells along this root-soil interface and can move through the cortex either along the cellulose cell walls or through the protoplasm and vacuoles. However, a layer called the *endodermis* lies inside the cortex but outside the vascular tissue of xylem and phloem, which has radially and horizontally oriented walls that are impregnated with suberin forming a band referred to as the *casparian strip* (Fig. 3.13). Water cannot diffuse through the suberin of the casparian strip, so it can move through the endodermal tissue only by diffusing into the protoplasm and through the membranes of the endodermal cells. Thus, the endodermis acts as a partial barrier through which water can diffuse only by osmosis. If transpiration is slow and the osmotic potential (Equation 3.1) is low, the root can generate a positive turgor pressure because the endodermis blocks all movement along the cell wall. When transpiration is rapid the water potential gradient steepens, the positive pressure disappears and a negative pressure begins to develop.

As roots mature, the epidermis is ruptured and destroyed by the activity of a cork cambium which develops from the outer part of the cortex. The growth of the cork cells, which are also impregnated with suberin, increases resistance to diffusion of water, but it does not render the mature roots completely impermeable to water (Kramer and Kozlowski, 1960).

Most trees have roots which not only extend well beyond the spread of the branches but which can also penetrate deeply into the soil and into fissures and cracks among rocks. The roots are most dense in soil layers where both oxygen and soil moisture are abundant, because these factors favor high physiological activity and root growth (Kramer, 1969). A tree with a growing crown is able to maintain adequate water absorption only by continual branching, rebranching, and extension of young, unsuberized root tips into moist soil. Thus, rapid absorption of water by a root system depends in part upon the total extent and in part on the continual growth of roots. If root systems were to completely cease growing by extension, the soil mass occupied by the existing roots could become so depleted of available water that the trees would suffer from serious lack of water, particularly if transpiration were high (Kramer and Kozlowski, 1960).

As was mentioned earlier in the chapter, water absorption is dependent upon the water potential gradients both in the soil and in the plant, that is, the more the available soil moisture, the steeper the free energy gradient of the water around the root and the faster the movement of water into the root. Any salts or substances dissolved in the soil solution can reduce the free energy gradient and affect the movement of water into the roots. However, salts are rarely sufficiently concentrated in the soil solution to become a major factor limiting to tree growth. Low soil temperatures may also decrease the permeability of roots to water by increasing the viscosity of water and by reducing capillary movement. Since the condition, rate of growth, and extent of roots also affect water absorption, any factor that retards root elongation such as low oxygen availability, varying amounts of growth regulators, and little available food will correspondingly decrease the area of contact between the young roots and soil water and thereby cause water absorption to decrease (Kramer and Kozlowski, 1960).

VIII. Internal Water Relations of Trees

The water in a plant forms a continuous hydrostatic system (MacDougal, 1924). As water molecules diffuse out of the cells of the leaf, additional water molecules take their place. Usually the absorption of water in the roots lags behind water loss, so that the volume of water in the xylem decreases, the cells shrink slightly, and the water becomes subjected to increasing tensions without breaking the continuity of the water within the stem. If the water potential in the xylem becomes markedly less than the water potential of the cortex of the root, water may diffuse out of the cortex faster than it enters, in which case the turgor of the cortical tissues would decline, and tensions that were first confined to the xylem would extend to the surface of the root. As a result of these tensions, water is pulled by mass flow not only through the xylem but also from the epidermis through the outer portions of the root (Kramer and Kozlowski, 1960; Meyer et al., 1973).

The continuity of the water status in a tree is maintained because there are exceedingly strong cohesive forces among water molecules confined in the cells, and strong adhesive forces between the water and the cell walls. If one of these water columns breaks, it contracts like a released rubber band, and the cell or segment of the water-conducting system no longer functions (Kramer and Kozlowski, 1960; Plumb and Bridgman, 1972). However, the water columns apparently remain continuous in a sufficiently large number of cells so that resistance to movement of water from the roots to the leaves is not in itself variable enough to produce significant variation in water stress.

Frequently the amount of water absorbed during the day lags behind the water lost from the leaves even in plants growing in moist soil (Fig. 3.12; MacDougal, 1924). As the pressure becomes increasingly negative, the adhesive and cohesive forces of water exert a greater pull on the xylem cells, which is transmitted throughout the hydrostatic system causing further reduction in cell volumes through the plant. Such volume changes in the stem have already been described in Chapter 2 (see Figs. 2.12, 2.14, 2.15).

The rate of water absorption increases during the day until the rate of water uptake equals the rate of water loss, at which time no further shrinkage in volume of the plant occurs. Cessation of shrinkage occurs in the early afternoon if the soil is moist and transpiration low. However, if the soil is dry and transpiration high, the actual rate of water absorption may continue to lag behind transpiration until sunset, when stomates close completely and cause transpiration rates to markedly decline. Water then begins to enter the roots faster than it is lost, the volumes of cells in the stem begin to increase, and the stem swells.

Detailed studies of a 110-year-old *Pinus ponderosa* in Arizona (Brown, 1968; Budelsky, 1969; and Drew et al., 1972), show how the hourly changes in stem size are associated with the natural water relations, transpiration rate, variations in solar radiation, temperature, and the exchange of carbon dioxide (Figs. 3.14 and 3.15). An air-conditioned, transparent, polyethylene tent was placed over the tree during six separate 48-hour periods through a single growing season, as shown by the six t's marked on Fig. 2.15 (Fritts, 1966).

Figure 2.15 shows the general growth of the tree and Fig. 2.21 portrays the environmental regime of the site. April 15 was the approximate beginning date of growth in 1966. On that day buds opened, cambial activity began, and soil moisture was relatively high (Fig. 2.21). On April 15 (Fig. 3.14) transpiration rates continued to rise until around noon as the water potential gradients between the leaves and the atmosphere increased. Water absorption apparently lagged behind water loss and the volume of the stem decreased. Clouds that formed from 1:00 p.m. to 2:00 p.m. reduced water potential gradients and caused transpiration to decline. This decrease in water loss was accompanied by a declining rate of cell shrinkage, especially in the upper portions of the stem. Clear weather returned at 2:30 p.m., transpiration began to increase again, and the basal portions of the stem showed accelerated shrinkage. Shadows were cast by neighboring trees after 3:00 p.m., light levels and air and

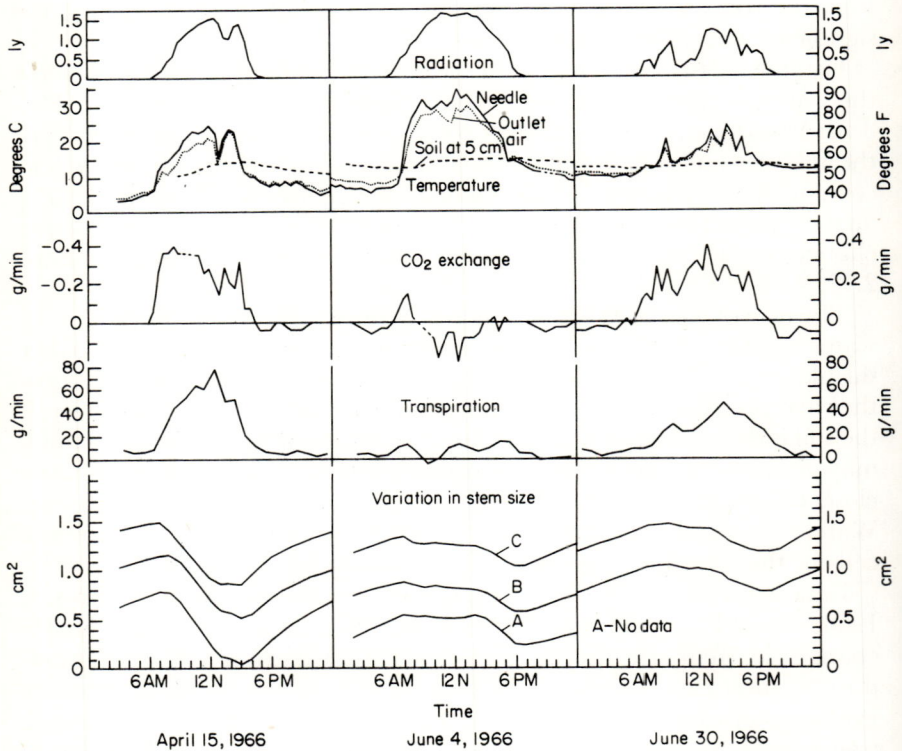

FIG. 3.14. Variation in stem size compared with variation in solar radiation, temperature, carbon dioxide exchange, and transpiration for an entire tree of *Pinus ponderosa* on a semiarid site in southern Arizona. Measurements are for three days of drought prior to the onset of the summer rains. A shower between June 4 and June 30 replenished some soil moisture, and greater cloud cover on June 30 decreased the severity of the drought. The tree was enclosed within an air-conditioned tent. (See periods marked with t in Fig. 2.15.) Dendrographs measured stem size at the base (A), middle (B), and upper (C) portions of the tree bole.

needle temperatures began to decline, transpiration decreased, water absorption exceeded water loss, and swelling in the stem commenced. The most rapid swelling occurred in the evening, when transpiration was low but water potential gradients between the plant and soil were steep. As pressure potentials and the turgor of cells in the tree increased throughout the night, water potential gradients between the roots and soil diminished and the rate of water absorption and the amount of stem swelling diminished.

By June 4, soil moisture was low and water deficits had begun to inhibit enlargement of growing cells in the stem (Fig. 2.15). Air temperatures had risen (Fig. 3.14) and water potential gradients between the leaf and air were undoubtedly steep. However, water content of cells was probably relatively low and leaf resistances were high, so when

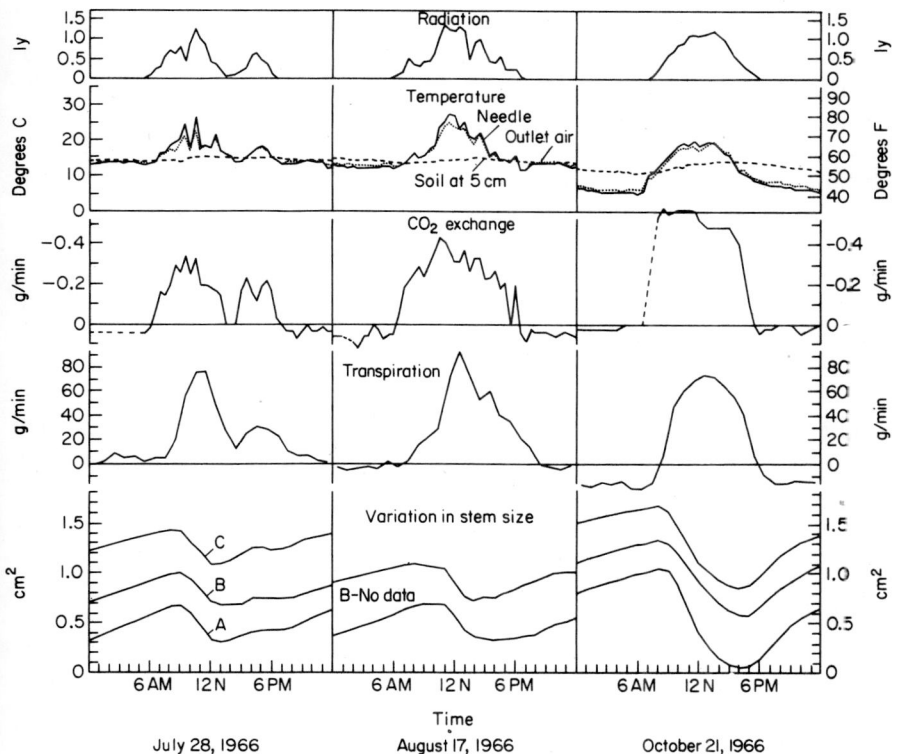

FIG. 3.15. Variations in stem size compared with variation in solar radiation, temperature, carbon dioxide exchange, and transpiration for an entire tree of *Pinus ponderosa* on a semiarid site in southern Arizona. Measurements are for three days after the onset of summer rains. Instrumentation was the same as for Fig. 3.14.

transpiration commenced after 6:00 a.m., its rate was markedly lower than on April 15 (Fig. 3.14). Dendrographs recorded an initial decrease in stem size, needle temperature soon became excessive, and the stomates (not monitored on this day) apparently closed. Even though water potential gradients outside the leaves became very steep, a high internal resistance, probably from stomatal closure, caused transpiration rates to remain low. This low rate of transpiration approximated the rate of water absorption from 8:00 a.m. through 3:00 p.m., so that the water balance and size of cells in the stem (as measured by the dendrographs) remained relatively unchanged. Stomates probably reopened around 3:00 p.m. as needle and air temperatures declined. The rate of transpiration exceeded the rate of water absorption, and the size of the stem decreased until 6:00 p.m., when low light levels caused stomates to close again.

Measurements for three summer days, June 30, July 28, and August 17, are shown in Figs 3.14 and 3.15. The weather was partly cloudy during all three days, a common condition for southeastern Arizona during the summer season. Showers associated with the periods of cloudy weather replenished soil moisture, water was absorbed more rapidly by the roots, and water potentials in the plant increased (Fig. 2.15). Stomates apparently remained open during the day, except possibly for the most cloudy weather; the internal resistance to transpiration within the needles was low, and even though the water potential gradients between the needles and air were less than in June, daytime rates of transpiration were high (Figs 3.14, 3.15). Transpiration greatly exceeded water absorption during the cloudless portions of the day and the stems exhibited a marked decline in size. As clouds built up late in the morning, the water potential of the air increased, and light levels became sufficiently low to cause stomatal closure, which in turn caused transpiration rates to decline. The water lost by the tree became less than the water absorbed by the roots, and the stem began to swell.

The rains of summer ceased prior to October 21 and soil moisture began to decline (Fig. 2.21). Measurements on October 21 (Fig. 3.15) show that transpiration increased rapidly after sunrise and continued at a high rate throughout the day so that the water loss exceeded water absorption and stems shrank rapidly throughout most of the day. The measurements of transpiration during the night exhibit negative values, implying that there was more water in the air entering than leaving the enclosure around the tree, and suggesting that some water vapor in the air was absorbed directly by the tree. However, the *dewpoint* of the air was below the temperature of the tree, so that the air water potential appeared to be lower than the water potential of the leaves. Even though no error could be detected in the measurements or in the design of the enclosure, our present knowledge of water relations offers no suitable explanation for the apparent absorption of moisture by this tree, and the possibility of negative transpiration must remain an open question.

IX. Moisture Stress and Tree Form

Competition for water among the various plant tissues is a major factor in tree development and form. Those tissues that develop the lowest water potentials obtain water at the expense of those having high water potentials (Kramer and Kozlowski, 1960). Young leaves often obtain water at the expense of older leaves; shaded leaves may become dry and die before leaves in full sun because they cannot develop sufficiently low water potentials; and growing tissues usually obtain water at the expense of other tissues. For example, stem tips may continue to elongate and the

FIG. 3.16. Long-lived spike-topped trees of *Pinus longaeva* which contain rings that vary greatly in width and which correlate highly with variations in climate. (Courtesy of C. W. Ferguson)

cambium may remain active at a time when other tissues are shrinking (Fritts *et al.*, 1965c; Budelsky, 1969).

The height to which water must be lifted in a tall tree can also affect the tensions that are developed in various tissues and competition for water within the tree. The top branches of a tall tree may be subjected more frequently to high water stress than are other parts of the same tree. Therefore the apices of many old trees often grow very slowly, and in extreme cases the tops die while lower branches may continue to grow. With time, the conical crown of a young tree becomes spherical as the tree matures, and then it becomes distinctly flattened at the top as it reaches advanced age. Damage to a root system as well as drought can result in dieback of the tops. As erosional processes remove soil from around an aging tree, roots may become exposed and die, causing a greater reduction in water absorption and increased water stress. The dead terminal branches may remain intact for many years, giving rise to the feature referred to as a *spike top*, an important characteristic useful in locating extremely old, dry-site trees (Figs 1.12 and 3.16).

X. Soil Factors Affecting Root Growth, Ring-width Sensitivity, and Longevity

Sensitivity, as defined in Chapter 1, is the amount of relative variability in width from one ring to the next. Often the longest, most homogeneous, and most sensitive series of annual rings are from old trees growing on cliff faces, talus slopes, and rocky sites where there is little soil space for the growth of roots. The soil is relatively coarse textured and can include humus mixed with partially weathered parent material lodged in crevices and among large broken rock fragments.

The roots of a young tree growing on such a site gradually extend throughout the available soil mass. As the root volume increases, the amount of water available to the tree increases, allowing for increases in the growth of stems and leaves. When the roots have fully permeated the available soil mass, however, further increases in the volume of the roots cannot occur. Any growth of new leaves and branches which increases the transpiring surfaces will result in increased competition for water among the existing tissues of the tree. As competition increases extreme conditions of water stress will develop in those branches that are least vigorous, excessively exposed, or shaded, and they will die. The mass of living shoots is pruned and maintained in proportion to the absorbing surface of the roots, especially during years of extreme water stress.

Unless there is an enlargement of the available soil environment, little increase in growth rate is possible. Except for the variability in ring width that occurs from one year to the next in proportion to amount of moisture available to the tree, no major change in stem size and in foliage or root mass occurs because of the limited space in the environment. The volume of the annual wood increment remains relatively constant so that there may be a slow decline in maximum ring width associated with the increasing circumference of the cambial sheath around the stem. Partial dieback of the cambium may occur in very old trees so that eventually only a strip of bark connects the living crown with the living roots (Figs 1.12, 3.16). Such individuals can live for several centuries or, in certain species, for several thousand years, exhibiting variability in ring widths that is highly correlated with variations in precipitation and temperature, without exhibiting any marked changes in the foliage, the roots, and in the mean growth rate (Fig. 1.12).

The crevice which supports an old tree may retain an extremely small proportion of any precipitation that falls. Large amounts of water may be lost to runoff, especially during intense summer storms or during winter storms which are prolonged, and only the small amount that is retained in the crevice can become available to the plant. Rocks often serve as collecting basins, and moisture is channeled into or away from certain sites. In addition, only a small volume of the crevice soil may be exposed to direct evaporation at the soil surface, so most of the available water that is trapped by the soil is subsequently absorbed by deeply penetrating roots. Since the total amount of available soil moisture is small, it becomes depleted over a relatively short period of time. The water potential of the plant decreases relatively rapidly between rains as the wilting point is approached, and water stress usually persists until soil moisture is replenished. The water table is commonly too deep to supply significant amounts of moisture to roots, but local areas of soil moisture seepage can sometimes provide a small but reliable moisture supply enabling the tree to survive. High mortality may be observed in those young trees whose roots are unable to penetrate to a more or less reliable moisture supply.

On crevice sites in semiarid western North America, water stress limits the growth of trees for longer periods of time than on sites with deeper or better developed soils (Fig. 3.17). As a result of these limiting conditions, ring widths from crevice-site trees are not only more variable from one year to the next, but they are on the average narrower and may be more highly correlated with variations in the number of precipitation days throughout the year than with the total precipitation amounts.

The tree growing on a crevice site may be a valuable study tree since the crevice environment can lead to longevity, high year-to-year ring-width variability, and little interdependence of ring widths upon conditions in prior years. This occurs because the crown and the xylem

FIG. 3.17. A crevice site *Pinus edulis* from Nothern Arizona in which the most exposed and least vigorous branches have died as a result of extreme water stress. Note the stunted and contorted shape of the tree and the large amount of exposed rock in the site.

increment cannot increase substantially over a period of several years of unusually high precipitation because the excess precipitation drains from the operational environment.

Trees on more optimum growth sites, on the other hand, are better able to use the high precipitation. Their crown can expand in height and aerial extent in the "luxury" of a favorable environment and there is a marked increase in ring width corresponding to the crown increment. However, a subsequent drought may be highly detrimental because the increase in transpirational area of the crown causes the tree to be more subjected to water stress. A several-year dry spell following years of high moisture can cause a large reduction in the crown area, the rings for several years can become very narrow (and sometimes absent), and in extreme cases death of the tree can result.

Since the crevice site tree will have much less moisture available to it during a wet period, the area of its crown, the xylem increment, and the tree vigor do not respond to the favorable climate as much as they did in the optimum site tree. During the subsequent drought the ring may become narrow without being associated with a large reduction in crown area and a decrease in growth potential that persists for several years as occurs in the optimum site trees.

XI. Uptake of Mineral Salts

Salts dissolved in the soil water or on the surface of the soil particles may also move into the roots, be transported through the tree, and be utilized in processes affecting growth. Molecules and ions of salt, like water, often move by diffusion along gradients of decreasing free energy. Factors affecting the free energy gradient, such as permeability of the root and root growth, therefore, can affect the uptake of salts. Salts may be accumulated by the plant against a free energy gradient, however, through expenditure of metabolic energy. This accumulation is called *active absorption*. The mechanism by which the accumulation occurs is not completely understood, but one popular theory involves a carrier. This carrier becomes bound to the mineral salt, and the combination can move across membranes which are impermeable to the salt alone (Kramer and Kozlowski, 1960).

Individual cells can accumulate mineral salts in concentrations considerably higher than those occurring in the environment. The process is selective so the concentration of certain ions is controlled independently of the concentration of others. Since active absorption is influenced markedly by the physiological activity of the roots, factors such as low soil oxygen, low temperature, and low amounts of food and growth-promoting substances can also limit the absorption of salt.

Mineral salts are absorbed at the soil-root surface by an exchange of ions between the root and the soil. The root produces ions of H^+ and HCO_3^- and these ions are exchanged for mineral ions such as Ca^{2+}, K^+, and Fe^{3+}. Exchange may also occur between the root and minerals in water solution both outside and inside the root.

As is the case for many other soil factors, the absorption of salts rarely varies sufficiently from year to year to be important to the relative variations in ring width. However, changes in the mineral cycles of the soil, which may result from changes in the soil or from pollution of the environment, may alter the mineral content of plants. When other factors are not limiting, the availability of minerals can be the most limiting one and may influence ring characteristics such as the maximum ring width. There is growing interest in the possibility that microelements or isotopes in the air, dissolved in rain water or carried as dust, might be absorbed by the tree through the leaves or roots and utilized in the structure of the ring. For example, carbon-14 from the rings has been shown to have varied markedly from one decade to the next over the past 10,000 years (Olsson, 1970), and certain stable isotopes of hydrogen, oxygen, and carbon, resulting from differences in sources of atmospheric moisture, could conceivably vary in measurable concentrations from ring to ring.

It has been suggested by some workers (Jonsson and Sundberg, 1972) that acidity of rain water due to present-day industrial pollution of the air may, over a period of time, alter soil factors which affect forest growth. If this occurs, and if it becomes a world-wide phenomenon, trees which are now limited by climatic factors may become more limited by changes in acidity and pollutants. This could influence calibration of ring width with climatic factors and limit future applications of tree rings to problems of climate.

XII. Translocation

The long-distance movement of organic and inorganic solutes from one part of the plant to another is referred to as *translocation*. The term is not ordinarily applied to the diffusion of solutes from cell to cell, but it does include the movement of mineral salts dissolved in water which usually are transported upward from the roots to the leaves through the xylem. The salts may then be translocated out of leaves through the phloem and become distributed among various tissues of the plant. Organic substances are generally translocated through the phloem. Some salts absorbed from the soil, such as those of nitrogen, phosphorus, or sulfur, can react with organic substances that have been translocated to the roots. The salts which are then a part of the organic compounds are translocated out of the root through the phloem (Kramer and Kozlowski, 1960). However, the pathway for upward translocation of mineral salts, more specifically ions dissolved in water, is usually the xylem. High rates of transpiration, high concentrations of mineral salts in the soil, and favorable aeration of the soil favor rapid translocation of ions through the xylem.

Translocation through the phloem, unlike translocation in the xylem, may be bidirectional, that is, different substances may be translocated in opposite directions simultaneously. Translocation in phloem can occur at substantial rates through the living sieve cells which are metabolically active (Fig. 2.1). Lateral movement of organic molecules in the stem occurs in the radially oriented and living rays.

Several hypotheses have been proposed to explain the mechanism of translocation in the phloem, but no one hypotheses is entirely satisfactory

(Kramer and Kozlowski, 1960). The hypotheses involve simple diffusion, movement along the protoplasmic interfaces of the phloem, active streaming of the protoplasm throughout the elongated sieve tubes contained in the phloem, and mass movement developed by the water potential gradient throughout the plant.

Downward translocation of organic molecules from large branches occurs in a vertical direction in straight-grained wood so that the substances are used by the cambial tissues and roots directly below them. Often trees exhibit spiral grain and translocation occurs in a spiral direction. Annual rings formed by cambia in the path of the translocation stream from a heavily foliated branch or from the crown often are wider than those not in the path. Changes in the growth and shape of the crown which are due to changes in competition or exposure of the tree can alter the translocation of organic substances down the stem and can cause corresponding changes in ring widths.

The variability in translocation of food and growth substances from irregularly shaped crowns of aging *Juniperus osteosperma* (Fig. 3.18), *Pinus longaeva*, *Pinus flexilis*, and *Artemisia tridentata* may be partly responsible for the asymmetric growth of stems, partial dieback of the cambium, and the formation of thin strips of live bark that connect the live tops with the live roots (Fig. 3.18). If the strip of cambium continues to grow for many years, it will produce flattened slab-shaped stem form. Sometimes the wood grain is twisted and the strips of living bark frequently spiral around the stem of such strip-barked trees (Wright and Mooney, 1965; Ferguson, 1964, 1968; Fritts, 1969).

Injury of the bark may destroy the phloem. If a segment of bark completely encircling the stem dies, transpiration of organic food to the roots will be blocked. If the injury is permanent, the stem and roots below the injury die after all food reserves are exhausted. If injury is not permanent or a portion of the bark remains intact, a reorientation of the conducting tissue occurs and new xylem and phloem are formed around the wound, which function effectively in both upward and downward translocation (Kramer and Kozlowski, 1960; Meyer *et al.*, 1973). Serious injury of the stem by fire can kill as much as 90 per cent of the bark circumference in some trees without causing a noticeable reduction in growth in the stem below the injury. However, severe fire injury can reduce ring-width growth (Jemison, 1944) or induce anomalous growth in the injured area. Fungi, insects, and bark-eating animals can also injure the bark and effectively girdle the stem of a tree. Some diseases infect phloem cells and thereby interfere with food conduction in the tree.

Natural root grafts may establish connections between root systems of adjacent trees (Graham and Bormann, 1966; Eis, 1970; Stone, 1974), and

FIG. 3.18. A gnarled crevice site *Juniperus osteosperma* from Northern Arizona in which the leaf surface has been limited by the small soil volume available to the roots. Dieback in the branches causes partial die-back of the cambium so that only narrow strips of living bark remain intact along the branches, main stems, and roots of the tree.

sizeable quantities of food and other substances may be translocated from one tree to another of the same species through such grafts. A group of stumps in a dense forest may be connected so effectively through root grafts that the root systems will enable the stump to survive on carbohydrates supplied by neighboring standing trees (Kramer and Kozlowski, 1960). LaMarche and Wallace (1972) report on a *Sequoia sempervirens* stump in which the rings continued to form after the tree was cut, with no major change in mean width. This indicated there was substantial translocation through roots grafted to neighboring trees and that translocation through such grafts must have been substantial even before the tree was cut.

Young sprouts from roots of grafted seedlings grow more rapidly than the nongrafted neighbors, since they can obtain carbohydrates from the mother tree (Laufersweiler, 1955). It is likely that the translocation of food through root grafts is an important factor in the survival and growth of some suppressed subcanopy trees (Graham and Bormann, 1966). The effect of root grafting and subsequent translocation from neighboring trees on a ring-width chronology has not been ascertained. It is possible that root grafting would have little effect on the climatic information in rings because all neighboring trees would be essentially responding to the same variation in macroclimate. However, root grafting could conceivably reduce some of the differences in ring-width response among trees because each tree would be utilizing the same common root system and drawing upon a communal source of moisture.

The movement of organic solutes over long distances throughout the plant is basically controlled by factors affecting diffusion of substances into and out of the translocation stream as well as by the physiological activity of the phloem. Food seems to move toward regions of high metabolic activity, and certain growth regulators may at least partly control the direction of translocation because of their effects on the physiological activity of various tissues in the plant. Kramer and Kozlowski (1960) point out that changes in temperature may affect certain metabolic processes and influence the rate of transport in phloem. Oxygen is also necessary for high metabolic activity, and presumably poor soil aeration may hinder or prevent translocation of organic substances through some roots.

Studies on translocation rates and times of peak movement show the following results. Huber et al. (1937) reported that maximum rates of translocation in oak stems follow the daytime peak of net photosynthesis in the crown. Maximum rates occurred in the early evening first near the stem top and last at the base, while minimum rates occurred in early morning. The movement of carbohydrates out of leaves is reportedly more rapid during the day than night (Meyer et al., 1973), since during daytime photosynthesis is producing high concentrations of carbohydrates in the leaves. Zimmermann (1969) reports translocation rates for sugar in stems of *Fraxinus americana* of approximately 30–70 cm per hour. Lateral transport of organic molecules in the stem may occur more slowly than vertical transport due to smaller free energy gradients in that direction.

According to Wardlaw (1968) it is probable that the effect of environmental factors on actual conduction of foods and other organic substances is of minor importance in determining ring widths as well as their pattern and distribution within the stem. The environmental factors are more likely to critically limit the food-making or food-using processes directly than the rate of food movement. The various processes and limiting factors affecting them are described in the following chapter.

Chapter 4

Basic Physiological Processes: Food Synthesis and Assimilation of Cell Constituents

I	Introduction	156
II	Photosynthesis and Respiration	157
III	Synthesis of Foods and Assimilation	159
IV	Measurement of Photosynthesis and Respiration	162
V	Factors Affecting Photosynthesis and Respiration	163
	A. Light	163
	B. Temperature	165
	C. Water	170
	D. Components of the atmosphere	175
	E. Physiological factors	176
VI	The Annual Net Photosynthetic Regime for a *Pinus ponderosa* on a Semiarid Site	178
VII	Some Implications of these Physiological Measurements . . .	182
VIII	The Distribution of Foods and Interactions with Growth . .	184
IX	Essential Mineral Salts	189
X	Growth-Regulating Substances	190
XI	Physiological Preconditioning and Correlating Systems . . .	193
	A. Photoperiod	194
	B. Dormancy	195
	C. Drought resistance	196
	D. Frost hardiness	197
	E. Frost rings	198
	F. Fire scars	200
XII	Changes in the Physiological Seasons with Varying Elevation of the Tree Sites	202

I. Introduction

The life activity of a plant is dependent upon a continual source of energy which is usually derived from food. When the plant is actively growing, large amounts of food must be available, not only as a source of energy, but also as a source of raw materials from which the new tissues are built. It has been shown in previous chapters that the various growth processes

4. FOOD SYNTHESIS AND ASSIMILATION OF CELL CONSTITUENTS

throughout the tree occur at different rates which can vary from one minute to the next, one month to the next, and from year to year throughout the operational history of the plant. The availability of foods can have a variety of effects on plant growth, depending upon which growth processes are involved, their rate of activity at the particular time of year, and age of the plant.

Chapter 3 describes those processes involved in the movement of materials into and out of the plant and among its living cells. This chapter will deal with the most important biochemical processes involved in (1) the making of organic compounds classed as foods, (2) the release of energy within living cells, and (3) the synthesis of complex organic and inorganic substances which become living protoplasm, nonliving cell parts, and other substances that are necessary for life and growth.

Near the end of the chapter, consideration is given to some of the preconditioning systems within the plant which influence the relative activity of processes which in turn affect growth. Finally, the seasonal variations in climate are described as a function of elevation and are related to variations in the physiological activity of the plant.

II. Photosynthesis and Respiration

Photosynthesis, sometimes referred to as *carbon assimilation*, is the process by which sugar (glucose) is manufactured from carbon dioxide and water by the chlorophyll-containing tissues (chlorenchyma) in the presence of light, with oxygen being formed as a by-product. Light energy is absorbed by the chlorophyll-containing plastids called *chloroplasts* and converted to chemical energy in the glucose molecule. The resultant glucose manufactured in the process is sometimes referred to as *photosynthate*.

The simplest chemical equation for photosynthesis may be written as follows:

$$6\,CO_2 + 6\,H_2O \xrightarrow[\substack{\text{chlorophyll} \\ \text{(in living cells)}}]{673 \text{ kg-cal light energy}} C_6H_{12}O_6 + 6\,O_2\uparrow \qquad (4.1)$$

Actually the process is a highly complex one where the carbon from carbon dioxide becomes a part of a number of intermediate substances before glucose is formed (Meyer et al., 1973). There are two basic steps involved, the first requiring light and the second involving chemical reactions which depend upon favorable temperatures but not light. It is satisfactory for our purposes, however, to regard the process as a single reaction in which glucose ($C_6H_{12}O_6$) and oxygen are the end products.

In higher plants two different chlorophylls (identified as chlorophylls *a* and *b*) may be involved in photosynthesis. Basically, each chlorophyll molecule is built from carbohydrates and salts of nitrogen and magnesium. Salts of iron are also essential to certain biochemical reactions that are involved in the chlorophyll synthesis (Meyer *et al.*, 1973). Photosynthesis is an important process because the glucose molecule, or chemically altered fragments of it, form the building blocks from which all other organic substances are made.

The energy that is stored in the glucose molecules can be released by a series of biochemical reactions, referred to collectively as *respiration*. Part of this energy is lost as heat, but the remainder is transferred to high energy molecules called *adenosine triphospate* (ATP). The energy is stored as ATP molecules until it is utilized directly by energy-using processes occurring in the cell. The simplest chemical equation for respiration may be written as follows:

$$C_6H_{12}O_6 + 6\ O_2 \longrightarrow 6\ CO_2 + 6\ H_2O\ +\ 673\ \text{kg-cal energy} \qquad (4.2)$$

Such respiration which utilizes oxygen and glucose and releases carbon dioxide, water, and energy is *aerobic respiration*. The glucose molecule is chemically altered and this substance is split into smaller fragments. Some of these fragments, in turn, form basic structural units for the synthesis of the various foods and other cell-building substances. Respiration results in a loss of dry weight and occurs continuously in all living cells, both during the day and night. Respiration may be extremely slow in dormant seeds and dormant plants, while it is most rapid in the active meristems, including the cambium, growing shoots, root tips, young leaves, and growing fruits.

If atmospheric oxygen is absent, *anaerobic respiration* which does not require free oxygen can occur in cells of certain species. Anaerobic respiration can occur in roots when soils are poorly aerated and in tissues where diffusion of oxygen is restricted by impermeable structures such as the seed coat (Kramer and Kozlowski, 1960). Unlike aerobic respiration described above, anaerobic respiration releases only a portion of the energy chemically bound in the sugar molecule. Various chemicals such as aldehydes and alcohols are by-products, and if these substances accumulate, they may become sufficiently concentrated to be toxic to the tissues in which they are formed (Kramer and Kozlowski, 1960). When oxygen is available but in limited supply, both aerobic and anaerobic respiration may occur.

III. Synthesis of Foods and Assimilation

If one is to properly model and analyze ring-width growth as a function of climatic factors, it is important that he be aware of the role of glucose in the making of foods and other substances which are necessary for growth and development (Fritts, 1971). The following section describes the three basic classes of organic materials that are synthesized from glucose, and their functions in the growth of the plant. The term *food*, as used in this text, refers only to organic molecules classed as carbohydrates, fats, and proteins which contain chemically bound energy. Mineral salts which may be essential for growth are technically not foods, because they are not utilized as an important energy source. The best-known carbohydrate foods include sugars and starches, to be distinguished from carbohydrate cell constituents such as cellulose and lignin, which are synthesized from food but cannot be utilized again as a food or energy source. Sugars such as glucose are utilized in respiration as well as in the synthesis of the other two food classes, fats and proteins. Any glucose that is not so utilized may accumulate or be converted to starch or some other type of reserve food.

Fats and fat-like substances are present in every living cell. They are important constituents of protoplasm and may accumulate in large quantities in storage tissues or may form the waxes, such as cutin or suberin, that are commonly deposited in the cell walls of the endodermis, epidermis, and cork.

Proteins also are important constituents of protoplasm, and in some cases they can serve as food storage material as in seeds. In addition, most enzymes which catalyze the chemical reactions within cells are proteins. Protein molecules are comprised of long chains of amino acids, and the chemical characteristics of each protein are determined by the order of the amino acids along the chain.

The biochemical reactions involved in the synthesis of foods from glucose are extremely complex. These reactions basically involve two types of chemical processes. The first, referred to as a *condensation-hydrolysis reaction* represents a change in solubility of reactants in water but involves little change in energy.

$$\text{glucose} \underset{\underset{\text{(amylase)}}{\text{hydrolysis}}}{\overset{\text{condensation}}{\rightleftarrows}} \text{starch} + \text{water} \qquad (4.3)$$

When *condensation* occurs molecules of soluble substances such as glucose combine with one another to form an insoluble and chemically more complex substance such as starch. *Hydrolysis* is in a sense the reverse of condensation in that relatively complex insoluble substances are converted to smaller soluble ones as occurs in *digestion*. Both types of reactions are catalyzed by enzymes produced by the plant, but the direction and rate of the reaction may be controlled by a combination of the specific enzymes, the concentrations of reactants, acidity of the solution, and other conditions within the plant. The presence of the enzyme *amylase* is required for both the synthesis of starch from glucose and the reverse reaction, hydrolysis, which converts starch back to glucose.

The rates of these particular condensation–hydrolysis reactions are markedly influenced by the rates of production, consumption, and translocation of glucose throughout the tissues of the plant. Under daylight, when photosynthesis is rapid, much of the glucose that is produced by the leaves diffuses out of the green cells of the leaves and is translocated to areas of the tree where glucose concentrations are lower. The glucose then can be utilized for growth or can accumulate in the living cells. The accumulation of glucose drives the condensation reaction (to the right in Equation 4.3) which converts glucose to starch. During the first hours of darkness, when photosynthesis has ceased but respiration continues, glucose becomes most evenly distributed around the plant, and its concentration declines until the reaction is forced in the opposite direction initiating hydrolysis. By early morning starch is being digested back to glucose.

The significance of the condensation–hydrolysis reaction is that starch, unlike sugar, is insoluble in water, and its presence in a cell has no effect upon the water potential. Changes in the rates and direction of condensation and hydrolysis, therefore, affect the amounts of dissolved substances in cells, influence the cell–water relationships, and as a result play an important role in many processes such as the opening and closing of the stomates, translocation, and growth.

The second basic type of biochemical reaction in a plant is *oxidation-reduction*, which involves an energy exchange in the reactants. The energy is usually derived from ATP. The molecule which gains energy is said to be *reduced* during the reaction, and the molecule which loses energy is *oxidized*.

Figure 4.1 diagrams the interrelationships among the basic food-making processes. Arrows in the vertical direction represent oxidation-reduction reactions, while arrows in the horizontal direction represent condensation-hydrolysis reactions. The synthesis of many foods involves both types of reactions. For example, in the synthesis of fats, carbohydrates are first reduced in oxidation-reduction reactions to form glycerol and high energy-containing fatty acids, and then fats are synthesized by condensation of glycerol and three fatty acids. Because fatty acids contain considerably more chemically bound energy than the carbohydrates, they are more efficient food storage molecules than carbohydrates. Before the insoluble fats can be utilized in processes such as respiration and the manufacture of cell parts, they must be hydrolized to the more soluble fatty acids and glycerol.

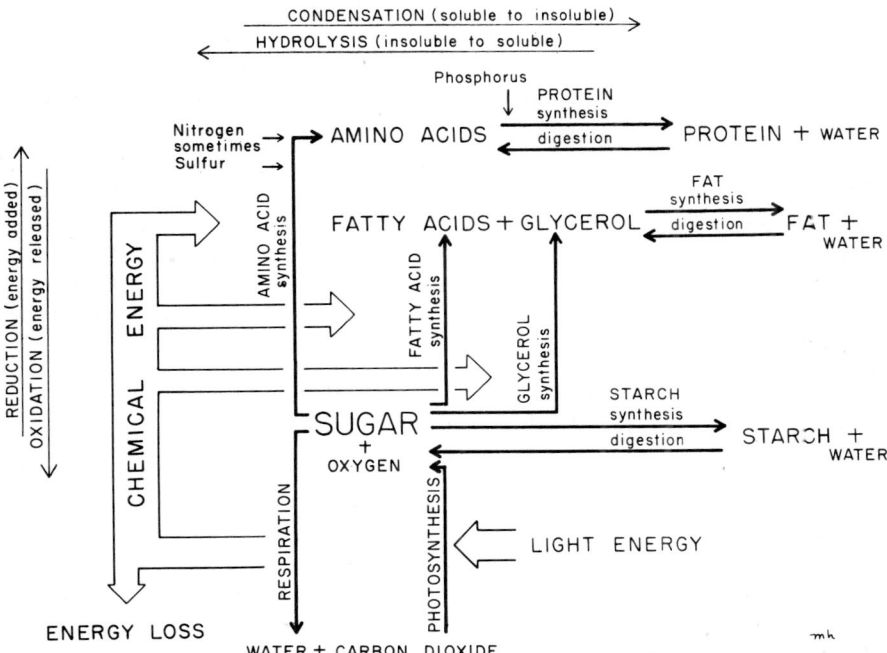

FIG. 4.1. Diagram representing the basic processes of food synthesis in green plants. The downward pointing arrow represents an oxidation reaction in which energy is released, and the upward arrow represents a reduction reaction in which energy is added. Horizontal arrows to the right represent condensation and to the left, hydrolysis. Organic substances on the left are soluble, those on the right are insoluble. Light energy is utilized in photosynthesis, as shown by the input arrows at lower right. Respiration releases this energy which was stored in the sugar (glucose) molecule. The energy is either released or transferred to reduction reactions, such as glycerol synthesis, fatty acid synthesis, and amino acid synthesis (arrow marked CHEMICAL ENERGY, on left). By means of condensation amino acids are converted to insoluble proteins, fatty acids to fat, and glucose to insoluble carbohydrates, such as starch. Insoluble substances are converted to soluble substances in the process of digestion. All reactions require specific enzymes. Nitrogen, phosphorus, and sulfur are necessary for the formation of certain amino acids and proteins. (Drawing by M. Huggins)

Amino acid synthesis (Fig. 4.1) involves reduction of certain mineral salts as well as changes in organic molecules. For example, reduced nitrogen and sometimes sulfur are combined with carbon compounds derived from carbohydrates. The synthesis of the soluble amino acids requires expenditure of chemical energy. The amino acids can then be linked to one another by a series of complex condensation reactions to form insoluble protein molecules.

The term *assimilation* is used in this book as an extension of food synthesis. In its simplest aspects, assimilation may be considered to be the utilization of carbohydrates, fats, and proteins to synthesize the protoplasm, cell walls, and numerous other substances making up the enzyme systems, pigments, and structures of plants. Assimilation involves the same sort of reactions involved in food synthesis, except that the end products become functional constituents of a cell. Assimilation requires the energy released by respiration, it utilizes soluble foods and minerals, and it depends upon numerous enzyme systems found in living cells. All the parts of a growth ring are essentially produced by assimilation, which may in turn be dependent upon any factors that limit photosynthesis, respiration, synthesis of food, and the general level of metabolism in the plant.

IV. Measurement of Photosynthesis and Respiration

Since the processes of photosynthesis and respiration involve a number of biochemical reactions and both processes can occur simultaneously, it is not an easy task to obtain accurate measurements of their respective rates. Both processes affect the amount of glucose in the plant and the exchange of carbon dioxide and oxygen between the atmosphere and the plant. At night respiration can be measured directly from the changes in carbon dioxide, oxygen, and dry weight, since photosynthesis does not occur in the dark. In the presence of light during the day, changes in the same variables provide a measure of the net effect of photosynthesis minus respiration, which is referred to as *net photosynthesis*. *Gross photosynthesis* is the total amount of glucose formed without regard to its consumption in other processes. It can be estimated by assuming that daytime respiration rates equal nighttime respiration rates and then adding the estimated daytime respiration to the net photosynthesis measurement. It has been found in recent years that such calculations may underestimate gross photosynthesis, because in certain plants respiration in light is more rapid than respiration in the dark (Ray, 1972; Decker, 1957).

Many estimates of photosynthesis and respiration involve measurements of carbon dioxide as it increases or decreases in an airstream passed over the plant material. The rates of the processes are expressed as grams of carbon dioxide utilized or lost per gram of dry weight of the plant. Other approaches utilize measurements of the amounts of oxygen absorbed, or changes in dry weight. For a general discussion of various techniques for measuring these processes, see Kramer and Kozlowski (1960) and Meyer et al. (1973).

V. Factors Affecting Photosynthesis and Respiration

Since all foods and cell constituents are derived from glucose molecules, tree growth can be markedly affected by conditions influencing photosynthesis and respiration rates. Both external factors and internal conditions of the tree affect the rates of these two processes. The important environmental factors involved are light, temperature, moisture, available gases, and soil fertility. The internal conditions include such things as physiological state of the tissues, the condition of the stomates, accumulation or availability of foods such as glucose, and any structural modification of functional plant tissues resulting from prior growth, increasing tree age, and injury of some plant part.

A. Light

Light is an important factor in photosynthesis, because it is the ultimate source of energy that can be utilized in the making of glucose. At low intensities, light is usually the most limiting factor, and the photosynthetic rates vary directly as a function of light intensity. At high light intensities, however, a further increase in intensity may not increase the rate of photosynthesis, because chloroplasts may be "light saturated" and some other factor becomes most limiting to the process (Fig. 4.2). Often clouds can reduce light intensities sufficiently during midday to cause photosynthetic rates to be low. (See Figs. 3.14 and 3.15 for June 30, July 28, and August 17, 1966.)

The length of the daylight period can affect the length of the photosynthetically active period. This may be an important factor for the growth and survival of trees at high latitudes where the growing season is short, because days are long and photosynthesis occurs at high rates almost continually throughout cloudless periods in summer (Tranquillini, 1964a).

All leaves or needles within a tree crown are not equally exposed to light. Those on the east side become light-saturated early in the morning, while those on the west may be light-saturated by late afternoon. Leaves in the lower crown may rarely reach light saturation, due to shading, and as a result there is a tendency for net photosynthesis, as measured for the entire tree, to increase to some extent with any increase of light intensity. As a result, the maximum rate of photosynthesis for a single leaf occurs at lower light intensities than the maximum for the entire crown of the tree.

The intensities at which light saturation occurs may vary because of differences in structure and optical density of leaves. Leaves that have grown in full sun are generally thicker and darker green than leaves grown in the shade (Kramer and Kozlowski, 1960). These thick leaves absorb light more efficiently and become light-saturated at higher intensities than leaves grown in shade. Photosynthesis is also reported to utilize certain frequencies of light more efficiently than others (Gates, 1965). However, the various light frequencies in a moderately dense forest are probably not sufficiently altered from these of full sunlight to affect photosynthesis to any substantial extent (Kramer and Kozlowski, 1960).

As a tree grows and the upper crown extends over and shades the lower leaves, photosynthesis in the most shaded branches can become so limited that they produce less food than they consume. These lower branches will eventually lose their vigor and die (Kramer and Kozlowski, 1960). Since a significant amount of food reserves can become consumed by these

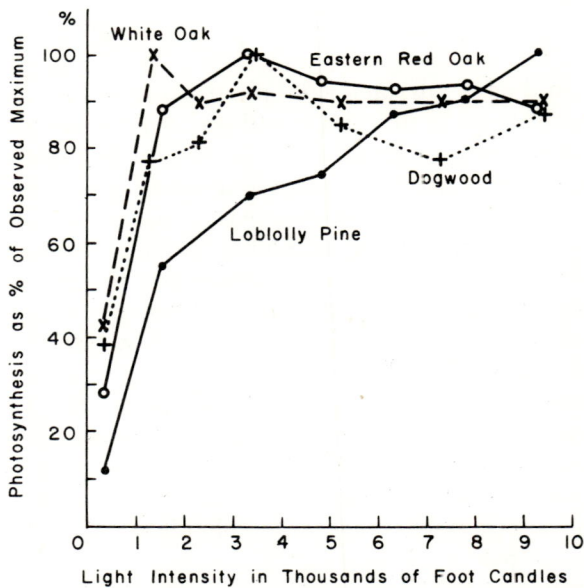

FIG. 4.2. The rates of gross photosynthesis occurring in tree seedlings for one-hour periods at various light intensities. Values are expressed as percentages of maximum observed rates. Temperature was maintained at approximately 30°C. (Redrawn from Kramer and Decker, 1944)

lower branches, it is possible that their death in a stress year can actually increase the overall accumulation of food in the tree during the following years.

Different species require different minimal light intensities to carry on photosynthesis and are said to differ in their tolerances to light. Seedlings of species which require high light intensities can become established in a site only after disturbance and opening of the canopy or in early successional stages when there is sufficient light to carry on the photosynthesis necessary for growth. The more shade-tolerant trees can establish themselves at any time, though they usually become established later in the successional sequence (Daubenmire, 1959).

There are two distinct groups of biochemical respiration reactions, one of which is called *photorespiration* and involves light (Ray 1972). Those plants which carry on photorespiration are referred to as C-3 plants; those which carry on respiration unaffected by light are called C-4 plants. Even though light does not affect respiration itself in C-4 plants, it can have indirect effects on the reaction because it enters the energy budget and affects plant temperature (see Chapter 5). Light can also influence photosynthesis, which in turn increases the amount of oxidizable substances available for respiration; light can influence chemical reactions involving the state of food reserves; and, as mentioned earlier, light can also influence stomatal opening and the exchange of gases involved in respiration. Brown (1968) reports high rates of net carbon dioxide production indicating high rates of respiration in *Pinus ponderosa* under high light levels during the day, which he attributes to high daytime temperatures of the needles due to heating by the sun.

B. *Temperature*

Photosynthesis and respiration occur over a wide range of temperatures, and temperature can affect each of the processes differently. Temperature also may affect conditions and processes such as dormancy, meristematic activity, reproduction, and growth, all of which interact with and modify respiration and the net photosynthesis of the plant.

The temperature minimum for net photosynthesis generally lies between $-2°$ and $-5°C$ (Tranquillini, 1964a), while respiration has been measured at temperatures as low as $-12°C$ (Freeland, 1944). Brown (1968) reports that he could not identify measurable exchanges in carbon dioxide from needles of *Pinus ponderosa* below temperatures ranging from $-2°$ to $-4°C$. The persistence of low temperatures can prevent significant amounts of respiration and photosynthesis from

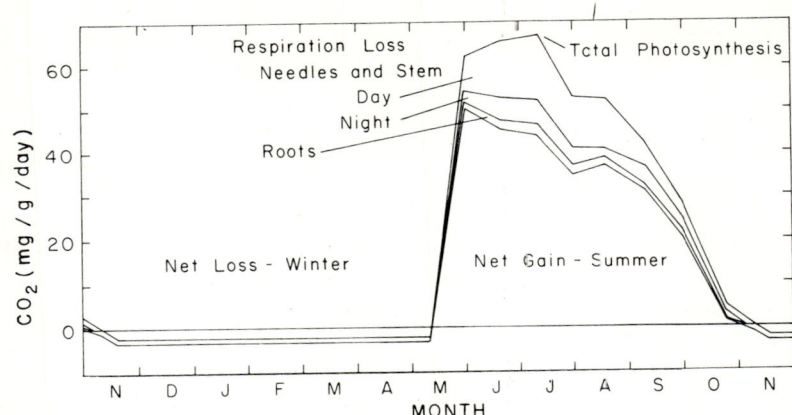

FIG. 4.3. The carbon dioxide balance throughout the year for young pines at tree line. The top line represents the total daily carbon dioxide used in gross photosynthesis. The area between the top and second highest line represents the carbon dioxide loss during the day due to needle and stem respiration. The area between the second and third highest lines represents loss of carbon dioxide due to respiration in the needles at night, and the area between the third fourth highest lines represents loss of carbon dioxide in the roots. The area between the lowest line and zero represents the daily net gain in carbon within the plant. During the winter little photosynthesis occurs and there is a net loss of carbon dioxide due to respiration of the plant. (From Tranquillini, 1964b)

occurring in conifers during long periods throughout the winter (Fig. 4.3; Tranquillini, 1964a, b). However, during brief periods of elevated temperatures within a long cold period respiration may increase much more rapidly than photosynthesis so that there is a loss of stored foods (Tranquillini, 1964a). Photosynthesis can also occur during winter in the twigs of deciduous trees (Perry, 1971).

Wintertime respiration of the high elevation *Pinus longaeva* in California, is reported to remain at the summer rate. Schulze *et al.* (1967) estimate that for *Pinus longaeva* at least half the summer season is needed to manufacture enough carbohydrate to equal that consumed during the brief warm periods in a normal winter. The probability of a negative net photosynthesis balance occurring throughout a year would increase with elevation and latitude, because the photosynthetic season becomes shorter and the inactive winter period longer. To a certain degree, the location of alpine and arctic tree lines may be influenced by the balance between winter respiration and summer net photosynthesis. Summer photosynthesis above the tree line may be reduced to such an extent that there is not enough food surplus to compensate for the loss due to respiration during the long winter (Tranquillini, 1964b).

When daytime temperatures are below 0°C during clear winter

4. FOOD SYNTHESIS AND ASSIMILATION OF CELL CONSTITUENTS 167

weather, high elevation conifers have been shown to be especially sensisitive to high radiation and to suffer considerable loss of chlorophyll, causing the foliage to turn yellow (Tranquillini, 1964a). In late winter, respiration in the yellow-colored foliage utilizes sizeable amounts of stored carbohydrates at a time when the absence of chlorophyll prevents photosynthesis from occurring. Normal photosynthesis resumes in spring after chlorophyll is resynthesized in response to warm weather. If shoots of *Pinus cembra* that are in subfreezing temperatures are brought to $+10°C$ in midwinter, about three days are required before net photosynthesis can be measured (Tranquillini, 1964b). If the shoots are placed in a warm temperature toward the end of winter, however, net photosynthesis can be measured immediately, although the rates are markedly lower than those that occur in the summer (Fig. 4.3). Tranquillini reports further that differences between the photosynthetic capacity of trees in the high mountains and those in the valleys are due primarily to differences in the frequency of severe frost which can destroy and thus deactivate chlorophyll in the leaves.

On warm sites such as those at low elevations or on south-facing exposures, there may be little deactivation of chlorophyll during winter. For example, Brown (1968) found little reduction in wintertime photosynthetic capacity and no decrease in chlorophyll content for *Pinus ponderosa* near its lower forest limits in southern Arizona. He did, however, find daytime release of carbon dioxide following winter nights when the trunk of the tree was frozen (Fig. 4.4). In these instances, carbon dioxide was released during the morning until noon when the tree had thawed and photosynthesis began. Carbon dioxide was absorbed during the afternoon until light became limiting at the end of the day (Fig. 4.4). When nighttime freezing of the trunk was less extreme, less carbon dioxide was lost by the branches and net photosynthesis reached maximum rates earlier in the day. In contrast, no carbon dioxide was lost during winter days which followed warm nights when the stems did not freeze. Under these conditions, net photosynthesis began at sunrise and continued at a high rate throughout the day except for periods when light, temperature, or moisture became limiting to the process (Fig. 4.4).

High temperatures can be limiting to both photosynthesis and respiration. In general, gross photosynthesis increases rapidly with increases in temperatures in the $0°$ to $15°C$ temperature range (Fig. 4.5). At higher temperatures there may be little change in the photosynthetic rate. Respiration, on the other hand, increases at a less rapid rate in the lower temperature range (Fig. 4.5). Respiration in many plants increases 2·0 to 2·5 times for each $10°C$ rise in temperature within the $10°$ to $30°C$ range (Meyer *et al.*, 1973). The optimum temperature for net

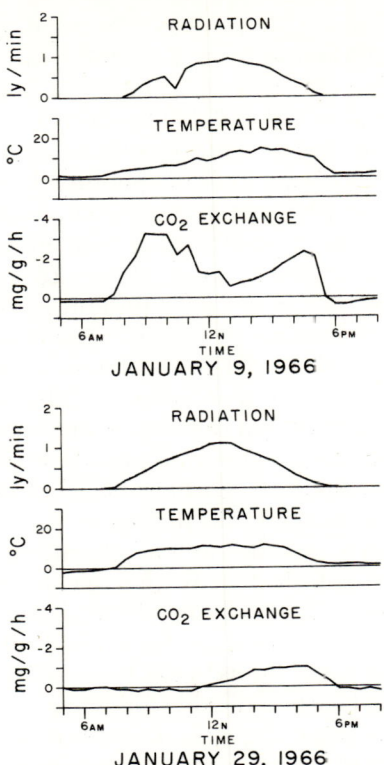

FIG. 4.4. Net carbon dioxide exchange from a branch of an arid-site *Pinus ponderosa* during two winter days. January 9 was preceded by a relatively warm night when no freezing occurred within the main stem of the tree, while January 29 was preceded by a cold night when freezing caused marked shrinkage in the tissues of the main stem. (Redrawn from Brown, 1968)

photosynthesis of temperate-region plants lies between 15° and 30°C. As temperatures increase above the optimum, net photosynthesis declines, because the rate of respiration increases faster than the rate of gross photosynthesis. Eventually more food may be consumed than is manufactured, and if the conditions persist, death due to starvation can result. The thermal death point of most living plant cells is between 50° to 60°C. Death due to extreme heat is caused by the coagulation of protein components in the protoplasm (Meyer *et al.*, 1973). Optimum temperatures for net photosynthesis may also vary among different species, throughout the season, and from site to site, depending upon prior light intensities, moisture availability, and factors which precondition the tree. Optimal temperatures for net photosynthesis in tropical and subtropical species often lie between 25° and 30°C, and in

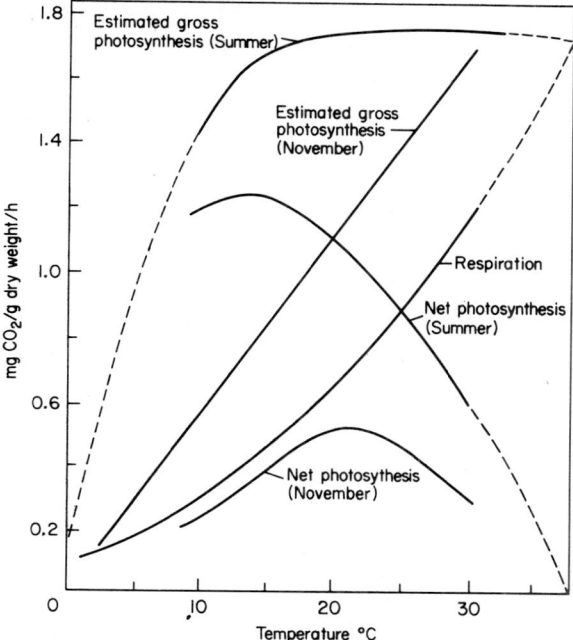

FIG. 4.5. The apparent relationship among temperature and estimated gross photosynthesis, net photosynthesis, and respiration in *Pinus longaeva* during summer and the month of November. Dashed lines represent extrapolated data, and respiration is assumed to occur at the same relative rates throughout the day and throughout the year. (Redrawn from data presented by Mooney et al., 1966, and Schulze et al., 1967)

species of the cooler temperate latitudes they may lie between 15° and 20°C. At the upper tree line, however, the optimal temperature may be 2° or 3° lower than 15°C (Tranquillini, 1964a).

The specific relationship among temperature, estimated gross photosynthesis, net photosynthesis, and respiration in *Pinus longaeva* for summer and fall months obtained by Mooney and his colleagues is summarized in Fig. 4.5. The rate of gross photosynthesis in summer appears to rise rapidly from temperatures near freezing to 15°C, but further increases in temperature are not accompanied by marked changes in gross photosynthesis. On the other hand, respiration, as inferred from measurements made in the dark, increases exponentially as a function of temperature between 0° and 30°C. The maximum rate of net photosynthesis measured in summer for high elevation *Pinus longaeva* occurs around temperatures of 15°C. Young plants grown under greenhouse conditions, however, exhibit lower rates of net photosynthesis at 10°C and higher rates at temperatures of 20° and 30°C (Mooney et al., 1966). Seedlings measured in the field during November exhibit low rates of net photosynthesis with optimum values occurring around 20°C (Fig. 4.5; Schulze et al., 1967).

High respiration rates due to high temperatures of nongreen tissues, such as those in the stem and root, may result in deleterious conditions within the plant. For example, soils that are compact and waterlogged may reach critical temperatures in summer. Oxygen can be depleted by aerobic respiration of soil microorganisms to such an extent that continued metabolic activity and life can be continued only by anaerobic respiration in the roots. Thus, prolonged flooding in the summer months can be injurious to the deeper roots, but if there is a series of wet and dry cycles or if flooding occurs during the colder winter period, respiration may not exhaust the available soil oxygen sufficiently to cause death or a reduction in the growth rates of roots.

C. Water

Slatyer (1967) defines two main modes of action of water deficits on photosynthesis: (a) the indirect effect of water deficits on stomatal closure and reduced diffusion of carbon dioxide into the leaf, and (b) the direct effect of water deficits on the biochemical reactions involved in photosynthesis. Slatyer goes on to point out that photosynthesis is dependent on four main groups of processes all of which can be affected by water stress: (1) diffusive processes associated with the supply of carbon dioxide to photosynthetic sites, (2) photochemical processes associated with the utilization of light energy, (3) chemical processes associated with the chemical reduction of carbon dioxide, and (4) transport processes affecting the translocation of photosynthate away from the production sites. In general, the rates of net photosynthesis can begin to decline when the water potential is one to three bars below zero and continues to decline in a more or less linear manner as cell turgidity decreases. The carbon dioxide exchange between the leaf and air reverses from a negative to a positive flux as respiration exceeds photosynthesis.

The opening and closing of stomates is a mechanism that regulates the water balance and maintains a favorable water balance within the plant. As mentioned in Chapter 3, when the guard cells surrounding a stomate increase in size, they arch and the stomate opens. The processes involved in this opening are relatively complex. It will suffice for this discussion to point out that the guard cells are the only cells of the leaf epidermis that contain chloroplasts. Photosynthesis commences when the guard cells are illuminated in the morning and carbon dioxide concentrations in the cell sap are reduced below night concentrations. Since carbon dioxide forms a weak acid in water, the cell becomes less acid as carbon dioxide is removed. This shift in acidity forces starch in the cell to be hydrolyzed to glucose. The increase in glucose lowers the osmotic potential of the cell, osmosis occurs, and water diffuses from the neighboring nongreen cells. The associated increase in pressure potential and swelling of the adjacent guard cells causes the stomates to open. In the evening, the reverse conditions occur and cause stomates to close.

It was mentioned in Chapter 3 that when the stomates close, leaf resistance to the diffusion of water vapor increases and the rate of transpiration declines. Although the diffusion of carbon dioxide may also be reduced, stomatal closure can be expected to influence water vapor transport to a greater extent than the carbon dioxide flux because of differences in the sizes of the two kinds of molecules (Slatyer, 1967). As water stress is imposed, factors other than low concentrations of carbon dioxide appear to become progressively more limiting to net photosynthesis.

Water deficits can affect respiration rates and thus produce changes in net photosynthesis. If water stress is imposed suddenly, an increase in respiration rate may be observed first, but after a period of time a reduction in respiration rate occurs, representing a sort of physiological adjustment. If stress is imposed gradually, however, no initial increase occurs and the rate of respiration gradually declines.

Wright and Mooney (1965) subjected three 20- to 40-year-old trees of *Pinus longaeva* to various soil moisture levels and measured net photosynthesis and respiration. Fig. 4.6 shows only a slight observable reduction of net photosynthesis and respiration while soil moisture levels were higher than −15 bars, the approximate water potential of wilting point. Severe reduction in net photosynthesis occurred at water potentials lower than wilting point.

Slatyer (1967) describes the effects of water deficits resulting from the drying of soil on the diurnal variations of metabolic processes. A base level of stress is imposed by declining water potential in the soil. Superimposed upon this is the water stress due to the diurnal lag of water absorption behind transpiration. Initially, metabolic processes are suppressed only during the afternoon period of maximum water deficit, but this period will become longer each day. Stomatal closure at this time can retard the transpiration, causing leaf temperatures to rise above the temperatures of the surrounding air. The elevated leaf temperatures can, in turn, cause increased respiration and reduced net photosynthesis. In addition, closed stomates can conceivably reduce photosynthesis by reducing the diffusion of carbon dioxide into the leaves. Finally, net photosynthesis may be affected adversely by reduction in turgor of the chlorenchyma.

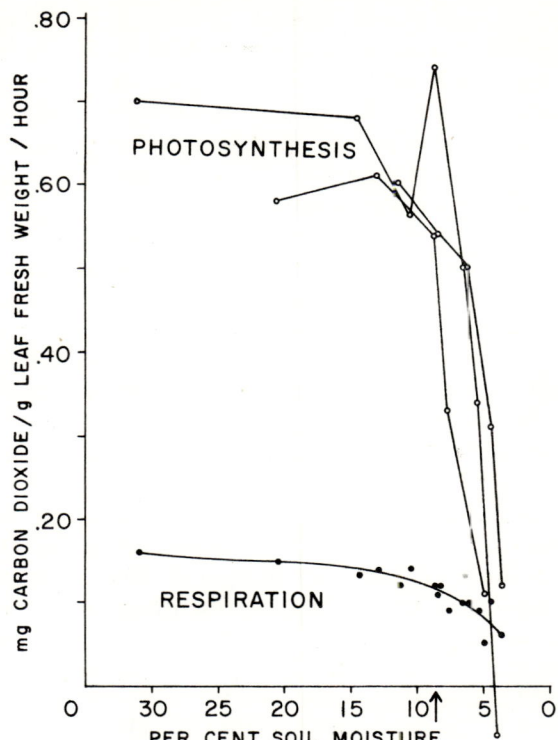

FIG. 4.6. Net photosynthesis and nighttime respiration of *Pinus longaeva* as a function of soil moisture. Lines connect rates of net photosynthesis for individual plants. The line for respiration is the mean for the three plants. Arrow designates soil moisture percentage at wilting point where soil water potentials are approximately −15 bars. (From Wright and Mooney, 1965)

The effects of soil moisture on photosynthesis have been studied by Brown (1968) for *Pinus ponderosa* in Arizona. As soil moisture decreased from April through June (see Fig. 3.14), rates of net photosynthesis declined and periods of net carbon dioxide production during the day increased in frequency and duration. Whenever replenishment of soil moisture occurred, water potentials increased and rates of net photosynthesis increased so that the net carbon dioxide flux became more negative (carbon dioxide was absorbed by the leaves) during the day as long as light intensities were adequate.

Figures 4.7 and 4.8 show measurements of carbon dioxide exchange from branches of *Pinus ponderosa* in a transparent enclosure during six days of different environmental regimes; three represent days when soil

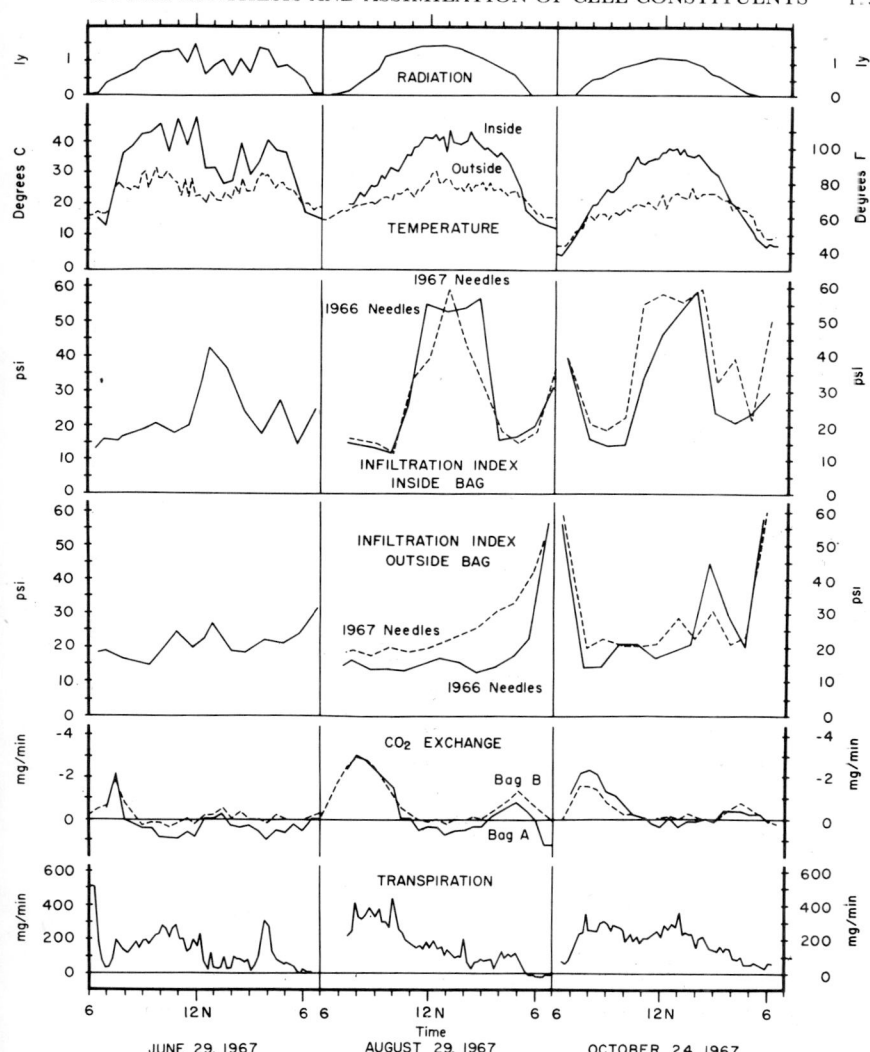

FIG. 4.7. Measurements of solar radiation and temperature, an index to stomatal size, carbon dioxide exchange, and transpiration as measured from one- and two-year-old needles on branches of *Pinus ponderosa* enclosed and unenclosed in a polyethylene bag during three days of apparent water stress. Carbon dioxide exchange is negative when photosynthesis exceeds the rate of respiration and positive when the relative rates of the two processes are reversed. Transpiration represents water lost from branches in the enclosure. Radiation is measured in langley units; the stomatal size is expressed as pounds of pressure per square inch required for infiltration; and carbon dioxide exchange and transpiration are expressed as milligrams per minute. Stomates are closed at pressures of 60 psi. As stomates open, the infiltration pressures become lower.

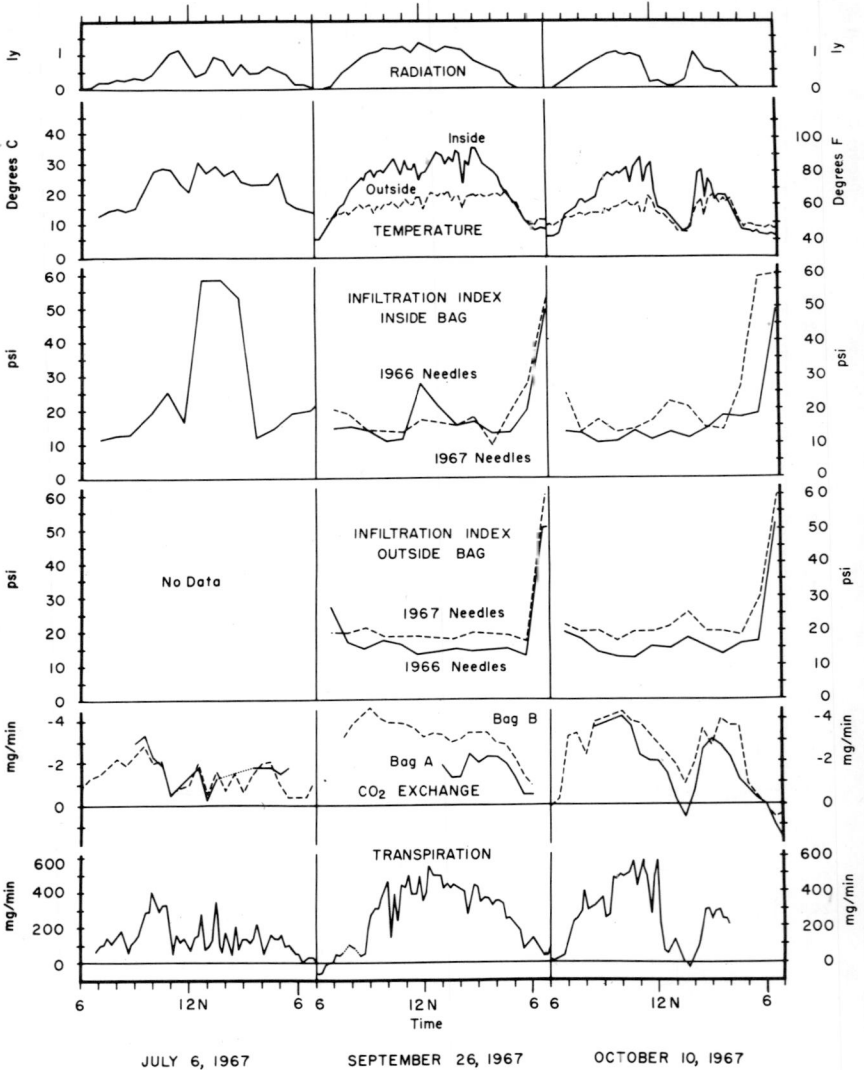

FIG. 4.8. Same as Fig. 4.7 except measurements are for three days of no apparent water stress.

moisture was low and three when soil moisture was ample. The rates of respiration, the amounts of solar radiation, temperatures, and relative size of stomates measured by infiltration techniques are also shown (see Chapter 3) (Drew et al., 1972). During the midday period of the driest and hottest days, stomates closed within the polyethylene enclosure, but stomates on unenclosed needles remained relatively open. Since complete stomatal closure has also been observed for branches in the open during dry, warm summer days (Drew, 1967; Drew *et al.*, 1972), it may be inferred that conditions within the bag truly represent natural conditions resulting from drought, although in the particular instance shown, the enclosure became warmer than the outside air, which accentuated stress and raised respiration rates. Under dry conditions illustrated by Fig. 4.7, about two hours after net photosynthesis commenced, its rate began to decline with increasing radiation and air temperatures, while stomates remained open and transpiration was relatively high. Needle temperatures that were above 25°C (the optimum for net photosynthesis) were often recorded, and water stress was increasing as indicated by declining stem size (see Fig. 3.14). Typically stomatal closure occurs when respiration approximates gross photosynthesis and there is no net carbon dioxide exchange between the foliage and the surrounding atmosphere. In this example from an arid-site *Pinus ponderosa*, the maintenance of open stomates until the carbon dioxide flux reaches zero suggests that high temperature and high water stress, rather than low carbon dioxide, must have caused the declining rates of photosynthesis to occur.

When soil moisture was high, such as on September 26 and October 10 (Fig. 4.8), stomates remained open, needle temperatures were lower, and rates of both transpiration and net photosynthesis remained high, except during cloudy weather which both lowered light intensities and reduced needle temperatures (Brown, 1968).

Slatyer (1967) states that as stress becomes severe and water potentials approach the permanent wilting point, the effect throughout the day of the diurnal water deficit will become less important (as in Fig. 3.14, June 4, 1966). Stomates will close for most of the day and only small amounts of water diffuse through the cuticular surfaces of the leaves. Daytime temperatures of leaves can rise well above air temperature, and net photosynthesis, cell division, and overall plant growth are likely to decline.

D. *Components of the Atmosphere*

Photosynthesis in trees with adequate water and light is often limited by low concentrations of carbon dioxide. In such cases the rate of photosynthesis can be increased simply by raising carbon dioxide concentrations in the surrounding air. Carbon dioxide in an actively photosynthesizing forest canopy is most limiting early in the afternoon when it reaches its lowest concentration. Carbon dioxide is more likely to limit photosynthesis in trees at high elevations than at low elevations, because the pressure of gases decreases with increasing elevation, and the rate of diffusion declines (Gale, 1972; Kramer and Kozlowski, 1960).

Ordinarily, oxygen is not limiting to tissues in organs above the ground. However, as was discussed in the last chapter, the oxygen concentrations in soils may decline to less than 10%, which is sufficiently low to limit respiration and root growth in species not well adapted to wet soils. It is sometimes suggested that in poorly aerated soils carbon dioxide accumulates to toxic levels, but a deficiency of oxygen rather than excess carbon dioxide appears to be the more important factor.

Sometimes gases other than those normally present in the atmosphere become limiting due to their toxic effects upon plants. For example, industrial sulfur dioxide or ozone may adversely affect plants. Most species of plants are injured by exposures of only one hour to an atmosphere containing as little as one part in a million of certain toxic gases (Meyer *et al.*, 1973).

E. *Physiological Factors*

A number of conditions related to age and physiological activity can influence respiration and photosynthesis. Respiration occurs in all living cells, but the rates are considerably higher in younger tissues than in older tissues because the former usually contain a greater number of growing cells. Rates of respiration of maturing cells are inversely correlated with the increasing vacuole size and with increasing proportion in the amount of cell wall material to the amount of protoplasm. High respiration rates in the stem are confined largely to the apical and cambial meristems and adjacent young tissues. In apical meristems, respiration may decrease abruptly in August when the buds become dormant and increase abruptly in the spring at the time of bud break (Kramer and Kozlowski, 1960).

Photosynthetic rates generally increase with increasing age of leaves or needles up to a level representing leaf maturity, and then they decline (Freeland, 1952; Kramer and Kozlowski, 1960). Mooney *et al.* (1966) measured photosynthesis in needles of *Pinus longaeva* at the end of a growing season and found that the needles formed in the course of the summer were somewhat less efficient than the older needles. This may be a result of mutual shading among the five needles of the young fascicle, for they remain in a tight bundle until some time after the autumn of their first year. Mooney *et al.* (1966) report that the photosynthetic rates in needles from one to five years in age were markedly similar. Fritts (1969) reports that in some trees of *Pinus longaeva* needles are green and appear capable of photosynthesis for 15 to 30 years.

In many other species the photosynthetic efficiency drops rapidly and needle numbers decrease with increasing needle age greater than one year. O'Neil (1962) reports that there is little reduction in growth of *Pinus banksiana* when two- or three-year-old foliage has been removed from the trees, which suggests that the photosynthesis in the two- and three-year-old needles is inefficient. When the one-year-old foliage is removed, however, there is a marked reduction in shoot and diameter growth, as well as a high mortality of apical buds. Silver (1962) states that 90% of the total foliage produced in a given year by *Pseudotsuga menziesii* is lost at the end of five years. The percentages of foliage usually lost at ages one to five years are 28%, 23%, 17%, 13%, and 10%, respectively in this evergreen species.

Photosynthetic efficiency sometimes diminishes with increasing age of trees, regardless of needle age. Results of Wright and Mooney (1965) show that old trees of *Pinus longaeva* have lower ratios of photosynthesis to respiration at temperatures of $10°$, $20°$, and $30°C$ than do younger trees. According to Möller et al. (1954), about 46% of the food produced by photosynthesis is used in respiration of eight-year-old *Fagus silvatica*, while at 25 years, respiration consumes only 40% of the photosynthate. However, with increasing age thereafter, the percentage of photosynthate consumed by respiration increases to about 50% in 95 years. This change is a function of the ratio between the amount of respiring tissue and the amount of photosynthesizing tissue. Young trees generally have a higher ratio of photosynthetic surface to respiring tissue than older trees (Kramer and Kozlowski, 1960).

The actual accumulation of glucose in the leaves, particularly during the latter part of the daylight period, can become limiting to photosynthesis. Sweet and Wareing (1966) report that growth in *Pinus radiata* appears to affect photosynthesis directly by reducing the amount of photosynthate that can accumulate. Therefore the young, fast-growing seedling may have a more efficient system than a dormant or old tree simply because there is less accumulation of photosynthate in the leaves and concentrations of glucose are less likely to become the most limiting factor to photosynthesis.

Möller et al. (1954) present evidence that a decrease in net dry-matter produced by aging stands is caused partly by a decreased rate of photosynthesis in the leaves and partly by a small increase in respiration accompanied by a greater loss of dry weight in roots, branches, and twigs (Fig. 4.9).

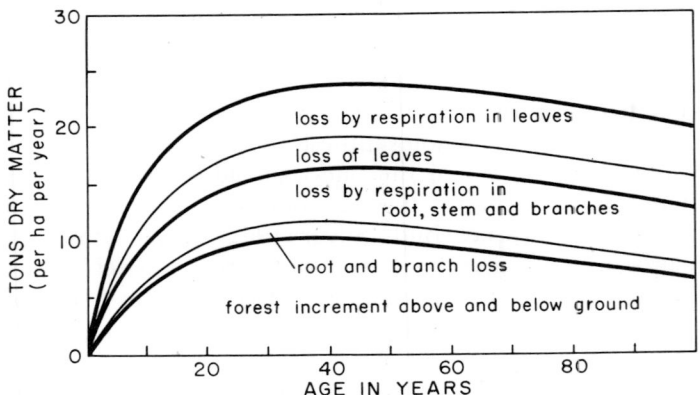

FIG. 4.9. Dry matter production in *Fagus silvatica* plotted as a function of tree age. The upper heavy line is dry matter production representing gross photosynthesis; the middle heavy line is net production translocated to other organs by the leaves; the heavy line at the bottom represents the annual accumulation of dry matter both above and below ground. The area between the upper and middle heavy lines represents the loss of dry matter through respiration in leaves and through shedding of leaves, while the area between the middle and lower heavy line represents the loss of dry matter by respiration and loss of roots and branches. (From Möller *et al.*, 1954)

Mooney (1972) points out that the differences in light sensitivities of respiration in C-3 and C-4 plants may help explain distribution and behavior of certain plants. In general, C-4 plants which cannot carry on photorespiration are less severely affected by extremely high temperatures and seem to occur more frequently in hot, bright, and dry environments. The C-3 plants occur in cooler habitats or are active in the cooler seasons of the year. While differences in these two types of metabolic pathways could account for differences in ring-width and climatic relationships, all trees so far used in dendroclimatology appear to be the C-4 type.

VI. The Annual Net Photosynthetic Regime for a *Pinus Ponderosa* on a Semiarid Site

It will be shown in Chapter 5 that the ring widths in warm and arid-site trees are statistically related to climatic conditions during a nine- to ten-month period prior to growth, as well as to the three to four months of the

4. FOOD SYNTHESIS AND ASSIMILATION OF CELL CONSTITUENTS

growing season from April through July. These statistical relationships can result from limiting effects of climate on the annual net photosynthetic regime of the plant. Some of the biological evidence for this effect is described below.

The carbon dioxide exchange for branches of *Pinus ponderosa* enclosed in a ventilated polyethylene bag was monitored during several years of contrasting climate (Brown, 1968). The enclosure was mounted on branches of a tree growing in an extremely arid site in southern Arizona at 8,500 ft (2591 m). The carbon dioxide in the enclosure, along with a number of other environmental variables affecting the trees, were monitored from May, 1964, through October, 1967. An important part of the study involved the application of energy balance equations (Chapter 5) to assess the environmental conditions that were affected by the particular carbon dioxide measuring technique being applied. For example, it was shown by Brown (1968) that the polyethylene enclosure actually reduced the heat load, but that needle temperatures during the day inside the enclosure were frequently higher than needle temperatures outside the enclosure, because of differences in air movement. The enclosed tissue was continually subjected to very low wind velocities (approximately 0·1 mph, or 0·045 m/sec), while the unenclosed portions of the tree crown were commonly exposed to higher wind velocities.

During summer days when transpiration rates were low and the branch was subjected to high solar radiation, air temperatures in the enclosure were sometimes 10°C, and in exceptional cases 20°C, higher than temperatures on the outside (see Figs 4.7 and 4.8). Air temperatures within the enclosure at night were from 2° to 7°C below temperatures of the outside air, depending upon wind conditions (Brown, 1968). It was concluded from this analysis that the measuring technique used in the study did somewhat modify the actual temperature environment by essentially excluding the influence of wind. The modification remained within the natural limits of the plant, however, and simply caused a bias toward warmer-than-average temperatures. Since none of the years of study was extreme, the particular conclusions drawn about the patterns of carbon dioxide exchange are considered valid and comparable to natural conditions in extremely warm and dry years which climatic records indicate have occurred in the past.

Figure 4.10 is a plot of climate and carbon dioxide exchange of the enclosed branches expressed as three-day averages throughout two years. Precipitation and temperature during the winter of 1964–1965 were near the average of the five previous years. The frequent occurrence of a carbon dioxide uptake during the warmer winter days indicated net photosynthesis could occur almost any time when temperatures were

G

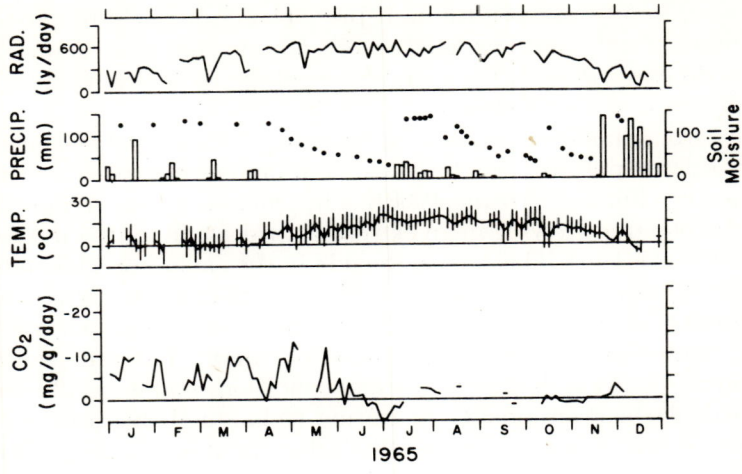

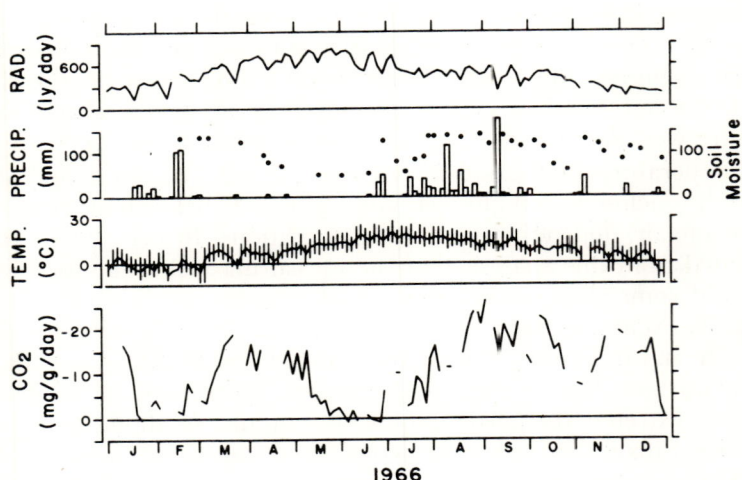

FIG. 4.10. The climate and carbon dioxide exchange during 1965 and 1966 for a branch from an arid-site *Pinus ponderosa* in southern Arizona. In 1965 less precipitation fell, soil moisture was lower, and the amount of net photosynthesis was more limited by moisture deficits than in 1966. The radiation, temperature, and carbon dioxide exchange are shown as three-day averages. Precipitation is the total for the three-day period, and the soil moisture values (shown as dots) are corrected dial readings from nine Colman soil-moisture units. The vertical bars indicate the range of temperatures during each three-day interval. (Redrawn from Brown, 1968)

near or above freezing. However, net photosynthesis ceased whenever daytime temperatures were less than $-2°C$, and it was low during days following nighttime freezing of the stem (see Fig. 4.4). Net photosynthesis generally increased with increasing temperature above freezing.

Soil moisture began to decline in April of 1965, after the snow had melted. Air temperatures for the months of March, May, and June were 1° to 2°C below the five-year average for these months, and during March the rate of net photosynthesis varied directly as a function of temperature and light intensity. However, later in the season as soil moisture was depleted and air temperatures rose, high water stress and excessively high needle temperatures became more limiting and caused the rates of net photosynthesis to decline. During the warmest and driest periods, daytime respiration exceeded gross photosynthesis, and carbon dioxide was released from the enclosed branch. Fig. 4.10 shows that the mean carbon dioxide exchange for the entire 24-hour period was positive for many days in June. As drought and air temperatures increased in June, dendrograph records, stomatal infiltration pressure, and relative turgidity measurements indicate that water stress increased throughout the tree. For example, infiltration rates (Drew et al., 1972) showed that stomates closed during midday in unenclosed portions of the tree, while those within the enclosure often closed at an earlier hour. Days of high temperature resulted in little net gain of photosynthates. Growth at the base of the stem, which began in April (see Chapter 2), must have utilized primarily stored foods because there was little net accumulation of photosynthate in the leaves.

Precipitation during July and August of 1965 was below the five-year average for those months. Virtually no precipitation occurred from September until late November, and the occasional rains replenished moisture only in the upper levels of the soil. Even though net photosynthesis was not monitored continually due to equipment failure, sufficient data were collected to show that net photosynthesis for the dry autumn of 1965 was markedly lower than the following year, when there was more moisture in the soil.

A storm in November, 1965, was followed by a cold winter with snow persisting for several months (Fig. 4.10). Where data are available for the winter of 1966 and when day temperatures were below $-2°C$, little or no net photosynthesis occurred. When day temperatures were above freezing, net photosynthesis occurred at higher rates than in 1965, probably because of the deep snow and its effect on insulating the soil and recharging soil moisture reserves. Net photosynthesis was rapid in March and April because soil moisture was abundant and air temperatures did not greatly exceed the optimum for the process. Growth started early in

April (see Chapter 2), and in May and June net photosynthesis of the enclosed branch declined with rising needle temperatures. Air temperatures in June, 1966, were $3.3°C$ warmer than in June of 1965, but net photosynthesis was more rapid, apparently as a result of the greater soil moisture supply and higher transpiration rates which maintained lower temperatures in the leaves.

The 1966 summer rains started late in June, which was earlier than the previous year, and precipitation in both August and September was high. Consequently, the average rate of net photosynthesis was markedly greater than during the previous summer. High net photosynthesis continued until late in October, when soil moisture began to decline. A storm in November replenished soil moisture, and high rates of net photosynthesis were measured until daytime temperatures dropped below $-2°C$ near the end of the year.

In summary, the rates of net photosynthesis were lower during 1965 than in 1966, as it was a year of average winter precipitation and below average summer precipitation. In contrast, the higher net photosynthesis in 1966 occurred because of high precipitation during both the winter and summer periods. Even though air temperatures were higher during June of 1966, net photosynthesis was higher than in the prior year because of greater soil moisture reserves.

Brown's measurements indicate that low-elevation, arid-site conifers, unlike those at high elevations described by Tranquillini (1964b), Mooney et al. (1966), and Schulze et al. (1967), can carry on net photosynthesis almost any time in the year if leaf temperatures during the day are not $-2°C$ or below. At low elevations low air temperatures are less often limiting to photosynthesis during the winter than at high elevations, but soil moisture in winter is more likely to be insufficient and internal water stress is more likely to occur.

VII. Some Implications of these Physiological Measurements

Measurements of net photosynthesis, respiration, and transpiration in arid-site conifers confirm that these processes vary sufficiently during the autumn, winter, and spring periods to have a substantial effect upon food reserves and subsequent ring growth. To be more specific high soil moisture and low air temperatures (except for those below $-2°C$) are shown to favor rapid net photosynthesis and production of glucose and derived stored foods, which are then available for growth in the spring. Moisture recharge of the soil which occurs during the autumn and winter can favor high photosynthesis the following spring when temperatures

4. FOOD SYNTHESIS AND ASSIMILATION OF CELL CONSTITUENTS

are favorable and growth begins. High growth, however, may not always accompany high winter temperatures if moisture is the most limiting factor, because the potentially favorable effects of high winter temperatures are counteracted by the unfavorable effects of high temperatures on transpiration and depletion of soil moisture reserves. In such arid sites, the amount of net photosynthesis is proportional to the amount of water which is available and is transpired whether the transpiration occurs in winter or is delayed until the spring (Budelsky, 1969).

Trees in colder sites, on north-facing slopes, or at higher elevations and latitudes may have more moisture available in the soil than those on arid sites, so that the increased moisture loss due to high transpiration during warmer periods in winter may not substantially affect the moisture available later in the year. In such cases, high temperatures in winter and early spring can favor rapid net photosynthesis and increased physiological activity, which could lead to either early initiation of cambial activity, or rapid growth, and favor the formation of a wide ring.

The measurements of net photosynthesis, respiration, and transpiration during the growing season also indicate that these processes can exert an immediate effect on growth occurring at the same time by their effect on water relations, food reserves, and the making of growth-controlling substances derived from them. However, during a brief period early in the growing season low temperatures may become limiting to growth directly unless moisture stress is severe and more limiting. If so, high temperature accelerates transpiration, and water stress may limit growth and alter growth-regulating processes.

Temperatures during late spring and summer may at times exceed the optimum for net photosynthesis and perhaps for other growth-controlling processes. If water is also scarce, the deleterious effects of high temperature can be accentuated by the increased water stress. In the most arid sites high temperatures accompanied by high water stress may be the primary limiting factors during summer, except during cloudy weather and for periods during and after a rainfall when temperatures are lower and water stress is not as great.

As the cambial activity slows down during midsummer the unfavorable effects of high temperature and low moisture begin to impinge more on the accumulation of stored foods than on the current growth, which in turn will affect the growth of the following year. When ring growth in August or September ceases, any limiting temperature and moisture conditions which affect processes and food reserves cannot affect growth until it is resumed in the following spring. Sometimes environmental conditions which hasten the cessation of growth during late summer may have the reverse effect upon the growth the next year, for photosynthate which would have been used for building cell parts is shunted into stored food reserves which are then available for more growth the following year.

During a warm and relatively dry summer season at high elevations, high latitudes, or on cool sites, prolonged water stress accompanied by excessively high temperatures may be rare, though they can cause a decrease in the rate of net photosynthesis and affect growth if water is in limited supply (Hari and Luukkanen, 1973). As on warmer and more arid sites, with the passage of time the effect of climatic factors on the ring that is being formed declines as the growth processes subside, while the effects on the width of the next year's ring increases through this control of food production, accumulation of stored food, and other conditions important to growth in the following spring.

In early autumn temperatures in cool high-elevation sites are not only lower but they may reach critically low levels and become directly limiting to processes long before they are low enough to be critical in warmer low-elevation sites. Thus, at high elevations and latitudes temperature rather than moisture is best correlated with the width of the following year's ring.

During the winter, high elevation or high latitude conifers may be deeply frozen so that the chlorophyll is deactivated. Periods of unusually warm weather may thaw out the tops, produce extreme water stress, favor excessive respiration, and hence bring about depletions of food reserves. This can lead to an inverse correlation of winter temperatures with food reserves and with the width of the subsequent season's ring.

VIII. The Distribution of Foods and Interactions with Growth

The relative distribution and consumption of food varies among the structures and organs of the tree and from one season to the next. The areas of the circles and squares in Fig. 4.11 illustrate the average amounts of glucose that can become involved in the various tissues of a typical middle-aged deciduous forest tree. Approximately 10% of the total glucose produced in leaves is consumed by leaf respiration. The remainder becomes stored or is involved in assimilation of other parts of the tree, with the aerial parts utilizing 81% and the roots utilizing 9%. Within the aerial parts of a tree, each year's growth of new leaves utilizes approximately 36%, the bole 36%, and the branches 8%. Fruit production utilizes about 1% of the total glucose. Allocation to the various structures in the diagram includes the glucose consumed in respiration as well as the portion that is converted to the structure of each tree part.

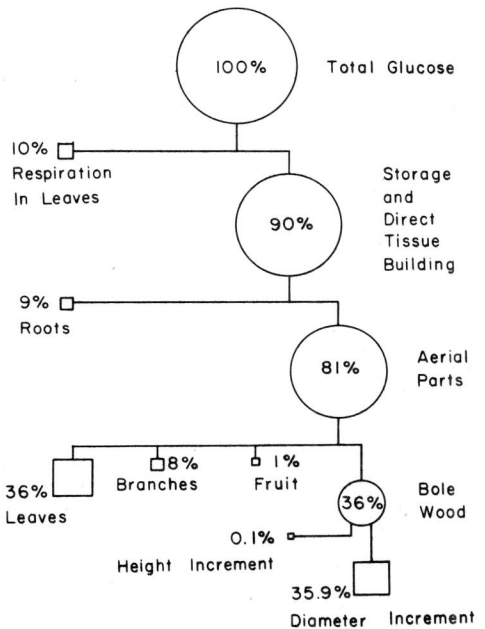

FIG. 4.11. Diagram showing the disposition throughout the tree of glucose made in the leaves. Ten percent of the glucose is consumed by respiration in the leaves and the remaining 90% goes to storage and tissue building. Of this 90%, 81% goes to tissues in aerial parts and 9% is used for growth of roots. The aerial distribution includes 36% which is used in growth of leaves, 8% for branches, 1% for fruit, and 36% for bole diameter growth. The squares represent ultimate dispositions and the circles intermediate allocations. (Modified and redrawn from Baker, 1950)

Figure 4.9 is actually another representation of the same phenomenon plotted as a function of tree age and expressed as respiration loss. It shows that up to two-thirds of the dry matter production is eventually consumed by respiration in the various organs of the tree. The remaining one-third represents the increase in dry matter in the above-ground and below-ground portions of the tree.

The processes of photosynthesis, respiration, and assimilation vary throughout the season and affect food accumulation and depletion throughout the tree. Respiration and assimilation occur continuously in living tissues but are most rapid during the most active period of growth. Photosynthesis can be active as long as there is light, moderate temperatures, sufficient moisture in the plant tissue, and green leaves. Ordinarily photosynthesis occurs at sufficiently high rates to replenish stored foods used by respiration and assimilation. At night, however, when respiration and assimilation are especially high, or during periods of low photosynthetic activity, more food may be consumed than is produced by photosynthesis and a net decrease in reserve foods may result. If the food consumption continues faster than food manufacture, the supplies become exhausted and the tree dies. In years of drought, the stored foods may become low and become the most limiting process to cambial activity, especially during the *grand period of growth* (see Chapter 2).

The reserve of food may be used at different times and by different plant tissues, depending upon various conditions in the tree. The crown in an arid-site tree is likely to utilize most of the photosynthates produced during the growing season, with the result that relatively small amounts may be translocated to the lower stem and roots during the period of active crown growth. As a result, cambial activity in the roots and in the lower stem may occur largely at the expense of stored foods accumulated during the prior dormant season. The low correlation between ring width at the stem base of arid-site trees and the climate of the growing season and the high correlation with the climate of the prior autumn, winter, and spring is consistent with such a dependency of the cambium on food reserves derived from photosynthesis in the prior autumn, winter, and spring.

As a tree ages, the efficiency of photosynthesis may decline, and a lack of food available from photosynthesis may become increasingly critical to growth in the lower stem and roots. At the same time, the circumference at the cambium continues to increase and the available foods must be distributed to a larger growing area. The ratio of roots to shoots may decline; less water may be absorbed, with water loss more often exceeding water supply and branches suffering repeated drying, resulting in narrower growth rings with more variable widths. As these conditions become more limiting, the tree becomes weaker, so that disease or a year of extreme climate is more likely to cause destruction and death of the tree.

In aging *Pinus longaeva* and *Pinus flexilis* the amount of roots relative to the amount of shoots referred to as the *root/shoot ratio*, is essentially constant. But as the stem grows, the cambial tissue becomes distributed around a larger and larger circumference until those portions farthest from food, hormone, and water sources die. In very old individuals, the portion of living bark may be reduced to a narrow strip which connects a small group of living branches with the few remaining living roots (Fig.

4.12, also see Figs. 1.12 and 3.18). The total living tissue remains relatively constant and the ratio of food-producing and food-consuming tissues are kept in balance by the gradual change from "full-bark" to "stripped-bark" trees. Sometimes the width of the strip will increase and decrease to some extent with changes occurring in the crown and roots; but in *Pinus longaeva* the average width of the annual rings in these "stripped-bark" trees may remain remarkably constant even though the arc of the strip may change (Ferguson, 1968; Fritts, 1969).

FIG. 4.12. *Pinus longaeva* may live to great ages. As the tree reaches old age, a portion of the cambium dies, leaving a strip of bark which connects the flagged top with the live roots. After the strip is formed the average ring width ceases to change with increasing tree age. (Courtesy of James Harsha and Valmore LaMarche)

Various growing regions of the plant, including root and stem tips, cambia, fruits, and seeds, compete for food and other growth substances. The success of the growing regions in competition seems to vary with their age and stage of development. The apical branch of a conifer seems to have competitive advantage over the side branches. Young leaves may compete for and receive substances that originate in older leaves. After the young leaves mature, they lose some of the competitive advantage and begin to lose substances to the areas of the plant that are still growing

(Rangenekar and Forward, 1973). Flowers, fruits, and seeds often obtain food at the expense of roots and other vegetative structures. Competition for food in *Fagus silvatica* between growing fruit and the cambia may cause narrow rings to form during years of especially abundant fruit set (Holmsgaard, 1962). Changes in the translocation of food substances and in the vigor of the crown can influence the availability of food and other growth substances to the cambium at the stem base and cause the climatic information recorded in the ring widths and cell structures to change in a systematic pattern throughout the life history of the tree (Chapter 2). For further discussion of this topic also see Chapter 5.

Other types of interactions occur as a result of environmental factors affecting production and expansion of the foliage, which in turn influence the capacity of the tree to photosynthesize at a given light level during subsequent years (Fritts *et al.*, 1965c; Fritts, 1969; Garrett and Zahner, 1973; LaMarche, 1974a). The lengths and possibly the numbers of needles in many conifers are affected by conditions during the growing season (Fig. 4.13). As long as the needles persist on the stem and are photosynthetically active, they can influence photosynthetic capacity and food accumulation, and hence can affect subsequent ring-width growth.

Fritts (1969) and LaMarche (1974a) report that needles are retained by *Pinus longaeva* and appear to be photosynthetically active for 15 to 30 years, so that an extreme climate of one year which affects the photosynthetic area and food production in that species can influence growth for the following 15 to 30 years. However, for arid-site trees this provides a type of stability to the photosynthetically active area, because one or two years of severe climate can affect only a small percentage of the total needle mass. LaMarche (1974a) observed that at higher elevations, however, when the spring and summer temperatures are low for a period of 10 to 15 years they can reduce the photosynthetic area substantially and cause a marked reduction in ring width. Furthermore, recovery may take 10 to 15 additional years of near-average spring and summer temperatures because the needles that were affected are held for such a long time on the trees.

Other conifers on stress sites show similar effects of climate on needle number and size, but they retain their needles for only a few years. One or two years of extreme climate can have a marked effect on the needle mass in these species, which results in a marked change in growth lagging one, two or three years behind the occurrence of the climate. Such differences in needle retention, as well as defoliation, changes in root mass, and fluctuations in fruit set, can result in variations in size and other characteristics of tree rings (see Chapter 6).

4. FOOD SYNTHESIS AND ASSIMILATION OF CELL CONSTITUENTS

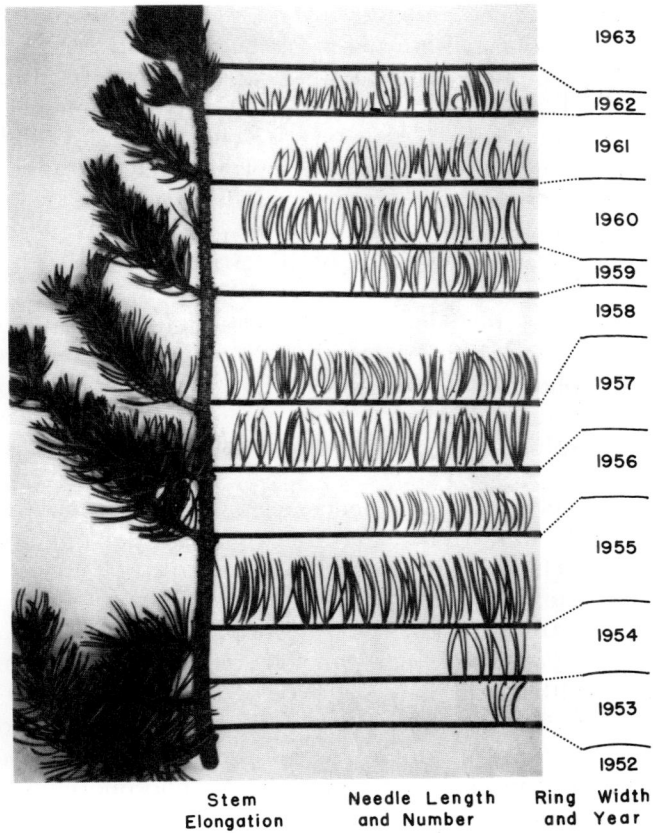

FIG. 4.13. Needle lengths and numbers, stem elongation, as well as ring widths of arid-site *Pinus edulis* may vary from one year to the next depending upon controlling factors of the environment. The needles which persist up to 10 years for this species and for longer intervals for other species can affect photosynthetic capacity, food accumulation, and the ring width. The needles shown have been stripped from the corresponding segment of the main stem and arranged to display the variation in numbers and length. The heavy horizontal lines mark the positions of the apical buds for each year and the distances between them correspond to the amount of shoot growth. The distance between the arcs drawn on the right is proportional to the area of the rings on the section cut from the stem base shown in the figure for 1952. (From Fritts *et al.*, 1965c)

IX. Essential Mineral Salts

The term *nutrient* used in a botanical context refers to mineral salts which are essential in small amounts for the normal growth and survival of the plant. As has been mentioned earlier, these nutrients should not be considered true foods because they are not significant sources of energy. Their actual functions within plants are varied. Nitrogen, sulfur, and phosphorus are components of amino acids and protein; nitrogen and magnesium are components of chlorophyll; and iron, while not a part of the chlorophyll molecule, is necessary for chlorophyll synthesis. Other elements, including potassium, calcium, boron, manganese, zinc, copper, molybdenum, and chlorine can also influence rates of important processes in trees. In addition, mineral salts influence water potentials within plant cells, and the balance of the various ions affects membrane permeability.

Soil minerals also may control the distribution of some tree species. Wright and Mooney (1965) report that *Pinus longaeva* is relatively more tolerant of low phosphorus levels in soils than some of its competitors and suggest that this tolerance may explain the greater abundance of *Pinus longaeva* on phosphorus-deficient dolomitic soils in the White Mountains of California.

The limiting effect of mineral deficiencies is most apparent when other environmental factors are not limiting. If a particular mineral such as phosphorus is deficient enough to influence growth processes, it will restrict the capacity of a tree to respond during years when climatic factors are favorable for growth. For example, Fritts (1969) noted that water deficits in *Pinus longaeva* growing on dolomitic soils account for the variations in width for rings which are below the average size, but when ample water is available the rings are all about the same width. It is conceivable that the uniformity in size of the larger *Pinus longaeva* rings could be due to nutrient deficiency (perhaps phosphorus) setting the upper limit to ring width.

Naturally occurring minerals are of limited importance in dendroclimatology, however, because they tend to be constant from year to year and therefore do not cause variations in ring widths. On the other hand, substances arising from environmental pollution or other disturbances may affect soil minerals and produce systemic changes in availability of minerals, which in turn may result in changes in growth (Jonsson and Sundberg, 1972). Certain elements reaching high concentration in polluted habitats can have direct toxic effects on the trees of the area in terms of cell structure and ring width (Vinš, 1965).

X. Growth-Regulating Substances

Plant *growth regulators*, or *hormones*, are organic compounds produced in small amounts which promote, inhibit, or qualitatively modify growth. The better known growth regulators include *auxins*, compounds which are responsible for cell enlargement; *antiauxins*, compounds that inhibit action of auxins; *kinins*, substances which promote cell division; *gibberellins*, compounds affecting stem elongation; and *inhibitors*, substances that reduce or stop growth and induce dormancy. For details on kinds of substances that act as growth regulators see Leopold (1964) and Giertych (1964).

If water, temperature, and food supplies are not limiting, tracheid diameter may be governed primarily by concentrations of certain hormones produced elsewhere in the tree. There is general agreement among plant physiologists that auxins are the hormones most important to the enlargement of cells, and that they are produced in the stem apices and are transported normally in a basipetal direction to the tissues in the stem (Giertych, 1964). Those factors limiting vegetative growth of the crown, therefore, can affect the auxins reaching the cambial regions and influence tracheid size (Larson, 1964). For example, the very rapid rate of the first flush of apical growth in spring produces a large quantity of auxin which migrates down the stem and stimulates production of large earlywood cells throughout the cambial regions in the stems and roots. As the season progresses, the rate of shoot elongation declines until shoot growth finally ceases. This causes auxin synthesis to decline, tracheid enlargement to be less rapid, and small, thick-walled latewood cells to be produced.

Larson (1962) proposes that auxin synthesis and transport in conifers account for the varying rates of growth and cell enlargement along the main axis of the stem. Therefore, both the width and structure of rings can be highly related to vigor of apical growth, auxin synthesis and transport, and the proximity of the cambium to the auxin source. This would help explain the presence of thin-walled tracheids making up low density latewood sometimes called *juvenile wood*, formed in the upper stems nearest the auxin source. With increasing distance from the stem apex, tracheid size decreases, latewood cells are smaller and thicker-walled, and there is an increasing difference in density between earlywood and latewood.

Digby and Wareing (1966b) note that the method of spread of cambial activity in spring among nonconiferous trees is different in ring-porous and diffuse-porous species. They show that in diffuse-porous trees (in which earlywood and latewood vessels are similar in size) growth promoters such as auxin are produced first by the swelling apical buds and later by tissues at the base of the tree as the gradient concentration of the growth regulator migrates down the stem. In ring-porous trees (those with large vessels in the earlywood), a gradient of auxin is not apparent in the stem at all. Rather, it is believed by some that an auxin precursor is present throughout the stem in early spring which is converted to auxin at all levels just before the buds swell. Since the auxin is in the same concentration throughout the stem, cambial activity in ring-porous species is initiated simultaneously throughout the stem.

The transition from the large cells of earlywood to the small cells of latewood in ring-porous species may be related to a decline in auxin level resulting from cessation of apical growth in the crown or simply to the completion of pore enlargement and the beginning of active division of the cambium (Phipps, 1967). However, Digby and Wareing (1966b) show that after cessation of shoot growth the mature leaves of ring-porous species continue to produce enough auxin to maintain continued cambial activity and the formation of the latewood.

Westing (1968) points out that various aspects of xylem differentiation, such as radial enlargement, longitudinal enlargement, wall thickening, and lignification are under separate, more or less independent, hormone controls. Wodzicki (1964) presents evidence that growth substances found in cortical tissues of *Larix decidua* influence the thickening of tracheid cells independent from the effect of auxins. Digby and Wareing (1966a) show that the relative levels of both auxin and gibberellin may be important in determining whether the cambium differentiates xylem or phloem.

Giertych (1964) states that some of these features of growth may be governed by a balance between growth promoters and inhibitors rather than by the concentration of any one growth-controlling substance. For example, the breaking of bud dormancy may be associated both with a decline in inhibitors and an increase of promoters, which in turn initiate cambial activity.

In the case of certain trees, such as *Quercus robur*, which have two phases of shoot growth during a season, one in April and May and the other in August separated by a rest period, the dormancy of summer is temporarily characterized by moderately high concentrations of growth inhibitors in the buds. This temporary dormancy can be broken by temperature alteration or by certain chemical treatments. The more permanent dormancy of autumn is accompanied by higher concentrations of inhibitors and is not broken easily until the following spring.

Larson (1962) points out that in *Pinus resinosa* false rings invariably result from second flushes of shoot growth which produce auxin and stimulate a change in the cambium from production of latewood to production of earlywood. The stimulus is of limited intensity and reaches the stem base only under exceptionally favorable conditions. Thus, the occurrence of false rings near the growing tips of the stem without false rings at the base of the stem (see Fig. 2.23) may be attributed to variations in auxin gradients within the tree (Fritts *et al.*, 1965a). Westing (1968) states that the production of compression wood in an inclined gymnosperm stem may be attributed to lateral redistribution or asymmetric production of a growth-regulating substance, a growth inhibitor, or an auxin-destroying enzyme.

Larson (1962) also proposes that differences in earlywood and latewood in trees with large crowns grown in the open and those with small crowns grown under more restricted conditions may be attributed to variations in the production of growth regulators resulting from differences in crown vigor. The absence of certain rings at the stem base of suppressed trees or of those trees growing in a very unfavorable climate may result from reduced activity in the shoot and an auxin stimulus which is too weak to initiate cambial activity by the time it reaches the stem base. Giertych (1964) reports that effects of day length on flowering and other phenomena may be observed by a change in concentrations of growth regulators before there is any visible change in growth.

Future studies will undoubtedly reveal many other ways environmental factors can affect the production or movement of growth regulators, which in turn affect growth. It is possible that the delayed effect of autumn and winter climate on growth in the following spring could result from an interaction between climate and the formation of some growth regulator. The anatomical changes in ring structure associated with its position along the stem, tree age, and crown vigor are undoubtedly related to production, movement, and distribution of growth regulators as well as the amounts of available foods.

Considerable differences exist in points of view regarding the relative importance of growth regulators, internal water stress, and food to cambial growth. All three factors are undoubtedly interrelated and important. Each may be of primary significance in certain situations, species, and sites, and all growth models should recognize that any one or a combination of these factors may become limiting under certain circumstances to ring-width growth.

XI. Physiological Preconditioning and Correlating Systems

The growth of a tree is, of course, directly influenced by its prevailing environment, but often there are many lingering effects on current growth from previous environmental conditions, as was suggested in Fig. 1.10. *Physiological preconditioning* is the induction of internal metabolic conditions which eventually influence the growth or biochemical reactions of a tree during a later stage in its life history. Such conditions may be variable and complex. For example, Lowry (1966) noted meteorological factors over a 27-month period that appeared to precondition *Pseudotsuga menziesii* and influence its cone crop. Climatic conditions were important during periods of activity such as the differentiation of buds, the formation of pollen, and fruit set.

Several examples of the effects of physiological preconditioning on ring growth have been described in earlier sections. Some of these are: (1) the accumulation and storage of food and other substances by arid-site conifers during the dormant period; (2) the production of root and shoot masses and variation in the root-shoot ratio; (3) the changes that occur with increasing tree age and size which can influence efficiency and rates of various processes; (4) the production of growth-controlling substances within the tree; and (5) the deactivation of chlorophyll and loss of photosynthetic capacity in conifers during the winter when temperatures are low. Additional examples of physiological preconditioning which possibly affect the formation of xylem rings are the influences of photoperiod, the induction and breaking of dormancy, the development of drought resistance and frost hardiness, and the presence of frost rings and fire scars resulting from injury to tissues of the tree.

A. Photoperiod

Changes in length of exposure to light (*photoperiod*) may influence both vegetative and reproductive phases of plant growth. Shoot growth, radial growth, breaking of dormancy, leaf abscission, frost resistance, seed germination, and flowering are processes known to be influenced by length of day. The growth of woody species so far investigated with respect to day length responds to short days by growth inhibition and to long days by growth promotion.

Photoperiod does not vary from year to year and therefore does not directly produce variable features in successively formed rings. The response to photoperiod can, however, be important as a genetically controlled factor that prevents unusually early bud break during a warm winter or the flushing of buds late in the season, which would produce young tissue susceptible to injury from frost. Although the genetic factors are constant for a given plant and cannot produce variable ring width from year to year, genetic factors do vary from plant to plant and are probably responsible for different photoperiodic responses among individual trees and among different populations or races, which result in differences in the date of bud break, in the duration and rate of cambial activity, and in the time of dormancy in autumn. Day length is most likely a significant factor in tree growth at high latitudes, also, where large differences in daylight occur seasonally and the length of the summer is short. As was discussed earlier, the very long summer days permit photosynthesis to proceed at high rates. For further discussion of photoperiod as a possible factor in tree growth see Downs (1962) and Kramer and Kozlowski (1960).

B. Dormancy

Trees do not grow continually, no matter how favorable the environment, but rather periods of active growth will alternate with periods of inactivity and dormancy. Tropical species in a continuously favorable environment may make several growth flushes during the year (Kramer and Kozlowski, 1960). In some temperate trees that at times produce multiple rings, growth can occur in two or three periods, separated by brief intervals of dormancy. Dormancy of this type is classified as *temporary*, as it lasts a few days or a few weeks. Growth may begin again spontaneously or it may be stimulated by a change in the environment. Dormancy which occurs for periods of many weeks or months is classified as *permanent*. In temperate regions, permanent dormancy usually cannot be broken until plants have been exposed to low temperatures for one or two months, normally during winter months. While temporary or permanent dormancy can be induced by environmental factors, it can be caused by internal conditions (Kramer and Kozlowski, 1960). For instance, the tree itself may bring about the conditions for dormancy by producing an inhibitor or by exhausting some substance essential for growth. The age of tissues also seems to affect the rate and degree of entrance into the period of rest (Levitt, 1956). For a more detailed description of these types of phenomena see Samish (1954).

Dormancy may be viewed as a type of "safety factor" which has evolved through the death of all individuals of a population that failed to become dormant and did not survive in a given climate. Dormancy is important for dendroclimatic consideration because it is a natural factor that modulates the response of growth to the environment at particular phenological stages of the tree. Climatic factors in the autumn may affect conditions which hasten or increase the depth of dormancy, while others may govern the rate and timing of the breaking of dormancy in the spring. When dormancy is broken, climate may directly affect the production of growth-regulating substances and influence the various growth processes, as well as affect net photosynthesis, accumulation of food, and water relations of the tree. During the dormant period when climate cannot affect growth directly, it can affect subsequent growth through its influence on soil moisture, photosynthesis, and food accumulation, or other processes that are still active in the plant. A variety of effects of this nature delayed for one or more years may be seen in many statistical results of tree ring–climatic relationships and are a part of the models developed in Chapters 5, 7, and 9.

C. Drought Resistance

Different species can vary in their ability to withstand drought, and individuals of the same species can also vary depending upon conditions which have preconditioned the plant. There are four basic strategies or adaptations which enable plants to survive conditions of drought: (1) escaping the drought; (2) evading the drought; (3) enduring the drought; and (4) resisting the drought (Parker, 1968). Examples of drought-escaping plants are the annuals which survive the dry period as dormant seeds. Drought evaders are able to economize on water use by structural or other modifications such as thick cutin or dense hairs on the leaf which allow them to evade internal water stress. Drought-enduring plants, rather than conserving water by some permanent feature, may lose their leaves, modify their root structure, or change in some other way that enables them to survive during periods of drought. On the other hand drought resistors possess certain physiological adaptations that allow their tissues to continue life processes under the stress of drought. Drought resistance in trees involves primarily the latter two types.

Most workers view drought resistance as a characteristic largely under genetic control and hence not a phenomenon that can change from one year to the next. For example, *Juniperus virginiana* and *J. communis* are apparently more resistant to desiccation than many common species of *Pinus, Picea, Abies,* and certain broad-leaved species of trees (Parker, 1968). Hence, these two species of *Juniperus* can survive at lower elevations and on drier sites than the other less well-adapted plants.

Certain conditions of particular plant tissues may cause physiological changes or structural modifications which can affect a tree's ability to grow or survive through periods of drought or water stress (Parker, 1968). The extent and conditions of root systems can modify water absorption or water loss. For example, trees in the sun which undergo high photosynthesis can develop larger root systems than those in the shade. The roots of the Mediterranean tree *Pinus halepensis* cease growth in times of drought, the outer root cortex turns brown and collapses, and a heavy suberin layer develops which retards dehydration of the roots and enables the species to better withstand drought (Parker, 1968).

Leaves developing in the sun or those in the upper portion of a tree crown may withstand more desiccation than those in the shade or those formed in the less vigorously growing areas of the crown, because the former may have a thicker cutin on their leaves, the osmotic potentials in the plant may be lower because of higher concentration of sugar and other dissolved substances in the cells, or their stomates may be able to exert a stronger control over water loss. Similarly, vigorously growing trees in open habitats can often survive more severe drought than those in the shade of a dense forest canopy, or a period of unusually sunny and cool weather may enable a tree to develop physiological conditions, enabling them to better withstand conditions of water stress.

In general, if water stress increases slowly in a drought-resistant plant, the physiological conditions in its tissues are more likely to keep pace with and compensate for the water stress. However, if the onset of water stress is sudden, a reaction may occur during which time water may be withdrawn from the plant tissue and adverse effects on various processes may be evident. With the passing of time a period of restitution may follow in which the cells become better able to develop the necessary water potentials so that water diffuses back into some of the desiccated tissues of the plant. Very often drought appears to be more damaging if it is preceded by a long wet period of vigorous growth than if it is preceded by conditions of more reduced growth and occasional periods of water stress.

The ability of *Pinus ponderosa* and *P. edulis* to continue growth during the dry period before the summer rains as described in Chapter 2 may represent a type of drought resistance which could in part result from some preconditioning of the plant. Some of the statistical correlation of temperature and precipitation in one year or season with the growth of the following spring could well arise from preconditioning leading to improved drought resistance of the tissues within the plant.

D. *Frost Hardiness*

If temperatures remain unusually warm late in the autumn, an abrupt onset of cold weather may injure many temperate trees; but if temperatures decline in a more gradual manner, hardiness or resistance to frost injury develops. A period of unseasonably warm weather in winter can decrease frost hardiness, but a return to colder temperatures will usually reestablish it. In the early spring if the weather turns unseasonably warm, some trees may lose their frost hardiness, thereby becoming more susceptible to freezing injury. Plants that had survived the most extreme freeze during midwinter may be killed or injured by very slight freezing in the spring after dormancy has been broken (Levitt, 1956).

Low soil moisture, high levels of light, and a scarcity of certain minerals may interact with low temperature to increase frost hardiness (Levitt, 1956). For example, *Pinus longaeva* at its upper tree line in the White Mountains of California does not form the contorted and stunted *krummholz* form, as it does in climates of the Mt. Washington area, Nevada, and the Rocky Mountains where summer drought is lacking. LaMarche and Mooney (1972) believe that drought conditions, which always occur during the growing season in the White Mountains but not in the krummholz sites, make the trees frost hardy so that winter conditions do not dry the foliage and kill the branches. This preconditioning by drought may be one reason why the dry-site trees maintain full stature right up to their high elevation limits. Because of the effects of frost hardiness (and possibly drought resistance), dry and cold conditions late in autumn or early in the spring may have a favorable effect upon subsequent growth, because it assures that any subsequent low temperatures or dry conditions cannot damage the foliage of the tree. Thus, if they acted in the above manner, both low moisture and low temperature during these periods, prior to winter dormancy would be expected to be correlated with wide rings.

If frost hardiness or drought resistance is lacking and the foliage is killed, large variations in the remaining living tissue may occur from one year to the next. Small differences in relief and wind exposure may control the amounts of living tissue that remain at the end of winter, and there is often little or no synchronization in ring widths or crossdating from tree to tree or from branch to branch in the same tree. The coupling of growth with climatic factors becomes insufficient to provide meaningful analysis of past climate from the contorted upper tree line trees. However, by analyzing rings of these contorted krummholz trees, interesting studies may be made on age and movement by vegetative propagation. The interested reader can refer to Levitt (1956) and Parker (1968) for a detailed discussion of frost and drought resistance in woody plants.

E. Frost Rings

Freezing is harmless to tissues that are frost hardy, because crystallization of water is restricted to the intercellular spaces. When freezing occurs very rapidly or the tissues are not hardened, ice may form in the cytoplasm or vacuole of living cells (Meyer *et al.*, 1973) or ice may crush young growing cells. The active cambium in branches is especially susceptible to injury when temperatures drop below freezing. Such injury produces distorted xylem tissue that is referred to as a *frost ring* (Fig. 4.14).

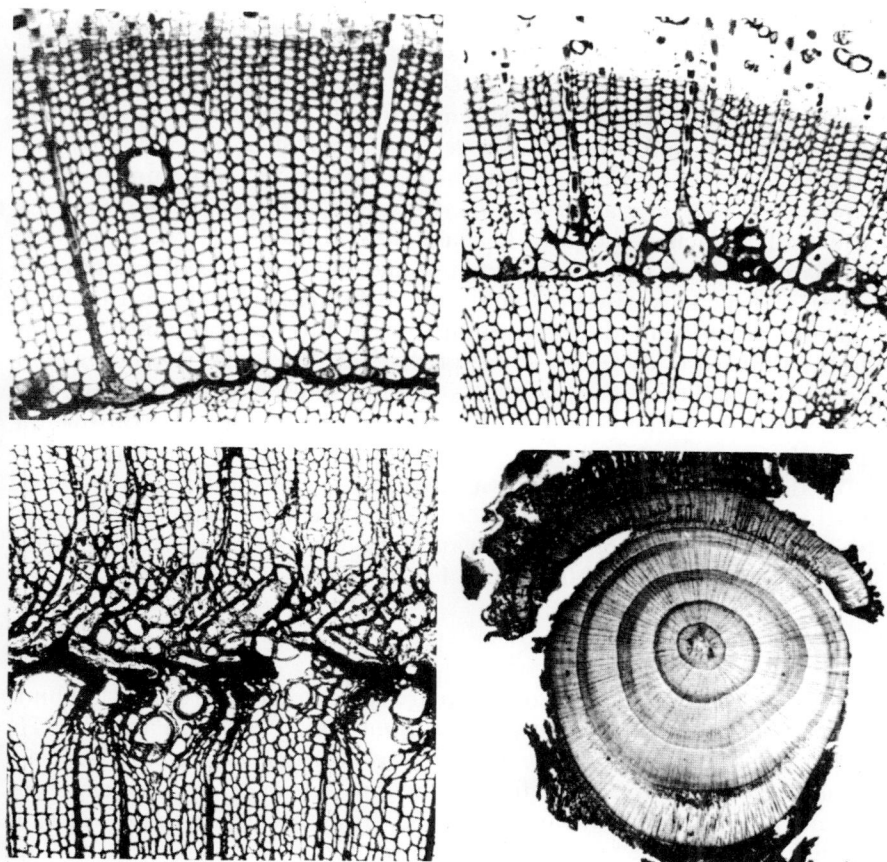

FIG. 4.14. Transverse sections of tree stems showing frost damage. The distinct dark line within the ring is a layer of crushed cells. Deformed and incompletely lignified tracheids often lie inside the crushed cells, while parenchyma and abnormal tracheids can lie outside. These abnormal cells form lines of weakness which may cause the wood to break, and ruptures may occur in the stem (see lower right photo). The upper sections are *Picea glauca* (from Glerum and Farrar, 1966), and the lower are *Pseudotsuga menziesii* (left) and *Larix decidua* (right). (From Day and Peace, 1934, courtesy of Oxford University Press)

Mature tracheids become crumpled and the cells that are undifferentiated at the time of the freeze become crushed (Glock, 1951). Rays are commonly offset at the line of injury. In the case of a severe frost, the cambium and therefore the tree may be killed outright. For injured but surviving trees, the first regenerated cells are enlarged and deformed, but successively formed cells gradually assume normal size and form. The effects of a severe frost may also seriously defoliate the tree and weaken stems. Glock (1951) reports that mechanical breakage and mortality of branches and trees may occur for several years following a severe frost.

Glock (1951) and LaMarche (1970) associate an injury along the inner portion of a ring with frosts occurring in spring after initiation of cambial activity. Frost injury near the outer boundary of the ring is associated with a freeze late in the growing season before the cambium becomes dormant. Frost rings are less frequently found in the larger stems than in the small branches, probably because of the insulation of the thick bark, the late initiation of cambial activity, and the early growth cessation in portions which are distant from the growing stem apex. Thus, frost rings may be less frequent in older portions of stem, but this does not necessarily imply that frosts were less severe at that height on the tree or in more recent years. Thus, it follows that one must sample trees of comparable size and age to compare the absence or presence of frosts during the period of growth. Glock successfully uses frost rings to label and study growth layers in species such as *Cupressus arizonica* that form several distinct rings in a year. Frost rings can be important markers which aid in the dating of trees of high elevation or high latitude environments (LaMarche, 1970; LaMarche and Fritts, 1971b).

F. Fire Scars

High temperatures resulting from fire can defoliate a tree and kill buds and living tissue exposed to the heat (Jemison, 1944). J. Parker (1971) reports that trees are more resistant to heating in winter and least resistant in summer as indicated by the rate of respiration after exposure to one to ten minutes' heating at 57°C. *Juniperus virginiana* possesses an ability to withstand both extreme heating and low moisture levels. The thick bark of species such as *Pinus ponderosa* often insulates the large trees from injurious effects of small ground fires. Fig. 4.15 is an especially interesting example of a fire scar in *Pinus sylvestris*. A fire apparently killed all of the cambium except for two small areas, where growth rate accelerated after injury and the new tissue expanded around the stem. Eighteen years after the fire, the bark had fully encircled the original scarred stem.

Fire injury may cause a marked reduction in the total growth of a stem if a substantial part of the foliage is destroyed, but in the area of wounding, the ring may be anomalously large. Fire may destroy competition in a stand of trees, remove the shade, release minerals, and make more soil area available to the roots. Thus, after a forest fire, rings of some trees become wider, the ring-width variations are virtually unrelated to climate, and there is often very little year-to-year width

4. FOOD SYNTHESIS AND ASSIMILATION OF CELL CONSTITUENTS

variability common to all trees of a stand. What ring-width variability exists is mostly due to local site factors, which often vary slowly through time so that there are long periods of several years with enhanced or sometimes reduced growth. As time passes, the rings become narrower (or wider) and the year-to-year variability shared by all trees increases as climatic factors once again become increasingly limiting to ring width.

FIG. 4.15. A cross section of a stem of *Pinus sylvestris* which was severely damaged by fire when the tree was about 25 years old. The cambium remained alive at two locations. Rapid growth occurred after the fire, and in 18 years the new growth encircled and buried the scarred stem. (Specimen courtesy of Jon Pilcher, photo by James Harsha)

XII. Changes in the Physiological Seasons with Varying Elevation of the Tree Sites

The physiological processes affecting tree rings and their most important climatic controls have been described in the preceding pages. It can be seen from the discussion that there are certain periods of time within the year when climatic factors are generally favorable to a process, other times when factors are limiting, and still other times when the process may be essentially inactive. These various periods are usually synchronous with the climatic seasons. For example, during winter the growth processes may be dormant and physiological activity of the tree may be low for long periods of time. Spring and autumn are transition periods during which temperatures and moisture may change and become less limiting to certain processes and more limiting to others. In summer high temperature, and water stress may become limiting when temperatures exceed the optimum for certain processes or when there is insufficient soil moisture. The varying periods of physiological activity throughout the year may be described as the *physiological seasons*. The season may vary depending upon the processes, the site, and the species of plant. Several generalizations are apparent, however, when mean monthly temperature regimes are examined for sites of various elevations and certain temperature or moisture thresholds are identified.

Figure 4.16 is a plot of several thresholds for temperature and water stress as they vary with elevation throughout the year. Mean climatic data were obtained for various western North American sites between latitudes 34° to 38°N. The elevations of climatic stations ranged from 5,500 ft (1677 m), which approximates the lower forest border, to 10,400 ft (3171 m), which approximates the upper elevational tree limits. The monthly temperature data for each record were scanned and the mean data for five temperature thresholds were estimated as follows: The monthly minimum, mean, and maximum temperatures were each plotted as occurring on the 15th day of each month. The seasonal regime for each of the three means was obtained by drawing lines through the plotted points. The plots of temperature were examined and whenever the lines intersected any of the chosen thresholds, they were entered at the appropriate elevation in Fig. 4.16. Lines were then drawn through these points for dates when (1) the mean maximum temperatures intersected 0°C, (2) the mean temperature intersected 0°C, (3) the mean minimum temperature intersected 0°C, (4) the mean maximum temperatures intersected 25°C, and (5) the mean maximum temperatures intersected 30°C.

These particular temperature thresholds are in part arbitrary, for they do not represent the results of any particular physiological study. The rationale for their use is as follows:

(1) When maximum air temperatures remain below 0°C it can be assumed that the trees in general are physiologically inactive. Chlorophyll in coniferous trees probably has been deactivated and the only processes of importance are respiration and transpiration, which may occur for short intervals of time when temperatures rise

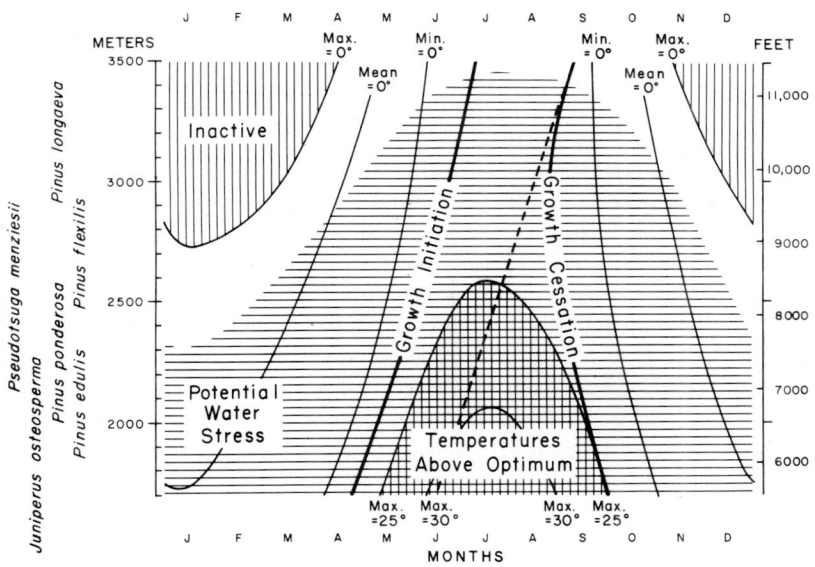

FIG. 4.16. Diagrammatic representation of elevations and approximate dates when maximum, mean, and minimum air temperatures are at 0°C; when maximum temperatures are at 25°C and 30°C; and when the growing period begins and ends. The elevations and times when water stress is most likely to be limiting are shaded with horizontal lines. The vertical lines indicate the photosynthetically inactive period when maximum temperatures are generally below 0°C. The area shaded with both horizontal and vertical lines indicates the dates and elevations when maximum temperatures are on the average above 25°C. The heavy dashed line designates the earliest date that growth would be expected to cease because of severe water stress. The approximate elevational ranges of certain trees are indicated on the left margin. Data are from latitudes 34°-38°N in western North America and are based in part upon LaMarche and Mooney (1972), Erdman et al. (1969), Fritts (1969), Fritts et al. (1965c), and U.S. National Weather Service. (Drawing by M. Huggins).

above freezing. The plot in Fig. 4.16 shows that temperatures are below this threshold for substantial periods of time only at high elevation sites.

(2) When mean maximum temperatures are above 0°C but the mean temperatures remain below this value, certain processes may occur during the daytime period. Chlorophyll regeneration may begin during warm periods when night temperatures are not extremely low, and net photosynthesis can occur during the warmest intervals. Temperatures are likely to be directly correlated with the accumulation of photosynthates, but they are not high enough to cause serious water stress except in habitats that are exceptionally arid.

(3) As mean air temperatures rise above 0°C but minimum temperatures are below that value, daytime temperatures are frequently above freezing and processes such as net photosynthesis can occur at substantial rates. Nighttime temperatures are low and there is little consumption of food by nighttime respiration. If moisture is adequate, periods with higher than normal temperatures are associated with high productivity and accumulation of foods. Except for extremely high elevations, internal water stress may occur at midday if soil moisture supply is inadequate. It is possible that this increasing water stress can counteract the favorable effects of the moderate temperatures on processes such as net photosynthesis.

(4) When mean minimum temperatures remain above 0°C but the maximum temperatures are lower than 25°C, the physiological activity is highest and many of the growth processes may be initiated. Temperature is rarely limiting except for unusually cold or warm days. Moisture reserves can become low in dry years, however, and water stress can become limiting.

(5) At low elevations maximum temperatures during the summer season may rise above 25°C and sometimes above 30°C. These temperatures are above the optimum for many processes, especially net photosynthesis, and during years of low precipitation the inhibiting effects of heightened temperatures are accentuated. As the temperatures begin to decline in late summer, the thresholds are passed in reverse order, and the processes reverse in like manner. However, growth may cease well before temperatures become limiting to the process. If moisture is available, net photosynthesis can occur rapidly and foods can accumulate. As temperatures approach 0°C most growth processes cease, and winter hardening may occur while net photosynthesis remains active. When maximum temperatures fall below 0°C chlorophyll may be deactivated,

photosynthesis then ceases, and the winter inactive period commences. Different species, of course, vary in their tolerance to various temperatures and degrees of water stress, and the microenvironmental conditions can accentuate or counteract the particular climatic effects.

On the left margin of Fig. 4.16, each species name spans its approximate elevational range. When these ranges are compared to the diagram of physiological seasons, several features become evident. The physiologically inactive period is long in duration throughout the range of *Pinus longaeva*, but relatively short for other species, except at high elevations and cold exposures. For most arid-site species the average maximum temperatures in winter are above freezing.

Mean temperatures in winter reach 0°C levels at all elevations considered in the diagram. At the high elevational tree line, mean temperatures average freezing or below for more than half the year, while for the lower forest border temperatures as low as 0°C may last for no more than a few weeks to a few months. During April minimum temperatures at low elevations can rise above 0°C and growth may commence, but at tree line this threshold is not attained until June or early July. The period of above-optimum temperatures occurs during midsummer only for species at low elevations, where the effects of high temperatures are likely to be most severe if water is unavailable and the plants are subjected to water stress. In winter water stress is generally restricted to sites below 7500 ft (2287 m). In summer water stress can occur at all elevations, although it is rare at the upper elevational tree line.

It can also be noted from the diagram that the growing season at low elevations is potentially longer than at high ones. However, early cessation due to water stress and elevated plant temperatures can occur there as early as June, and is less likely to occur with increasing elevation. Above 9,800 ft (2988 m) the length of the growing season does not vary substantially, although both the dates of growth initiation and cessation occur later with increasing elevation.

Figure 4.16 represents changes associated only with elevation, but modifying factors of the site such as exposure, slope, and soil characteristics may further modify the microclimate and affect the timing of the physiological seasons. The marked differences in the elevational ranges of the listed species and the wide variations in the physiological seasons suggest that there will be a variety of ways that climatic factors can affect the plant processes. Since the processes interact to negate or accentuate each other's effect, it is reasonable to expect that the net effect on growth will vary not only among the different species but also among the various macro- and microenvironments of the sites. The following chapter deals with the important factors of the macroclimate and the environment as they may influence growth-controlling processes which affect ring width; and it describes some models which express the most important growth–climatic relationships.

Chapter 5

The Climate–Growth System

I	Introduction	207
II	The Energy and Water Balances	208
III	Site Factors Which can Modify the Energy Balance . . .	213
	A. Topography	214
	B. Substrate	216
	C. Elevation and orographic factors	218
IV	Biotic and Other Nonclimatic Factors: Dendrochronological Examples	219
V	Modeling Relationships in the Ring-width and Climatic System .	223
	A. Rationale	224
	B. Approach	224
	C. Model constraints	226
VI	A Model for Factors Affecting Cambial Activity and hence Ring Width .	226
	A. Temperature	227
	B. Water	229
	C. Growth-controlling substances	230
	D. Amounts of building materials	230
VII	Modeling the Effects of Temperature and Precipitation on Ring Width .	231
	A. Model part A	232
	B. Model part B	233
	C. Model part C	235
VIII	The Concept of the Climatic "Window"	238
IX	The Concept of the Response Function	240
X	Suitability and Limitations of the Growth Model	242

I. Introduction

It was pointed out in Chapter 1 that an important and basic step in dendroclimatological research is modeling the climate–growth system. The biological portions of the system have been described in Chapters 2, 3, and 4. It is now necessary to consider how the biological system is linked to climate, how the variables of the site may affect the linkages, and how the essential structure and components of the system may be modeled to assist in the identification of the important dendroclimatological components. As mentioned in Chapter 1 such modeling can be accomplished by means of word diagrams, logical statements, mathematical equations, or other analytical schemes which portray the important linkages between the initial causal phenomena, which in this case include variables of climate and site, and the resultant end product, which is ring width.

This chapter begins with a description of the two major components of the climatic portion of the system, the energy and water balances which link the macroclimate to the operational environment of the plant. Consideration is given to site factors such as topography, substrate, and elevation, as well as to the biotic and other factors which can modify or nullify certain components of these energy and water balances and thereby alter, complicate, or obscure the tree growth–climate relationship.

The discussion of the climate–tree growth system as a whole begins with a model of those factors or conditions that can become operationally limiting to cambial growth. This rather simple word diagram is then expanded and generalized to a more comprehensive one including a variety of processes occurring throughout the tree, as well as those occurring in the cambial region, which are in turn linked with two physical factors of the macroclimate, temperatures and precipitation. The possibility for both direct and inversely correlated relationships between growth and these two physical factors is accounted for by the model. It also allows for those variations that may be attributed to changes in the season, differences in the microsites, and differences among tree species.

II. The Energy and Water Balances

The amount of energy and water that a plant utilizes depends upon the amounts that are available to it and the portions which are absorbed and lost. Availability of both energy and water is very much dependent upon the conditions of both the atmosphere and the site in which the plant is growing, and the relative importance of these conditions can be best portrayed by mathematical equations. The inflow, outflow, sources, and sinks, and transformations of energy are referred to collectively as the *energy balance*, while similar components for water are referred to as the *water balance*. All energy and water balance equations are based upon conservation of energy and matter, which is to say that any change occurring within a system will affect that system's overall balance and be associated with a corresponding addition or loss of energy or water. The energy balance for a column of atmosphere and a soil is described by Sellers (1965) as

$$G = (Q+q)(1-a) + I\downarrow - I\uparrow - H - LE + F_i - F_o \qquad (5.1)$$

5. THE CLIMATE–GROWTH SYSTEM

where G is the net transfer or flux of heat (energy) into or out of the system; Q is the direct solar radiation incident on the exposed surface of the soil; q is the diffuse solar radiation reflected, or scattered from atmospheric dust and clouds incident on the same surface; a is the *albedo* (reflectivity) of the earth's surface; $I\downarrow$ is the absorption of longwave or infrared radiation from the atmosphere into the ground at the surface; $I\uparrow$ is the loss of longwave radiation from the soil surface to the atmosphere; H is the *sensible heat* exchange between the soil and air; LE is the heat loss by evaporation where L is the *latent heat* of vaporization (cooling by evaporation of 590 cal/g) and E the amount of moisture evaporated; F_i is the horizontal transfer of heat from surrounding soil; F_o is the horizontal transfer of heat to the surrounding soil.

The first five items govern the flux or transfer of incoming radiant energy that is delivered to a given soil surface. The next three items, $I\uparrow$, H, LE, govern the dissipation of this energy back to the atmosphere and space after it has been absorbed, and the last two items deal with small amounts of energy that are transferred laterally into or out of the column of soil.

This energy balance equation is an approximate one, but in general it applies over any time period from seconds to many years where the relative importance of items can vary as a function of the weather, the latitudinal and elevational location of the site, its exposure and slope, and other characteristics such as soil color, water content, and the amount of plant cover. Components other than those included in the equation are usually small and can be neglected (Sellers, 1965).

The water balance for a column of soil can be portrayed in a similar fashion,

$$g = r + D + f_i - E - f_o \tag{5.2}$$

where g is the net rate of change in moisture content in a column of soil; r is the moisture added by precipitation; D is the moisture added by condensation of dew; f_i is the horizontal flux of moisture into the column of soil; E is the loss of moisture from the soil by *evapotranspiration*; f_o is the horizontal and downward flux of moisture out of the column of soil.

Usually dew formation is negligible (Sellers, 1965). The water balance is tied to the energy balance through the term E, because large amounts of energy are dissipated by evaporation from the soil and plants.

A soil moisture accounting system, such as the one developed by Thornthwaite and Mather (1955, 1957) utilizes daily precipitation, mean daily air temperature, and potential radiation as the moisture and energy input, and considers moisture storage along with the depletion of moisture from evaporation (E) and run off (f_o). Thornthwaite and

Mather's accounting system was used to compute total available and lower-layer available soil moisture in Fig. 2.21 and evapotranspiration deficits shown in Fig. 2.24. Such accounting systems are thought to be useful in that they convert weather data into new variables that hopefully simulate growth-influencing environmental conditions (Zahner and Stage, 1966). Palmer (1965) uses a water balance approach to convert temperature and precipitation data to an index of drought. This index is relatively well correlated with variations in ring width (Julian and Fritts, 1968) and is a climatically derived factor that may be calibrated with and reconstructed from ring-width indices.

Large discrepancies arise, however, between actual and calculated moisture conditions for arid sites. These discrepancies have two possible causes: First, the use of mean temperature may not be appropriate, for it does not take into account the extreme diurnal range of temperatures that occurs on arid sites, nor does it take into account the radiational heat load as it affects the daytime energy balance of the plants. Second, the run off from the high-intensity storms that occur during summer is not adequately considered by the water-accounting procedure.

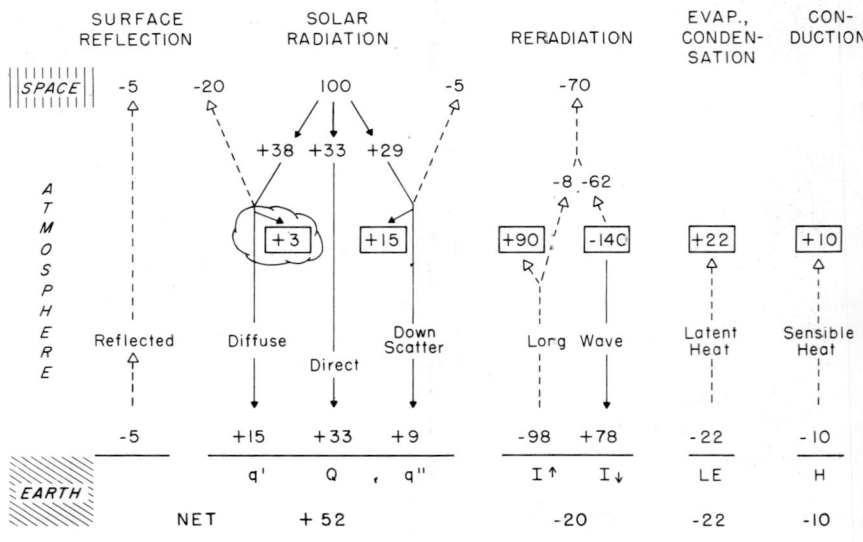

FIG. 5.1. The mean annual energy balance of the earth–atmosphere system for the Northern Hemisphere. The 100 units are equivalent to 0·485 cal/cm^2/minute, which is based on a solar constant of 1·946 cal/cm^2/min. Downward fluxes are indicated by solid lines and arrows, outward fluxes by dashed lines and open arrows. Symbols refer to Equation 5.1. (After Gates, 1962, as modified by Sellers, personal communication) (Drawing by M. Huggins)

Gates (1967) discusses the water balance equation and describes an aerodynamic method which he believes is more suitable for accurate estimation of the water balance. The method has not been applied to dendroclimatic work because insufficient data are available for a statistical comparison with annual ring widths.

Figure 5.1 shows the components of mean annual energy balance starting with 100 units per minute of incoming solar radiation at the top of the atmosphere. An average of 38 units are affected by clouds, and of these 38 units, clouds reflect 20 units back into space, absorb 3 units, and transmit 15 units to the ground surface. Other atmospheric substances deal with 29 units, reflecting 5 back to space, absorbing 15 in the atmosphere, and scattering 9 downward. Direct solar radiation accounts for 33 of the 100 units which are added to the diffuse radiation and down scatter, making a total of 57 units reaching the earth's surface, 5 of which are reflected back to space.

The surface of the earth radiates heat very much like a black body. The other columns in Fig. 5.1 indicate that 98 units are reradiated outward from the ground surface, 22 units are transferred as latent heat by evaporation and condensation, and 10 units as sensible heat. The total amount of energy transferred from all sources to the atmosphere is 140, as shown by summing the positive numbers inside the boxes. A total of 78 units of these 140 are reradiated back to the earth while 62 units, plus the 8 units reradiated directly from the earth are lost to space. This energy is the longwave infrared radiation included in the term, I, in Equation 5.1, and it occurs day or night, summer or winter, sunny or overcast.

The condition of the atmosphere is always changing and the components of the energy balance may vary somewhat in response to yearly or decadal changes as well as shorter-term fluctuations. For example, the net rate of exchange in heat, G (Equation 5.1), may be altered by changes in the energy received from the sun, Q, by changes in atmospheric carbon dioxide, dust, or pollutants which intercept some of the energy. Snow cover may alter the albedo, a, causing a significant portion of the incoming radiation to be reflected back to space, and it may reduce the sensible heat transfer between the atmosphere and the ground. Oceans may absorb large quantities of heat during the periods of warming and may dissipate the energy during periods of cooling, so that they contribute to long-term variations in climate.

Such long-term variations in the energy balance can result in changes in weather and climate that produce correlated variations in ring widths for certain trees and sites. In such cases the appropriate variations in the earth's energy and water balance can be calibrated with and inferred from past variations in the ring widths. The reconstruction of sea surface temperatures (Douglas, 1973) and of the levels of Lake Athabasca (Stockton and Fritts, 1973) are good examples of reconstructing elements of the energy and water balance by tree-ring analysis.

Every living organism is exposed to and lives in this stream of energy expressed by the energy balance equation. The temperature of an organism is a function of how rapidly it absorbs and dissipates the energy it receives, as well as any additional energy resulting from metabolism. Its temperature will fluctuate as the hourly, daily, seasonal, and long-term variations in the energy balance cause changes in the organism's operational environment.

As for all organisms, plants are coupled to the energy and water balances depending upon certain properties and characteristics they may possess (Gates, 1968a, b). As mentioned in Chapter 1, sometimes the coupling is tight and the plant absorbs and dissipates energy or moisture readily, or the coupling is loose so that plant temperature and moisture may lag substantially behind those of the surrounding environment. The coupling factor related to the energy balance is referred to as *absorptivity* and may depend upon the composition of the plant surface, spectral quality and geometry of radiation on the plant surface, and perhaps time (Gates, 1968a, b).

The plant is coupled to air temperature across a boundary layer of adhering air surrounding the organism, and variations in the depth of this boundary layer may influence the rate of energy exchange between the organism and the atmosphere. The depth of the boundary layer depends upon wind speed, the temperature gradient between the plant's surface and the air, and the size, shape, and orientation of the plant's surface.

The boundary layer surrounding the plant along with the epidermis also offers resistance to transpiration, which is the most important coupling factor for exchange of moisture between a plant surface and the atmosphere. The details of transpiration as it relates to the water balance have already been described in Chapter 3.

Gates (1962) expresses the energy balance for plants in a form similar to the energy balance at the soil surface in Equation 5.1:

$$G_o = Q_s - (a+t) Q_s + \varepsilon R - LE - H_g - H_a \pm s - P + M \quad (5.3)$$

where G_o is the net transfer or flux of heat (energy) between the atmosphere and the plant; Q_s is the solar radiation on the plant surface; a is the reflectivity or albedo of the plant; t is the *transmissivity*, or the ability to transmit the energy through the plant; ε is the *emissivity* of the plant, its power to emit the radiation from the surface which often is less than that of a black body; R is the thermal longwave radiation flux from the ground, atmosphere, and objects surrounding the plant; LE is the heat lost by evaporation from the plant; H_g is the transfer of sensible heat by conduction from the plant to the ground; H_a is the transfer of sensible heat by conduction and convection from the plant to the air; s is a term for storage of energy by heating or cooling of the plant mass over short periods of time; P is the energy used by the plant for photosynthesis; M is the heat released from the plant by respiration.

It can be seen from the above equation and prior discussion of the energy balance that the environmental factors which are most relevant to the energy transfer process of the aerial parts of a plant are sunlight and skylight; thermal radiation from the ground, atmosphere, and nearby objects; air temperature; wind; availability of water for transpiration; and water vapor in the air. If the energy absorbed is greater than or less than energy lost, plant temperatures will rise above or decline below the temperatures of the ambient air.

Evaporative cooling by transpiration is usually sufficiently rapid for well-watered plants to maintain the plant temperatures near or below air temperatures if the absorbed energy does not greatly exceed the energy dissipated by convection and reradiation. However, under high heat loads during the day, especially when water stress occurs, the energy absorbed by exposed leaves usually exceeds the energy that is dissipated, and aerial portions of plants can rise to temperatures in excess of 10° to 20°C above those of the surrounding air (see Fig. 3.6). Wind increases convection and helps to maintain plant temperatures near those of the ambient air, but wind may have detrimental effects in that it removes the vapor boundary layer and increases the rate of water loss to the surrounding atmosphere (see Fig. 3.7).

The brief description of the energy and water balance as outlined here is included to call attention to the many factors other than air temperature and precipitation which affect the actual temperature and water status of a plant. Unfortunately, detailed data are often unavailable for most components of the energy budget for a sufficient number of years to be calibrated with variations in tree rings. However, correct interpretation of statistical correlation of ring width with mean monthly temperature and precipitation involving trees at various elevations, on various exposures, and in various climatic regions requires an understanding of the differences in microclimates that can result from differences in the energy and water balances.

III. Site Factors which can Modify the Energy Balance

Many physical phenomena are not a specific part of the operational environments because they do not directly impinge upon the organisms' physiological processes. However, such phenomena may have indirect effects on plant processes because they play an important role in the development of a particular site. For example, the topography and soil are the result of geological, physical, and biological processes operating through past geological ages creating the land mass supporting the site. The plant community on a site is a result of past plant migrations, and the establishment, growth, reproduction, and death of individuals on the site. The death of a tree frequently results in more light and moisture available to neighboring trees while the growth of a neighbor may increase competition for water and light. These factors are important to the operational environment in that they create features that modify the energy and water balances.

A. Topography

Topography is one of the most important factors of the site which affects the energy and water balance by controlling the incident

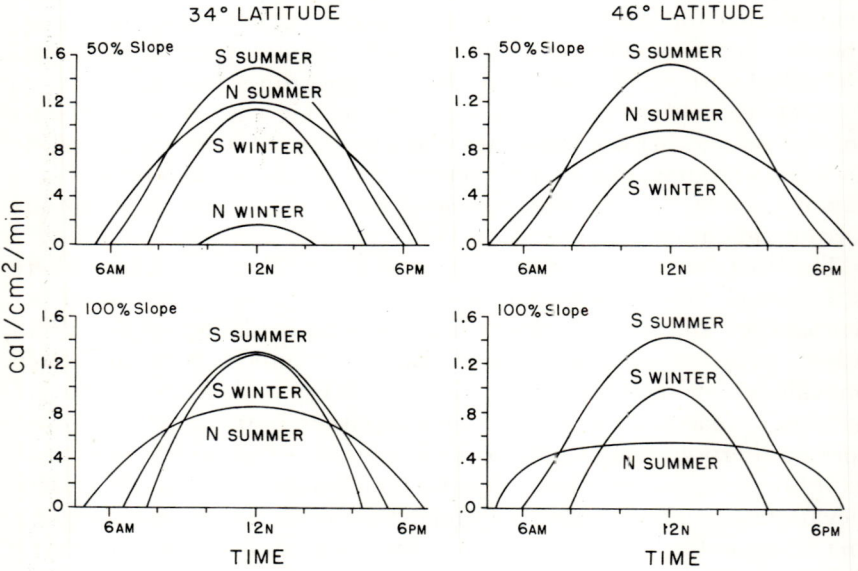

FIG. 5.2. The theoretical regime of solar radiation reaching a site on a clear day for 2 slopes, 2 latitudes, 2 seasons, and 2 exposures (N = north-facing and S = south-facing) Calculations are for December 3 or January 10 (winter) and for June 1 or July 12 (summer). Sun declination from the Equator is assumed to be 22; solar constant, 1·946 cal/cm²/min; and transmission coefficient, 0·8, to account for atmospheric thickness at 22° latitudes. (Data from Fons et al., 1960) (Drawing by M. Huggins)

radiation, Q, received on the site and by influencing the amount and disposition of moisture. South-facing slopes in the Northern Hemisphere on the average receive more annual solar radiation per unit area and are drier than other slopes (Haase, 1970). Fig. 5.2 shows the theoretical radiation, Q, on north- and south-facing 50% (26·57°) and 100% (45°) slopes at latitudes of 34° and 46° North (Fons et al., 1960).

It is apparent from this figure that the 100% south-facing slope at 34°N latitude (the approximate latitude of southern USA) receives almost as much radiation in winter as in summer. The 50% slope at 34° latitude is the only north-facing slope shown which receives any direct radiation in winter. Because of the angle of sun and the day length in summer, both south-facing slopes at 46° latitude (northern USA) potentially receive more radiation than similar slopes at 34° latitude. In general, the north-facing slopes in summer receive as much as or more total radiation during the day than the comparable south-facing slopes in winter. The contrast in radiation between summer and winter is generally greater at a higher latitude than at a lower latitude site.

Because arid, south-facing 100% slopes at low latitudes can receive as much solar radiation in winter as in summer, winter temperatures are high and trees on such sites are more likely to remain physiologically active even during the middle of winter (Fig. 4.16). During the day, temperatures of plants in direct sunlight can rise much above those of the air; but soil temperatures can be relatively low, so that high daytime temperatures can cause transpiration to exceed water absorption, thus creating water stress and desiccation of plant tissue even when there is moisture available in the soil. As a result, net photosynthesis in such low latitude trees may be limited more by high than low temperatures in winter. Thus, at low latitudes on south-facing slopes, temperatures at any time during the year can be high enough to be detrimental to physiological conditions in trees, subsequently affecting ring-width growth the following spring. For more northerly sites, such as 46° N latitude, there is a more marked difference in the energy received during the summer and winter on 100% south-facing slopes, and elevated plant temperatures in winter are not as likely to become high enough to limit processes in the plants. This phenomenon, involving elevated temperatures on south-facing slopes, has been discussed in Chapters 3 and 4 and can be substantiated by field measurements as well as inferred from numerous statistical analyses of environmental–growth relationships described in Chapter 8 (Fritts, 1974).

Haase (1970) reports that in southern Arizona the southeast-facing slope is the most arid exposure in the summer, due to cloudless mornings. By noon or early afternoon, when radiation is potentially high on the south- and southwest-facing sites, clouds develop which reduce net solar radiation. Erdman *et al.* (1969) measured forest environments in several sites at Mesa Verde, Colorado, and they report that a southwest-facing slope is drier than a northeast-facing slope. The rings of *Pinus edulis* at Mesa Verde substantiate these findings in that high ring-width variability, which indicates the effects of drought, is characteristic of the more arid southwest-facing exposures.

Fritts *et al.* (1965c) state that ring-width variability is greater in trees on steep, well-drained canyon walls than in trees on the mesa tops, indicating greater aridity for the canyon-wall sites. However, the extremely old *Pseudotsuga menziesii* from Mesa Verde are not only on steep slopes but the exposure is north-facing so that these trees receive little direct solar radiation in winter (Fig 5.2). As might be expected, the ring widths from these trees do not correlate as well with winter climate as do trees on south-facing exposures, probably because the north-facing trees remain frozen during the cooler winter season and are unable to respond to certain variations in the winter environment.

Fritts (1969) used ring-width variability and other statistical characteristics in *Pinus longaeva* to assess the aridity of sites on different exposures in the White Mountains of California. He found that substantial variations in ring-width characteristics were related to topographic differences. Trees on the south-facing slopes, which receive the highest incident radiation, exhibited the most ring-width variability (sensitivity) and were inferred to be on the driest sites. The trees on the west-facing, steepest, and most exposed slopes also indicated aridity by their high ring-width variability. It was suggested that in the latter case snow is removed from these exposures by the prevailing winds.

Trees on a southeast-facing slope near a protecting ridge had wide rings and low ring-width variability. Drifting snow was observed to accumulate in the lee of the ridge, adding to the moisture of the site. Thus, it could be assumed that the trees on the southeast-facing and protected slope would grow more rapidly and uniformly than those on slopes exposed to the sweep of the wind because of the additional moisture from drifting snow. Both ring characteristics and stand density indicate that growth is most favored on north-facing slopes and valleys, where snow can accumulate but the incident radiation is low and where the trees are protected from the sweep of the wind.

B. *Substrate*

The parent material, or soil type, may modify the environment of a site in several ways and therefore must be taken into account when modeling or interpreting tree-growth conditions. *Pinus longaeva* in the White Mountains is largely confined to dolomitic substrates. Mooney *et al.* (1962) and Wright and Mooney (1965) demonstrate that this is in part due to the lighter color of the dolomite which has a 15% to 20% greater albedo than sandstone. The greater reflectivity results in lower soil temperatures and lower evaporation rates, so that a dolomitic soil remains consistently wetter and cooler than sandstone soil when other environmental factors are equal. In addition to the above conditions, Wright and Mooney (1965) point out that there are lower levels of certain minerals on dolomite than on sandstone. *Pinus longaeva*, which is more tolerant and therefore less limited by the low mineral availability than other species, is dominant in the plant community on dolomitic soils, but occurs infrequently on sandstone and granite where it apparently cannot compete with other plants.

Fritts (1969) reports that rings of bristlecone pine are wider on granite and, to a lesser extent, on sandstone than on soils derived from dolomite. This may indicate that minerals may also be limiting to *P. longaeva* on the dolomite and may be at least in part responsible for their slow growth on dolomitic sites. Dendrochronologists have frequently noted that the oldest trees with the most datable and varied rings are often found growing on limestone or dolomitic substrates.

Erdman *et al.* (1969) and States (1968) note that trees in semiarid regions often grow best on soils with a heavy charge of rock fragments. Such soils hold less moisture per volume of soil than fine-textured soils, but the penetration of moisture is deeper and less water is lost by evaporation. As a result, more of the moisture that falls as precipitation can be utilized by deeply rooted trees than by shallow-rooted grasses. Run off is also less on the coarse-textured than on fine-textured soils. Erdman *et al.* (1969) note that interbedding of shale between sandstone produces seeps and springs at the contact, which serve as sources of moisture for trees. For more discussion of similar phenomena see Chapter 3.

Fritts and Holowaychuck (1959) describe soil profile differences (Fig. 3.11), and Vinš (1970) and Stockton and Fritts (1971c) describe changes in water tables which control root distribution and water relations of trees and influence their growth. Other factors of substrate, such as the amount of leaf litter, the percentage of surface covered by rock, and soil permeability, may markedly affect the operational environment of trees on particular sites. However, these soil and site factors do not change measurably from year to year and therefore are not usually factors governing the year-to-year ring-width chronology of a site.

C. Elevation and Orographic Factors

Elevation or altitude is a site factor affecting mean annual precipitation and temperature regimes (Fig. 4.16). Disregarding the influence of surface features, air temperatures decrease with increasing elevation. This decrease in temperature is referred to as the *lapse rate*. An average value for the lapse rate is 0·5°C per 100 m elevation (Geiger, 1965).

The density of the air column also decreases with altitude, so that there is less scattering and absorption of direct solar radiation by the atmosphere (Fig. 5.1). As a result, plants at high elevations are often subjected to high radiation loads and high daytime temperatures during cloudless days. At night the exposed surfaces lose heat rapidly because there is less reradiation by the atmosphere back to the earth surface, and plant temperatures can fall well below the temperatures of the air.

Mountains may serve as barriers to moisture-laden winds, so that high-elevation sites receive more precipitation than the sites at lower elevations. Since temperatures are lower in the mountains, moisture evaporates more slowly and a greater proportion of the precipitation may fall as snow. Snow in turn reflects the solar radiation and reduces that portion of energy absorbed in the site. Snow also can insulate the soil, keeping it relatively warm, and reducing the outgoing longwave radiation so that night-time temperatures in the above-ground tissues drop well below freezing.

Sites in the lee of mountain ranges usually receive significantly less precipitation than comparable sites on the windward slopes. Little moisture falls on *Pinus longaeva* sites in the White Mountains of California even though they are at high elevations, because the Sierra Nevada mountains to the west are an orographic barrier removing a large portion of the moisture from the prevailing westerly winds. Although available moisture generally increases and temperature decreases with increasing elevation in this area (Wright and Mooney, 1965; Fritts, 1969), there is less change in tree-ring characteristics associated with elevational difference than with topographic exposures of the tree sites. More details of this analysis are presented in Chapter 6.

Additional examples of elevational, orographic, or other types of physiographic factors influencing microclimate are the *foehn* or *chinook winds*, cold air drainage, temperature inversions, and mountain-valley winds (see Geiger, 1965). Any correlation between variations in macroclimates and ring widths can be influenced by the microclimates as they control the ranges of environmental factors that limit the processes in plants. As will be discussed in Chapter 8, there are many differences in the response of ring width to climatic factors which can be attributed to microclimatic conditions in the respective tree sites.

IV. Biotic and Other Nonclimatic Factors: Dendrochronological Examples

The location and condition of surrounding plants may influence the operational environment of a tree by casting shade, by influencing the drifting or blast of wind-carried snow, by competing for moisture or essential minerals, or by producing substances which favor or inhibit growth. Certain details of these influences have been described in Chapters 3 and 4. Because none of these factors is directly climatic in origin, the dendrochronologist needs to be able to both recognize them and remove them from tree-ring data being used for climatic analysis.

Changes in the operational environment of a tree due to the growth and senescence of neighboring trees usually occur slowly and in more or less systematic fashion. Because the resultant variations are nonclimatic in origin, they will differ from tree to tree and from site to site and they can be at least partially removed by the standardization of ring widths. For the same reason this variability from tree to tree can be used quite successfully for ecological analyses such as the assessment of stand history, site disturbance, and other ecological conditions and events. Although not the focus of this text, it is interesting to look at a few examples of studies concerning nonclimatic factors affecting the ring-width variations in trees.

There is extensive literature involving the use of ring counts to determine the age of trees following a catastrophic event, to determine the age structure of a forest, or to identify the age of a scar on a tree stem. Lawrence (1950), Lawrence and Lawrence (1961), Sigafoos and Hendricks (1961), and Alestalo (1971) have used ring counts to estimate tree age to determine the minimum length of time since a substrate surface was exposed and available for establishment of plants. These studies assume that existing trees on the site represent the original invaders after the given substrate was produced. Sigafoos and Hendricks (1969) show that age estimates may be improved by establishing the interval of time between stabilization of the substrate surface and establishment of the tree seedlings used in age determination.

Tree-ring crossdating, though more time consuming than ring counting, usually produces a more reliable determination of age because it takes into account any missing rings or double rings. Roughton (1962) and Alestalo (1971) review some of the literature which discusses the use of crossdating to obtain dates for past occurrences of fires, frost, insect infestations, release from competition, mortality, animal use, vegetative reproduction, glacial movements, and other geomorphological changes that have left a record in the rings of trees. Vinš (1970) describes similar studies on forest environments in Czechoslovakia.

The dendrochronologist can often reconstruct the past history of many events by detailed study of stems and anomalous structure of the rings (Sigafoos, 1964; Sigafoos and Sigafoos, 1966; Potter, 1969; Alestalo, 1971; Helley and LaMarche, 1973). Landslides, floods, earthquakes, and glacial advances can bury, injure, or tilt a tree stem and may cause changes in the growth which are recorded in the rings (Lawrence, 1950; Lawrence and Lawrence, 1958b; Potter, 1969; Page, 1970; Alestalo, 1971; LaMarche and Wallace, 1972). Tilting of angiosperms can be

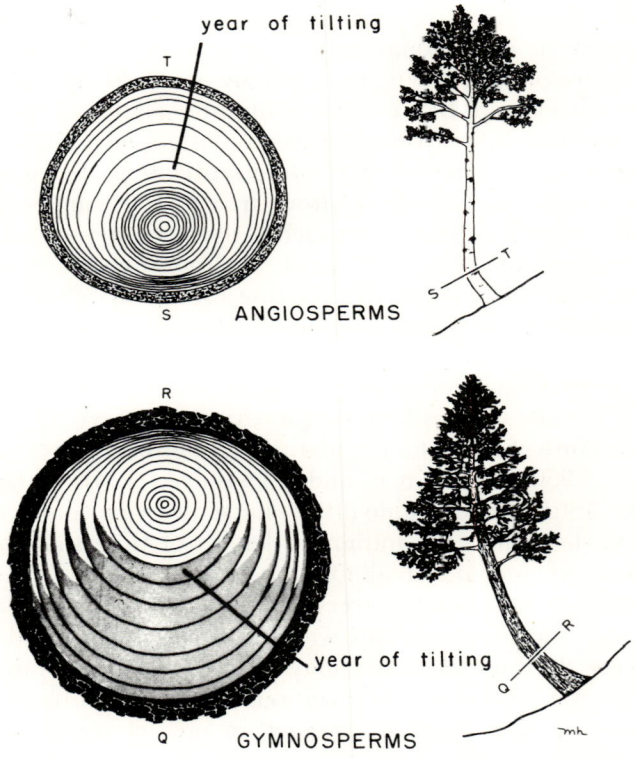

FIG. 5.3. Tilting of trees may be dated because after the year of tilting growth rings are formed asymmetrically in the stem. In angiosperms, the widest rings are formed in the stem on the upper side of the tilt, producing tension wood, while in gymnosperms the widest rings form on the lower side of the tilt, producing compression wood. (Modified from Lawrence, 1950) (Drawing by M. Huggins)

recognized by the presence of tension wood on the upward side of the tilt (Chapter 2). In gymnosperms, compression wood producing wider rings is found on the downward side of the tilt (Fig. 5.3).

Direct injury can produce scars on a tree stem, alter the ring structure, or affect crown form which itself produces a change in ring width (Alestalo, 1971). Scars at the base of trees are commonly used to reconstruct the past frequency of wildfire (Fig. 5.4) (Weaver, 1951; Erdman *et al.*, 1969; Kilgore, 1970; Billings, 1969; Heinselman, 1969, 1973). Abrupt changes in ring width have been used as evidence for snow slides (Potter, 1969), for burial of trees from deposition of rock debris or volcanic ash (Lawrence, 1950; Smiley, 1958; Sigafoos, 1964; Druce, 1966; Helley and LaMarche, 1973), for injury due to smoke pollution (Fig. 5.5) (Bakke, 1913; Clevenger, 1913; Vinš, 1965, 1970), for

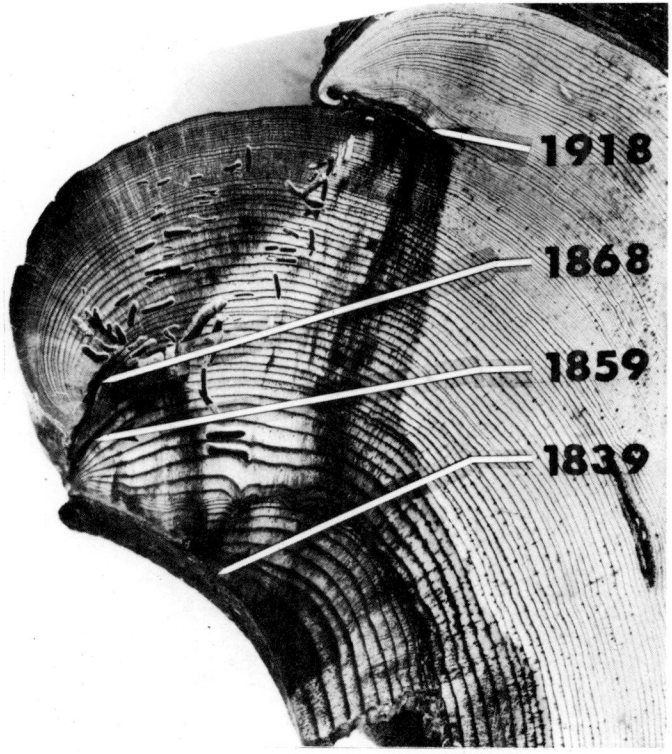

FIG. 5.4. Cross section of a tree stem showing scars from fires for 1839, 1859, 1868, and 1918. A part of the cambium has been killed and the living portion of the stem has grown over and buried some of the scarred tissue. (Specimen courtesy of David F. Aldrich and Robert W. Utrech, U.S. Forest Service)

disturbance of a site by rupturing of faults during earthquakes (Page, 1970; LaMarche and Wallace, 1972), for reduced soil aeration by rising lake levels (Antevs, 1939), and for damage by browsing herbivores (Lawrence, 1952; Vinš, 1970). Spencer (1964) reconstructed past variations in populations of the porcupine (*Erethizon epixanthum*) in Mesa Verde, Colorado, by using tree rings to date browsing scars on *Pinus edulis*.

Sometimes radiocarbon dating of ancient wood can be used to establish the relative age of a buried forest while the precise relative dating of tree-ring patterns is used to establish the histories of the individual trees. Although such relative dates are not accurately placed in time, the trees are dated accurately relative to one another, and such information can lead to considerable understanding of past geological changes and climatic events (Lawrence and Lawrence, 1958a; Munaut, 1966).

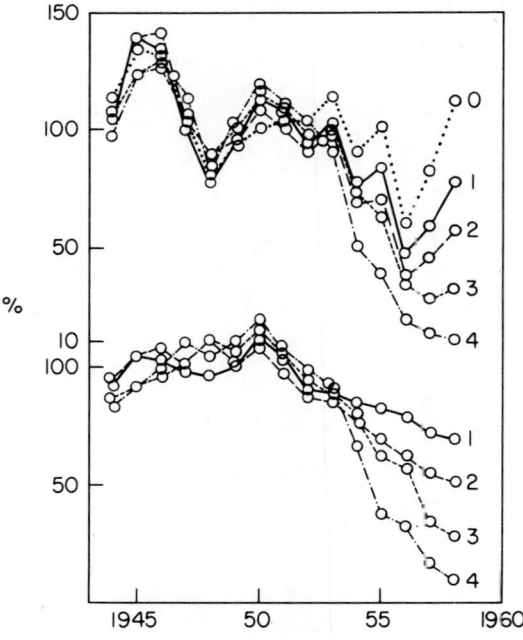

FIG. 5.5. Ring-width variation present in forest stands in Czechoslovakia subjected to various degrees of air pollution. The plot labeled 0 in the upper part of the figure represents growth indices for an untreated control stand, as compared to stands 1 to 4 representing increasing degrees of smoke pollution. The lower set of curves shows the growth of stands 1 to 4 as a percentage of the unaffected one labeled 0 at the top. (Vinš, 1965)

Koerber and Wickman (1970) discuss the use of ring-width measurements to evaluate the impact of insect defoliation on forest growth. They state that an increase in insect populations will increase defoliation of the host trees and reduce their photosynthetic capacity, which results in marked decreases in growth for a number of consecutive years following the insect attack. They also develop a model for the effect of weather factors on insect populations (Fig. 5.6) and show that the same climatic factors of low precipitation and high temperatures which are correlated with narrow tree rings also favor the growth of insect populations, increasing defoliation, and reducing further the width of tree rings.

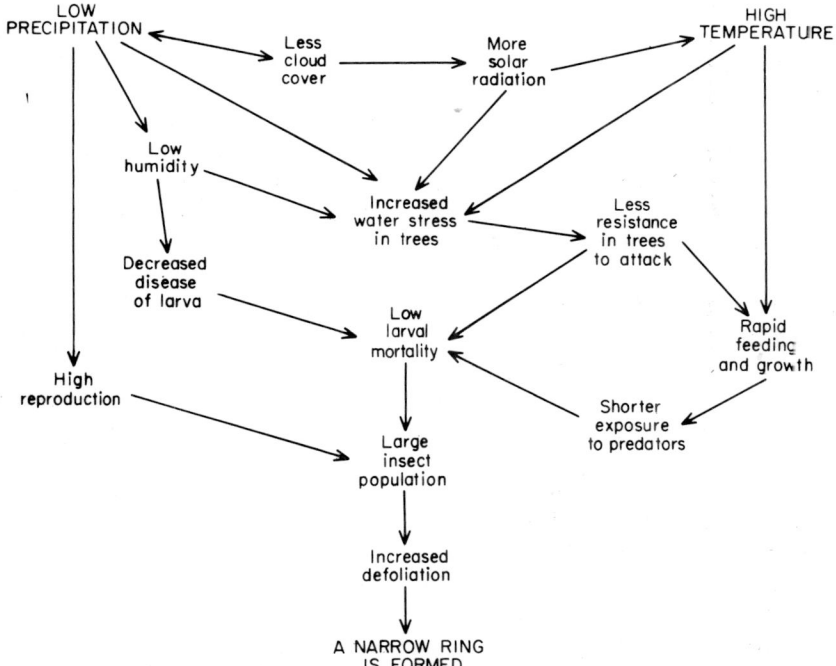

FIG. 5.6. A model of the relationships between low precipitation and high temperatures on insect populations, which decrease tree foliage and reinforce the effect of unfavorable climate on tree growth. (Redrawn from Koerber and Wickman, 1970)

V. Modeling Relationships in the Ring-Width and Climatic System

The foregoing portions of the book include descriptions of the most important plant processes affecting ring-width growth. Many of the factors that limit these processes are shown to be linked to macroclimatic conditions through the energy and water balances. However, none of these processes and climatic linkages operates in a simple and singular manner. All are part of an ecological system in which there are almost endless numbers of interrelationships and complexities affecting the linkages between climatic factors and ring width (Billings, 1952).

However, the system can be simplified and described by developing models, which are represented in the following pages as a network of linkages (some unidirectional and some bidirectional) among specific states or conditions in the climate–plant system.

A. Rationale

The general rationale behind these process–network models is to express the essential nature of a complex system in the most intelligible form possible, a form that is also amenable to hypothesis testing and conceptual development. At the outset the hypotheses tested may principally involve the model itself, such as the identification of any variables included in the model that are irrelevant to it, the search for variables omitted from the model that should be a part of it, identification of any inaccuracies in the model, evaluation of alternatives to the model structures, and recognition of improperly formulated constraints. As a model increasingly resembles the natural system, it becomes more and more useful in the testing of inferences or hypotheses about the system, anticipating unusual conditions and situations, identifying new problems and possible solutions, as well as developing new models for further analysis.

This last use is perhaps the most important for these particular models because they have led to the identification and development of multivariate statistical models which in turn are used to quantify and test the process–network models themselves. They have also led to the development of other multivariate statistical models which allow precise calibration of climate and ring-width data and objective statistical estimations of past climatic variations from the corresponding variations in ring widths.

B. Approach

Because of the many complexities among the specific biochemical reactions and the physical linkages in the system, it is extremely difficult to model the details of individual physiological processes. In cases where such process-by-process modeling has been accomplished, the results, although very interesting and revealing, usually pertain to particular circumstances and deal with simplified estimates for subsystems rather than the entire system complex.

For the time being it is more practical, as well as sufficient for dendrochronological purposes, to generalize the processes in a nonquantitative fashion emphasizing the major pathways in the system, including the range but not all the possible physiological components. Such generalizations must be consistent with relevant biochemical and physiological experimental evidence, while allowing for the fact that almost all such experiments must deal with only one aspect of the whole system, and therefore may be incomplete. It is most important that the generalizations be consistent with the basic concepts, observations of the system as a whole, and field evidence from habitats and species to which the models are relevant.

Such modeling is not without its own type of experimentation and hypothesis testing. Field measurements on the actual trees to be modeled may be obtained, stratified sampling of various sites can be used to study natural changes in specified variables, and actual manipulation of particular environmental factors in the field is possible to test certain kinds of hypotheses.

The models used are in a sense always tentative in that they are based on the best inferences and the best data available at the time. Each one may be considered to be suitable, however, until the emergence of new evidence. When this happens, the models should be reexamined, tested, and revised to incorporate the new information. As more and more evidence and experience accumulates, the models will improve, and in time they may become considered adequate representations of basic principles and concepts of the growth–climate relationships.

Such an approach to modeling is probably more justifiable for dendroclimatic study than for biological research, because the primary object of dendroclimatology is to use the structure of the ring to statistically reconstruct past climate. It makes little difference to the statistical result whether the relationships arise because a winter drought affects photosynthesis and food reserves which alternately limit growth, or whether a winter drought affects soil moisture, hormone precursors, and differentiation of buds which control the concentrations of growth regulators affecting subsequent growth. Either model can explain the lag between the climatic cause and the effect on tree growth. New evidence will undoubtedly alter the details of the models described in this volume, but it cannot alter the statistical relationships. On the other hand, new information can improve understanding of the relationship in such a way as to suggest a new model which could improve statistical analysis and change the results. For this reason, physiological and ecological study, as well as statistical analysis of correlated relationships, are important to continued development of dendroclimatic and dendroecological research.

C. Model Constraints

It has been shown that a climatic variable can limit a wide variety of plant processes, which in turn can become limiting to growth and ring width. While the growth process at a particular instant in time is essentially controlled by that process which is most limiting, there are variable interactions within the plant and nonlinear situations that can make a model structure intricate. Arrows are used to join and to specify the direction of such interacting situations, and the sign of the effect is indicated, but scales or numeric quantities are not assigned to any of these relationships.

Different segments of the model are used to distinguish between cases where a set of climatic variables may have a positive as opposed to an inverse effect, and the importance of each segment is considered to vary and be additive from one month to the next.

VI. A Model for Factors Affecting Cambial Activity and hence Ring Width

A convenient place to start is the model of cambial activity described by Wilson (1964) and Wilson and Howard (1968) (see Chapter 2 and Fig. 2.7). They look on the cambium as a kind of factory to which building materials, energy sources, hormones, and water are fed from the nearby phloem and xylem. Any factor limiting the supply of these materials can limit growth. The complex cell structures of the ring are manufactured and added in the processes of assimilation which are dependent on favorable temperature. Whenever temperatures in the cambium are low, assimilation can be limiting and growth can be reduced. Cell differentiation also occurs but does not need to be included in our model because once the cell is fully enlarged, changes due to cell differentiation will not affect ring width.

There are four primary limiting conditions that must be considered in such a model. They are temperature of the growing tissues, water stress in the tissue, concentrations of growth regulators, and the amounts of building materials including foods and mineral salts. Fig. 5.7 visually outlines the relationships among these factors with appropriate words and arrows. Any limiting situations in any one of these factors can affect

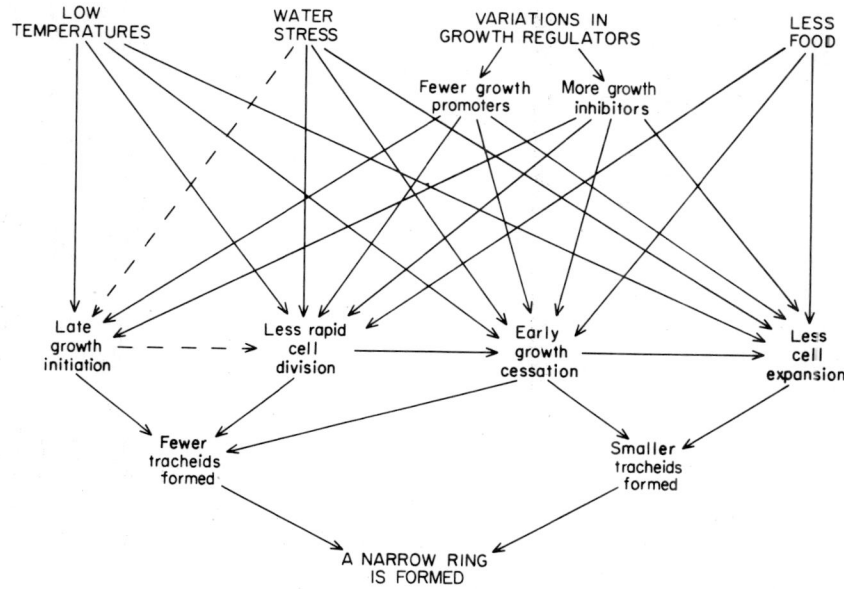

FIG. 5.7. A model describing the four major factors causing a reduction in growth of the cambium and production of a narrow growth ring. The arrows indicate cause and effect and include various types of interrelations among the processes and variables. It is implied from the diagram that the effect of the opposite extreme will increase both cambial activity and ring width.

the beginning and ending of the actively growing period, the rate of cell division, and the expansion of cells. These in turn affect number and size of the tracheids that are produced and thereby the ring width.

A. Temperature

Temperature of the cambium can be a directly limiting factor through its control of respiration and assimilation processes. The direct effects of temperature on radial growth are most frequently observed at the beginning of the season when unusually cold weather causes a delay in the breaking of dormancy. This favors a narrow ring by decreasing the length of the growing period. Once growth is started, any period of extremely cool weather near 0° C may limit growth, but as temperatures rise to those of summer other factors usually become more limiting. At high elevations or high latitudes, where the potential growing season is short and plant temperatures can be low while the cambium is still active, radial growth may be greatly reduced by cold weather during any time in the growing season (Hustich, 1945; Mikola, 1952, 1962; Giddings, 1953; Eklund, 1957). Except at such extremely cold sites, cambial activity in late summer may cease due to internal conditions long before temperatures are low enough to be limiting to the assimilation and cell division processes (Fritts, 1969) (also see Fig. 4.16). Of course, if temperatures during the growing season become low enough to freeze the cambial tissues, much damage can result which will markedly affect the appearance and subsequent ring growth.

The energy budget of trees may be modified by the low temperatures and influence cambial activity in the stem. For example, trees at high elevations or high latitudes on south-facing slopes may receive direct solar radiation during the daytime causing temperatures to be higher than those in trees on north-facing slopes. On such exposures, temperature is most likely to become limiting to cambial activity during cloudy weather, at night, or when high wind dissipates the energy and reduces the cambial temperatures to levels too low for assimilation processes. Also on southern exposures, certain portions of trees near the ground may be warmed by longwave radiation from nearby objects and from the ground surface, stimulating early initiation of growth, before activity in the upper parts of the tree is underway.

Soils in the shade or on north-facing slopes may remain frozen and snow may persist for long periods of time, causing initiation of growth in the spring to be delayed by the persistence of low temperatures. Because of the limiting effect of low temperature on the initiation and continuation of cambial activity in trees on north-facing slopes, ring-width variations are more likely to be directly correlated with air temperatures of the early spring than for trees on south-facing exposures.

The microhabitats in polar regions may vary in a different fashion from those at lower latitudes because of the low sun angle and the long summer days near the poles. Radiation and heat loads on the plant are low, temperatures are moderate, and the diurnal range of temperatures is often less than at lower latitudes. Temperatures can become low enough during daylight hours to limit both respiration and photosynthesis, but in alpine habitats at lower latitudes, high rather than low temperatures may be limiting during the day and low temperatures are more apt to be limiting at night.

Much work on the growth-environmental relationships of trees at high latitudes and upper tree line is in progress. Evidence so far supports the generalization that unusually low air temperatures coupled with cloudy weather during the growing season can directly limit growth and cause rings to be narrow. Temperatures may also be limiting to photosynthesis, but in the model under discussion this would affect the amounts of building materials and hence not be a factor directly limiting the growth processes.

In general there is little evidence that high temperatures are directly limiting to cambial activity in most trees used for dendroclimatological purposes, except for extreme heating resulting from such conditions as forest fires which can damage the cambial tissues. As in the case of frost damage, only when cambial tissues are killed outright or injured sufficiently to alter subsequent cambial activity is there a direct effect of such extreme temperatures on ring width.

B. *Water*

Water stress has been found to account for a large amount of ring-width variation. During a dry summer, water is obviously likely to be deficient enough to be limiting to the cell division and cell enlargement processes. Although much experimental evidence can be cited for a direct effect of water stress on cambial activity, the evidence is in large part circumstantial. This has led to considerable controversy over the question of whether water actually affects the growth process directly or affects some other factor such as photosynthesis or auxin synthesis which in turn influences the growth process. Zahner (1968) and Glerum (1970) present and describe some of the growing evidence supporting the hypothesis that moisture content of the cambium can become a direct limiting factor. This evidence includes numerous measurements and counts of cells that have repeatedly demonstrated that reduced rates of cell division and cell enlargement occur when the cambium is subject to internal water stress. The gross expansions and contractions of the stem diameter throughout the day and throughout the growing season, as documented in Chapters 2 and 3, represent some of the more obvious evidence.

However, unusually low solute and matrix potentials within the cambial tissue may enable the growing cells to compete favorably for water within the tree. As a result, cambial initials may remain relatively turgid and continue to divide, while other tissues such as the phloem and xylem are shrinking and undergoing water deficits. As the growing cells differentiate and mature, they become more vacuolated than the cambial initials, their solute potentials may rise and they may begin to show increasing amounts of shrinkage due to plant water deficits. These water deficits first affect the fully enlarged cells, and then the enlarging cells before they become limiting to the cells that are dividing. Thus, the first noticeable effects of water stress may be a reduction in cell size or a change in cell-wall thickening. These effects often vary from time to time, depending upon the degree of water stress and the life span and condition of the cell derivatives (Zahner, 1968).

If the drought persists long enough to induce dormancy, the last latewood tissue formed in the ring under severe drought may be markedly less dense and the latewood thinner than the latewood formed in a year with a more moist climate. A number of variations in tracheid structure and size, apparently due to water stress in a drought-subjected *Pinus ponderosa*, are shown in Figs. 2.19, 2.20, 2.22, and 2.23. The reader is also directed to a paper by Zahner (1968) for a detailed and informative discussion of the many indirect effects that water deficits can have on certain processes which in turn affect ring width.

C. Growth-controlling Substances

The concentrations of growth regulators may also be factors directly affecting radial growth. The influence of growth promotors and growth inhibitors on initiation, cessation, and rate of cambial growth has been described in Chapters 2 and 4.

In general the initiation of cambial activity is a result of a decline in growth inhibitors and an increase in growth promotors. If other factors are not more limiting the amount of cambial activity will increase to an optimum rate because of changes in the growth-regulating substances. The large earlywood cells formed during spring and early summer are associated with high auxin concentrations, while the shift to lower cambial activity and the formation of latewood is associated with a decline in auxins and possible changes in other growth regulators. The cessation of cell division and cell enlargement at the end of the growing season can be attributed to decreases in growth promotors and increases in growth inhibitors.

The rate of shoot growth may be controlling the production of these growth regulators which must then move to the region of the cambium before they affect growth rates (Larson, 1960). Also, water stress or low temperatures can reduce shoot growth or affect translocation which in turn governs the supply of growth regulators at the base of the stem. These factors of course are indirect controls, so they cannot be included in the particular model under discussion.

D. Amounts of Building Materials

Much circumstantial evidence has been accumulated which indicates that rates of growth in arid-site conifers are affected by changes in the amount of foods and related growth-controlling substances. If conditions are severely limiting to food manufacture and accumulation for long periods prior to the growing season, the rate of cambial activity in the spring and summer can be most limited by food availibility. In such cases, the ring is narrow even though the temperature, moisture, and growth regulators are not limiting.

It is also possible that available minerals could be the most limiting factor to cambial activity. The cambial growth in *Pinus longaeva* on dolomitic soils may be limited by the low availability of certain minerals in the soil as some ions enter into the manufacture of protoplasm and certain proteins within the growth tissues. Minerals may also be indirectly limiting through their effects on the manufacture of foods and growth substances elsewhere in the plant.

Thus, the model of factors affecting cambial activity, as outlined by words and arrows in Fig. 5.7, includes four major limiting conditions in the operational environment surrounding the cambial tissues of the plant. It should be noted that temperature of the cambial tissues is the only climatic variable mentioned in the model and the model can account for only the direct temperature ring-width relationship. Although water stress is closely associated with variations in the water balance, it depends upon a variety of plant processes affecting water absorption in the roots and the transpiration from the leaves. These processes in turn can be affected by a variety of climatic factors including precipitation, humidity, temperature, wind, and light.

The growth-regulating substances are manufactured elsewhere in the plant so that the amounts in the cambium are governed indirectly by the conditions affecting growth and the activity of the stems, leaves, and roots as well as the translocation of the growth substances from their site of manufacture to the cambium. Mineral nutrition cannot be tied to any particular climatic factor and can be eliminated from a model of climate-growth relationships. However, the limiting effects of stored foods which become the building materials and energy source are highly dependent upon a variety of climate-affected processes, including photosynthesis, respiration, food manufacture, and assimilation.

VII. Modeling the Effects of Temperature and Precipitation on Ring Width

It was shown in the previous section that the climatic factors affecting ring width could not be adequately described in a model dealing only with the cambial tissue. It is necessary to consider the entire operational environment of the tree and the various effects of factors on processes occurring at different times in the year, and in different parts of the tree. Figs 5.8 through 5.10 diagram three segments of a model which describes these relationships. The variables of temperature and precipitation were chosen as input variables because actual measurements of these variables could be used to test statistical models describing the same effects. In addition, these two climatic variables are easy to interpret in terms of other variables such as solar radiation, water vapor, wind, and atmospheric pressure; and they are variables most often used to describe changes in past climate. For the sake of simplicity, the linkages between macroclimate and microclimate have been omitted from the diagram. The following discussion treats the model in segments representing (A) the manner in which low precipitation and high temperature during the growing season lead to the formation of a narrow ring, (B) the manner in which the same two factors occurring prior to the growing season lead to the formation of a narrow ring, and (C) the ways and circumstances in which high precipitation and low temperatures can lead to the formation of a narrow ring.

A. Model Part A

The diagram referred to as Part A (Fig. 5.8) represents relationships

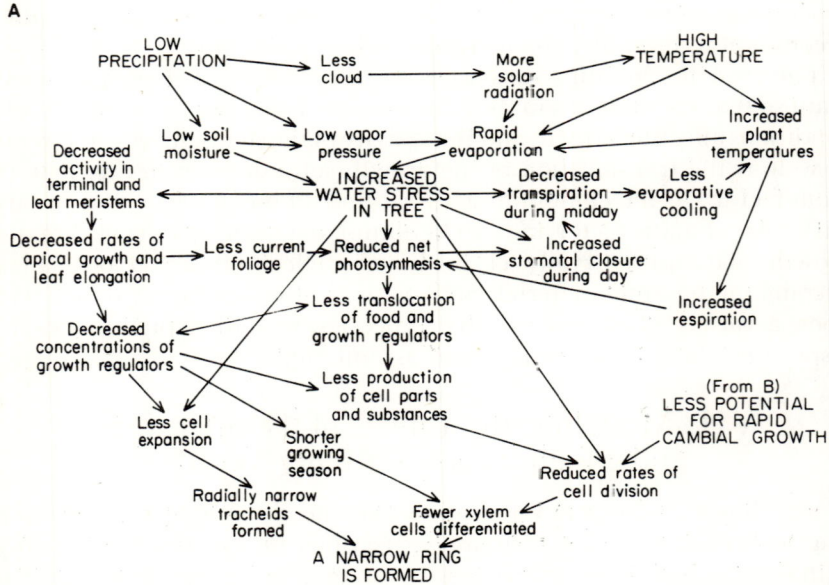

FIG. 5.8. Model Part A. A diagram representing some of the relationships that cause climatic factors of low precipitation and high temperature *during the growing season* to lead to the formation of a narrow ring in arid-site trees. The arrows indicate the net effects and include various processes and their interactions. It is implied that the effects of high precipitation and low temperature are the opposite, that is, ring width will increase.

that occur during the interval of time when the cambium is active. Low precipitation is generally associated with high temperatures, and high precipitation with low temperatures, because precipitation is usually accompanied by increased cloud cover which reduces the solar radiation flux (Q, Equation 5.1). Low precipitation and higher radiation, of course, increase environmental temperatures, and both climatic conditions accentuate water stress. Low precipitation results in low soil moisture recharge, which leads to reduced rates of water absorption. High temperatures increase the water potential gradients surrounding the tree and favor rapid water loss. As shown in Fig. 4.10, net photosynthesis in *Pinus ponderosa* is reduced during periods of high water stress and high temperatures. Depending on the situation, net photosynthesis may be reduced because water stress lowers photosynthetic rates directly, because increased temperatures raise respiration rates, or because of a combination of these two. With less net photosynthesis there is less food available for growth.

The left-hand portion of Part A describes the effects of water stress on activity in the apical meristems and growing leaves, which are the source of important growth regulators affecting cambial activity in the stem, as was discussed in the prior model and in Chapter 4. Increased water stress can limit the activity in the meristems so that there are lower concentrations of growth promoters (or higher concentrations of growth inhibitors) and reduced cambial activity in the main stem.

Increasing water stress can reduce cell expansion directly (see Fig. 5.7 and Zahner, 1968), and the production of new cells and the length of the growing season may be shortened as a result of water stress. Interactions are shown where decreased apical growth may cause less foliage to be formed, which in turn reduces net photosynthesis late in the season or in the following year. Interactions between growth regulators, translocation, and utilization of food are also shown.

B. Model Part B

Studies on ring widths and climatic relationships for conifers on arid sites repeatedly demonstrate that the climate during the growing season (Part A) is often less important to ring-width growth than the climate of the autumn, winter, and spring prior to the growing season (Fritts, 1965, 1969; Fritts *et al.*, 1965c; Julian and Fritts, 1968).

Part B, shown in Fig. 5.9, diagrams relationships which appear to be responsible for the preconditioning of growth or which, in statistical terms, account for the lag in the growth response behind the occurrence of climate. As in Part A, much of the biological evidence for the

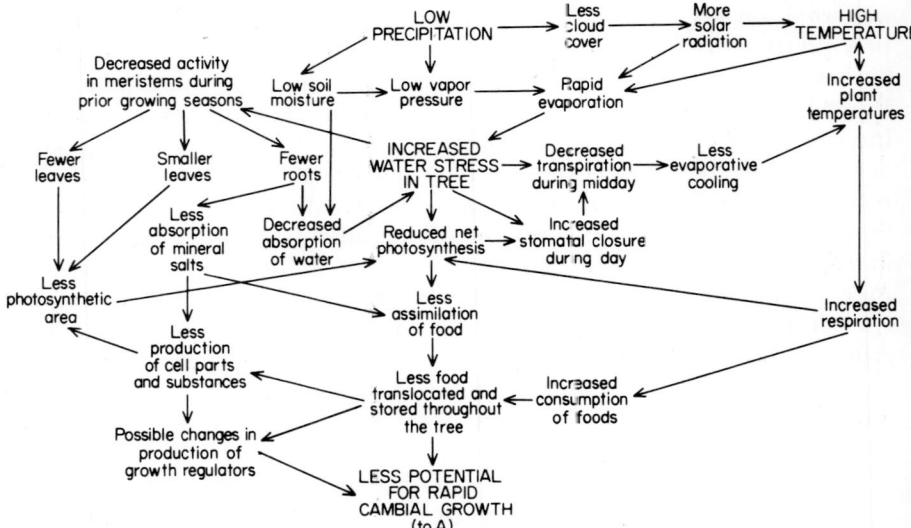

FIG. 5.9. Model Part B. A diagram representing some of the relationships that cause climatic factors of low precipitation and high temperature occurring *prior to the growing season* to lead to the formation of a narrow ring in arid-site trees. Compare with Fig. 5.8.

relationships expressed in this model is described in Chapters 2, 3, and 4. Of the three model segments Part B is most applicable to arid-site conifers on low-elevation sites near the lower forest border of the species. Temperatures there in autumn, winter, and spring are often above freezing (Fig. 4.16) and precipitation is low enough that the combined effect of low precipitation and high temperatures can create conditions of water stress during any time of the year. Because the trees can be photosynthetically active all year long as well, such water stress can limit net photosynthesis. High temperatures often reduce net photosynthesis by increasing the rate of respiration over the rate of photosynthesis and thereby reducing the accumulation of photosynthetic products as stored food. In such cases, when cambial activity commences in spring, the growth of trees can be most limited by a deficiency in the products assimilated from stored food.

Interacting relationships can also occur with the growth of prior seasons. Water stress during prior years may result in fewer and smaller leaves, buds, and roots, This reduces the total photosynthetic and absorptive areas of the tree, so that it is less able to photosynthesize or absorb moisture when favorable climatic conditions occur. In this manner, climatic conditions during prior years interact with those occurring immediately before and during the growing period to affect the ring width.

Chapter 2 describes some of the variations in apical and radial growth on arid sites. Growth commences in late spring after much of the soil moisture from winter snow has been lost by evapotranspiration. This is a time when the remaining soil moisture in arid, forest-border sites in southwestern North America is diminishing and net photosynthesis has reached its peak rate and is beginning to decline. A large part of growth that takes place early in the season (especially in the regions of the roots and at the stem base) is dependent upon foods accumulated and stored in the tree during the autumn, winter, and spring, because photosynthates that are currently produced by the leaves are utilized primarily by the growing crown. This situation remains even after the summer rains commence. Net photosynthesis usually increases, but shoot growth, needle enlargement, and growth of the reproductive organs within the crown also increase. The tissues in the crown continue to receive most of the currently produced photosynthates. The cambia at the stem base and in the roots are the last to receive translocated food and are most likely to remain dependent upon the foods accumulated throughout the winter and stored in tissues of the stem and root.

Therefore, the coincidence of the first half of the cambial growing season with dry conditions and water stress and the coincidence of the latter half with rapid shoot and fruit growth, which compete with the cambium for large amounts of the currently produced photosynthate, could cause the ring width at the base of the stem of mature trees to remain dependent upon stored food reserves accumulated over the autumn, winter, and spring. This situation would explain why ring-width variation from the stem base of mature trees is correlated so well with moisture and temperature conditions during the autumn, winter, and spring preceding the beginning of growth. It will be shown later in the chapter and in Chapter 7 that, where the spring season is on the average more moist than on these arid sites, ring width is less well correlated with the winter climate and more highly correlated with the spring and summer climates.

C. Model Part C

Some trees used for dendrochronological studies grow on sites where temperatures in the spring, summer, and autumn are low, and temperatures in the winter are well below freezing for long intervals of time. Such conditions are often found on north-facing slopes, low areas where cold air may accumulate during calm nights, and high latitudes or elevations where the annual net energy received by a tree may be low (Equation 5.3). At certain times throughout the year, conditions of extremely low temperatures and high precipitation may influence environmental conditions which directly or indirectly limit growth. A few of the important relationships leading to these conditions are shown in Part C (Fig. 5.10).

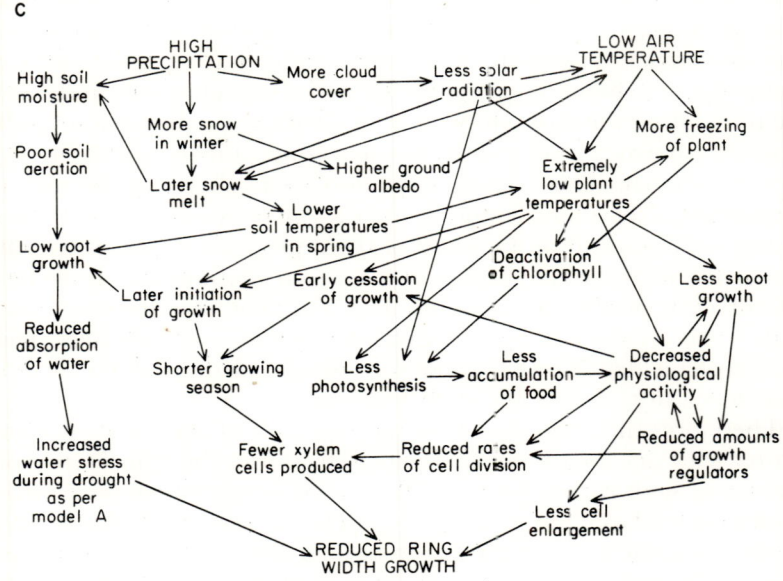

FIG. 5.10. Model Part C. A diagram representing some of the relationships that cause the climatic factors of high precipitation and low temperature to lead to the formation of a narrow ring. Compare with Figs 5.8 and 5.9.

As in Parts A and B, low temperatures are coupled with periods of high precipitation because increased cloud cover reduces solar radiation and the heat load. High precipitation in winter may result in increased snow cover, which raises ground albedo (a, Equation 5.1) and helps to maintain low air temperatures. Soil temperatures are determined by the temperature prior to the onset of insulating winter snows. Severe conditions in autumn, winter, and spring may ultimately curtail growth through freezing of the plant tissues and deactivation of the chlorophyll (see Chapter 4).

Cool temperatures during the growing season may also reduce physiological activity of the plant. When cambial activity is initiated it may be limited by low respiration rates, low concentrations of growth regulators, or low amounts of food (Fig. 5.7). Cool temperatures and deep snow cover may delay the onset of spring and cause initiation of growth to be late and slow. Unusually cold weather in late summer or in autumn may induce early cessation of growth. Both a shortening of the growing season and a reduction in the rates of cell division by cool temperatures can reduce the width of a ring. High precipitation during one season may saturate the soil in poorly drained sites, contribute to low aeration and low root growth, which in turn can induce water stress later in the year.

The growth–climate relationships shown in diagrams A, B, and C can vary in their relevance to a particular site throughout the year. Part C represents the most common relationships that occur in trees on temperate sites during abnormally cold months, while Parts A and B portray the most important relationships governing growth during the warmer seasons of spring, summer, and early autumn. With increasing elevation or latitude, Part C becomes applicable for an increasing number of months of the year. This varying relevance of the models to a particular site may itself be modeled mathematically in terms of the following equation:

$$W_t = G_t \left[\sum_{j=1}^{m} (\alpha_j A + \beta_j B + k_j C) \right] \qquad (5.4)$$

where W_t is ring width and G_t the growth function which varies inversely with tree age t. The values of α_j, β_j, and k_j are weights representing the importance of each part of the model, A, B, and C, and they assume varying positive values for m months throughout the year. The limiting effect on growth of low precipitation and high temperature, relative to the effect of high precipitation and low temperature, may be expressed for the jth month of the year by adding α_j and β_j and subtracting k_j. These values would represent the tree-growth response to combinations of dry–warm conditions versus cool–moist conditions throughout the year.

There are also some very important relationships involving a positive association of both temperature and precipitation which have been omitted from the diagrams. Low temperatures and low precipitation occurring late in the summer or early in autumn together can prolong cambial activity in one year with the added utilization of stored foods, which reduces the amounts of food available for the next year's growth. These conditions can also affect the formation of vegetative and reproductive buds which can exert an important control over cambial activity in the following year. There is also evidence (Chapter 4) for low temperatures and low precipitation in late autumn promoting winter hardening and hence favoring survival of living tissues through the cold winter season. There can also be limiting effects of both low temperature and low moisture occurring together as they can delay breaking of dormancy and shorten the length of the growing period for that year. Some other relationships not shown in any part of the model are the effects of climate on flower initiation, fertilization, fruit set, and maturation, and the interrelationships between the amounts of growth for the current season with growth for several prior years.

VIII. The Concept of the Climatic "Window"

The net effects of environmental factors on ring width can vary from species to species and from site to site. In addition, the climate during certain months of the year may have a greater influence upon growth than the climate of other seasons. The tree may be thought of as a "window" or filter which, by means of its physiological processes, passes and converts a certain climatic input into a certain ring-width output. Different species pass different amounts of information through their "windows" at different times of the year. These differences could be expressed by different absolute values of α, β, and k for different months within the year (Equation 5.4).

Figure 5.11 shows diagrammatically the mean climatic "windows" calculated from correlation analyses made on five different species or species complexes (Fritts, 1974). The diagrams are smoothed to express the relative effect of variability in climate at different times of the year upon the variability in width of the annual ring. The physiological seasons for these species are shown in Fig. 4.16. The ring-width variations for *Pinus longaeva* (P.l.) and *Pinus aristata* (P.a.), which grow at high elevations, reflect a large amount of climatic information from the summer and autumn prior to growth (25% and 26%). In the winter the climate accounts for 23% of the ring-width variation, while in spring it accounts for only 15%.

The ring-width variations of *P. flexilis* (P.f.), a species which has a larger ecological amplitude than *P. longaeva*, are influenced greatly by the climates of autumn (32%). The ring-width variations of *Pseudotsuga menziesii* (Ps.m.) and *Pinus ponderosa* (P.p.), two species which occur at lower elevations (see Fig. 4.16) and on drier and warmer sites, show the largest response to autumn and spring climate. *Pinus ponderosa* emphasizes the climate of late spring and early summer (13%) more than does *Pseudotsuga menziesii* with only 8% growth variation determined by summer climate. The ring-width variability of *Pinus edulis* (P.e.) and *Pinus monophylla* (P.mn.), species growing at the lowest elevations of the five groups shown in the figure, is more influenced by winter environment than the ring-width variability of the other four groups. The window for this particular group passes a large amount of variation in climate of autumn, winter, and spring (29%, 27%, and 20%). During the summer concurrent with growth, the window allows climate to affect only 6% of the variance in growth.

The area under each curve in Fig. 5.11 included in the period of cambial activity (g) is proportional to the relative effect on ring-width variability of the climate during the growing season. The area may be compared to the remaining area under the curve which is proportional to the effect of climate for the period preceding growth. Such comparison clearly illustrates that the climate during the growing season (partly described in Part A, Fig. 5.8) is less important to growth than is the

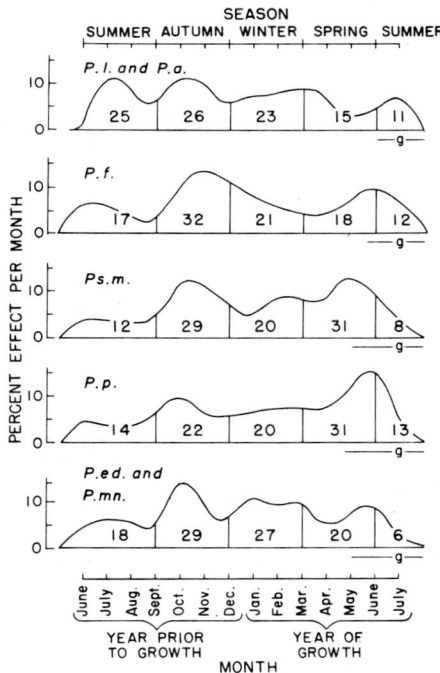

FIG. 5.11. The mean climatic "window" for five different coniferous species from a variety of arid sites in western North America. The plot shows the percent correlation of *mean* temperatures and *total* precipitation for each month with variation in the width of annual rings. Vertical lines designate the seasons and the numbers are the percentages of growth affected by climate during each season. The line marked g designates the portion of the climatic window included in the period of growth. The data are smoothed by averaging data for adjacent months. P.l. = *Pinus longaeva* and P.a. = *P. aristata* [25 sites analyzed]; P.f. = *P. flexilis* [4 sites analyzed]; Psm. = *Pseudotsuga menziesii* [40 sites analyzed]; P.p. = *Pinus ponderosa* [35 sites analyzed]; and P.e. = *P. edulis* and P.mn. = *P. monophylla* [15 sites analyzed].

climate prior to the growing season (partly described in Part B, Fig. 5.9). In general, the climate prior to the growing season exerts the dominant control over ring-width variation for the trees and sites used for dendroclimatic analyses from semiarid North America.

IX. The Concept of the Response Function

The correlation between climatic factors and growth may show striking changes from one month to the next. This relationship can be expressed as *response functions*, a set of 28 statistics (to be further described in Chapter 7) which represent the separate relative effects on ring width due to variations in mean temperature and total precipitation during each month of a 14-month period prior to and concurrent with the period of growth. A positive value for the response function statistic for a given month indicates a direct relationship (the greater the value of the climatic variable, the greater the growth); a negative value indicates an inverse relationship (the less the value of the variable, the greater the growth). An inverse effect may be just as important as a direct effect, only the sign of the coupling relationships is reversed. Figure 5.12 shows smoothed plots of the response functions for the same sites and species included in the analysis of the climatic "window" shown in Fig. 5.11. Three additional statistics shown on the right in the plots represent the mean effect of prior growth on ring width for lags of 1, 2, and 3 years.

It is evident from the mean response function shown in Fig. 5.12 that the ring-width variations in the lower elevation species, those plotted at the bottom of the figure, are more influenced by monthly precipitation than by monthly temperature. Precipitation is directly related to growth, especially in autumn, winter, and spring, while temperatures are often inversely related to growth in autumn, late spring, and summer. The growth of the two upper elevation species, those near the top of the figure, are affected equally by precipitation and temperature, although the signs of the relationship are different. There are also more months when high precipitation and low temperatures, as shown in Fig. 5.10, may limit growth (Fig. 5.12).

Ring width on the average for *Pinus edulis* (P.ed.) and *Pinus monophylla* (P.mn.) is largely dependent on variations in precipitation for autumn and winter. Ring widths for *Pinus ponderosa* (P.p.) and *Pseudotsuga menziesii* (Ps.m.) are dependent more on varying precipitation in autumn and spring. At the same time, variations in temperatures are less important to ring-width growth of these two species than to growth of the other four species. High growth in *Pinus flexilis* (P.f.), is dependent upon high precipitation and low temperatures in the prior summer, autumn, early winter, and late spring. High growth of *Pinus longaeva* (P.l.) and *Pinus aristata* (P.a.) is dependent upon high precipitation and high temperatures in the prior summer, high precipitation in autumn and early winter, low temperatures in autumn, winter, and spring, and low precipitation in early summer at the time when the trees begin to grow (Fritts, 1969).

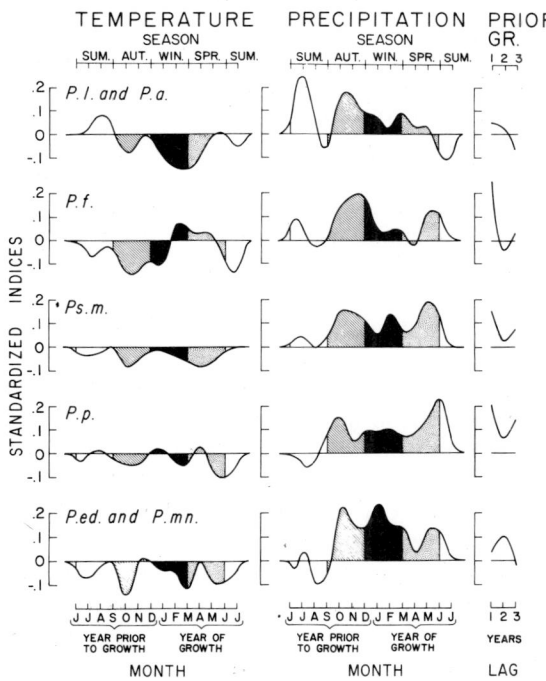

FIG. 5.12. The mean response functions for five different coniferous species from a variety of arid sites in western North America. The plotted values show the relative effect of variations in mean temperature and total precipitation for each month on the variations in ring width. A value above zero indicates a direct relationship (an increase in the variable is associated with an increase in growth), and a value below zero indicates an inverse relationship (an increase in the variable is associated with a decrease in growth). Differences in shading mark the various seasons. Sample size, species symbols, the growth period, and smoothing procedure are the same as for Fig. 5.11. Computational procedures for deriving response functions are described in Chapter 7.

The results shown in Figs 5.11 and 5.12 express the *average* climatic "window" and the *average* response function for a species growing on many sites. Large variations in climatic "window" and response functions also occur as a function of site factors and microclimate. The details of these variations are described more fully in Chapter 8.

X. Suitability and Limitations of the Growth Model

It is obvious that no single model can resolve all the complexities of the growth–climate system. However, the above process-network models along with Equation 5.4. allow for both direct and inverse effects, for variations in the climatic "window" and response function from one month to the next, and to some extent for interactions among the physiological environmental processes. The most useful attribute of the model is that the arrows do not represent an individual plant process but rather the net effect of the climatic factors on all relevant processes.

For example, it has been mentioned (Chapter 4) that higher than normal air temperatures can affect ring width positively or negatively, depending upon the particular processes which are most active. It can elevate plant temperatures so that growth rates increase; it can hasten root growth and favor absorption of soil moisture; or it can increase leaf development, exposing to light more tissue which can carry on photosynthesis. However, the elevated plant temperatures also can increase respiration rates; they can decrease the rate of net photosynthesis; and they can increase transpiration which can accentuate conditions of water stress. The first three conditions are usually beneficial to growth, while all the latter conditions are usually detrimental to growth. Early in the spring when the beneficial effects dominate, the model could indicate that the net effect of elevated temperatures is to increase growth. Later in the season when the latter conditions may dominate, the model could indicate that the net effect of elevated temperatures is to decrease growth.

The process–network model is especially suited for cases where variables can also compensate for each other's effects. For example, low soil moisture can be offset by the favorable effects of low temperatures on reducing transpiration, or the reverse might also occur with high soil moisture offsetting the effects of high temperature. Such situations have been observed for low-elevation and arid-site trees, where subfreezing temperatures can limit net photosynthesis and hence the production of food (Fig. 4.3), but low temperatures cause freezing and hold moisture on or within the soil in the form of ice and snow. As long as it is moderately cold, little evapotranspiration and soil moisture depletion can occur. When temperatures are less severe and rise above freezing, the increase in net photosynthesis is accompanied by an increase in evapotranspiration so that soil moisture declines. A return to cold temperatures stops photosynthesis but also converts the moisture back to ice so evapotranspiration ceases as well. As a result of the association between these two conditions, the unfavorable effects of low winter temperatures on photosynthesis are offset by the favorable effects which hold the moisture in the soil.

If a site on average is colder than the one described above, the detrimental effects of low temperatures may dominate because of deactivation of chlorophyll, so that low temperatures may be associated with low food reserves and the model would indicate low growth. However, if the site on average is warmer than the one described above, the effects of higher than normal temperatures may lead to increased evapotranspiration, so that higher temperatures rather than lower ones become associated with less available moisture, low reserves, and low growth.

The effects of many factors are often cumulative, so if they are limiting during many days throughout the year, they are modeled to have a marked net effect on growth. However, variable actions can sometimes involve short-period effects, such as the occurrence of freezing temperatures for one night during the growing season which can damage the cambium and produce a frost ring. The model used in this chapter does not handle the effect of single but severe events efficiently, but events which occur with some regularity are reflected in the mean and therefore *are* handled well.

When there is interaction among variables, their combined effects can be greater than the sum of their independent effects. For example, the combined effects can be many times more limiting if precipitation is low when air temperatures are high, than if either occurs separately. The model treats such interactions by combining the effects of high temperatures with those of low precipitation and the effects of low temperatures with those of high precipitation, which is a compromise short of using a logarithmic or other similar transformation.

Some factors may condition the plant or its environment so that the plant responds in a certain way at a later time. The rate of warming and the temperature regime in the spring is an example. Spring temperatures not only control evapotranspiration, but they also may control initiation and rate of root growth. In a normal spring when warming is gradual, root growth can maintain increasing absorptive surface to match the increasing water demand by the crown. If soil moisture is adequate, water stress is unlikely to be limiting to the tree. An unusually late and cold spring could delay the initiation of root growth and perpetuate low soil temperatures. If the weather suddenly turns hot and dry, transpiration may exceed the abilities of the roots to absorb water, both because of low growth rates of the roots and because the soil temperatures are low. Water stress would result until enough time had elapsed for soils to warm up and for root growth to reestablish enough active surface so that water absorption equaled water loss.

Another example of delayed response is the growing crown which expands and intercepts both more sunlight and precipitation, and casts greater shade. Thus, the regimes of soil moisture and temperature may be gradually altered, causing possible changes in the timing of initiation and cessation of growth. Model Part B attempts to account for some of these interacting relationships, but changes from one month to the next and one year to the next are assumed to be additive; that is, the effect in one year or month is added to the effect in other years and months. Any relations of these characters are averaged to obtain the net effect; interactions involving multiplicative effects between months and years are not reflected in the model.

There may be other kinds of effects of variables on growth rates over long periods of time. For example, insects (Fig. 5.6), hail, and frost can defoliate the tree and produce an immediate and marked change in a variety of processes affecting growth. Fire, death, or windthrow of a neighbor, or a landslide might open the forest canopy, increase the available light and food, and affect other conditions which could cause a rapid increase in the growth rate. A change in the soil moisture regime arising from an alteration of a river channel or fluctuation in lake level may drastically change the amounts of water and air in the soil and produce an abrupt change in tree growth.

Depending upon the circumstances, the growth rate may accelerate or decelerate rather abruptly, and may then be associated with a gradual return to the original unaltered state, or the tree may remain at the altered growth rate for the remaining portion of its life. There are also some factors that become increasingly or decreasingly limiting very gradually over long periods of time. Variations caused by changing tree height, changes in soil, slope degradation, and certain kinds of competition may be indistinguishable from the biological changes associated with increasing age because they occur so slowly through time.

The same climatic factor can vary in its effect on growth depending upon the time scale of the reaction. A drought during a given year can reduce ring width for that year and for several years thereafter. However, the effects of the drought on growth may become less if the drought persists for a long period of time, for the forest may gradually change in structure making the tree more drought resistant. For example, weak neighbors die, the forest stand becomes less dense, and the roots and crowns of the remaining individuals can expand and occupy the vacated sites. Each tree remaining in the stand is less limited by comparable moisture conditions at the end than at the beginning of the prolonged period of drought, because competition from neighbors is reduced. In this manner, the reduction in density of the forest community counteracts the deleterious effects of prolonged drought on the survivors. Thus, the ring widths in arid-site trees may become proportionately wider over a long period of time, even though drought conditions persist. In such cases, the ring-width variation will not record the long persistence of a drought. However, this may not apply to the most arid sites where trees are so widely scattered that they do not influence each other very much.

The generalities of the process–network model in Figs 5.8 through 5.10 appear flexible enough to handle a large number of the complicating and interacting relationships of growth and climate. While the models are representative of the most important relationships, they most certainly do not include all the individual processes that operate in the system, and they cannot be used in their present form to obtain quantitative estimates. However, they do adequately summarize the important physiological processes described in Chapters 2 through 4 which link precipitation and temperature to variations in ring width.

The following chapters will describe statistical and other methods for further study of this system and for testing the importance and significance of particular relationships. Techniques will also be introduced for calibrating specific ring-width variations with climatic variations in factors such as temperature and precipitation, and for reconstructing the same or similar climatic variations using ring-width variations for a number of species in a variety of locations and sites.

Chapter 6

The Statistics of Ring-Width and Climatic Data

I	Reliability of Measurements	246
	A. Collecting adequate field data	247
	B. Ensuring adequate replication	248
	C. Preparing the specimens	249
	D. Crossdating	249
	E. Careful measurement	250
	F. Testing for inhomogeneity of climatic data	252
II	General Statistics	254
III	Standardization	261
IV	Filtering Techniques	268
V	Other Methods for Assessing the Growth Curves	277
VI	Analysis of Variance	282
VII	Analysis of Chronology Error	290
VIII	Correlation Analysis	293
IX	Power Spectrum and Cross Power Spectrum Analyses	295
X	Variability in Statistical Characteristics of Ring Widths among Sites	300
XI	Statistical Characteristics of Ring Widths within a Tree	304

I. Reliability of Measurements

The discussion in Chapters 2 through 5 has focused on the general problem of modeling the ring-width/climatic system. Attention will now be directed to the problem of objectively quantifying characteristics of both the tree-ring and climatic data which are to be used for statistical modeling, calibration, and climatic reconstruction.

The basic data under consideration in this chapter are ring-width measurements and climatic records. Each data set is made up of measurements which are averaged and transformed in certain ways to facilitate study and analysis. The numbers that are generated from these data are collectively referred to as *statistics* and may represent the means (averages), the variability, or some other feature of the data. However, statistics are no more reliable than the data from which they are derived.

6. THE STATISTICS OF RING-WIDTH AND CLIMATIC DATA

The following paragraphs describe some routine but important procedures that ensure that the original tree-ring and climatic data are collected and processed correctly, maintaining the maximum amount of useful information as well as data reliability.

A. Collecting Adequate Field Data

The first procedure of importance is an obvious one: to obtain complete field notes and information about the sampled trees and sites, not only as a matter of record but for use in subsequent analysis. The samples should be labeled carefully and clearly, and the date, species, location, elevation, exposure, slope, and any other noticeable characteristics should be recorded. Exact specification of location in terms of latitude and longitude is preferable to the use of local names since the latter often lead to ambiguity. Without proper site location on a map, there can be no assessment of distances to recording weather stations and to other stands, no evaluation of topographic factors, and little possibility of returning for subsequent collection. It may be helpful to cite the name

Table 6.I A checklist of collection and site information

1. Name and address of submittor. Name and address of collector(s).
2. Date collected.
3. Source of collected material (living trees; historical; archaeological; geological; or other source).
4. Name of site. Single or merged site?
5. Site elevation. Latitude. Longitude. Country. State. County.
6. Collected genus. Species. Variety (optional).
7. Number of trees collected. Total number of radii. Number of radii per tree.
8. Type of measurement (total ring width; earlywood; latewood; density).
9. Unit of measurement used (100ths mm; 100ths inch; other).
10. Beginning year of dated chronology. Ending year of chronology.
11. Name, type, and scale of map used. Is map available? Are site photos available?
12. Slope angle of site. Direction slope faces.
13. Predominant soil characteristics (organic, clay, silt, sand, loam, gravel, rock fragments, bedrock).
14. Soil drainage. Soil chemistry. Parent material of soil.
15. Was the collected species dominant (in crown area coverage)?
16. What percent of total forest was the collected species?
17. Associated species. Vegetation zone or life zone.
18. Canopy position and competition.
19. Is forest growth natural, successional; natural, stable; cultivated, native species; or cultivated, introduced species?
20. If there is site disturbance is it man-caused; natural; or of unknown cause?
21. Are climatic data available for locale of collection site?
22. Additional comments.

and edition of the map which was used to identify the precise location. If one is likely to return to the area at a later time for recollection, distances from neighboring towns and roads are helpful, as well as permanent labels which can be attached to the trees that are sampled. Records of slope, exposure, soil, plant community characteristics, and obvious site disturbance can be used to account for unusual results or to select data to test future hypotheses.

Table 6.I is a checklist of site information requested from contributors to the International Tree-Ring Data Bank being established at Tucson, Arizona. Replicated well-dated ring widths are being placed on file along with a certain amount of site information. The first 10 items are required in order to identify the contributor, the character of the tree-ring materials, and the location of the site. The remaining items are listed as optional but are considered to be important for maximum scientific usefulness.

Other kinds of data may also be recorded, depending upon the nature of the problem and the goals of the project. The important point is that objective records must be kept in some way and that they should be summarized and made available in a usable form along with the ring-width measurements. Well-defined and complete records of site information assures that carefully dated and processed tree-ring materials will be of maximum value to future scientific analysis.

B. Ensuring Adequate Replication

A second important procedure is to ensure data reliability by sampling adequately. As mentioned in earlier chapters, two cores from each of 20 or more trees may be a typical sample, although smaller samples can be used where there is good crossdating and a large amount of ring-width variability. However, a sufficient quantity of materials should be obtained to allow for data loss when certain samples are found to be anomalous or to yield questionable dating. After a long, tiresome hike into a remote area, it is a temptation to stop collecting after the minimum number of trees have been sampled. Yet insurmountable difficulties are often encountered back at the laboratory which can invalidate a portion of the sample; and in such circumstances it is comforting to have collected enough extra specimens to compensate for the data loss due to the difficulties.

Some of these difficulties can be avoided by examining the sample in the field immediately after collection, before it is labeled and packed for transit. The rings for the last few decades can be surfaced using a sharp knife and examined with a hand lens to check whether there are any obvious growth anomalies, whether the rings are sufficiently variable, and whether crossdating is evident. This information can be used to judge the potential quality of the record and to help decide whether the site and species are appropriate and how many trees should be sampled. A procedure using analysis of variance statistics, described later in this chapter, can be utilized to evaluate past decisions concerning choice of sites, sample size, and design of analysis as a basis for planning future improvements.

C. Preparing the Specimens

A third important precaution, also seemingly obvious, involves mounting and preparing specimens so that the rings are clearly visible and easy to examine. Some workers have floundered because their samples were incorrectly mounted and the rings could not be seen well enough to be dated. The Laboratory of Tree-Ring Research uses pregrooved sticks in which the individual cores are mounted to expose the transverse (not the radial) surface. If a razor rather than sandpaper is to be used in surfacing, the best visibility of the rings is obtained if the cores are rotated approximately 30° from their vertical position (Stokes and Smiley, 1968). Bulk samples should be cut along the transverse plane, trimmed to facilitate ease in handling, and surfaced with knife or sandpaper.

If the materials are sanded, coarse paper is used first, followed by finer grit sizes. For example, a sequence of sandpaper grit sizes might be 180 or 220, 320, 380, and 400 or 600, depending to some extent upon the quality of the cut surface and upon the hardness of the wood. Finishing with soft felt, lambs wool, steel wool, or very fine emery paper can produce an exceptionally fine surface, the choice again depending upon the character of the wood. The interested reader should refer to Stokes and Smiley (1968) for details on specimen preparation.

D. Crossdating

As described in Chapter 1, crossdating is perhaps the most crucial procedure in tree-ring analysis, although one commonly finds tree-ring and climatic studies where only ring counts are made. Chronologies resulting from such studies can contain errors due to counting, mistaken identification of ring features, and ring absence. A surprising number of absent sets or false rings occur, even for sites that are relatively temperate. As emphasized earlier, crossdating is of vital importance in assuring that each ring width and climatic value is placed in its proper time sequence.

In certain cases where few rings are missing and troublesome false rings rarely occur, the rings can be measured first, certain computations made, and the data plotted to facilitate the crossdating process. Since this practice may encourage excessive reliance on statistics and on ring-width measurements as opposed to other helpful ring characteristics, it is recommended that all rings be examined, under magnification if required, and that features other than ring width be used to confirm any decisions that were made using the ring-width data and graphical analysis.

The crossdating obtained in most North American studies by professional dendrochronologists at present involves visual examination of every specimen. The Laboratory of Tree-Ring Research requires that a routine dating check be performed by a second person, and if serious discrepancies are found, the dates for the entire set must be verified by yet another worker. No specimen is selected for climatic analysis if there is *any* question as to the dating, even if the problem involves only one ring. Not only are the specimens checked against one another, but the entire sample is checked against chronologies from neighboring regions.

It is highly likely that some of the tedium involved in the visual crossdating operation will be increasingly facilitated by computer analysis. However, at present there is no fool-proof shortcut to the procedure of careful visual comparison (see Chapter 1, and Table 1.II).

E. Careful Measurement

A fifth reliability check involves verification of the precision of measurement. After all rings are dated and measured, a number of samples are selected at random, and one to three intervals of 20 successive rings are remeasured by an independent worker. These remeasurements are subtracted from the first measurements to obtain a set of 20 differences. Each of the differences is squared, the 20 squares summed with the limit of acceptability set at a sum of 0·10 for arid-site conifers and at a sum of 0·23 for *Quercus alba* (Fig. 6.1). The acceptable limits were obtained by plotting the frequency of occurrences of the summed squared differences obtained by expert measurers (see Fig. 6.1). The figure shows that the sums equal to or less than the limit are easily obtainable by a careful worker, while those greater than the limit may indicate the possibility of unacceptable measurement. In the latter case, the cause of the errors is determined and all past work of the particular measurer may be checked more closely. The Tree-Ring Laboratory of the U.S. Geological Survey measures each specimen twice and calculates an error term from the differences. If the standard deviation (see Equation 6.2) of

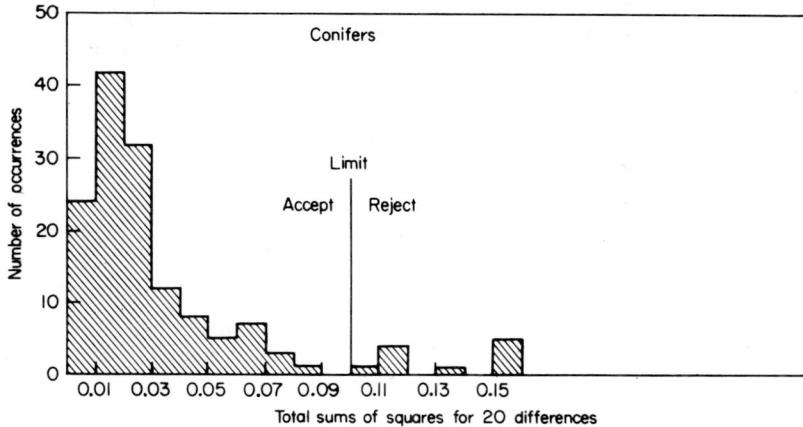

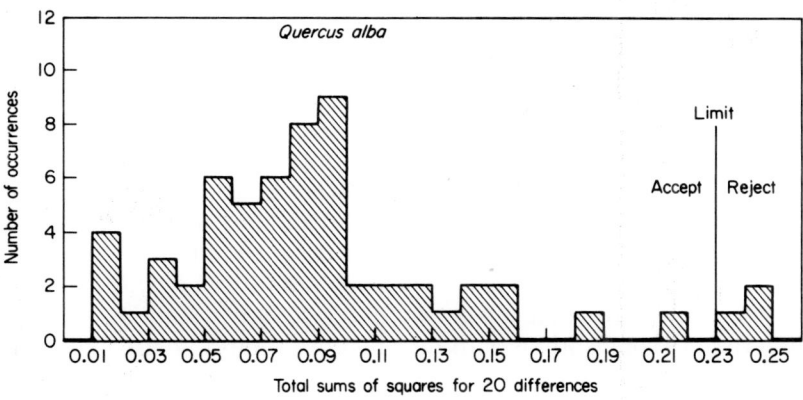

FIG. 6.1. A test of measurement accuracy can be made by comparing measurements of particular operators to those of experts. The above graphs are the frequency distribution for sums of squares of differences between replicate measurements of 20 successive ring widths obtained by experts on (1) arid-site conifers and (2) temperate forest *Quercus alba*. Each set of 20 measurements (expressed in millimeters) was made by different experts, and the differences between the first and second measurement for a given set of rings were computed, squared, the 20 squares summed, and the result tabulated in one of the frequency classes. Using the results from the figure, if the sum of 20 squared differences for a particular remeasurement from the first measurement is equal to or less than the limit, the measurements are judged to exhibit acceptable error, but if it is greater than the limit, the measurements are judged to be unacceptable and are sent back for reassessment and further checking.

the differences exceeds 0·10 mm the sample is remeasured (Phipps, personal communication).

The ring widths are also checked after they are key punched for analysis. This function is accomplished by running a computer program, *RWLST*, which checks for certain types of errors, calculates 20-year means, plots them, and lists all data. The errors in the computer listing are easily spotted and the plots scanned before choosing the curve-fitting option that is to be used in subsequent analysis (see Section III of this chapter and Table 1.II for time estimates).

F. Testing for Inhomogeneity of Climatic Data

Climatic data should also be carefully examined for missing information and for inhomogeneities. Records with large numbers of missing data, anomalous changes attributable to instrument relocations, or trends caused by nonclimatic changes such as urban growth, must not be used in dendroclimatic analysis unless they are corrected.

When less than 5% of the climatic record is missing, the missing values can often be estimated from those of a neighboring station, if they are available. Not only should the climatic record from the neighboring station be complete, but its location should be as similar in elevation, exposure, and topography as possible. A method for estimating precipitation data at one station from that at another utilizes the ratio of mean monthly precipitation for an interval of time that is common to both records (McDonald, 1957). The ratio is multiplied by the data that are present to estimate the missing value. At least two stations within 100 miles (161 km) of a station with missing data should be used, and these data should be averaged to obtain the best estimate.

Missing temperature data are usually estimated by a simple linear regression as described in standard textbooks on statistics. The monthly data common to neighboring stations are used for calibration, and the appropriate data are substituted into the equation to obtain a statistical estimate for the missing values. Separate equations are developed for each pair of stations and for each month to be estimated.

All climatic records should be examined and tested for inhomogeneities, because if anomalous trends or abrupt changes attributable to instrument relocations or other nonclimatic factors are present, they are likely to introduce significant error in the calibration relationships. Station histories which are published along with the climatic record may be checked for changes in either the instrumentation or geographical location of the station. Kohler (1949) describes one method for testing homogeneity of precipitation records (Fig. 6.2).

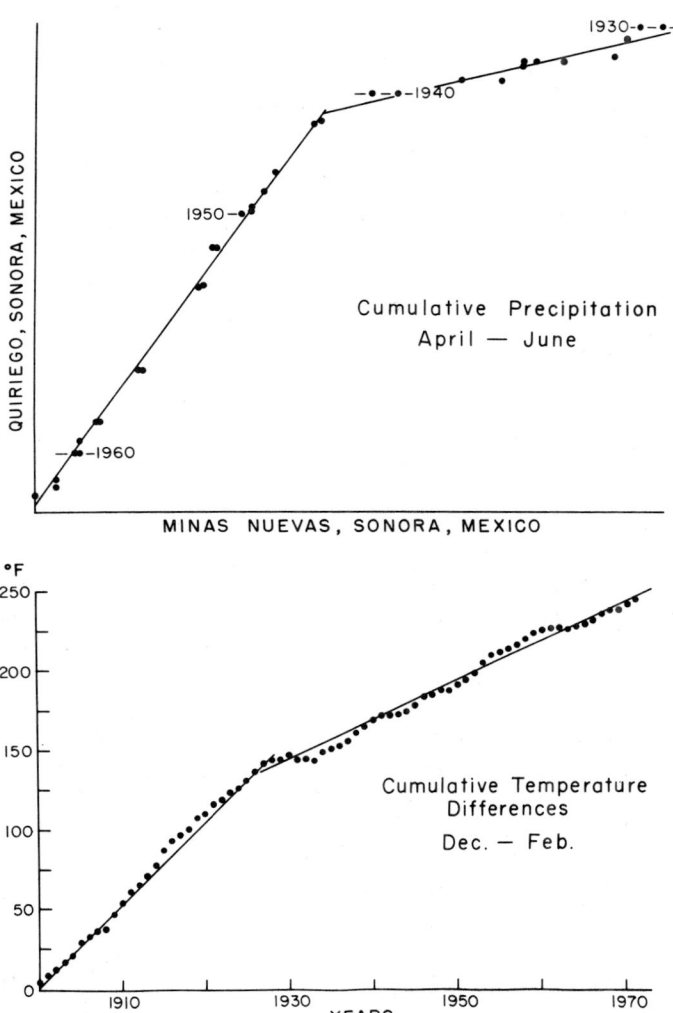

FIG. 6.2. Tests for homogeneity of precipitation and temperature at two stations in Mexico. Analysis of precipitation utilizes cumulative totals for April through June. The totals for one station are plotted as a function of the totals for the other station starting with the most recent record; the axes were scaled (not shown in figure) to facilitate comparisons. Analysis of temperature utilizes the cumulative differences between mean December through February temperatures for each year which are plotted as a function of the year. Inhomogeneity is indicated in both types of analyses by abrupt changes in slope. Straight lines are drawn by eye to accentuate the inhomogeneity. See text for explanation. (Data courtesy of Arthur Douglas)

Monthly precipitation data from each of several neighboring stations are summed to obtain totals for each season, and these data are compared. This procedure is facilitated by summing the seasonal data (monthly data are often too variable), starting with the most recent year, to obtain cumulative totals, which are plotted as a function of the equivalent totals for another station (Fig. 6.2). If the plotted totals fall along a straight diagonal line, the data for the two stations are judged to be relatively homogeneous. If there is a change in the slope, only one of the records is homogeneous. If both stations have been moved or affected by changes at different times, several variations in slope appear in the plot.

It may be necessary to compare several records to identify which one exhibits a particular inhomogeneity. One record already found to be homogeneous can be used to check a variety of neighboring stations. This technique, sometimes referred to as *double-mass analysis*, also provides a means of adjusting the data (Kohler, 1949). The precipitation associated with the shorter segment or the earlier record is multiplied by the ratio of the two slopes. Kohler (1949) provides a more complete explanation of this procedure.

The test for homogeneity of a temperature record requires somewhat more elaborate computations (Mitchell *et al.*, 1966). Sometimes, however, inhomogeneities can be identified from a simple plot of the temperature data averaged for a particular season or from the *first differences* of the averages (the value of each datum subtracted from its successor). The data are plotted as a function of year and the plots compared to those of other stations to identify errors as spikes or anomalous values in a particular record.

A shortcut method developed by LaMarche involves subtracting the seasonally averaged value for temperature in one record from the value for the same season in another record. These differences are then summed over time to obtain a cumulative difference (Fig. 6.2). As in the analysis of precipitation, a discontinuity in one of the records is accompanied by a change in slope of the cumulative temperature differences, and the ratio of the two slopes on each side of the change can be used to correct the faulty record. See Mitchell *et al.* (1966) and Panofsky and Brier (1968) for a detailed discussion concerning analysis and correction of climatic data.

II. General Statistics

Ring widths which are limited in large part by variations in the operational environment vary from one year to the next in a more or less irregular or quasi-random manner. It is customary to express these

6. THE STATISTICS OF RING-WIDTH AND CLIMATIC DATA

variations as a series of values ordered chronologically according to year of formation, the resulting sequence being referred to as a *time series*. As has been discussed in earlier chapters, the size of growth rings in trees is the result of a number of causal factors acting within the tree and within the tree's environment. The manner in which these causal factors act to form a time series is referred to as the *generating process*. The primary objective in the remaining portions of this chapter is to describe a variety of techniques for characterizing, analysing, and handling such time series data. Many of the techniques could be applied to climatic time series as well as tree rings.

The general characteristic of a ring-width time series can be described by a number of standard statistical parameters. The simplest of these is the mean m_x, which is defined as

$$m_x = \frac{1}{n}\sum_{t=1}^{t=n} x_t \qquad (6.1)$$

where Σ is the symbol for summations of the elements following it, x_t is the ordered ring-width sequence such that time (indicated by subscript t) varies sequentially from 1 (the calendar year of the earliest dated ring) to n (the last calendar year). The symbols above and below the summation sign indicate these two limits. The *mean* is the point that is closest to all values for the data set analogous to the center of gravity.

A second statistic typically generated from a ring-width time series is the *variance* (s_x^2), which is a measure of the scatter of values about the mean and is calculated as

$$s_x^2 = \frac{1}{n-1}\sum_{t=1}^{t=n}(x_t - m_x)^2 \qquad (6.2)$$

where the symbols are the same as in Equation 6.1. This variance is often converted to a different statistic, the *standard deviation* (s_x), which is simply the square root of the variance. The variance is used for certain statistical tests, while the standard deviation is easier to visualize as a measure of scatter of the data values from their mean expressed in the units of the original measurements. If the data in the time series are *normally distributed*, that is if their distribution closely approaches that of a statistically normal curve, two-thirds of the data are expected to lie within a distance of one standard deviation on each side of the mean. For large samples, approximately 95% of the data lie within a distance of two standard deviations on each side of the mean.

While these two statistics represent specific measurements obtained from a given set of values, they will be somewhat different if another data set is taken from the same population. Therefore each statistic should be regarded as an *estimate* of the true value which can only be computed from the entire population, or *statistical universe*. The following discussion will regard all statistical measurements as statistical *estimates*. It is possible to use the variance of the available data to obtain an estimate of how a particular statistic, such as the mean, varies from the mean of the entire population. This particular estimate is referred to as the *standard error* of the mean (SE_m) and can be calculated as

$$SE_m = \sqrt{\frac{s_x^2}{n}} \qquad (6.3)$$

where n and s_x^2 are the same as in Equation 6.2.

For large samples, the mean of the entire population, the *true value* in terms of the statistical universe, will lie within a distance of two standard errors from the sample mean in 21 out of 22 cases and within three standard errors of the mean in 369 out of 370 cases (Ezekiel and Fox, 1959). The possibility for error is greater for means of samples with less than 30 items because characteristics of different samples fluctuate by chance from those of the statistical universe, and the estimate of the standard error itself will vary from one sample to the next. The amount of this increased error or bias is a function of the sample size (Ezekiel and Fox, 1959) so that the bias or expected variation for a standard error based upon a sample of three trees is 50%; a sample of four trees is 41%; a sample of five is 35%; a sample of 10 trees is 24%; and a sample of 20 trees is 16%. Thus, in cases of a mean of four observations, the estimate of the standard error obtained from Equation 6.3 is itself in error, and its value should be adjusted upward by 41%. If the mean is based upon 10 observations, the value should be increased from the calculated estimate of the standard error by 24%. For explanation of this correction see Ezekiel and Fox (1959).

As a result of the above error, very small samples are subject to sizeable random variations; and increasing sample sizes from 5 to 10 trees, for example, can lead to considerable reduction in the expected random variation with improvement in the reliability of all its statistics. If samples are not available in sufficient numbers for a particular site, it may be necessary to avoid using the data or to combine them with those from a similar geographically related sample.

Because of the large error in small samples, data to be entered into the International Tree-Ring Data Bank mentioned earlier must be replicated with a minimum of 10 different trees, although additional replication including two samples per tree is recommended. Later in this chapter it will be shown that reliability of a ring-width chronology is not only dependent upon the number of items included in the collection but

6. THE STATISTICS OF RING-WIDTH AND CLIMATIC DATA

also upon the amount of heterogeneity or homogeneity of the materials from which the data were sampled.

The *correlation coefficient* is another statistic which measures interdependence of association between two data sets, x_t and y_t, where t again represents the time dimension. As such, the correlation coefficient is used in dendroclimatology to measure associations between two time series, such as chronologies from different trees or sites, or a chronology and a climatic sequence. The correlation coefficient (r_{xy}) is defined as

$$r_{xy} = \frac{\sum_{t=1}^{t=n}(x_t - m_x)(y_t - m_y)}{(n-1)s_x s_y} \tag{6.4}$$

where m_x, m_y, s_x, s_y are the means and the standard deviations of the two sets of data; and n is the number of items compared, namely, the sample size. The correlation coefficient between the two data sets being compared can range from an upper value of $+1$, which indicates perfect and direct agreement, to a value of -1, which indicates perfect and inverse agreement. If the two data sets are completely independent or random with respect to one another, the correlation coefficient takes a zero value.

The correlation coefficient is used in dendroclimatology not only to measure association between two different time series, but also to measure associations between items lagged in time. For example, the ring width for a particular year (t) can be correlated with the ring widths in the previous or following year ($t-1$ or $t+1$) or correlated with rings widths at lags greater than 1. Figs 1.10 and 1.11 diagram such relationships and Chapters 2 and 5 have described some of the biological causes of such time-related interdependence. To obtain the measurement the correlation coefficient (r_{xy} Equation 6.4) is calculated for terms x_t and x_{t+L}, where, in the case of tree rings, L is the lag in years of the second ring behind the first and n is reduced in number by the length of the lag, L. The resulting statistic is referred to as the *autocorrelation* (or *serial correlation*) *coefficient*. When $L = 1$ year, it is the first-order autocorrelation coefficient; while a lag of 2 gives the second-order autocorrelation coefficient; and a lag of m is the mth-order autocorrelation coefficient. Of course if the items in a series are completely random with respect to their positions within the time series, the values of the computed autocorrelation coefficients at all lags will be small and will vary in a random fashion about a mean value of zero. However, many time series dealt with in dendroclimatology, especially those derived from ring widths, are autocorrelated to a significant extent.

Another statistic designed especially for tree-ring analysis measures the relative difference in width from one ring to the next and is referred to as *mean sensitivity* (ms_x). Douglass (1936) describes this statistic as the "mean percentage change from each measured yearly ring value to the next." The average mean sensitivity for a series is calculated as

$$ms_x = \frac{1}{n-1} \sum_{t=1}^{t=n-1} \left| \frac{2(x_{t+1} - x_t)}{x_{t+1} + x_t} \right| \qquad (6.5)$$

where x_t is each datum and the vertical lines designate the absolute value (neglecting the sign) of the term enclosed by them. The denominator of the term scales the absolute values of the differences between adjacent ring widths, x_t and x_{t+1}, so that the differences are proportional to the average of the two widths. The values of mean sensitivity range from 0 where there is no difference to 2 where a zero value occurs next to a nonzero one in the time sequence.

In general, each series of ring-width measurements can be regarded as a time series made up, at least in part, of randomly varying components. Usually, the sizes of successive growth rings are statistically dependent on one another due to persistence, trends, cycles, or other nonrandom components produced by climate and the generating process of tree growth. The dendroclimatologist, therefore, is concerned with identifying as clearly and unambiguously as possible the precise nature and extent of both the randomness and nonrandomness in each ring-width time series, not only for accurate reconstruction of past climate, but also to provide information on the nature of the tree growth-generating process itself. Also, certain undesirable characteristics must be properly assessed and analyzed in an appropriate manner designed to meet important statistical assumptions.

For example, when successive values of a time series such as ring widths are found to be statistically interdependent, the *effective size* of the sample (the number of elements which are truly independent) is reduced. Therefore, special precautions are necessary when using familiar textbook methods for evaluating the characteristics and statistical significance of time series interrelationships (see Chapter 7 and Mitchell *et al.*, 1966).

One type of nonrandomness is *autoregression*. The term *regression* is used in cases where one value is affected by the condition of another value, and the prefix, *auto*, is used when the relationship involves values ordered in a single time series. *Correlation* refers to association of two variables without implying the direction of the dependence. However, autocorrelation coefficients calculated for a number of different lags can be used to estimate the nature of autoregressive linkage. The simplest type of dependence is first-order autoregression, representing a linkage of each value in a time series with only the condition of the item immediately preceding

6. THE STATISTICS OF RING-WIDTH AND CLIMATIC DATA

it. In such cases the *first-order autocorrelation* (L = 1) is all that is needed to estimate the degree of the relationship, and the autocorrelations for higher lags (L > 1) will decrease in the following fashion:

$$r_L = r_1{}^L \qquad (6.6)$$

where r_L is the correlation between the items at the specified lag L, r_1 is the first-order autocorrelation, and the superscript L is the power to which the value for r_1 must be raised in the calculation. Thus if r_1 has a value of 0·5, the correlation at a lag of 2 is $(0·5)^2$ or 0·25, and the correlation at a lag of 3 is $(0·5)^3$ or 0·125. The theoretical values for higher-order autocorrelations remain positive although they approach zero.

Growth cycles, long-term variations in ring width lasting for several years or decades, or trends in the data all affect autocorrelation which may be either positive or negative depending upon the lag and the nature of the relationship. In the case of cycles, the autocorrelation coefficients will be positive from a lag of 1 up to a lag that is approximately one-fourth of the cycle wavelength. The coefficients are negative for higher-order lags, with the largest negative value occurring at a lag equivalent to half the wavelength of the cycle. In the case of trends or very long-term persistence, the higher-order autocorrelation coefficients remain positive and higher at greater lags than in cases where only first-order autoregression is present (Equation 6.6). It is most common to encounter a mixture of several types of nonrandomness in time series of ring widths.

Autocorrelation coefficients calculated for a number of lags may be plotted to help identify the nature of the nonrandomness in a time series. However, as the lags become larger the reliability of the autocorrelation estimate decreases because the number of observations common to the lagged and unlagged set becomes smaller. This does not create serious difficulty in analyses if the number of lags does not exceed 20% of the number of observations (Jenkins and Watts, 1968). Thus, for a time series of 100 items, only the autocorrelations for lags 1 through 20 should be considered.

It is useful to deal with these types of nonrandomness as if they could be represented by cycles with varying wavelengths. For purposes of discussion this text will consider all nonrandom variations in ring width that last longer than eight years (represented by cycles with wavelengths greater than eight years) as *low-frequency variance*. Such long-term variations in ring widths can arise from changes in tree structure or the environment, including relatively long-term variations in climate. When wavelengths are greater than the length of the entire ring-width series, the variance is referred to as a *trend*. Trend results from such things as the changing growth potential of the tree resulting from increasing age, successional alterations in the forest community, geological changes, or very gradual variations in climate. Changes which are of shorter durations, representing cycles with wavelengths of eight years or less, will be considered as *high-frequency variance*. Included in this category would be year-to-year ring-width variations, fruiting cycles in trees, and short-term variations in climate.

In addition to the presence of autocorrelation, cycles, or trends, the variance of ring-width series is usually nonhomogeneous through time in that it decreases with increasing age of the tree (Fig. 6.3). Also, the mean ring width, the standard deviation, and the autocorrelation structure can vary markedly depending upon site conditions as well as tree age. When

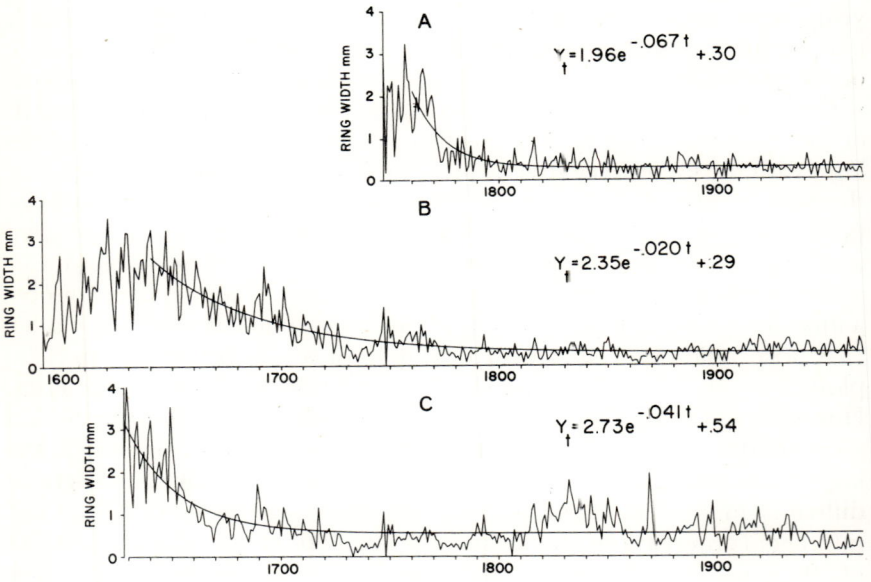

FIG. 6.3. Time series of ring widths from three Douglas-fir (*Pseudotsuga menziesii*) growing in New Mexico, plotted as a function of the dated year along with the exponential curve that was fitted to each data set and the equation that describes it. (See Equation 6.7 and text for explanation.)

such variations prevail, the data must be transformed by a particular procedure referred to as *standardization*, to form a new time series with trend removed and a mean and variance that are more homogeneous with respect to time (Matalas, 1962). Details of the standardizing procedure and its effects on statistics are described in the next section.

III. Standardization

Three unstandardized series of ring widths are plotted as a function of time in Fig. 6.3, and some of the statistics derived from them are shown in Table 6.II. Each series shows a general decline in ring width and in variance, both of which are associated with increasing tree age (for a biological explanation of this trend see Chapters 1 and 2). The A series in Fig. 6.3 has a mean of 0·53, a standard deviation of 0·42, a first-order autocorrelation of 0·75, and a mean sensitivity of 0·64 (Table 6.II). The large value of mean sensitivity indicates the presence of considerable high-frequency variance. A trend of declining growth with increasing age for the first 70 years in the life of the tree is apparent from the figure. The plot for series B shows a more gradual trend than that for A; the lower mean sensitivity indicates less high-frequency variance; and the presence of more low-frequency variance as evidenced by the high autocorrelation and the excursions in ring widths from the fitted curve that last for 10 or more years (see 1730's, and 1857–66 for negative excursions, and 1687–96, 1760's, and 1913–1922 for positive excursions). Series C exhibits less high-frequency variance as indicated by lower values of mean sensitivity,· trend is also apparent, and there is more low-frequency variance than in the other two, represented by several long excursions from the fitted curve persisting more than 20 years.

An examination of other statistics included in Table 6.II shows that the mean ring width is highest for series B and lowest for series A, and the standard deviation for ring widths is markedly lower in A than in B and C. As would be expected from the presence of trends, first-order autocorrelations are high in all three series of ring widths.

TABLE 6.II Statistics of three ring-width series and their indices[a]

	Series		
Statistic	A	B	C
Mean ring width – mm	0·53	0·80	0·73
Mean index	1·00	1·00	1·00
Standard deviation of ring widths – mm	0·42	0·60	0·60
Standard deviation of indices	0·54	0·45	0·55
1st order autocorrelation of ring widths	0·75	0·81	0·78
1st order autocorrelation of indices	0·10	0·31	0·60
Mean sensitivity of ring widths	0·64	0·43	0·39
Mean sensitivity of indices	0·64	0·44	0·39

[a] See Figures 6.3 and 6.7.

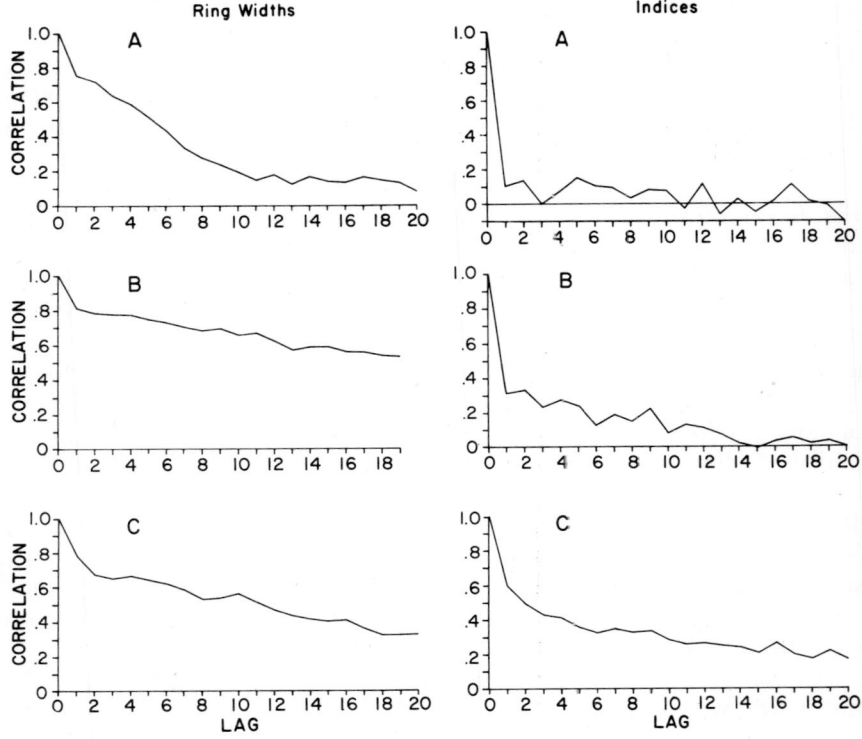

FIG. 6.4. Autocorrelations for lags from 0 to 20 years for each ring-width series shown in Fig. 6.3 and for each derived index series shown in Fig. 6.7.

Figure 6.4 (left) shows plots of autocorrelation coefficients for the three ring-width series in Fig. 6.3 for lags 0 to 20 years. At lag 0 each series is correlated with itself, so the value of the coefficient is a perfect 1·0. The values at lags greater than 0 decline from left to right with increasing length of the lag. For series A the plotted values most resemble the expected decline for a first-order autoregressive process (Equation 6.6), indicating the least amount of long-term or very low-frequency variance. The values of the autocorrelation coefficients in B decline in a more or less linear fashion after a lag of 1, although these values are somewhat larger than in A and C due to the greater length of the trend in the B series (Fig. 6.4).

Repeated experience has shown that the plots in Fig. 6.3 are typical of ring-width data derived from many coniferous species growing on drought-subjected sites. Such curves are generally satisfactorily estimated by fitting a curve to the data with the following *exponential* form:

$$y_t = ae^{-bt} + k \qquad (6.7)$$

where the values of a, b, and k vary from series to series depending upon the slope of the curve required to fit the data, e is the base of natural logarithms and y_t is the expected growth at a given year t. The values of t as in previous equations vary from 1 to n. The curves shown in Fig. 6.3 represent the best statistical estimate of this exponential form to the particular ring-width sequence. The values for the equation used to generate the curves are shown above each plot.

Often there is an increase in ring width for the first 10 to 30 years in the life of a tree and these portions cannot be adequately fitted by the exponential curve (see the early portions of A and B; Fig. 6.3 where no curve is drawn). These data may either be disregarded, fitted with an upward-trending straight line, or the entire curve fitted with a polynomial equation or other function capable of expressing an increase in width of the first-formed rings. Since these early rings often provide the least reliable climatic information anyway, they can be discarded without any substantial loss of information.

A variety of biological growth functions has been used for the curve-fitting process, such as parabolas, hyperbolas, logarithmic functions, polynomials, and moving averages. The use of some functions is defended on theoretical grounds, while the use of others, such as polynomials and moving averages, is justified because they can be applied to a wide variety of situations. However, the exponential function has been found to be adequate for many North American conifers because it approximates the various parabolic, hyperbolic, and logarithmic forms, and resembles the declining rate in the conifer biological growth function.

As a result of the many dendroclimatological studies now being attempted on deciduous species, in dense forests, and in other new regions, there are more complications which cannot be handled by fitting an exponential function. These complications result from a variety of changes that can occur throughout the life of a tree within a forest due to stand disturbance and the changing forest environment. The more versatile orthogonal polynomial curve, using coefficients at higher and higher powers (Figs 6.5 and 6.6), is now applied in such cases, although care must be taken not to use it where interesting climatic information might be approximated and removed by the polynomial function. There are many situations, as illustrated in Fig. 6.6, where the complex curves described by a polynomial equation with many coefficients are the only means of obtaining an adequate growth function. In such cases there is a risk that some climatic variation will be removed by the more flexible curve-fitting option. Because of this possibility a test is made as coefficients are added, and the procedure is terminated before the curve becomes too flexible when 5% or less of the variance is reduced by the addition of a coefficient.

An alternative to calculating a complicated statistical equation is the construction of a growth curve by hand using a graphical approximation, as was done by early dendroclimatologists. This method involves computing 10- or 20-year means at 10-year intervals, drawing a curve or straight line through the points, reading the curve value for every year, and adjusting the curve upward or downward so that its mean matches the mean of the ring widths to which the curve was fitted (Stokes and Smiley, 1968). When the graphical procedure is used, no values for the constants a, b, and k are available.

A computer program for ring-width series standardization developed by Fritts *et al.*, (1969) attempts first to fit the exponential form of Equation 6.7 to each ring-width series with the following restrictions: If the value of b in the equation is found to be positive rather than negative (that is if the slope of the exponential curve is upward and thus opposite

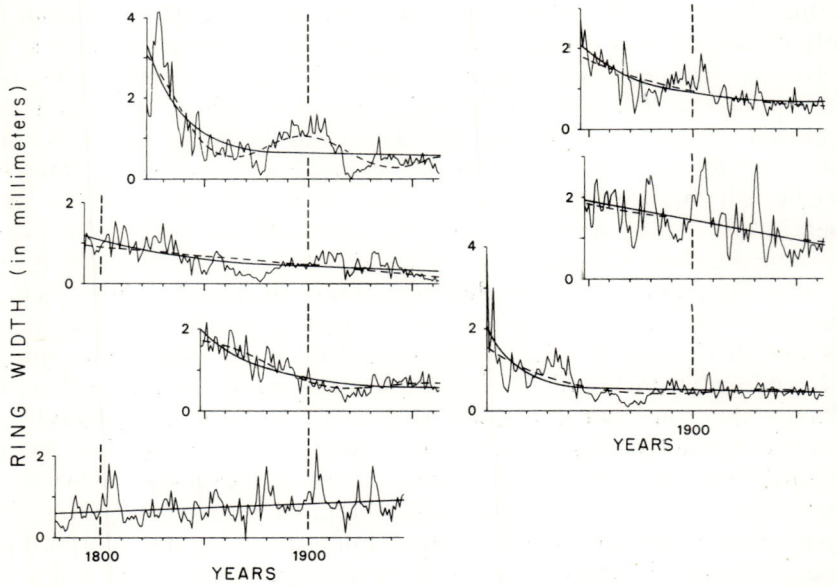

FIG. 6.5. Seven ring-width series which were fitted by an exponential or linear curve (solid line) and by a polynomial curve (dashed line).

the slope expected for normal trees) or if *a* and *k* are negative, the computer rejects that particular exponential equation as it is inappropriate to the negative exponential growth model (see Fig. 6.3). In its place a straight line is fitted to the ring widths as a function of time depending upon specifications provided by the investigator. In certain situations there is justification for using only a straight line with a negative or zero slope. In such cases if the slope is positive, the least-squares fitted line or curve is rejected and a horizontal line is fitted through the mean ring width as the expected growth at all ages. Since the program is designed to be flexible, it requires that the investigator choose the specific kinds of curves to be fit to the data, and that he specify whenever the growth model changes between one ring-width series and the next.

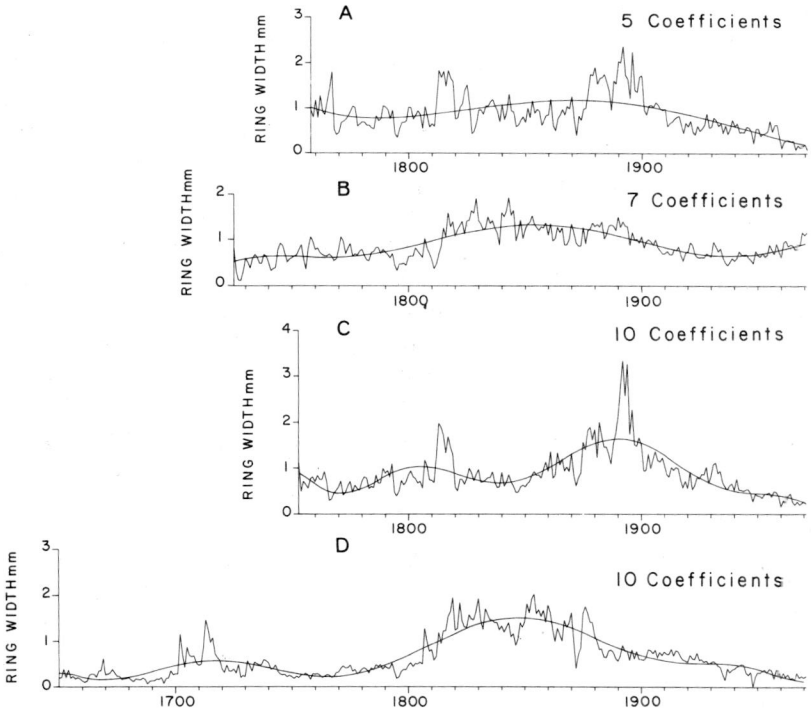

FIG. 6.6. Four ring-width series which required the fitting of high-order polynomials to remove the long-term nonclimatic variations in growth. Since the trees were from the same general area, it is likely that the differences among the polynomial curves reflect local conditions surrounding the trees and sites and not variations in the macroclimate. The greater the numbers of coefficients for each polynomial equation, the greater the degree of complexity in the shape of the curve.

Some trends and a large portion of the low-frequency variance in growth cannot be approximated by the exponential equation, by a straight line of negative slope, or by a polynomial equation. These changes are purposely left in the ring-width series if they are not excessive, because they can provide potential information on long-term climatic variations. In some European studies a 22-year *moving average* (Bitvinskas, 1974) is used for the curve function, but the procedure removes all growth variations including those due to climate that occur on time scales equal to or greater than a 22-year duration. Since information about this time scale is important to climatic analysis, the use of a 22-year moving average is not advisable for use in most dendroclimatic analyses.

After an appropriate curve has been ascertained for the growth changes that are associated with increasing tree age, the equation is solved for the expected yearly growth (Y_t). The measured ring widths (W_t) are then converted to ring-width indices (I_t) by dividing each width for year t by the expected growth (Y_t), as follows:

$$I_t = \frac{W_t}{Y_t} \tag{6.8}$$

Division by the expected growth both removes the trend in growth and scales the variance so that it is approximately the same throughout the entire length of the time series. The result of the division by (Y_t) for the data points graphed in Fig. 6.3 are plotted in Fig. 6.7. The latter represents a new time series referred to as *ring-width indices* or the *standardized ring-width chronology*. The indices from each of these ring-width time series (Table 6.II) exhibit increased similarity in their statistical properties. The means of the tree samples all approximate a value of 1, the differences in the standard deviations are less than those for ring widths, and most of the autocorrelation due to trend is eliminated. However, mean sensitivity estimates are the same as those for the ring-width data sets.

The standardized ring-width indices are then averaged to obtain a mean chronology as shown at the bottom of Fig. 6.7. Standardization, in Douglass' words (1936), equalizes or brings all ring-width curves to a uniform mean value (a value of 1, see Table 6.II) so that one tree record with a large average growth will not dominate other records of small average growth when the two series are combined into a mean chronology.

The scaling of all series to mean values of 1 is an important feature of standardization, for the slow-growing trees that are under climatic stress often provide more information on climatic variation than the fast-growing ones. Thus, if the ring widths rather than indices were averaged before standardization, the variance of the series would be dominated by the fast-growing individuals whose ring widths fluctuate the most but are least limited by climate. Standardization before averaging converts all series to the same relative variance so that each is given the same weight when the series are averaged. In addition the variances, standard deviations, and autocorrelations of indices are calculated so that the basic characteristics of each growth response are automatically computed in the standardization process.

If climate changes slowly over a long period of time, it may be impossible to distinguish long-term effects of climatic change from the effects of increasing age. Depending upon the curve-fitting option, standardization will inadvertently remove some of these long-term growth changes due to low-frequency variations in climate. Because the

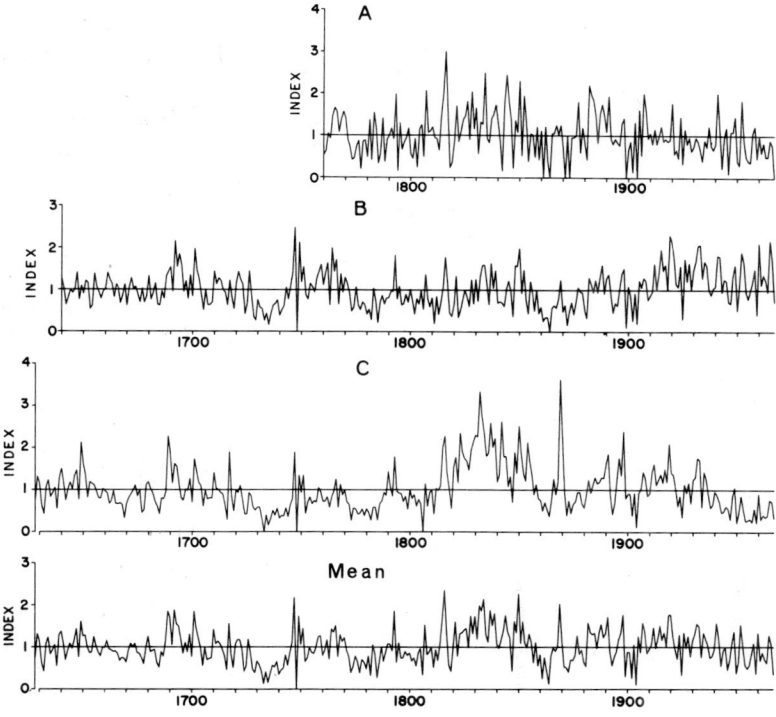

FIG. 6.7. Standardized indices calculated from ring widths by fitting a growth curve (Fig. 6.3) and dividing each ring width by the value of the growth curve. The indices from each year in series A, B, and C are averaged to obtain a mean chronology shown at the bottom of the figure.

polynomial curve-fitting option is more flexible, it is more likely than the exponential option to fit and remove some effects of long-term climatic changes. There are certain cases where an investigator may decide that there is important low frequency climatic information in the ring record (LaMarche, 1974a) and that it is better to live with the undesirable characteristics of ring widths than to risk removing or altering the information on climate by indiscriminately fitting an exponential or polynomial function.

The plots of autocorrelation coefficients derived from the indices may be compared to those for the ring-width time series (Fig. 6.4). The greatest differences between the autocorrelations of ring widths versus those of indices are most evident for series A and B (Table 6.II), in which most of the low-frequency variance was trend and was removed in the standardizing process. Since the autocorrelation coefficient at a lag of 1 for the indices of series A is near 0 and no clear systematic variations at higher order lags occur, it can be inferred that the information remaining in the indices of series A has more high-frequency than low-frequency variance. The first- and second-order autocorrelation coefficients for indices of series B (Fig. 6.3) are larger than those of A and are significant. This confirms the presence of more low-frequency variance. The high autocorrelations for series C up to lags of 20 years indicate that a significant amount of low-frequency variance remains in series C, although there is less than in the ring-width sequence (see Fig. 6.7 and Table 6.II).

Generally, there is an inverse relationship between mean sensitivity and autocorrelation in tree-ring chronologies, because the former measures the proportion of high-frequency variance, while the latter measures the proportion of low-frequency variance. The standard deviation, however, is a measure of variations in both frequency domains. As will be shown later, these three statistical measurements are useful in the selection of chronologies for dendroclimatic analysis.

IV. Filtering Techniques

Another method for studying variance at particular frequencies involves filtering out the undesirable variance by means of digital filters or moving averages, sometimes called *running means*. Moving averages are simply ring-width averages for a given number of successive rings, the sequence being moved ahead by one year each time the average is computed. In a three-year moving average the first is a mean of x_1, x_2, and x_3, the second x_2, x_3, and x_4, and the third x_3, x_4, and x_5, and so on. Each average is assigned to the year of the central ring in the sequence. Three-year and five-year moving averages were used in early dendrochronological studies (Douglass, 1936; de Martin, 1970) to smooth the year-to-year variability in proportion to changes occurring over periods greater than three years duration.

The common three-year moving average as described above places equal weight on all values being averaged, and this can produce peaks and troughs for the generated series in locations other than those which originally existed. However, the numbers to be averaged can be weighted, giving the central value twice the weight of the value on each end, in which case the peaks and troughs are the same as in the original series. In early studies a commonly used moving average weighted the central value by one-half and the end values by one-fourth. A variety of weighted moving averages have been applied, some including terms for 21 or more years.

The number and value of weights may vary, depending upon which frequencies of variation are to be retained and which are to be eliminated. The actual weights used to obtain particular types of moving averages are referred to as *filters*, which are analogous to color filters that change the optical spectrum of light passing through them. An unfiltered time series with equal variance at all frequencies is analogous to white light. The three-year weighted moving average described above blocks out the rapidly changing growth variations (high frequencies, analogous to blue light) and passes the slowly changing variations (low frequencies, analogous to red light). When the resulting averages retain only the long-term or low-frequency variations as in the case of the three-year weighted moving average, the weights are referred to as a *low-pass filter*.

Specially designed filters have been created for dendrochronological purposes which emphasize variance at certain prescribed frequencies. For example, 11-year and 22-year weighted means have been used to emphasize the 11-year and 22-year sunspot cycles. *First differences*, mentioned earlier in the chapter and calculated by subtracting the value of each item in a series from its immediate successor, is a type of *high-pass filter*. Mean sensitivity also acts as a high-pass filter, since it measures the absolute value of high-frequency variations observable in the differences between adjacent widths or indices. Mitchell *et al.*, (1966) present a brief discussion concerning the use and design of filters for climatological analysis.

Table 6.III includes the weightings for two specially designed filters of particular value to tree-ring analysis. They are *reciprocal filters*, which is to say that each of the two filters is designed to pass variance at opposite extremes of the frequency spectrum. The high-pass filter transmits or passes high frequencies (those with short wavelengths) and blocks most of

TABLE 6.III Weights of two reciprocal digital filters

Weight number[a]	Weight value	
	High-pass filter[b]	Low-pass filter[c]
+6	−0.0003	0.0003
+5	−0.0030	0.0030
+4	−0.0161	0.0161
+3	−0.0537	0.0537
+2	−0.1208	0.1208
+1	−0.1933	0.1933
0[d]	0.7744	0.2256
−1	−0.1933	0.1933
−2	−0.1208	0.1208
−3	−0.0537	0.0537
−4	−0.0161	0.0161
−5	−0.0030	0.0030
−6	−0.0003	0.0003

[a]Weight number expressed as position relative to central values of the data to be summed. [b]Passes variance with approximate wavelengths < 8 years. [c]Passes variance with approximate wavelengths > 8 years. [d]Central weight.

the variance with wavelengths greater than eight years, while the low-pass filter transmits or passes low frequencies (those with long wavelengths) and blocks the variance with wavelengths less than eight years. Since they are reciprocal, the filters can be applied to the same ring-width or ring-index series to separate the variance into its high- and low-frequency components. Neither filter is perfect and there is a certain amount of ambiguity in handling the variances at wavelengths near eight years. Consequently, a small portion of the variance is not passed by either one. A variety of filters can be developed to pass variations at a variety of different wavelengths (Craddock, 1957).

The weights of a filter, as in the case of a moving average, are multiplied by the appropriate ring-width values and the products summed as in the following equation (Mitchell, et al., (1966).

$$\overline{X}_t = \sum_{i=-n}^{i=+n} w_i x_{t+i} \qquad (6.9)$$

where $\overline{X}_t$ is the filtered value of the series corresponding to the tth term and w_i is the weight by which the value of the series i units removed from t is multiplied (Mitchell et al., 1966).

Using the three-year moving average mentioned above, $n = 1$ and i varies from -1 to $+1$. Both values on each side of the central value are multiplied by 0.25, the central value by 0.50, and the three products are

6. THE STATISTICS OF RING-WIDTH AND CLIMATIC DATA

summed. Using the weights in Table 6.III, n is 6, and weight numbers -6 to -1 and $+1$ to $+6$ are multiplied by the corresponding six items on either side of the central value and summed with the product of the central weight and the central value. The next filtered value is obtained by shifting the matching of weights and data one year ahead and repeating the operation. Note that the sum of all 13 filter weights for the low-pass filter has a value of 1. This feature preserves the mean and variance of a time series which has been operated upon so that the low-pass filtered data always have the same mean and same low-frequency variance as the original data. The sum of the weights in the high-pass filter equals 0 so that the high-pass filtered data have a zero mean. As in the low-pass filter, the values of the weights are chosen so that there is the same high-frequency variance in the filtered set as in the high-frequency component of the original data.

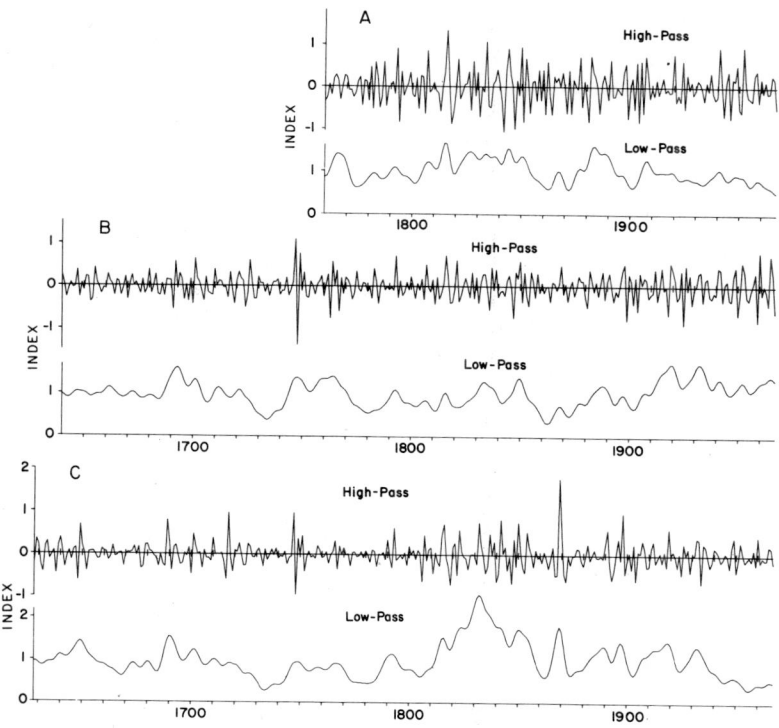

FIG. 6.8. Data from the three index series shown in Fig. 6.7 after they have been treated with the high-pass and low-pass filter weights included in Table 6.III.

Figure 6.8 includes plots of the filtered data from indices for A, B, and C (Fig. 6.7) which were passed by the two filters (Table 6.III). Table 6.IV includes the variances of the original indices and the two filtered sets, along with the percentages of variance passed in each case. Also shown are the correlation coefficients for comparisons (1) among the three index chronologies, (2) among the three high-pass data sets, and (3) among the three low-pass data sets. The data in Table 6.IV confirm the inferences derived from the measurements of mean sensitivity and autocorrelation, in that series A has more high-pass variance (66%) than low-pass variance (23%). The filtered series from C exhibit the greatest low-pass information or variance (59%) and the least high-pass information (31%). The filtered series from B have nearly equal amounts of information at high and low frequencies, respectively representing 48% and 41%. As mentioned earlier, a small amount of variance is not passed by either filter, due to rounding and smoothing by the filter weights, so that the percentages passed by the two filters do not add up to 100%.

Correlation analysis of the filtered sets (Table 6.IV) indicates varying associations among the three chronologies, depending upon the frequencies of the variances. The high-pass portions of choronology B are relatively well-correlated with A and C (0·695 and 0·619), but the low-pass components of B are not as highly correlated with the other two sets (0·198 and 0·252). The low-pass variances for A and C are more highly correlated (0·605), though the amount of low-pass information differs markedly between the two sets (0·176 versus 0·067). These results suggest

TABLE 6.IV Variances and correlation among the three chronologies before and after filtering[a]

	Ring-width index series		
	A	B	C
Unfiltered data			
Variance	0·239	0·202	0·299
Correlation with A	—	0·486	0·458
Correlation with B	—	—	0·379
High-pass filter			
Variance	0·190	0·097	0·092
% of unfiltered variance	66%	48%	31%
Correlation with A	—	0·695	0·403
Correlation with B	—	—	0·619
Low-pass filter			
Variance	0·067	0·083	0·176
% of unfiltered variance	23%	41%	59%
Correlation with A	—	0·198	0·605
Correlation with B	—	—	0·252

[a] See Figs 6.7 and 6.8.

that the processes generating the two frequency domains may be different among the three data sets.

Although these examples may be inconsequential in themselves, the data illustrate how the two reciprocal filters can be used to separate and study the variances at different frequencies with respect to time. When the variances are more poorly correlated at low frequencies than at high frequencies, it may be inferred that either the generating processes or the histories of the particular forests have differed among the trees and sites. Dissimilarities at high frequencies may be inferred to result from varying microclimates, or varying nonclimatic factors that create short-term (high-frequency) variations that persist for only a few years.

Julian and Fritts (1968) employed these filters to study chronology characteristics from 11 stands including trees of three different species in an area of similar topography near Denver, Colorado. The stands were separated from one another by varying distances ranging from less than 1 to 150 kilometers. The 11 chronologies for the interval 1860–1964 were correlated with one another and then were treated with the high- and low-pass filters shown in Table 6.III. Correlations were again calculated between sets of filtered data. Plot A (Fig. 6.9) includes the correlations for the unfiltered sets and shows a linear decrease in correlation with increased separation distance. However, when the variances in the chronologies were treated with the high-and low-pass filters (plots B and C, Fig. 6.9), it became apparent that the decline in correlation with increasing separation distance was mostly a feature of the high-frequency variance.

A search as to the cause for the lack of correlation among some of the low-frequency components led to the discovery that disturbance of the *Pinus ponderosa* sites could have been a factor. Actual plots of the *Pinus ponderosa* data revealed diverging trends in ring widths that could have been associated with disturbance from mining and grazing activity in the area, both of which had in fact occurred. Thus it was concluded that disturbance of the *Pinus ponderosa* sites produced the slowly changing variations in ring widths that were uncorrelated with the slowly changing variations in the chronologies from other sites. However, the correlations for the low-pass variance from stands of *Pseudotsuga menziesii* (as indicated by triangles in Fig. 6.9) were like those of the high-pass variance, so it was concluded that there was apparently less disturbance on these steeper and more moist sites.

Julian and Fritts (1968) also filtered data from three nearby climatic stations and calculated the correlation between precipitation and ring-width indices for what they judged to be the better tree-ring sites (Fig. 6.10). These results show a consistent and gradual decline in correlation

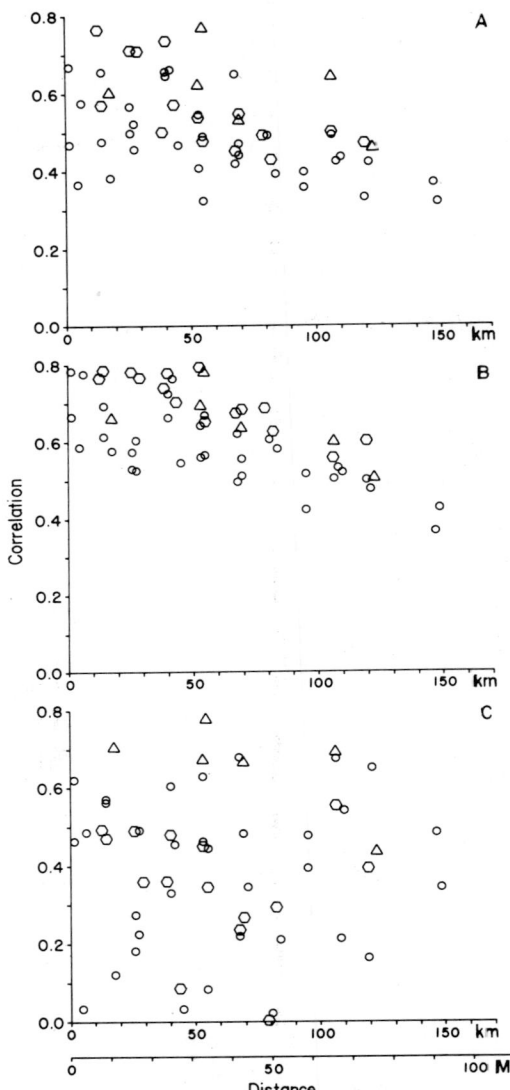

FIG. 6.9. The correlation among index chronologies from 11 stands of trees representing three different species plotted as a function of separation distance. Correlations are for (A) the mean indices, (B) the high-pass variance of the mean indices, and (C) the low-pass variance of the mean indices. Triangles designate those comparisons for chronologies of *Pseudotsuga menziesii*, hexagons designate those comparisons for chronologies of *Pinus ponderosa*, and circles designate comparisons for chronologies differing in species, one being *Pinus edulis*. For further discussion of these data, see text and Julian and Fritts (1968).

between the high-pass components of tree growth and climate with increasing separation distance, although considerable differences in correlations were noted for the low-pass components.

Stockton and Fritts (1971b) applied the same filtering and correlation techniques in a study of four chronologies from Arizona which are related to the average statewide climate (see section VII, Chapter 7). The tree-ring data were subjected to filtering and subsequent correlation analysis for varying intervals of 20 years from 1650 to 1957 (Fig. 6.11). There appeared to be certain 20-year periods of time when the low-frequency variations differed among the four chronologies within the state. It was first thought that the trees may have been affected by varying disturbance of the sites during these time periods. However, an examination of chronologies from neighboring sites and nearby states revealed similar low-frequency changes in growth, so it was concluded that in this case the differences among those particular chronologies were the result of climatic changes which produced a gradient across Arizona during the above-mentioned time spans. LaMarche (1974b) uses the same filtering techniques to test for differences in the generating processes between high-elevation and low-elevation *Pinus longaeva*.

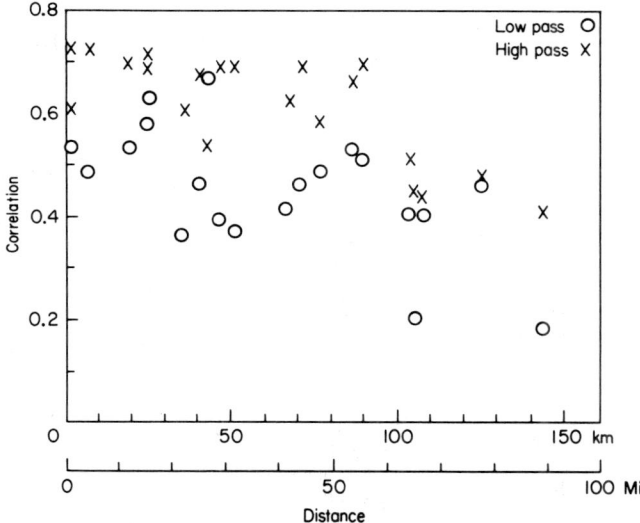

FIG. 6.10. The correlation between four ring-width chronologies and precipitation from three weather stations plotted as a function of separation distance. Monthly precipitation was transformed into a yearly precipitation index by use of a multiple regression relationship derived from analysis of the rings, and the data were subjected to the low-pass and high-pass filters before correlation coefficients were obtained. The first circle on the left represents two data points (see Julian and Fritts, 1968).

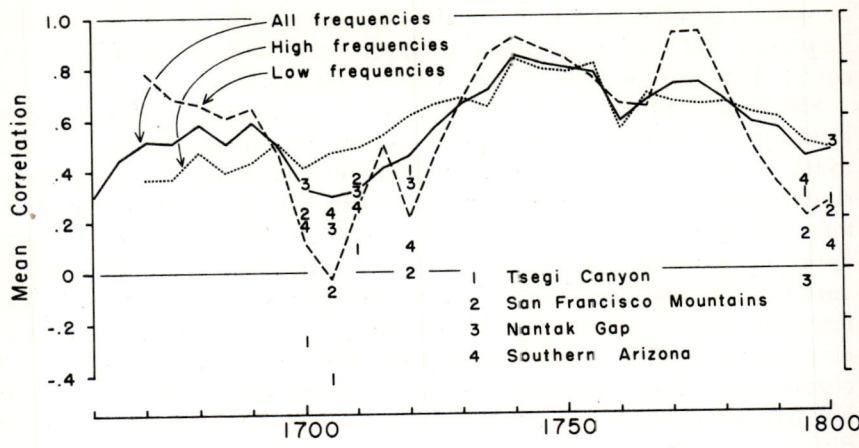

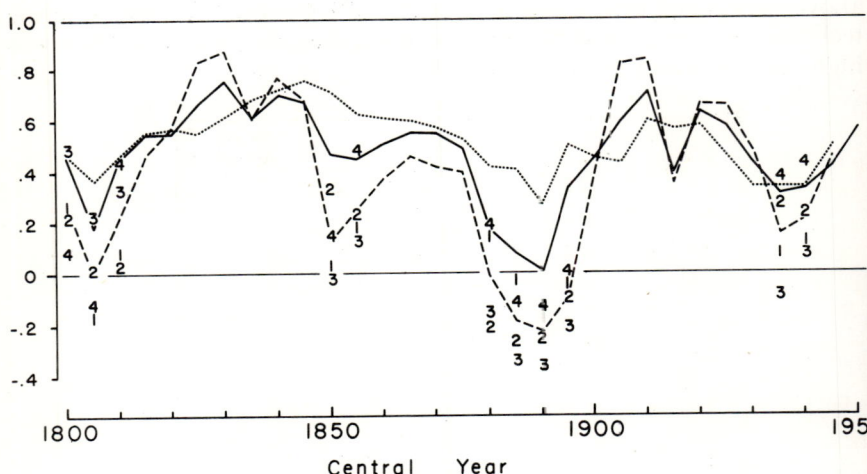

FIG. 6.11. The means of correlation coefficients plotted every five years for all possible correlations among indices of four Arizona ring-width chronologies calculated for overlapping 20-year intervals of time. Correlations were obtained for the unfiltered index data which include all frequencies, and for filtered data including either high frequencies or low frequencies as described in the text. The mean correlation for each of the four chronologies with the other three is plotted as the central position of numbers 1-4 only when the average correlation for the low-frequency data was low (less than 0·3). A negative value in the mean correlation indicates trends in the data that diverge from those for the other three chronologies. (Stockton and Fritts, 1971b)

The crossdating procedure involves a visual inspection and check for agreement which in essence utilizes the high-frequency variance in ring width. It is physically more dificult to scan a ring-width series by eye for the slowly varying changes in growth which may be nonclimatic in origin and which do not correlate from tree to tree and from site to site. Therefore, low-pass filters and correlation analysis can be of great assistance in identifying dissimilar low-frequency variations in growth. Chronologies to be used for climatic purposes may be screened by applying both the high- and low-pass filters and correlating the resulting series with similar data from neighboring sites. If the low-pass information in a particular chronology is poorly correlated with that of its neighbors, nonclimatic factors may be inferred to be excessively limiting and the particular chronology is deleted from the candidates for the climatic set. The ratio of high-pass to low-pass variance is also calculated and used to identify chronologies with disproportionate amounts of high- versus low-pass variance (Fritts and Shatz, 1975).

V. Other Methods for Assessing the Growth Curves

Systematic variations in ring widths may be assessed in a variety of ways other than those described above. Some workers regard systematic changes in ring width as purely geometric constraints, hypothesizing that the volume of wood added each year, in addition to being a function of ring width, is a function of the circumference of the cambium and height of the stem (Phipps, 1967). According to this model, ring width could be converted to a volume increment of growth for the entire stem, and a simple function of this increment would provide a good estimate of changes associated with tree age (Baker, 1950).

Other workers (Duff and Nolan, 1953; Smith and Wilsie, 1961) analyze ring growth on cross sections at every internode along the main stem and classify and plot ring features along three different gradients (Fig. 6.12). One gradient represents the usual horizontally arranged sequence of rings which is affected by climatic variation and the increase in tree age. The second represents the diagonal series of ring widths at various stem heights for the same year and expresses changes associated with increasing stem height and decreasing *cambial* age. The third series represents a vertically arranged sequence with each ring sampled at a given cambial age (that is, a constant ring number from the pith for all cases). This vertically arranged sequence often exhibits less systematic variation in ring width associated with increasing tree age than the other two sequences because cambial age was the same for all rings in the sequence. A number of internode-by-internode analyses using the Duff and Nolan approach have contributed considerable useful information about anatomical features and about certain growth factors and wood density changes (Richardson, 1961).

Theoretically, ring-width variability along the vertically ordered sequence should be the best dendroclimatic sequence, but the technique is generally impracticable for most dendrochronological studies, because it requires a cross section for every increment in the stem corresponding to a year's growth. However, a modification of this classification approach was used to study ring characteristics throughout the stem of a mature *Pinus ponderosa*. The data were classified into 20-year units rather than

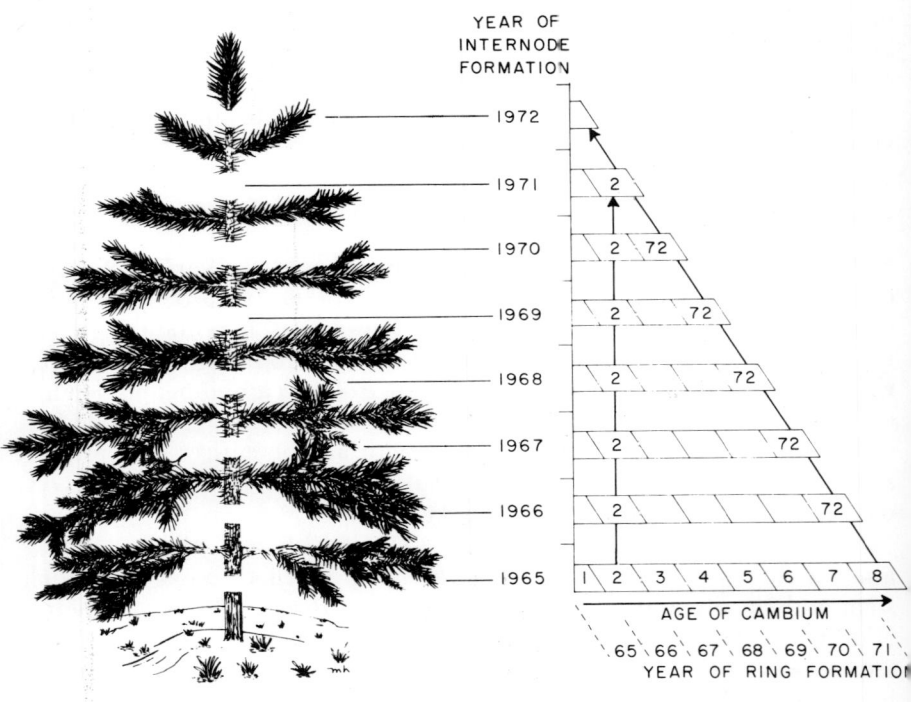

FIG. 6.12. The three classification schemes described by Duff and Nolan (1953) and Smith and Wilsie (1961). Each measurement corresponds to a parallelogram in the figure and these data for each year and season can be plotted as shown by the arrows in (1) the horizontal direction to show effects of changes in both year of formation and cambial age; (2) the diagonal direction to show increasing height in the tree, keeping year of formation constant; and (3) the vertical direction to show effects of changes in year of formation (holding cambial age constant). The horizontally averaged series is the one commonly used in dendroclimatology, but the vertical series is proposed by some to be a better representation of variations in climate. See text for further discussion.

one-year units and these data were compared to variations in ring characteristics associated with stem height, cambial age, and increasing tree age (Fritts et al., 1965a). See Figs. 6.20 and 6.21 and accompanying discussion on pages 304–310.

Some workers have proposed that the ring width from all trees of a given species and site can be expressed as a simple mathematical function of increasing age. For example, Mitchell (1967), following the lead of Scandinavian scientists, obtained a large sample of varying ages, classified each ring according to tree age, averaged the ring widths in each class, and plotted these averages as a function of tree age (Fig. 6.13).

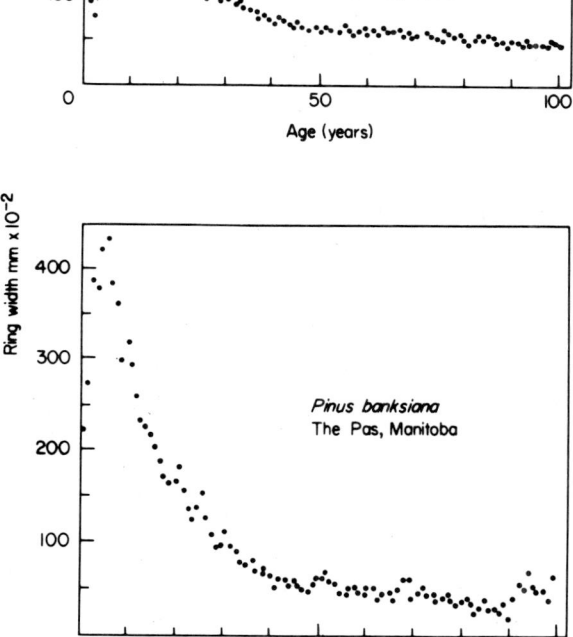

FIG. 6.13. The mean change in ring width associated with biological factors related to increasing age can be estimated by using trees of many age classes and averaging widths for rings of the same age. This procedure assumes that the environmental factors limiting growth in trees are randomly distributed through time and that the variations in these environmental factors become insignificant when large numbers are averaged from trees of all ages. A single smoothed curve fitted to the mean of all data shown in the figure is then used to standardize the ring-width series for all trees in the site. (Mitchell, 1967)

Variations in ring width due to climatic fluctuations are minimized in the averaging process, and the resultant curve presumably expresses the growth function that should be used for the sampled trees.

However, all individuals of a species rarely attain optimum growth at the same age, and individual trees differ in their growth rates because of differences in soil factors, competition, microclimate, and other factors governing the productivity of the site. Therefore, individual trees will deviate markedly and systematically from such a mean growth curve, and standardization by this method leads to more error in the chronology than is obtained by the empirical method of curve fitting described in Section III.

The use of a single growth curve could aid studies at the upper tree line or polar limits, where both increasing tree age and long-term changes in climate produce long-term variations in ring widths. It is possible that a single growth curve could be developed and applied to each measured set to distinguish between long-term changes due to increasing age from those associated with long-term variations in climate.

Ring widths are sometimes transformed to logarithms or they are plotted on a semilogarithmic scale. The general effect of this transformation is to lessen the variations in large rings and to increase the variations in small rings, but the effects of increasing tree age are not removed. An additional transformation (Jonsson, 1969) is one in which the logarithms of ring widths (lni_t) are averaged year-by-year and then a polynomial with two or more coefficients is fitted to the mean of the logarithmic series ($\overline{lni_t}$). Sometimes it is possible to simultaneously solve for the climatic relationships at the same time that the polynomial equation is obtained (Jonsson, 1969). Ring-width index values are then derived as follows:

$$x_t = \exp(lni_t - \overline{lni_t} - \sigma^2/2) \qquad (6.10)$$

where σ^2 is the variance of lni_t. This term is used to adjust for the difference in means between the arithmetic and logarithmic scales so that the mean of the indices has a value of 1·0. Jonsson and Fritts compared their two standardization techniques on identical data and found 77% variance agreement. Differences were attributed to variations within and between trees which were not removed by the polynomial curve that was fitted to the mean of logarithms. While the computational procedures are easier, there is much less flexibility in the Jonsson method since the length of record for all sampled trees must be the same; also there is no possibility for statistical analysis of variance and correlation, both of which are described in the following sections.

Hollstein (1975) has suggested a very promising and interesting

6. THE STATISTICS OF RING-WIDTH AND CLIMATIC DATA

alternative to standardizing procedures. He converts ring-width data to ratios by dividing each width by the width of the preceding ring and then takes the logarithm. These values for each year, which he calls "wuchswert," are averaged by year for all available samples. A simple linear regression is fit by least squares methods to the mean logarithmic series and yearly values of the regression subtracted to correct for the growth trend and to obtain a mean series which is zero.

One apparent difficulty with this method is that the resulting indices are logarithms of ratios which are dependent upon the climate of two successive years, so that the results are difficult to interpret and meaningful calibration of the values with climate is not easily obtained. However, this particular difficulty can be circumvented by taking the exponent of the mean ratios and converting the mean sequence of ratios back to mean yearly values which are comparable to the indices derived in the usual way. Also, it may be appropriate to allow for non logarithmic growth curves by fitting a straight line or polynomial to the logarithm of ratios for each ring-width sample and subtracting the value of the curve before the averaged *wuchswert* is obtained. A computer program to perform these calculations is now available.

No single method has been developed that is applicable to removing the growth functions for all species and sites. The computer-derived exponential function has proven satisfactory for a variety of coniferous species on arid sites and it appears to apply to many sites in Europe and the Arctic. The polynomial curve-fitting routine has been added to allow for cases with more complicated growth variations. Both types of curves are chosen for pragmatic reasons, that is to say they work for dendroclimatic analysis. While the using of a polynomial has no theoretical basis, it is the most convenient method for fitting a curve to accomplish standardization. Since the polynomial curve can approximate the exponential function, it is at present the best over all model for standardization, as long as care is taken to restrict the number of coefficients so that variations due to long-term climatic changes are not lost. If the simpler method of Hollstein, or some modification of it, proves feasible, it could be a suitable alternative to the more complicated standardization by division of a polynomial, computer-derived curve.

These standardizing techniques should never be applied blindly. It is routine practice at the Laboratory of Tree-Ring Research to plot all ring-width data as a function of tree age before computer analysis. Occasionally none of the available growth functions will adequately fit particular ring-width data. In such cases it may be inappropriate to apply the data to dendroclimatic study unless an alternative method can be derived for the growth-trend analysis.

VI. Analysis of Variance

The first five chapters have described a variety of factors that affect the tree ring-climate system and contribute to the ring-width chronology variance. Certain statistical measurements have been defined in this chapter which allow quantification of this variance. It is now appropriate to turn attention to analysis of variance, a very powerful technique which sorts out and measures the various sources in the tree ring-climate system that govern the chronology variance.

For example, the mean year-to-year pattern in ring-width indices from many trees in a region may be identified as variance resulting from limiting macroclimatic factors. Differences in these patterns from one stand to the next may be considered to result from a different source such as varying soil properties, different slopes, and the unique conditions of each plant community interaction with the macroclimate. The term *interaction* refers to the influence of one variable such as a site factor upon the effect of another variable such as macroclimate on growth. Differences also occur among individual trees within a stand which may be attributed to variations in tree vigor, competition, and the available operational environment. Variations in the growth within the tree stem may also occur as a result of growth in major branches and roots, local microclimatic variations, and other factors which control the operational environments for different tissues and organs. The relative importance of these sources of variation is assessed by a particular *analysis of variance* developed specifically for indices, derived from ring-width data which have been collected in a particular manner to represent several sources of variance. The following paragraphs describe the technique by means of a simple example. General descriptions of the technique may be found in Neter and Wassermann (1974), Simpson *et al.* (1960), and Steel and Torrie (1960).

A stand of *Pinus longaeva* in California was divided into two classes or groups of equal size, one of young trees and one of old trees. Since the age classes were intermixed within the same stand, differences in the mean chronologies of the two groups could be attributed to tree age as the source of variance. Differences in the chronologies between trees within each age class could be attributed to differences among trees as a source of variance. In addition, two cores were sampled from each of the trees in each group so that the chronology differences between them could be used to assess the sides of the tree stem as a source of variance. Four trees were sampled at random from each age class within the site, including two radii per tree which were subsequently placed in separate growth-rate classes, depending upon which core for each tree had the greatest average ring width.

6. THE STATISTICS OF RING-WIDTH AND CLIMATIC DATA

The samples were mounted, dated, widths of rings measured, and the indices derived for the entire length of each sampled stem radius. An interval of time common to all samples was specified to be 1950–1954. Usually this interval includes a common period of 100 or more years, but in this example only rings for the five years were analyzed. Table 6.V includes all indices for each age class, tree, core, and year.

The first required computations are the sums and the sums of squares for the five indices on each core, as shown in columns 9 and 10 of Table 6.V. (The individual data were analyzed by the computer but have been rounded in the table; therefore, small differences will occur in the least significant digits if recalculations are made from the data in the table.) The variances are calculated from sums and sums of squares statistics using the equations in Table 6.VI.

The indices from the different sides of each tree stem (Table 6.V, columns 4–8) are summed year by year to obtain five yearly values for each tree, and these values are summed, squared, and the squares summed (columns 11 and 12). The data for each growth-rate class within the young and old group are summed, these values squared, and the squares summed (columns 13, 14). In this example, all cores numbered as 1 are from the slower-growing sides of the trees and those numbered 2 are from the faster-growing sides. The sums and sums of squares for the slow-growing class are listed first. The two core classes between age groups are summed to obtain combined core-class chronologies for both groups (columns 15 and 16). All indices within groups are then summed year by year to obtain the sums and sums of squares for chronologies of young and old groups (columns 17 and 18), and finally all indices for all 16 cores are summed year by year to obtain the sums for the chronology for both groups as shown at the bottom of Table 6.V.

The next step in the analysis is a complicated procedure shown in (Table 6.VII). Rows 1–5 are the variance of group, core, and tree means. sums of squares as shown in Tables 6.V are adjusted (corrected) by subtracting the squares of several sums as shown in Table 6.VI working from row 1 to row 11. Each of these adjusted values is divided by the number of degrees of freedom to obtain the mean squares or variances (Table 6.VII). Rows 1–5 are the variance of group, core, and tree means. There is usually no significant variance accounted for by these values, because the standardizing procedure has forced all samples to have the same mean. Occasionally when the period of analysis is a small subset of a larger one, some of the variances for means are significant.

The *component variances* computed in Table 6.VI are obtained for the mean chronology (Y in row 6) and the other sources of variance that interact with it (rows 7–11, Tables 6.VI and 6.VII). They are the estimates

TABLE 6.V Sample data for analysis of variance[a]

Column				Years					Cores[b]		Trees		Core class within group		Core class		Groups	
1	2	3	4	5	6	7	8	9	10	11	12	13	14	15	16	17	18
								C× T/G	Y×C ×T/G	T/G	Y× T/G	C×G	Y× C×G	C	Y×C	G	Y×G
Group	Tree	Core	1950	1951	1952	1953	1954	Sums	Sums of squares[c]	Sums	Sums of squares	Sums	Sums of squares	Sums	Sums of squares	Sums	Sums of squares
Young	1	1	0·341	1·031	1·026	0·713	0·817	3·928	3·408	9·042	18·104	19·264	76·064	37·885	298·936	39·544	321·296
	1	2	0·415	1·251	1·395	0·928	1·125	5·114	5·810			20·281	84·768	38·071	300·488		
	2	1	0·479	1·291	0·814	1·105	1·088	4·777	4·963	8·953	17·210						
	2	2	0·581	1·226	0·785	0·960	0·625	4·177	3·769								
	3	1	1·116	1·385	1·164	1·205	0·931	5·801	6·837	11·667	27·602						
	3	2	1·069	1·308	1·159	1·349	0·980	5·865	6·977								
	4	1	0·844	0·921	0·980	1·095	0·918	4·758	4·563	9·882	19·668						
	4	2	0·882	1·295	1·000	0·971	0·977	5·125	5·352								
Old	1	1	0·706	0·974	1·355	1·336	0·935	5·306	5·942	10·825	24·346	18·621	74·132			36·411	281·329
	1	2	0·711	1·083	1·145	1·303	1·276	5·518	6·315			17·790	66·788				
	2	1	0·261	1·050	1·255	1·130	1·070	4·766	5·167	9·166	18·754						
	2	2	0·366	1·000	1·059	1·172	0·804	4·401	4·275								
	3	1	0·788	0·939	0·971	0·955	0·692	4·345	3·836	8·808	15·601						
	3	2	0·713	0·928	0·870	0·846	1·106	4·463	4·065								
	4	1	0·160	1·203	0·524	1·379	0·938	4·204	4·529	7·613	14·600						
	4	2	0·184	1·047	0·495	1·058	0·625	3·409	2·885								
Sum of all groups, trees, and cores			9·616	17·932	15·997	17·505	14·907	75·957	1198·572								

[a] Indices from *Pinus longaeva* for 5 years, 4 young trees, 4 old trees, 2 cores collected from each tree. [b] See Table 6.VI, left column, for definition of symbols, and equations for computations. [c] Each value is squared and the squares are summed. Key: groups (G) = 2; trees per group (T) = 4; cores per tree (C) = 2; rings per core (N) = 5.

of variances for each source which are calculated by separating the mean squares into their respective components and dividing by the number of elements in each part (Table 6.VI).

The analysis pools the variance in the appropriate ways and uses the differences to assess the importance of each source. The operation is described in conceptual terms in the following paragraphs. (For specific details consult Tables 6.V, 6.VI, and 6.VII.)

The pooled variances for the tree chronologies are subtracted from those of the individual 16 cores. There is always less variance in the former because divergent variations among the individual core chronologies are canceled out in the summing process. The greater the differences among the paired core chronologies within the same trees, the greater the reduction in variance resulting from the summing process and the greater the component attributed to the core chronology source (Table 6.VI, line 11).

The variances for the sums representing each of the two age groups are subtracted from the pooled data for the individual tree chronologies and the difference in variance is attributed to the tree chronology source (Table 6.VI, line 8). The variances in the summed chronology for both groups are subtracted from those for the sums of young and old groups and the difference is attributed to the age source (Table 6.VI, line 7). Likewise, the sums of the group variances are subtracted from the corresponding values of the core classes to derive the fast-growing versus slow-growing components (Table 6.VI, lines 9 and 10). Finally, the amount of variance remaining in the five totals for each year summed for all cores, trees, and groups represents the mean variance in the year-by-year chronology or that attributed to the macroclimatic source (Table 6.VI, line 6).

If a large amount of ring-width variability coincides in all cores, trees, and groups, only a small amount of variance is lost in the averaging process and a large amount remains in the year-by-year chronology of the site (the macroclimatic source). It may be inferred from this result that some large-scale factor, such as climate, has varied from year to year and has affected growth in the same way on all sides of all trees and in all groups. If, on the other hand, there are large differences in core and tree chronologies (with the result that the chronology summed for all trees and groups has little variance), it may be inferred that during each year a variety of factors have limited growth in a variety of ways in the different radii, trees, and groups diminishing the effects of macroclimatic variations on growth.

Sometimes there will be a greater reduction in variance when core chronologies are averaged into tree chronologies than when the tree

TABLE 6-VI

Calculations for Analysis of Variance of Tree-Ring Chronologies

Row	Source of Variation	Sum of Squares and Correction	Corrected sum of Squares
1*	Group means (G)	$\frac{1}{tcn}\left[\sum_{h=1}^{g}\left(\sum_{i=1}^{t}\sum_{j=1}^{c}\sum_{k=1}^{n} y_{hijk}\right)^2\right] - K$	$=G_c$
2	Core class means (C)	$\frac{1}{gtn}\left[\sum_{j=1}^{c}\left(\sum_{h=1}^{g}\sum_{i=1}^{t}\sum_{k=1}^{n} y_{hijk}\right)^2\right] - K$	$=C_c$
3	Tree means in groups (T/G)	$\frac{1}{cn}\left[\sum_{h=1}^{g}\sum_{i=1}^{t}\left(\sum_{j=1}^{c}\sum_{k=1}^{n} y_{hijk}\right)^2\right] - G_c - K$	$=T_c$
4*	Core means in groups (C × G)	$\frac{1}{tn}\left[\sum_{h=1}^{g}\sum_{j=1}^{c}\left(\sum_{i=1}^{t}\sum_{k=1}^{n} y_{hijk}\right)^2\right] - G_c - G_c - K$	$=CG_c$
5	Core means with trees in groups (C × T/G)	$\frac{1}{n}\left[\sum_{h=1}^{g}\sum_{i=1}^{t}\sum_{j=1}^{c}\left(\sum_{k=1}^{n} y_{hijk}\right)^2\right] - G_c - C_c - T_c - CG_c - K$	$=CT_c$
6	Mean indices in total chronology (Y)	$\frac{1}{gtc}\left[\sum_{k=1}^{n}\left(\sum_{h=1}^{g}\sum_{i=1}^{t}\sum_{j=1}^{c} y_{hijk}\right)^2\right] - K$	$=Y_c$
7*	Chronologies of groups (Y × G)	$\frac{1}{tc}\left[\sum_{k=1}^{n}\sum_{h=1}^{g}\left(\sum_{i=1}^{t}\sum_{j=1}^{c} y_{hijk}\right)^2\right] - Y_c - G_c - K$	$=YG_c$
8	Chronologies of trees in groups (Y × T/G)	$\frac{1}{c}\left[\sum_{k=1}^{n}\sum_{h=1}^{g}\sum_{i=1}^{t}\left(\sum_{j=1}^{c} y_{hijk}\right)^2\right] - Y_c - G_c - T_c - YG_c - K$	$=YT_c$
9	Chronologies of core classes (Y × C)	$\frac{1}{gt}\left[\sum_{k=1}^{n}\sum_{j=1}^{c}\left(\sum_{h=1}^{g}\sum_{i=1}^{t} y_{hijk}\right)^2\right] - Y_c - C_c - K$	$=YC_c$
10*	Chronologies of core classes with groups (Y × C × G)	$\frac{1}{t}\left[\sum_{k=1}^{n}\sum_{h=1}^{g}\sum_{j=1}^{c}\left(\sum_{i=1}^{t} y_{hijk}\right)^2\right] - Y_c - G_c - C_c - YG_c - YC_c - CG_c - K$	$=YCG_c$
11	Chronologies of cores with trees in groups (Y × C × T/G)	$\left[\sum_{k=1}^{n}\sum_{h=1}^{g}\sum_{i=1}^{t}\sum_{j=1}^{c}\left(y_{hijk}\right)^2\right] - Y_c - G_c - C_c - T_c - CG - CT_c - YG_c - YC_c - YT_c - YCG_c - K$	$=YCT_c$

*When g=1, these sources of variation do not exist, and the mean squares are zero.

$$K = \frac{1}{gtcn}\left(\sum_{h=1}^{g}\sum_{i=1}^{t}\sum_{j=1}^{c}\sum_{k=1}^{n} y_{hijk}\right)^2$$

6. THE STATISTICS OF RING-WIDTH AND CLIMATIC DATA

Degrees of freedom	Mean Square		Variance Component		Percentage
g-1	$G_c/(g-1)$	=MS(G)			
c-1	$C_c/(c-1)$	=MS(C)			
(t-1)g	$T_c/(t-1)g$	=MS(T)			
(c-1)(g-1)	$CG_c/(c-1)(g-1)$	=MS(CG)			
(t-1)(c-1)g	$CT_c/(t-1)(c-1)g$	=MS(CT)			
n-1	$Y_c/(n-1)$	=MS(Y)	$\frac{MS(Y) - MS(YT)}{gct}$	=VC(Y)	$\frac{VC(Y) \times 100}{TOTAL\ VC}$
(n-1)(g-1)	$YG_c/(n-1)(g-1)$	=MS(YG)	$\frac{MS(YG) - MS(YT)}{ct}$	=VC(YG)	$\frac{VC(YG) \times 100}{TOTAL\ VC}$
(n-1)(t-1)g	$YT_c/(n-1)(t-1)g$	=MS(YT)	$\frac{MS(YT) - MS(YCT)}{c}$	=VC(YT)	$\frac{VC(YT) \times 100}{TOTAL\ VC}$
(n-1)(c-1)	$YC_c/(n-1)(c-1)$	=MS(YC)	$\frac{MS(YC) - MS(YCT)}{gt}$	=VC(YC)	$\frac{VC(YC) \times 100}{TOTAL\ VC}$
(n-1)(c-1)(g-1)	$YCG_c/(n-1)(c-1)(g-1)$	=MS(YCG)	$\frac{MS(YCG) - MS(YCT)}{t}$	=VC(YCG)	$\frac{VC(YCG) \times 100}{TOTAL\ VC}$
(n-1)(c-1)(t-1)g	$YCT_c/(n-1)(c-1)(t-1)g$	=MS(YCT)	MS(YCT)	=VC(YCT)	$\frac{VC(YCT) \times 100}{TOTAL\ VC}$

where
  h = 1, 2, ..., g
  i = 1, 2, ..., t
  j = 1, 2, ..., c
  k = 1, 2, ..., n
  g = number of groups
  t = number of trees per group
  c = number of cores per tree
  n = number of years

TOTAL VC = VC(Y) + VC(YG) + VC(YT) + VC(YC) + VC(YCG) + VC(YCT)

TABLE 6.VII Analysis of variance results for *Pinus longaeva* sample[a]

Row	Source of variation		Raw sum	Corrected sum	Degrees of freedom	Mean square	Variance component	% variance component
1	Group means	G	72·243	0·127	1	0·127		
2	Core class means	C	72·117	0·001	1	0·001		
3	Tree means in groups	T/G	73·242	0·999	6	0·167		
4	Core means with groups	C×G	72·282	0·038	1	0·038		
5	Core means with trees in groups	C×T/G	73·518	0·237	6	0·040		
6	Mean indices in total chronology	Y	74·907	2·791	4	0·698	0·039	48
7	Chronologies of groups	Y×G	75·328	0·294	4	0·074	0·001	1
8	Chronologies of trees in groups	Y×T/G	77·944	1·617	24	0·064	0·025	30
9	Chronologies of core classes	Y×C	74·928	0·020	4	0·005	−0·002	0
10	Chronologies of core classes with groups	Y×C×G	75·438	0·051	4	0·013	−0·001	0
11	Chronologies of cores with trees in groups	Y×C×T/G	78·691	0·400	24	0·017	0·017	21
			K = 72·116		N = 80		Total Variance Components (disregarding minus values) = 0·082	

[a] From data in Table 6.V using equations in Table 6.VI.

chronologies are averaged into larger groups. Thus, a large variance is attributed to the core source and it may be inferred that the core chronologies within trees differ from one another because of factors such as irregularities in the tree crowns, injuries, or even carelessness in ring measurement. If there is a large reduction in variance when tree chronologies are summed, a high variance is attributed to the tree source, and it may be inferred that tree microclimates, local site variations, or other local factors have modified the tree responses to macroclimate.

In each analysis, the investigator collects his cores, trees, and groups in such a way as to analyze a particular component of the environment. Slope effects are studied by comparing groups of trees on contrasting slopes, soil effects by selecting groups on different soil types, and elevational effects by varying elevations of sites. Effects on ring-width variations around the tree due to the lean of the stem, exposure of slope, or crowding by neighbors is assessed by selecting the appropriate stem side that is used in the core class, while effects of drought years versus wet years are assessed by choosing appropriate years for analysis.

The sampling design for analysis of variance described here must be balanced and replicated like the example in Table 6.V. All samples must include the same years of growth, all trees must have the same number of replicated cores, and all groups must include the same number of trees. However, the trees are assumed to be randomly chosen so they are nested but not classed within groups. In other words, tree 1 of group 1 has no relation to tree 1 of group 2, but the same number of trees must be included in each group. The years, the core classes, and the groups are assumed to be fixed, that is, not selected in a random manner, but according to a conscious decision as to specific years, specific core classes, and specific groups.

The actual values of the variance components differ from site to site in proportion to the nature of the sampled materials, including the amount of autocorrelation in them. To make comparisons possible between chronologies with markedly different variance, the components are reduced to percentages as shown on the right of Table 6.VII.

The component variance for core classes (Table 6.VII, lines 9 and 10) frequently is near zero as there is rarely any contribution by that particular source. That is to say, the chronologies in the core classes are not significantly different. In the example, the variance components for core classes at both levels are negative (lines 9 and 10, Table 6.VII). Sometimes negative values do occur by chance or by round-off error, if the particular core classification is not important. As long as the negative values are small, they can be replaced by zeros in the calculation of the percentages. Since all variances are theoretically positive, no variance components should be large and negative. When such cases are encountered, they usually indicate that the wrong sums and sums of squares were used or some other computational error has been made.

The core classes were originally included in the analysis of variance to test hypotheses regarding chronology variability throughout the tree stem. Only a few cases with significant variance in a core class have been observed, and these cases usually represent samples from the upper and lower sides of leaning trees or from the up-slope and down-slope sides of trees on very steep slopes.

The results from the example in Table 6.VII show that the variance component in all cores, trees, and groups (the total group chronology variance Y was 48%. The variance due to differences in the chronologies of trees ($Y \times T/G$) accounted for 30% of the variance. The differences in the individual core chronologies ($Y \times C \times T/G$) accounted for 21% of the variance. The differences between the chronologies for the two groups ($Y \times G$) accounted for only 1% of the total variance.

Trees on dry sites can retain as much as 80% common variance in the total or group chronology for the site. The value of 48% in this particular example suggests that the sampled trees for the particular five years were only moderately limited by factors of the macroclimate. The high variance for tree components (30%), indicates a relatively heterogeneous site with varying factors affecting the ring-width chronology somewhat differently in each tree. If the component variance for core chronologies ($Y \times C \times T/G$) had been higher than that of the trees, it might be concluded that irregularities of crown size, competition, and distribution of food-and growth-controlling substances produced a larger amount of ring-width variability within the trees of the site. The 1% difference between age groups indicates that, at least for the five years and four trees that were tested, there are very few differences between the chronologies of the old vs young group. Additional examples of analyses of variance are presented later in this chapter.

VII. Analysis of Chronology Error

It was pointed out in Section II that statistics, such as the mean or variance, which are derived from small samples are likely to deviate from the true value of the particular statistical universe and that the error of these statistics can be estimated from the *standard error* statistic (Equation 6.3). In the case of a chronology developed from a number of cores and trees within several groups, the individual yearly values are in error depending upon the amount of error variance contributed by each source (cores, trees, and groups).

6. THE STATISTICS OF RING-WIDTH AND CLIMATIC DATA

The standard error for any yearly value of a chronology (SE_y) is calculated from the component variances (such as those shown in Table 6.VII) as follows:

$$SE_y = \left(\frac{VC(YT)}{t} + \frac{VC(YCT)}{ct}\right)^{\frac{1}{2}} \qquad (6.11)$$

where $VC(YT)$ and $VC(YCT)$ are the calculated variance components (Tables 6.VI and 6.VII) and t and c are the number of trees used and number of cores per tree in the chronology under consideration. Since core classes and groups are assumed to be statistically fixed (not randomly chosen), the calculations involve only the variances for tree chronologies and their interactions with the cores.

The components are unbiased estimates of these variances (that is, the results apply to the statistical universe and are independent of the sample size) so that different values for trees (t) and cores (c) can be substituted into Equation 6.11 to calculate the errors in samples of varying sizes. Table 6.VIII includes such calculations using the components in Table 6.VII and cases including from five to 20 trees per sample and from one to four cores per tree. Similar computations may use the component

TABLE 6.VIII Calculated mean standard error of estimates for various numbers for trees and cores[a]

Number of trees	Number of cores per tree			
	1	2	3	4
5	0·092	0·082	0·078	0·077
6	0·084	0·075	0·072	0·070
7	0·078	0·069	0·066	0·065
8	0·073	0·065	0·062	0·061
9	0·068	0·061	0·058	0·057
10	0·065	0·058	0·055	0·054
11	0·062	0·055	0·053	0·052
12	0·059	0·053	0·051	*0·049*
13	0·057	0·051	*0·049*	0·047
14	0·055	*0·049*	0·046	0·046
15	0·053	0·047	0·045	0·044
16	0·051	0·046	0·044	0·043
17	*0·050*	0·044	0·043	0·042
18	0·048	0·043	0·041	0·040
19	0·047	0·042	0·040	0·039
20	0·046	0·041	0·039	0·038

[a] Based upon variance components for tree chronologies of 0·025 and core chronologies of 0·017 in Table 6.VII and Equation 6.11.

variances of other samples to test for differences among sites (Fritts, 1969). Such analysis of variance statistics are especially useful for evaluating sampling strategies in terms of the information gained and the statistical reliability of the results.

For example, the data in Table 6.VIII confirm the generalization that the standard error decreases with any increase in the number of cores or trees. However, because of differences in the variance structure of the sample as indicated by the component variances, the error reduction is dependent more upon the number of trees in a sample than upon the number of cores sampled from each tree in the group. The data in Table 6.VIII indicate that if only one core is sampled per tree a sample size of 17 trees must be obtained to reduce the standard error to 0·050 or less (see values in italic type in the table). However, if two cores are sampled per tree, 14 trees are required; if three cores are sampled per tree, 13 trees are required; and if four cores are sampled per tree, 12 trees are required.

It could be inferred from the above results that a sample of one core per tree for 17 trees is the most efficient sampling scheme, but on a steep, rocky slope it may be easier to extract a second core from 14 trees than to find three extra trees. Since variance and other analyses require at least two cores per tree, the extra time required for sampling the second core is justified. In special circumstances where only a limited number of trees is available on a given site or where more information is needed from individual trees, the collector might decide to increase the sample to three or four cores per tree. On the other hand, where time is extremely limiting, the most efficient and desirable sampling scheme might be one which would utilize the time saved in collecting only one core per tree to obtain samples from additional trees.

Another feature of standard error computations was first noted in *Pinus edulis* from Mesa Verde National Park, Colorado. The standard error for a mean index of a chronology was found to be directly proportional to the average value of the index. That is, those years with high average indices generally exhibited higher standard errors than years with low average indices (Fig. 6.14). This manifestation is undoubtedly a consequence of the law of limiting factors described in Chapter 1. The cambia producing the narrow rings formed during years of severe climate are highly limited by similar characteristics of the macroclimatic factors, so that ring widths vary little from one tree to the next. The wide rings formed in less severe years are more limited by local site conditions than by factors of the macroclimate, and thus a larger variability in widths within trees, among trees, and among groups prevails. A test for this relationship is made routinely by checking the correlation between value of standard error and average index. The relationship varies in its importance among ring-width chronologies from arid sites, but at present no one has studied the problem sufficiently to explain its variance.

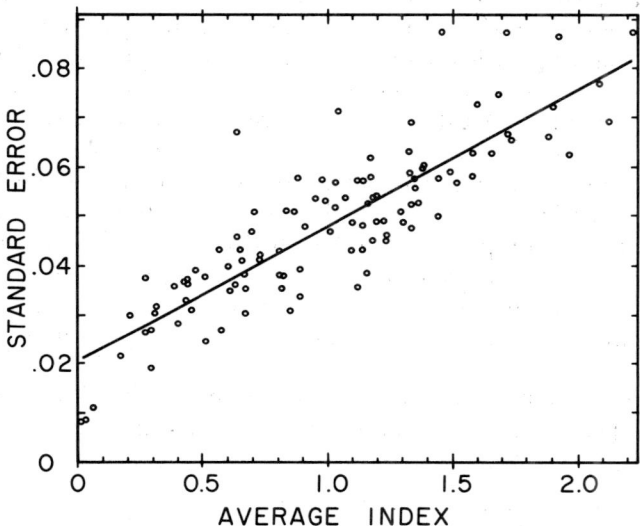

FIG. 6.14. The standard errors for average yearly values of some chronologies vary as a direct function of the average value in the index. The higher the value, the greater the standard error and the less precision of the index. The above sample is calculated for the average yearly index chronology from three groups of *Pinus edulis* from Mesa Verde, Colorado.

VIII. Correlation Analysis

It is not always possible to obtain adequate replication to perform an analysis of variance. In such cases correlation coefficients can provide information similar to analysis of variance. This information, however, is generally less useful than variance analysis, for correlation provides only a relative measure of variability in common between two data sets and it cannot be partitioned as easily into separate components. Nevertheless, because correlation is a simpler concept and easy to compute, it is sometimes favored over analysis of variance.

Correlation statistics are derived after standardizing the ring widths for individual cores. Correlation coefficients are obtained for all combinations possible within a group of chronologies. The correlations for cores paired within trees are tabulated separately from those paired between trees. The former measures the similarity within tree variance, while the latter measures the similarity between tree variance. In fact, the mean correlation for all pairs between trees is closely related to percent variance component for the group chronology in the analysis of variance [VC(Y), Table 6.VI]. That is, as the mean correlation increases between core chronologies in different trees, there is a parallel increase in the percent variance that is common among all cores, trees, and groups.

The relationship between the two measurements is so well correlated for arid-site conifers that the correlation results can be substituted into a simple regression equation to statistically estimate the corresponding variance component (Fig. 6.15). The equation shown in the figure is for the straight line fitted to the data points where Y is the group-chronology percent variance component and X is the mean correlation among core chronologies from different trees within groups. When analysis of variance is not possible, the correlation data can be substituted into the equation in Fig. 6.15 to estimate the percent common variance component.

Correlation analyses can be run for a number of time intervals such as for 20-year periods lagged every five years. A comparison of the results for

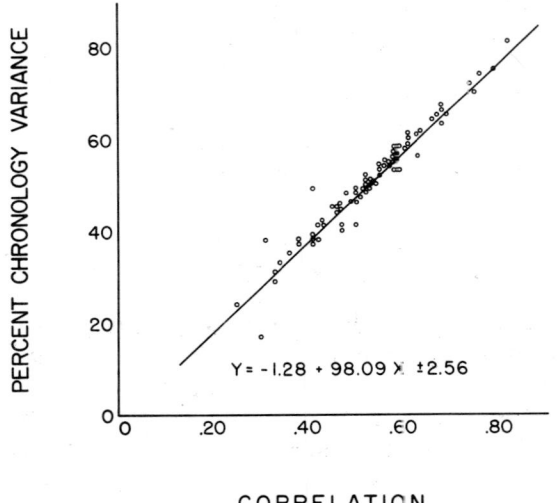

FIG. 6.15. The mean of correlation coefficients for all combinations of core chronologies paired between trees (excluding comparisons of cores from the same trees) is highly correlated with and may be used to estimate the percentage of variance in the group chronology [VC(Y) × 100/total VC] (Table 6.VI) as shown by the plotted points. The regression equation shown is for the line and is the least square fit to the data shown in the figure, where Y is the percent chronology variance and X is the mean of the correlation coefficients. The data are for semiarid North American conifers from a variety of species and sites.

different periods can be used to assess when ring widths were best correlated and hence most limited by climatic conditions as opposed to site factors. Correlation analysis can also be run on the mean chronologies for trees and the mean chronologies for groups to study the similarities and dissimilarities in these components. Analysis using correlation helped Charton and Harman (1973) to identify a decline in agreement among ring-width indices from *Quercus rubra* that could have been due to limiting effects of air pollution or possibly to anomalies in precipitation resulting from industrial pollution near the tree sites.

A dendrochronologist who selects cores from trees with the best crossdating and the highest sensitivity is simultaneously selecting for high percent variance in the group chronology, for high correlation among trees, and for a high proportion of climatic information in the chronology of the site. Analyses of variance and correlation allow the dendroclimatologist to evaluate his samples and to make decisions as to which ones are most likely to provide the best results for a particular experiment. The conclusions from these data can also guide future selections of tree-ring materials and sites and help in developing the most productive lines of research.

IX. Power Spectrum and Cross-power Spectrum Analyses

It is sometimes helpful to study nonrandomness in a ring-width time series using *power spectrum analysis*, which describes the data in a manner that presupposes nothing as to its nature in the frequency domain. For example, the ring-width index series may be regarded as an infinite number of oscillations or cycles described by an infinite number of wavelengths. The power spectrum analysis estimates the variance (the power) of each wavelength and expresses the results as a continuous distribution of wavelengths throughout the entire spectrum, ranging from the shortest wavelength that can be resolved by the annual values (a half cycle per year or one cycle every two years) to infinite wavelength (linear trend) (Mitchell, *et al.*, 1966).

The procedures for computing a power spectrum use the methods of Tukey (1950) and Blackman and Tukey (1958). First the autocorrelation coefficients are computed from one to a prescribed number of lags (Figs. 6.4, 6.16). These coefficients plotted against lag constitute the *autocorrelation function*. The power spectrum is then obtained by means of a Fourier transform of the autocorrelation function. The resolution of a power spectrum estimate is proportional to the number of lags that can be used to generate the autocorrelation function, which in turn is constrained by the record length. Very detailed estimates may be obtained if the time series is long, but they must be generalized to a few frequencies if the time series is short or few lags are analyzed (Jenkins and Watts, 1968).

The plotted autocorrelations and the power spectra of three ring-width chronologies are shown in Fig. 6.16 (Stockton, 1971, and Stockton, 1975). If the variations in the time series are purely random, the spectrum will approximate a horizontal line (all spectral estimates are the same for all frequencies) (Mitchell *et al.*, 1966). Since such a spectrum is analogous to white light, which includes equal amounts of light at all wavelengths, it is called a "white noise" spectrum. If a time series is a pure sine wave, the spectrum will contain a relatively sharp peak at the appropriate frequency for the sine wave. If there is a regular periodicity having a nonsinusoidal shape, the spectrum will contain not only a peak at the basic wavelength, but other peaks at wavelengths corresponding to one or more higher harmonics of its basic wavelength. If there is a quasi periodicity, or irregular rhythm, the spectrum will express it as a relatively broad hump spanning an appropriately wide range of wavelengths. Finally, if there is persistence whereby each value of the series is related to those values preceding it, the spectrum will be distorted across all wavelengths so that the greatest variance is at low frequencies and there is a decrease in power from long to shorter wavelengths. This spectrum is said to resemble that of "red noise" (Mitchell *et al.*, 1966). Most power spectra of ring-width series, like those shown in Fig. 6.16, exhibit "red noise" spectra indicating more persistence than would be expected if they were made up of completely random events. For more discussion of these and other techniques of time series analyses see Box and Jenkins (1970), Jenkins and Watts (1968), and Kisiel (1969).

The three power spectra in Fig. 6.16, in addition to showing "red noise," indicate that the chronology from site A has the least power (variance) of the three at high frequencies from 0.1 to 0.5 cycles per year while the chronology from site C has the most. The first is a cool and moist site and the chronology has moderately low mean sensitivity, while the latter is the driest and warmest site and the chronology has the highest mean sensitivity (Stockton, 1971, and Stockton, 1975).

The above techniques, in addition to being applied to single series variations, may also be used to study covariance between series expressed in the frequency domain (Jenkins and Watts, 1968). Studies of this type are referred to as a *cross-power spectrum analysis*. The computations utilize the autocorrelations of two series and the cross correlations between two series at various lags. Fourier transforms of the autocorrelation and cross correlation functions along with the phase relationships are obtained.

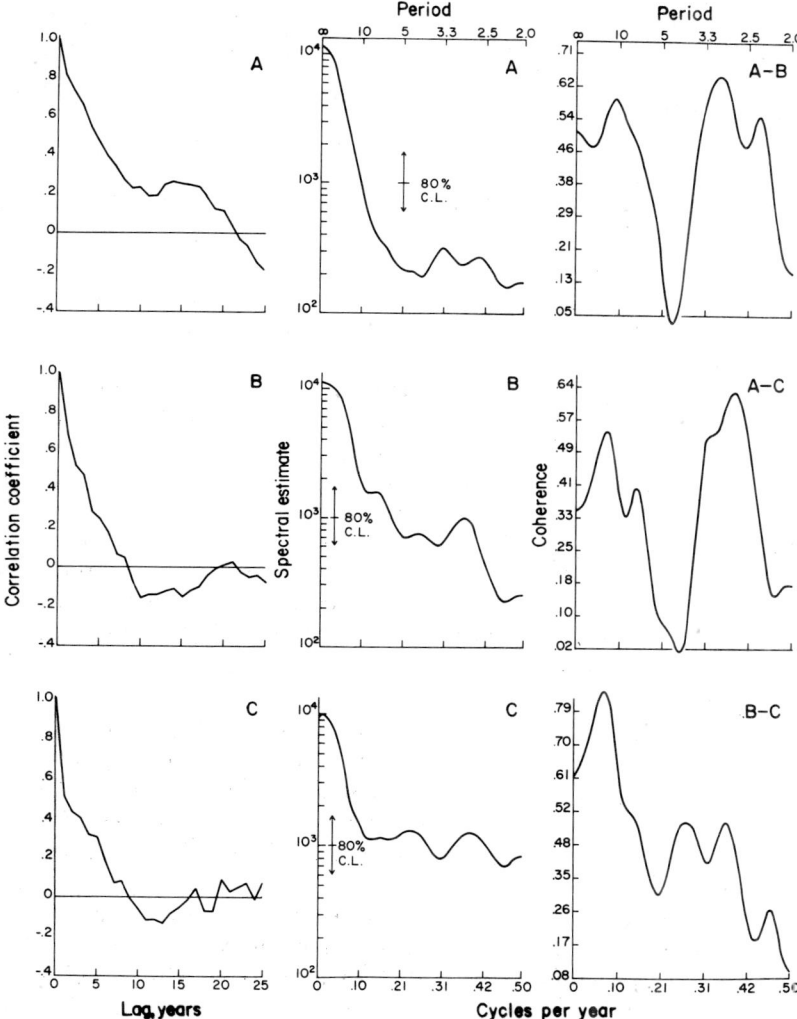

FIG. 6.16. An example of results from power and cross-power spectrum analyses of three site chronologies of *Pseudotsuga menziesii* in northern Arizona. The left-hand plots are the autocorrelations for lags from 0 to 25 years. The plots in the middle of the figure are the power spectra, while the right-hand plots show the coherence among the three chronologies obtained in the cross-power spectra analyses. Chronology A is from trees on a moist, cool site. Chronology C is from an arid site. Chronology B is from a relatively intermediate site. (Redrawn and reprinted by permission from "Long-term Streamflow Records Reconstructed from Tree Rings", Paper of the Laboratory of Tree-Ring Research at the University of Arizona, No. 5, by C. W. Stockton, Tucson: University of Arizona Press, © 1975.)

These data are used to measure *coherency* which represents the similarity between the two series at each frequency as in power spectrum analysis. Coherency is analogous to the square of the correlation between the two paired series at each frequency band.

The plots on the right of Fig. 6.16 are the coherencies among the same three chronologies. All pairs exhibit high coherency at 0·08 cycles per year or at periods of approximately 12 years. Pairs A–B and A–C exhibit higher coherency at 0·30 to 0·45 cycles per year than the pair B–C. The same two pairs exhibit little or no coherency at 0·24 cycles per year. This implies that the variance generated by series A near frequencies of 0·24 cycles per year is uncorrelated with the variance at the same frequency in the other two series.

Often power and cross-power spectrum analysis can lead to considerable insight into the biological and statistical relationships in regard to possible periodicities in factors controlling growth. For example, LaMarche (1974b) studied four chronologies of *Pinus longaeva* in the Snake Range of Nevada ranging from the low-elevation semiarid forest border to the high-elevation upper tree line. He noted that the chronologies at the lower and upper tree lines appeared relatively uncorrelated; but when the data were treated with the high- and low-pass filters shown in Table 6.III, there was high and direct correlation between the high-frequency variations and high but inverse correlation between the low-frequency variations.

The data were then subjected to power and cross-power spectrum analysis because LaMarche was concerned that the arbitrary spectrum selection of filters may not have divided the series into the "natural" frequency bands. He noted that the biggest difference between the two spectra occurred in the frequency range from about 0·1 to 0·5 cycles per year. As was the case for Stockton's results (1975) shown in Fig. 6.16, the arid forest border trees exhibited more variance at high frequencies than did the trees at high elevations on more moist and cool sites.

Cross-power spectrum analysis of LaMarche's data revealed that a portion of the variance in the chronologies was out of phase; that is, when a maximum in growth occurred at the lower tree line there was a tendency for it to be followed in three years by a minimum in growth at the high-elevation tree line. Also, the inverse correlation was found to be most pronounced at frequencies from 0·10 to 0·01 cycles per year.

LaMarche (1974b) attributed the high positive correlations at high frequencies to the direct effects of precipitation on ring widths for both the low and high elevation sites. The coherency at low-frequency and the out-of-phase relationships were attributed to the differing effects of climate on the growth of needles, which in turn affect photosynthesis and ring width (see Chapter 4). At the lower forest border short needles were formed during years when precipitation was low and temperatures were high, but at the high elevation short needles were formed when temperatures were low. Since the needles are retained and are photosynthetically active for 15 or more years, the growth response to temperature was not only different for the two sites but it was averaged over 15 years and it lagged behind the occurrence of climate.

The coherency between the two chronologies enabled LaMarche to identify four frequencies which were best correlated. These are shown in Fig. 6.17 as peaks from right to left at 0·45, 0·30, 0·14, and 0·037 cycles per year with periods corresponding to 2·2, 3·3, 7, and 27 years. LaMarche suggested that the tree-ring coherency at 0·45 cycles per year (with a period of 2·2 years) was probably linked to the well-known rhythmic variation in wind and pressure from equatorial regions known as the "quasi-biennial oscillation" which exhibits a period of 26 to 27 months (Landsberg, et al., 1963). He also suggested that coherency at periods of 27 years could be a result of solar-lunar tidal influences in the atmosphere as described by Brier (1968), which could control both warm-season temperatures in the central Great Basin and the tree needle lengths.

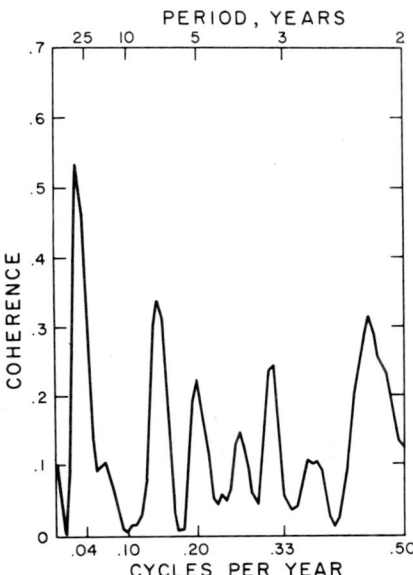

FIG. 6.17. An example of cross-power spectrum analysis showing the coherence between two *Pinus longaeva* chronologies in the Snake Range of Nevada. One chronology was from trees at the semiarid lower forest border and the other from the upper tree line. The peaks indicate coherency at about 0·45, 0·30, 0·14, and 0·037 cycles per year. (Redrawn from LaMarche, 1974b)

Power spectrum analyses also were employed by Bryson and Dutton (1961), LaMarche and Fritts (1972), and Sirén and Hari (1971) to study periodicities in ring-width data that might be associated with the 11-year and 22-year periodicity in sunspots. Although several apparent relationships were proposed, no ring-width series was proven to be significantly linked to sunspot numbers and none was confirmed to exhibit a significant spectral peak at the appropriate frequencies (LaMarche and Fritts, 1972).

X. Variability in Statistical Characteristics of Ring Widths among Sites

It has been shown that a number of statistical parameters may characterize the time series of ring widths and ring-width indices. These statistical parameters may vary because the generating functions of the time series differ among trees and sites.

Figure 6.18 portrays six tree-ring characteristics which vary with increasing aridity of a site and decreasing arboreal dominance of the stand. The area to the left of K in the diagram represents forest interior trees which exhibit little ring-width variability, while the area to the right of K represents trees of greater sensitivity and ring-width variability, generally those in semiarid forest border sites. To the left of J there is insufficient variability in common among trees to allow rings to be crossdated. To the right of M, there are so many partial or absent rings that crossdating is excessively difficult.

The changes occurring from left to right across the diagram are (1) decreasing average ring width, (2) decreasing arboreal dominance (density of the forest stand), (3) increasing correlation of ring widths within and among trees, (4) increasing correlation of ring widths with climatic variations, (5) increasing mean sensitivity and standard deviation, and (6) increasing percentage of rings that are locally absent. Sometimes there is a reduction in autocorrelation with increased aridity, except to the right of L where autocorrelation may begin to increase. The above characteristics change from left to right in proportion to decreasing amounts of effective precipitation and increasing variability in that precipitation. As these factors become more extreme, they result in more days when moisture becomes limiting to tree processes.

As mentioned in Chapters 1 and 4, such changes in ring-width characteristics are not only a direct result of macroclimatic variations but also of individual site factors which affect the degree to which climatic factors become growth-limiting. For example, the amounts of precipitation can vary due to topographic factors, or site-to-site

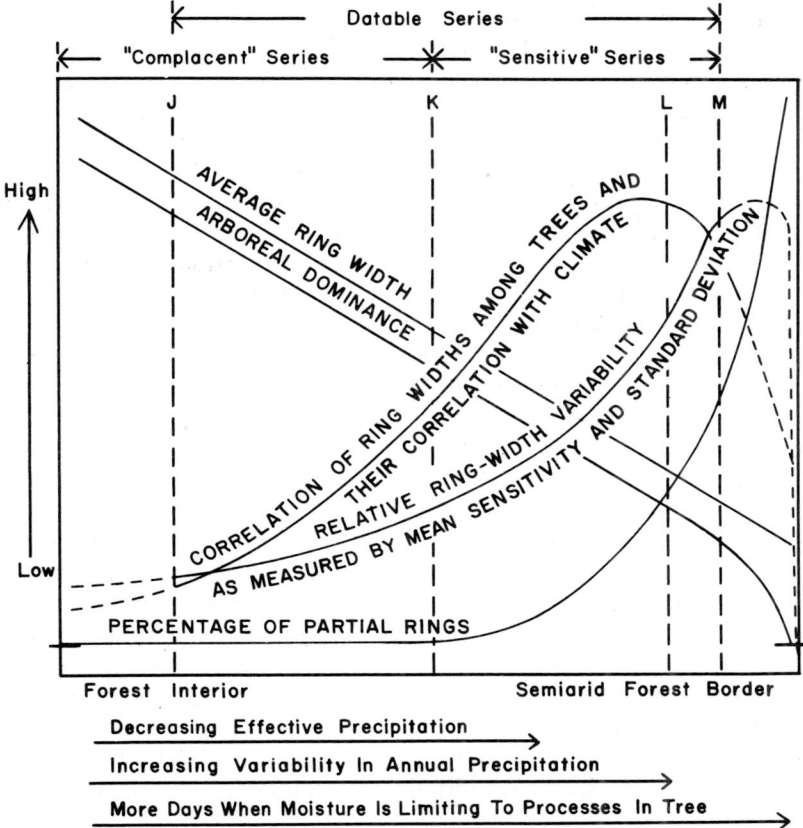

FIG. 6.18. Statistical and dendrochronological characteristics of ring-width chronologies from semiarid regions vary as a function of the proximity of the sampled site to the forest border. J is in the forest interior, where arboreal dominance (density of the stand) is high, rings are wide and "complacent," and there is little correlation in the ring-width variation among trees or between ring widths and climate. K marks the difference between "complacent" and sensitive series and L is near the forest border where arboreal dominance is low (the trees are more scattered), rings are narrow, mean sensitivity and standard deviation of the chronologies are high, and there is a high correlation of the chronologies among trees and with climate. Climatic and physiological conditions that change along the gradient are indicated. (Modified from Fritts *et al.*, 1965b)

variations in soil characteristics can cause different amounts of moisture to infiltrate the soil. Differences in site exposure can affect temperature, wind, and radiation, all of which affect the rate at which moisture is lost by evapotranspiration processes; and exposure of the site can affect the removal and drifting of snow, which modifies the soil moisture reserves.

Since the statistical characteristics of ring-width indices are generally related to the moisture availability on arid sites, their values can be used to estimate the amount of environmental stress on tree growth.

The utilization of tree-ring statistics to assess environmental differences at several tree sites is demonstrated in Fig. 6.19. A number of statistics were calculated from replicated ring-width indices for the last 100 years from *Pinus longaeva* in southeastern California. The statistics for each site included (1) the percentage of partial or absent rings, (2) mean sensitivity, (3) standard deviation, (4) mean correlation between cores within trees, (5) mean correlation among cores between trees, (6) the first-order autocorrelation (serial correlation), and (7) the percent variance components for cores [VC(YCT)], trees [VC(YT)], and for the entire group [VC(Y)]. The samples were obtained from sites representing differences in topography, elevation, and soil substrate. Differences among species and among trees of varying mean sensitivity were also considered. The data were arranged to represent particular environmental gradients, shown by the labels and diagram at the top of Fig. 6.19; the percentages of slope are plotted immediately below them.

The statistics for sites 6, 7, 8, 9, and 10, shown in Fig. 6.19, represented a topographic gradient from a steep and exposed SSW-facing slope across a small valley to a steep and exposed NNE-facing slope. Site 8 included the four young and four old trees used in this text to illustrate analysis of variance; but the results shown here include data for 10 trees of each age class and for 100 years.

Most of the statistics indicate that trees on the lower NNE-facing slope (site 9) were generally the least limited by climate. Few rings were absent, mean sensitivity and standard deviation were low, correlation within and among trees was low, the percent variances for cores was high, and the variances common to all trees were low. However, on the upper north-facing slope where the trees are exposed to wind (site 10) and on the south-facing slope (sites 6 and 7), the tree-ring statistics indicate that the environment was more limiting to growth.

Sites 11 and 12 in Fig. 6.19 face WNW and ESE near the sites discussed above. It was thought that the prevailing winds from an open meadow to the west might carry snow into the WNW-facing site although not into the opposite slope. However, only slight differences in statistical characteristics were apparent, and it was concluded that in this instance exposure had little effect upon ring-width characteristics.

An interesting topographic phenomenon is apparent from the data for sites 15, 13, 14, and 16 which were on a SSE-facing slope. The statistical characteristics of ring widths from trees on sites 13 and 14 indicate that relatively mesic conditions prevailed on the middle of the slope.

6. THE STATISTICS OF RING-WIDTH AND CLIMATIC DATA

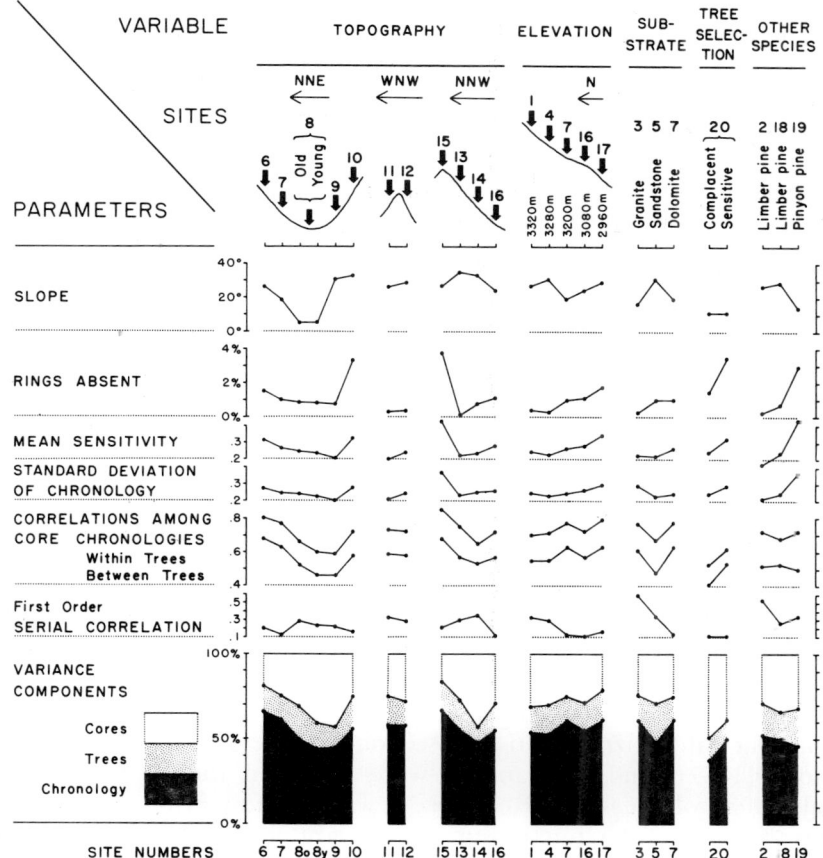

FIG. 6.19. The statistics of ring-width chronologies can be used to assess the stress of particular sites. The data are derived from analyses of replicated samples from *Pinus longaeva* in the White Mountains of California. Site variables of topography, elevation, and substrate are diagrammed at the top, as well as selection criteria of complacency as opposed to sensitivity, and differences among species. The variations in slope and values of statistical parameters are plotted below as a function of the variables shown at the top. (Reprinted by permission from "Bristlecone Pine in the White Mountains of California, Growth and Ring-width Characteristics", Paper of the Laboratory of Tree-Ring Research at the University of Arizona, No. 4, by H. Fritts, Tucson: University of Arizona Press, © 1969.)

However, the characteristics of rings for sites 15 and 16, which were above and below the two midslope sites, indicate more aridity. A greater number of rings were absent, and both mean sensitivities and standard deviations were higher. The correlations within and between trees and the percent variance of the group chronologies were highest for site 15. Aerial photographs made in the spring showed snow banks on slopes of this particular exposure. Therefore the decrease in ring-width variability and the decline in other arid-site characteristics of the midslope tree rings was attributed to increasing amounts of moisture from drifting snow.

The statistical characteristics along the gradient of elevation (sites 1, 4, 7, 16, 17) show changes associated with increasing aridity as a result of decreasing elevations. However, the magnitude of differences is less than those in the topographic comparisons (sites 6, 7, 8, 9, 10) depicted on the left of the figure. It was concluded that topographic variations in the White Mountains of California have a greater effect on arid conditions than elevation and hence a greater effect on ring-width characteristics.

Classifying trees into varying substrate, species, and sensitivity revealed differences in first-order autocorrelations, which were highest for *Pinus longaeva* on granite and for site 2 with limber pine, *Pinus flexilis*.

XI. Statistical Characteristics of Ring Widths within a Tree

The previous section described a number of statistical characteristics that can vary from site to site. The dendrochronologist may not only be interested in site variations, but he may also wonder about statistical variations in ring width and cell structure that can occur throughout the main stem and branches of trees. The following is a study of this within-tree variation.

The main stem of a 295-year-old *Pinus ponderosa* which was growing near Flagstaff, Arizona, had been sectioned at specified levels in 1935 by Glock (1937), and the samples were stored in the archives of the Laboratory of Tree-Ring Research. It was noted by Fritts *et al.* (1965a) that the dates of the innermost rings on eight of the sections differed by intervals of approximately 20 years. Thus, the rings within each section could be divided into 20-year intervals (Fig 6.20 and 6.21).

The first 20 rings on the lowermost section (section 1) represented approximately the first 20 years of growth in the life of the tree. The second set or segment of 20 rings on section 1 was formed during the second 20 years in the life of the tree and corresponded to the first 20 rings on the next section above it. The third segment of 20 rings on section 1 corresponded to the second one on section 2 and the first one on section 3. In a like manner, the eighth segment of 20 rings on section 1 corresponded to the first segment of 20 rings on section 8. Another way to view and classify the data set is to consider the outermost 20 rings in all eight sections which were formed during the last 20 years in the life of the tree (1915–1934) while tracing the rings inward; the seventh set of 20 year segments represented the first 20 years on section 8 and the seventh section from the center for section 1. The rings along four radii on each cross section were crossdated, examined by 20-year segments, and analyzed in terms of eight different statistics. The 80 rings for each group (20 on each of four radii) were used to calculate average ring width, average percentage of intra-annual growth bands (false rings), and average percentage of rings that were absent.

The ring widths for each of the four radii were converted to indices by standard techniques (Equations 6.7 and 6.8). Correlations were obtained by pairing the four radii on each section and segment. There were six possible paired combinations of the four radii, so six correlation coefficients were computed from the indices within each segment. The mean of these six coefficients is referred to as the intracorrelation for that particular time period and section.

The indices from the four radii were also averaged, and the mean index series was used to calculate the first-order autocorrelation (serial correlation), the standard deviation, and mean sensitivity for each 20-year segment. Correlation coefficients were also computed for the average indices on each 20-year segment paired with the average indices on other sections in the tree that were dated in the same 20-year time period. For example, the indices for the second 20-year segment of section 1 were correlated with those on the first 20-year segment of section 2. The indices for the third segment on section 1 were correlated with those for the second segment on section 2, as well as those in the first segment on section 3. In this latter case a third correlation coefficient was also obtained between the indices for the second segment on section 2 and the first segment on section 3. Extending the process to more segments and sections involves the computation of 84 correlation coefficients which express the intercorrelated variability in ring-width indices from section to section throughout the tree. All coefficients involving a particular 20-year segment were averaged. These averages are referred to as the *inter*correlation of indices among sections for each 20-year period, as opposed to the *intra*correlation of indices within a particular section described in the preceding paragraph.

Figure 6.20A portrays the 84 mean width values for the groups of 20 rings in their respective positions diagrammed as a half-radial section within the stem. The data for each of these blocks or segments were then pooled in four different classifications labeled as A, B, C, and D.

As shown by the emphasized lines in Fig. 6.20, the A classification included data which were pooled and averaged along columns of the 20-year segments aligned with the pith. The resulting averages which are plotted in column A of Fig. 6.21) are analogous to the horizontal sequence of Duff and Nolan (1953) (see Fig. 6.12). The first average, plotted on the left of column A, includes date from all rings formed by one- to 20-year-old cambia; the second includes data from all rings formed by 21- to 40-year-old cambia, and the last (plotted on the right of each graph) includes data from all rings formed by 121- to 140-year-old cambia. Thus, the averages express the variations associated with increasing cambial age and to some extent, tree age. Every average includes one segment from each section so that height is held constant.

The same data were utilized in classification B, but they were pooled and averaged in the direction of rows, each corresponding to one cross section. The resulting averages for ring width, which are also shown in Fig. 6.20 and plotted in column B (Fig. 6.21), are analogous to Duff and Nolan's vertical series (see Fig. 6.12). These data portray the variations within the tree stem (from left to right) associated with increasing height. However, cambial ages from one to 140 years are included in each average so that cambial age is held constant. As in the previous classification, the successive averages at increasing stem height also include some changes associated with the aging tree.

In classification C the data were pooled along the rows, but the segments represented by blocks in Fig. 6.20 were aligned with the outer wood surface, not the pith. The resulting averages and plots are analogous to the diagonal series in Duff and Nolan. From left to right they express the effects of increasing stem height, with tree age (the dates of the rings) held constant.

The averages for classification D were pooled in a diagonal direction

FIG. 6.20. Statistical data from eight sections of *Pinus ponderosa* are separated into 20-year segments and pooled in four classifications to study various patterns that occur throughout the stem. Data for mean ring widths for each 20-year segment are shown in 84 blocks in the upper left. The left margin of each diagram is aligned with the pith in the stem, the right with the bark, and the bottom and top rows represent the lowermost and uppermost sections. The mean values for each 20-year segment, shown in the boxes, are then pooled and averaged vertically in classification A in the direction of the emphasized columns. The resulting means express the effects of increasing cambial age starting with the first 20 years next to the pith. The same data are used in classification B (upper right), but the data are pooled in a horizontal direction shown by the emphasized rows. The resulting averages express the association of mean ring width with increasing height within the stem, but cambial age is held constant. For classification C the data are also pooled in a horizontal direction as shown by the emphasized rows in C, but the segments are chosen according to their alignment with the outer wood surface. The averages of this classification express changes in ring width associated with increasing height within the stem, but tree age, not cambial age, is held constant. Classification D uses the same block of data as classification C, but the data are pooled in a diagonal direction to obtain the mean ring width throughout the stem associated with each 20-year period of time. Plots of the average values obtained in each classification are shown in Fig. 6.21 (Fritts et al., 1965a).

6. THE STATISTICS OF RING-WIDTH AND CLIMATIC DATA 307

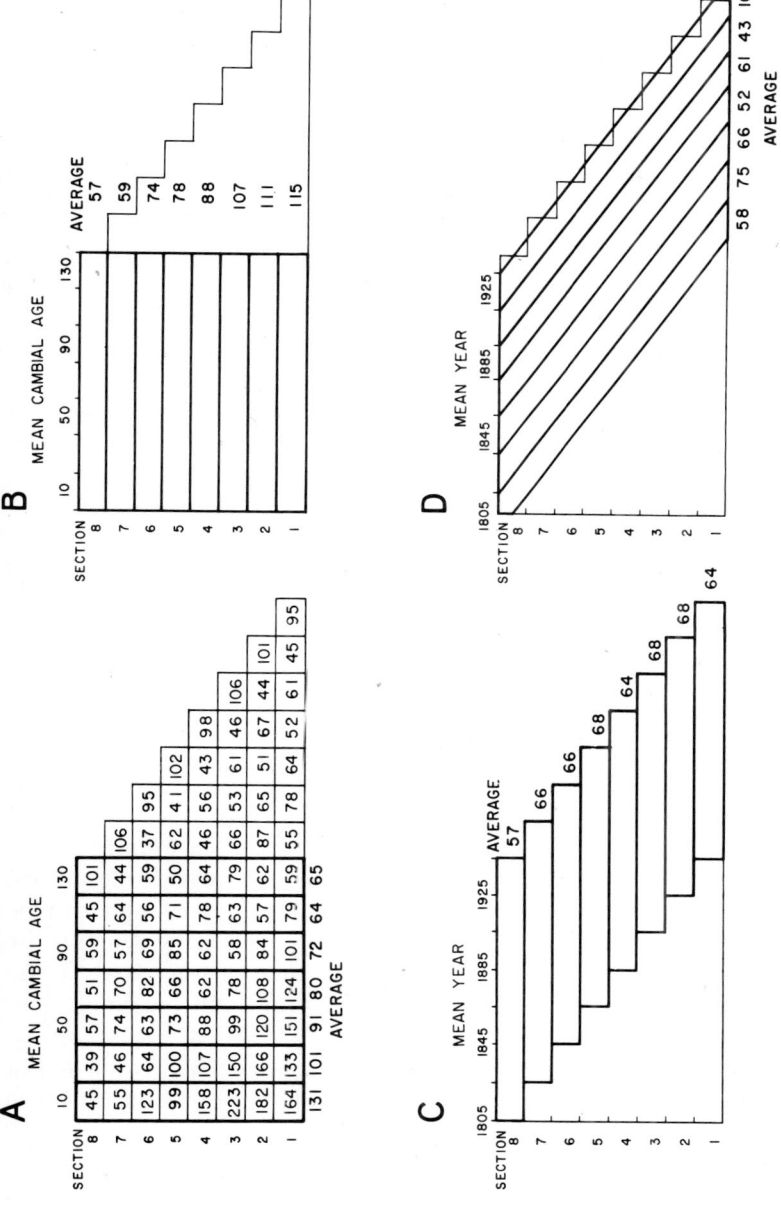

parallel to the outer surface of the wood. These data are analogous to the traditional dendrochronological series and the second horizontal series in Duff and Nolan, except that 20-year averages rather than individual indices are utilized. These averages express (from left to right) the effects of both increasing tree age and variations in climate. As in classification A, height within the stem is constant, but unlike A, cambial age decreases to some extent with increasing height; so cambial age does affect the plots to a limited extent.

The following discussion examines the variations in each statistic revealed by the four classifications plotted in Fig. 6.21 and relates them to material discussed in earlier chapters. Ring widths show the expected decline with increasing cambial age (A) and with increased height within the stem when cambial age is held constant (B). However, when tree age, not cambial age, is held constant (C), ring width does not change as a function of height except for a slight decline in width in the uppermost section. The sequences compared in this classification (as well as D) include rings formed after the tree had attained two-thirds its full height, so that the vigorous growth of youth is not considered. Thus, looking back at classification B the inverse correlation between ring width and tree height is probably more an effect of increasing tree age than of a height factor. The variations shown in classification D show a general decline in ring width with increasing tree age and a marked increase in width centered in 1925. This particular 20-year interval of high growth is associated with and undoubtedly caused by a period of exceptionally favorable climate.

The percentages of double rings are directly correlated with ring width, as shown in sequences A, B, and D. However, classification C indicates a tendency for more double rings in the uppermost sections. As mentioned in Chapters 2 and 4, double rings are often found in close proximity to the stem apex which is the source of growth regulators. Thus, the high frequency of double rings in the upper stem can be attributed to (1) the closeness of the rings to the stem apex which is the growth regulator source, (2) the vigor of the younger tissues, and (3) the greater propensity for water stress with increasing stem height.

FIG. 6.21. Changes in ring-width statistics in a *Pinus ponderosa* associated with the four different classifications using eight sections of the main stem and 20-year segments within the sections which are pooled and averaged as illustrated by the diagram at the top (also see Fig. 6.20). Every second section in the diagram and every other 20-year segment is shown in the diagram, and the solid lines connect the blocks of data that are pooled in each case. The plots express the averages of each statistic as a function of the 20-year segments from the inside toward the outside or the cross-section number, ordered from the bottom to the top. See text for further explanation. (Fritts *et al.*, 1965a)

6. THE STATISTICS OF RING-WIDTH AND CLIMATIC DATA

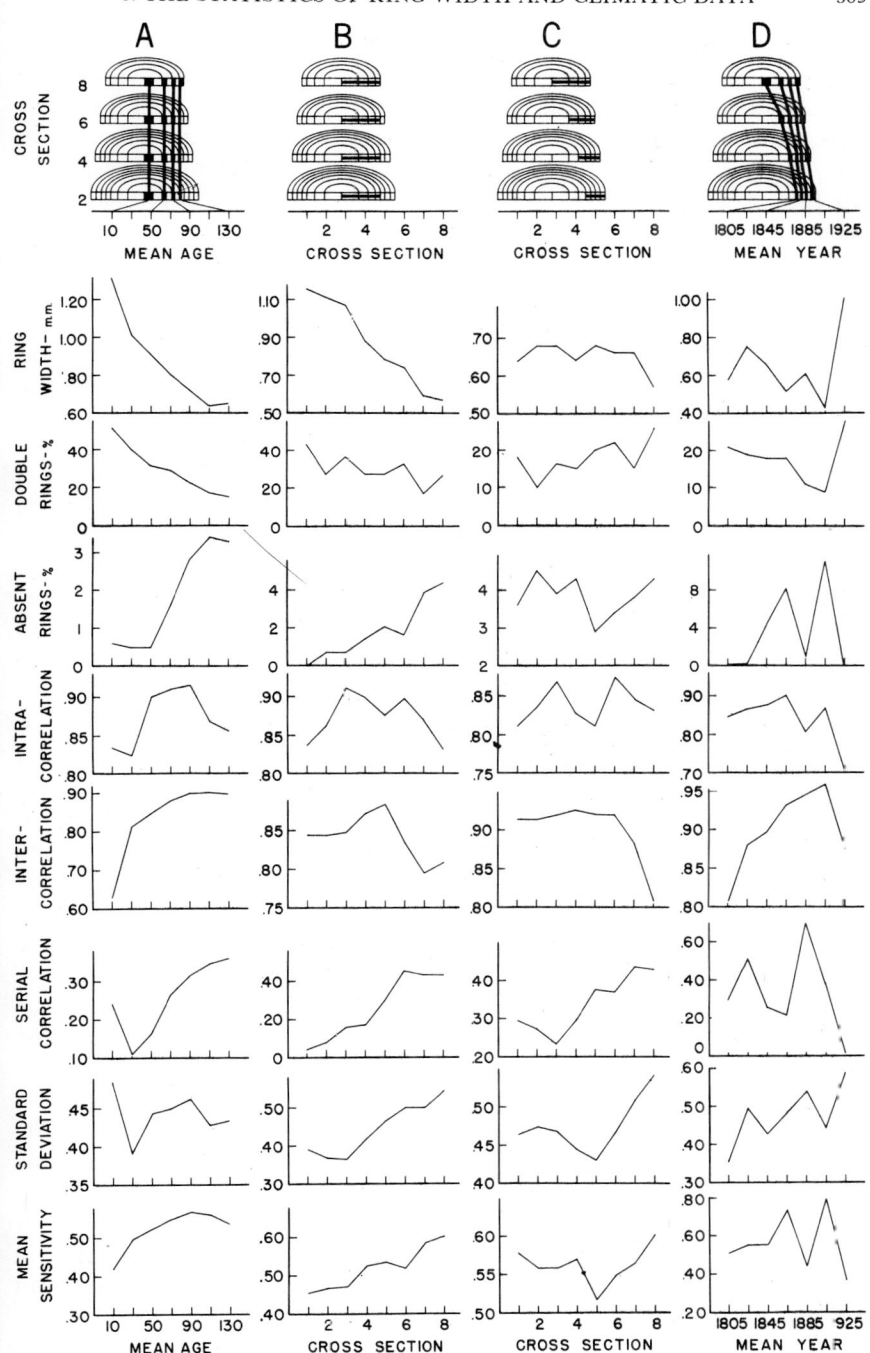

The percentages of absent rings show an expected inverse correlation with ring width in all four classifications. If the rings on the average are wide, there is a low probability that rings are absent. Cambial age (classification A) is also important in that absences are rare for tissues derived from one to 60-year-old cambia, but absences increase as a function of cambial age from 60 to 120 years; for greater cambial ages the percentages appear relatively constant. Classification C shows declining absence for those sections affected most by the lower crown (sections 5 and 6), a relationship attributable to a strong cambial initiation stimulus in the stem immediately below the major branches.

Both the intracorrelation (the correlation within a section and segment) and intercorrelation (the correlation among sections) show the poorest agreement among rings derived from one- to 40-year-old cambia (classification A). This result supports the contention that the first-formed rings on a stem yield the least information on climate. Intracorrelations are also low, indicating somewhat less agreement (1) in the uppermost portions of the stem, sections 7 and 8, (2) in sections near or just below the main branches, (3) at the ground line, and (4) in segments produced when rings were widest and climatic factors most favorable. Intercorrelation indicates that there is excellent agreement in ring-width patterns throughout the first six sections and for time periods when rings were narrow and climatic factors were highly limiting.

Autocorrelation, standard deviation, and mean sensitivity are all high in the uppermost and outermost stem segments. This is evidence of increasing variance at all frequencies with increasing tree height and age. These statistics also show marked variations associated with changing climatic conditions and with variability in other statistics for the same periods of time (column D). The anomalously high autocorrelations and standard deviations at cambial ages from one to 20 years are due to a brief period of increasing ring width in the first 10 rings in some sections. This increase in width was not adequately modeled by the negative exponential curve that was used in the standardization process, so that the divergent trend of growth inflated the autocorrelation and standard deviation statistics.

The above analysis illustrates how statistical characteristics can be objectively assessed throughout a tree stem. Classification by segments and section can be used to identify changes associated with various gradients representing height, age, or spatial variations within the stem. One highly significant conclusion from the study is that the ring-width variations at breast height (approx. 4·25 ft. or 1·3 m from the ground) in the stem provide good estimates of the ring-width variations throughout the entire stem and thus are probably the best estimates of variations in the macroclimate.

Many of the statistics dealt with in this chapter are readily available for analytical work, since they are routinely calculated in the tree-ring standardization programs. If used properly, the statistics can provide an objective and powerful tool for assessing variations from site to site, from tree to tree, and from one segment in the tree stem to the next. The statistics of new and old collections can be compared to obtain objective measures of their respective qualities. Inferences from such statistics can guide future collections, assist in model development, and suggest new approaches.

Statistical measurements will become increasingly important for assessment of site characteristics as tree-ring analysis is applied to new areas where the relationships between growth and climate are less apparent. Careful use of statistics should also make it possible to distinguish previously unrecognized phenomena in many and varied fields that lend themselves to dendrochronological analysis.

Chapter 7

Calibration

I	Introduction	312
II	The Procedure of Calibration	313
	A. The biological model	314
	B. The statistical model	316
	C. Verification	320
III	The Role of Statistics and Sample Size	321
IV	Degrees of Freedom and the Effective Sample Size	323
V	Selecting the Statistical Model	325
VI	The Diversity in Variable Selection	327
VII	Testing the Association between Variables	329
	A. The sign test	329
	B. Product means	331
	C. Simple correlation and reduction of error	332
	D. Conditional probability	333
VIII	Multivariate Techniques	340
	A. Multiple regression and correlation	341
	B. Regression after extracting the principal components	352

I. Introduction

Ring widths in trees on stress sites have been shown to vary from one year to the next in a more or less irregular manner, with a large portion of this variation being a function of fluctuating climatic conditions prior to and during the growing period when the rings were formed. This occurs because variations in the macroclimate for each year affect the operational environments of the trees which, in turn, influence the processes governing growth. The degree of relationship between climatic factors and ring-width variation is dependent upon (1) the ecological amplitude of the species, (2) the proximity of the sampled trees to their environmental limits, and (3) the range of variability in the limiting factors affecting growth.

As mentioned in earlier chapters, the growth-controlling processes in trees on moist sites with mild climates will be less limited by conditions associated with variations in climate than those on stress sites, but they will instead be limited by nonclimatic conditions such as shading by neighbors, lack of soil minerals, size of crown, and tree age. In such cases ring widths will vary widely from one tree to the next and only a small amount of variation will be common to all trees of a given site. This common variation, although it is less precisely defined, represents potential information on variations in the macroclimate.

From the point of view of a dendroclimatologist, the ring-width variation due to climatic factors would be considered analogous to the *signal* in an electronic system, while those due to nonclimatic factors would be analogous to *noise*. In such an analogy, the ring-width variations in trees experiencing moist and mild climates would have a small signal-to-noise ratio. The ring-width patterns in trees highly limited by climatically related environmental factors would have a large signal-to-noise ratio.

Not only does the ratio of signal to noise vary, but the specific values of ring width that are associated with specific values of climate can vary as a function of growth rates of the trees and the severity of the site. Furthermore, a lag may exist ranging from several months to 15 or more years between the climatic input of the system and the corresponding growth output. This chapter will describe procedures that can be used to identify the standardized ring-width variations that result from variations in climate and to transform them as well as other growth indicators into estimates of climate. Consideration will also be given to methodologies for determining the structure of a relationship including any lags in the growth response behind associated conditions of climate.

II. The Procedure of Calibration

The variations in ring widths might be considered analogous to the length of a column of mercury in a thermometer measured at regular intervals of time or to the dial reading on a scientific instrument which is sensing an environmental variable. The readings are compared to the actual conditions which the instrument is sensing to obtain a *calibration* between the two. New gradations can be made on the thermometer which take into account the association between the direct measurements of temperature and the observations on the instrument. In the case of dial readings, a calibration equation can be obtained with appropriate constants which converts the measurements on the instrument to units appropriate to the variable that is sensed. If the instrument produces a voltage, an equation may be found that converts voltage to a temperature measurement.

A *variable* is defined here as a characteristic, feature, or factor which can assume different values in successive individual cases. In each of the above examples there are two variables, those values observed on the instrument to be calibrated and those observed on a previously calibrated instrument which is now considered a standard. In the case of ring-width chronologies the indices are the values for one variable, and monthly precipitation or average monthly temperature represent the values of climatic variables with which the indices are calibrated. The success of such calibration is measured as the percent variance accounted for or reduced on one set of variables when another is used to estimate it. If the values of the chosen variables express important associations, the reduced variance is greater than that which could be obtained by chance.

The object of calibration as presented in this text is to establish the appropriate statistical or mathematical transformation which converts measurements of climate to those of growth or converts measurements of growth to those of climate, depending on the goals of the research. Such calibrations can be obtained in a number of ways, some of which involve relatively simple classifications and counts and others which involve rather complicated statistics. The simpler procedures often involve one factor at a time and require *a priori* judgments as to the nature of the relationships. Some of the more complex methods use statistics which help to identify the relationships, to eliminate unnecessary variables, and to assign appropriate weights. These complex methods often involve fewer *a priori* judgments.

All calibration procedures include a certain amount of modeling which describes or quantifies the biological–climatological linkage. A number of qualitative biological models already have been presented (Chapter 5). The task described in this chapter is to simplify these or similar models and to make them quantitative.

A. *The Biological Model*

Part A of Fig. 7.1 is a flow diagram which summarizes the task of developing biological models. First, some sort of meaningful biological–climatological linkage is hypothesized (step 1). A simple example might be that the width of a ring is controlled by available soil moisture and is therefore proportional to the precipitation over a several-month period ending when ring growth is complete. Another example might be that ring width is proportional to the hours of physiological activity allowed by favorable plant temperatures; thus, the length of time when temperatures are above a critical value during the biologically active period would correlate with ring width. A more complex model is described in Figs. 5.8, 5.9, and 5.10.

The researcher hypothesizes a particular model and then examines

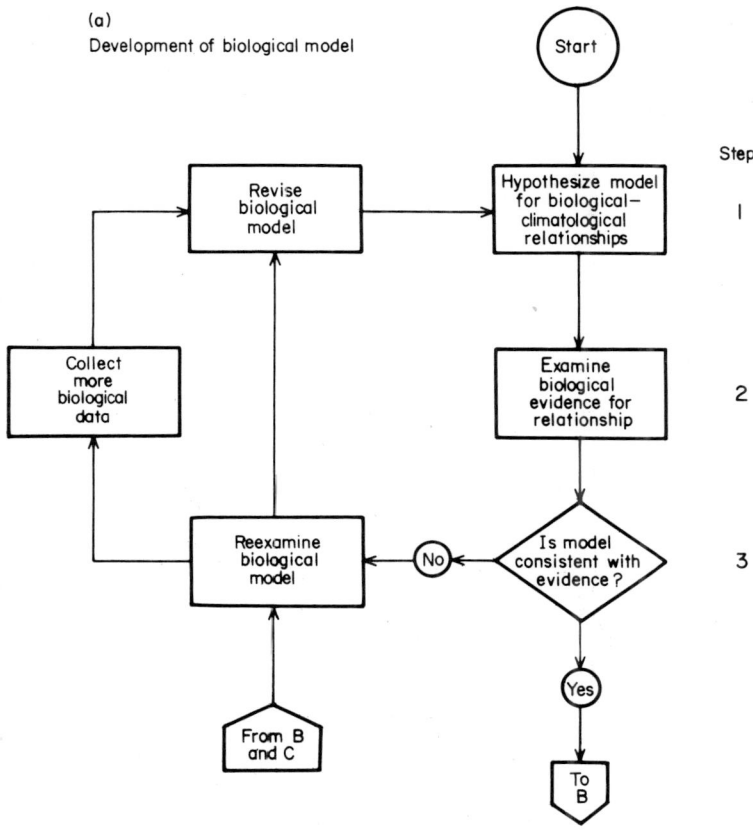

FIG. 7.1a

FIG. 7.1a, b, and c. Flowchart illustrating the steps necessary to obtain reliable calibrations between variations in ring widths with variations in climate. Steps are the same for both response functions and transfer functions except for the final use of the result.

the data, testing them against his experience or against evidence described in the literature of the field (Fig. 7.1a, steps 1, 2, 3). If there is insufficient biological information for the proposed relationship, he may collect more data before attempting to revise or test the model further; or if there are inconsistencies between the biological information and the model, he may reexamine the hypothesis, revise the model, and proceed to test the new relationship (steps 2 and 3). When sufficient information is consistent with the hypothesized biological model, he proceeds to calibration, part B of Fig. 7.1.

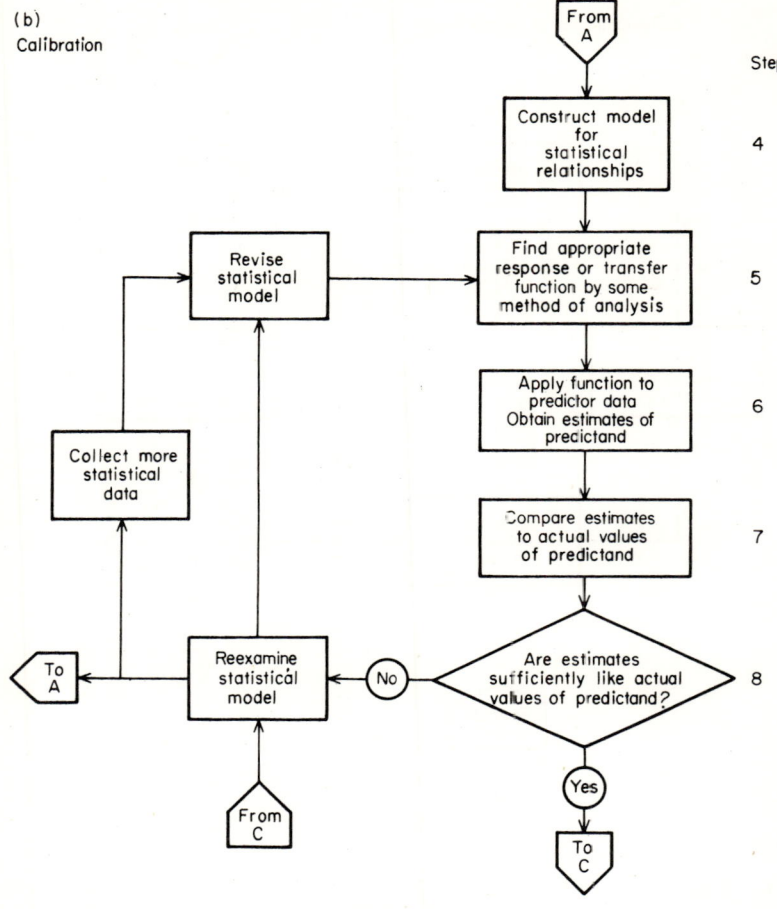

FIG. 7.1b

B. *The Statistical Model*

A statistical model is constructed which suitably portrays the general features of the biological one. Often this is accomplished by making simplifications and then by transforming the growth and climatic data in a manner that simulates the relationship. The transformations are compared to the actual data they are supposed to estimate (Fig. 7.1b, steps 4, 5, 6, 7). If the statistical model resembles the true biological relationship closely enough, the variation in computed values will closely resemble the variations in the observed values of the system output.

In the example using precipitation and growth, a linear relationship can be assumed between growth and total precipitation falling during the

(c) Verification

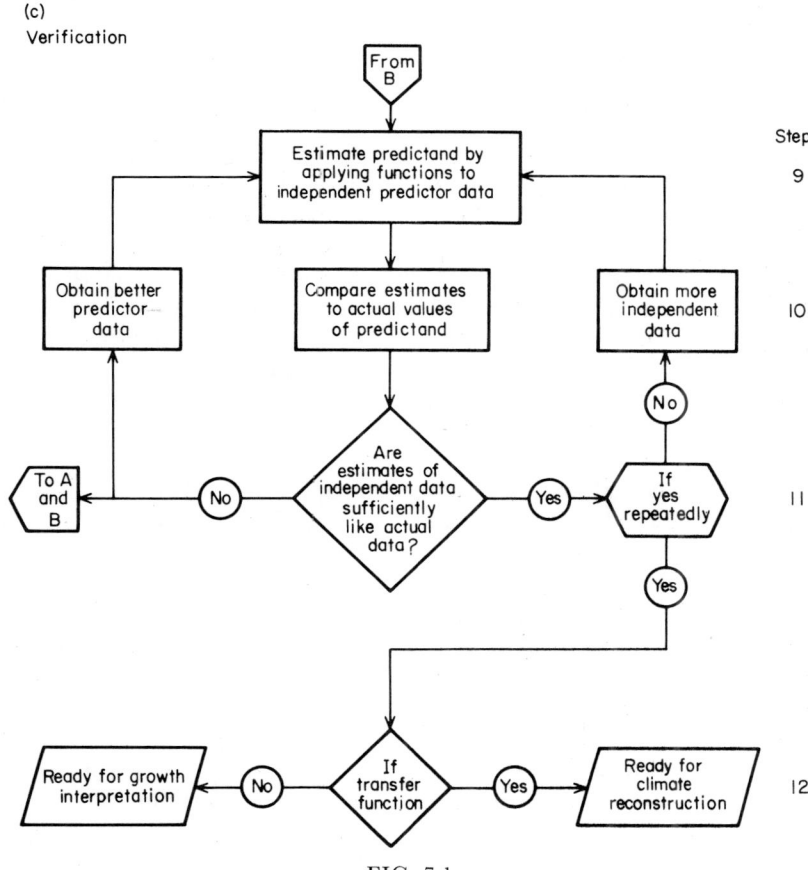

FIG. 7.1c

appropriate number of months. The amounts of monthly precipitation recorded at a nearby weather station are summed for each year and compared to yearly growth measurements. In the example using temperature records, the accumulated number of hours in each day above a chosen critical threshold are calculated and compared to the corresponding amounts of growth.

The calculated values represent the statistical estimates of growth. In these examples the climatic data are the *predictors*, and the actual growth the *predictand* (Fig. 7.1b, steps 6, 7, and 8). Some object to the term *prediction* because it implies the making of estimates forward in time. The term *retrodiction* has been proposed by Lawson (1972) to avoid this implication and is applied to statistical predictions involving estimates made backward in time. However, for the purposes of this volume the term *reconstruction* appears more appropriate. The terms *predictors* and *estimates* are used in their statistical sense to refer to the statistical input data and the statistical output results, while the *predictand* will refer to the actual data to which the estimates are compared. The differences between the predictand and the estimates are the *residuals* representing the noise of the system and any signal not accounted for by the statistical relationship.

The weightings given to the predictor variables and the statistical manner in which these variables are combined to make the estimates is referred to as a *response function* or *transfer function* (Fig. 7.1b, step 5). The term *response function,* introduced in Chapter 5, refers to weights or coefficients used to estimate growth from variables of climate. The weights describe how the tree "responds" to climate. The term *transfer function* refers to a different set of weights used to estimate climate from ring-width values. Growth records are "transferred" into reconstructions of climate.

The estimated values are computed by applying the predictor data to the model coefficients (step 6), and the resulting values are compared with those of the predictand (Fig. 7.2). If the values that are estimated via a straight-line relationship differ from the actual values more at the low extremes than at the high extremes, one may conclude that a curvilinear model is more appropriate. In such cases converting the data to logarithms, powers, square roots, or reciprocals may lead to improved results. If estimates differ markedly for particular cases, close examination may suggest how the model may be altered to fit the divergent cases more adequately. A variety of models may be constructed which weight and transform the predictor variables differently.

A simple method for comparing the predictand and estimates is to plot the year-by-year values one above the other or to plot one as a function of the other using a horizontal and vertical axis to represent the abscissa and the ordinate (Fig. 7.2). In the latter, the data points fall along a straight line if the relationship is linear, and they fall along a curve if the relationship is curvilinear (see plots on the right of Fig. 7.3). For the best-fitting linear relationships the cluster of plotted points on such a graph

---

FIG. 7.2. After calibration the estimated values can be compared to actual data by plotting the two, one above the other, as a function of time (left) or plotting the actual value as the ordinate (vertical axis) and the estimated values as the abscissa (horizontal axis) (right). If there is a linear relationship between the two series, the plotted points fall along a diagonal line. The data represent actual and reconstructed flow of the Colorado River estimated for Lee's Ferry, Arizona, using ring-width chronologies from the upper Colorado River Basin. Note that the left graph has two ordinate scales. (Redrawn and reprinted by permission from "Long-term Streamflow Records Reconstructed from Tree Rings", Paper of the Laboratory of Tree-Ring Research at the University of Arizona, No. 5, by C. W. Stockton, Tucson: University of Arizona Press, © 1975.)

7. CALIBRATION

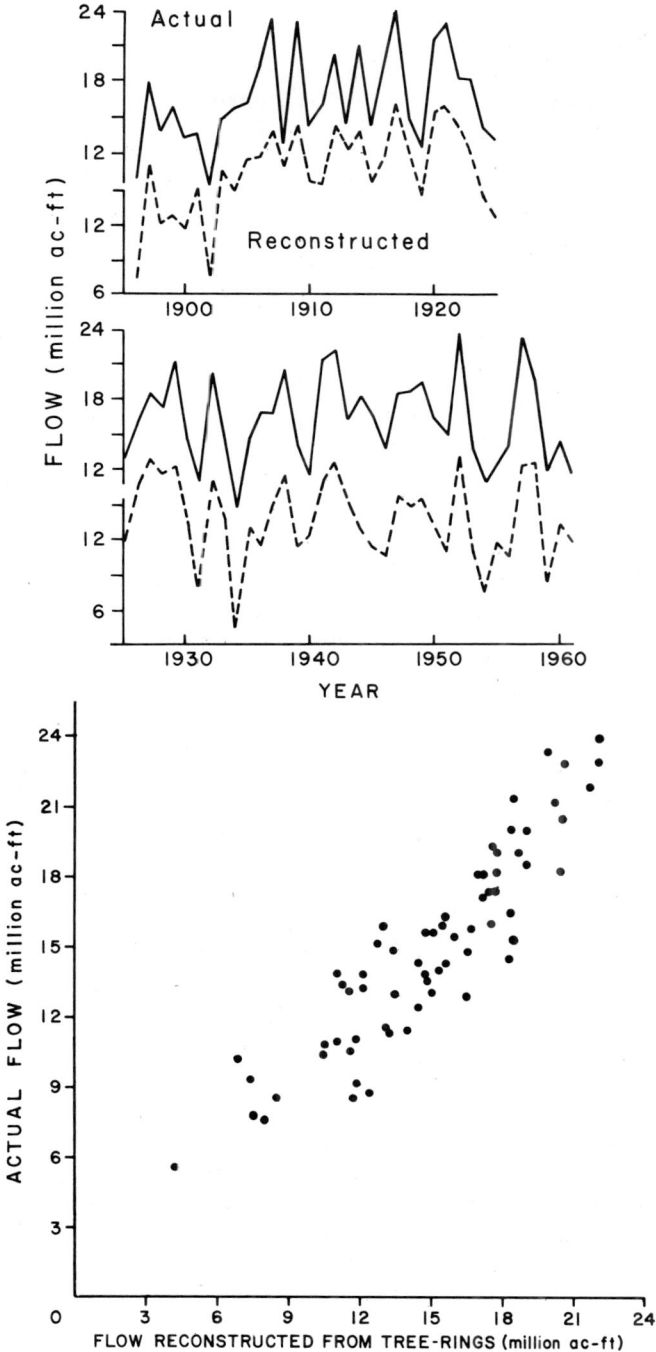

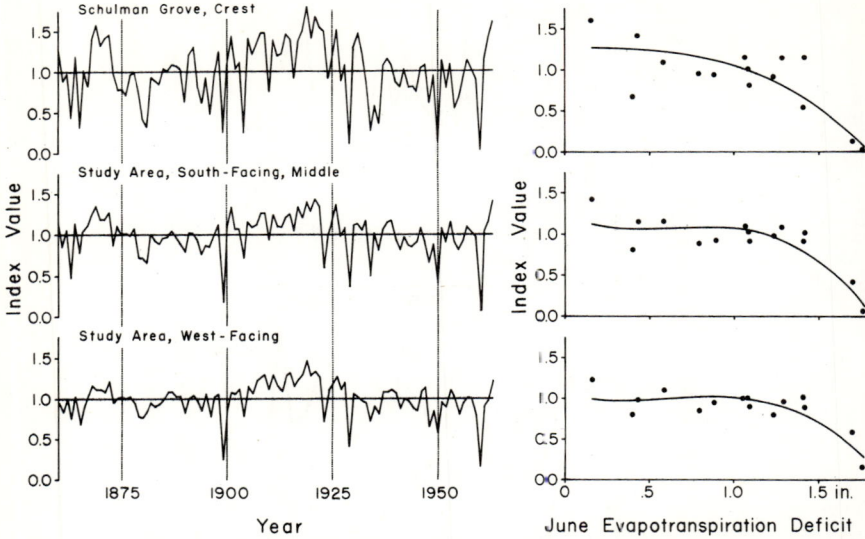

FIG. 7.3. Data from 1860 through 1963 from three ring-width chronologies for *Pinus longaeva* plotted against time on the left and against June evapotranspiration deficit on the right. In the latter only data for 1949–1963 are used. Growth is low for the years of most extreme evapotranspiration deficit, so that a curvilinear relationship results. The top chronology represents a limited site where the relationship is closest to being linear. The lower two are from more moist sites in which growth is limited by evapotranspiration deficit only for two extreme years. (Reprinted by permission from "Bristlecone Pine in the White Mountains of California, Growth and Ring-width Characteristics", Paper of the Laboratory of Tree-Ring Research at the University of Arizona, No. 4, by H. C. Fritts, Tucson: University of Arizona Press, © 1969.)

will resemble a narrow ellipse (Fig. 7.2). If the relationship is a weak one the data points will resemble a wider ellipse.

If the agreement between the estimates and the actual values is insufficient (Fig. 7.1b, step 8), the calibration is rejected and the statistical model, the biological model, or both, are reexamined. If the agreement is judged to be sufficient, the calibration is accepted and verification is then attempted (part C of Fig. 7.1).

C. Verification

Verification is obtained using data withheld from the calibration analysis. The coefficients of the calibration equation are applied to a set of predictor data for a different time period, for a different climatic station, or for different trees than those used to derive the calibration relationship. The data used in the original calibration are referred to as the *dependent set*, while those used for verification are the *independent set*. Verification is established when the estimates derived from the independent predictor set resemble the independent predictand set. The operation may be repeated several times, depending upon what data are available for the independent check (Fig. 7.1c, step 11). If the relationship repeatedly holds up when testing against independent sets the calibration is accepted as appropriate for the particular relationship (step 12).

In the past there have been few dendroclimatic verifications attempted. However, verification is as important as the reconstruction, especially when complex calibration equations are utilized and the size of the dependent data set is limited. As the number of variables in the model approaches the number of observations used for calibration, the more likely it is that the calibration will degenerate into "number matching" rather than quantification of a real relationship. Such calibrations often provide good estimates of those observations of the predictand variable which are included in the calibration set, but fail miserably when tested on observations from an independent data set. Also, autocorrelation reduces the number of truly independent observations in most tree-ring chronologies (see Chapter 6), and thus the likelihood of obtaining a relationship purely by chance is greater than one would expect from a randomly distributed data set.

Repeated verification on independent data is the most rigorous statistical proof that a particular relationship is not the result of pure chance, and that the reconstructions do in fact represent conditions that actually did exist. Verification is more commonly applied to transfer functions than to response functions, because the primary objective for the former is to obtain valid reconstructions while for the latter it is to examine the structural relationships. However, in certain cases response function analysis can be applied to an independent data set. For example, Ashby and Fritts (1972) tested a particular hypothesis about climatic variation and air pollution by applying a response function to an independent climatic data set.

III. The Role of Statistics and Sample Size

Certain statistical tests and procedures are described in the following pages which can help the researcher to sort through possible growth–climatic relationships and to decide whether his calibrations are significantly different from those occurring by chance. However, these statistical methods should never be accepted blindly, as they cannot substitute for clear analytical thinking, nor should statistical methods be distrusted unduly to the extent that they are resorted to after all other techniques have failed. Wallis and Roberts warn that, "He who accepts statistics indiscriminately will often be duped unnecessarily. But he who distrusts statistics indiscriminately will often be ignorant unnecessarily" (*Statistics: A New Approach.* Copyright, 1956, by the Free Press, a corporation). They go on to point out that "statistical interpretation depends not only on statistical ideas, but also on 'ordinary' clear thinking." They itemize several ways in which statistics can be misused. It is assumed that the reader is already familiar with such misuses, although it may be instructive to independently review examples described in their text. Included among other statistical texts which examine possible misuses concerning time-series analysis such as those dealing with tree-ring chronologies and climatic records are Mitchell *et al.* (1966) and Jenkins and Watts (1968).

Many statistical tests involve statements of probability and are therefore affected by sample size. The success or failure of a particular calibration can be expressed in terms of how much variance in the dependent variable is expressed by predictor variables and the probability that a particular result could have occurred solely by chance. Sometimes this is referred to as the *"goodness of fit"*, percent *variance calibrated*, or the *variance reduced* in a particular analysis. The tests usually involve a comparison of how good the fit is relative to the *residual variance*, a term including any noise plus relationships not accounted for in the analysis.

The *goodness of fit* is dependent upon the strength of the signal in the data set as well as the adequacy of the model in describing the relationship. For example, growth data from a single tree may include a signal of climate amounting to only 25% of the total variance, the remaining 75% representing randomly varying statistical noise which results from nonclimatic site factors and internal differences in controlling factors within the tree. In such a case, the maximum possible goodness of fit using only the single tree data in a valid model would be no more than 25%.

If several trees are sampled and the mean growth is obtained for each year, the variations in the yearly averages are less than the variations of ring widths from any one individual, because some of the noise in individual tree variation is canceled in the averaging process (see Chapter 6). The variance of the signal which is the same in every tree is not lost by averaging, so the potential variance attributable to the signal grows with increasing sample size while the total variance of the mean chronology is reduced. This relationship is described as the *Law of Large Numbers* (see Roberts, 1960). It is expressed in dendroclimatological terms as follows: As the number of individual trees sampled in a mean chronology increases, and as the number of radii used from each individual tree increases, the amount of variability due to noise in the mean chronology will decrease because there is less error in the means of the larger sample. At the same time there is an increase in the percentage of variation due to the climatic signal, the maximum possible "goodness of fit" attributable to climatic variation, and the ratio of signal to noise.

Another way to express the Law of Large Numbers is that the larger the sample, the less it is likely to differ from the statistical universe or from the true value for the mean. *True* is used here as the mean value of all trees and all radii in those trees available on a particular site. Of course, it is never possible to know the true value in this sense, so a portion of every chronology must include some statistical error or noise.

However, as was shown in Chapter 6, an increase in sample size from 5 to 10 trees will result in a far greater increase in signal and reduction in noise than an increase in sample size from 15 to 20 trees. Therefore, the relative improvement in a mean chronology resulting from adding one individual to the sample is greater the smaller the sample size.

No strict rule can be stated as to what is an adequate sample size for climatic analysis, since adequacy depends upon the degree of reliability one desires. Thus, the adequacy of 10 trees for a sample on arid sites is based upon a subjective decision that the error in such a chronology is tolerable for a particular study. However, as dendroclimatic analysis becomes more and more precise and is applied with greater frequency to trees in less extreme environments, it will probably become necessary to obtain larger sample sizes.

The investigator should evaluate the reliability that is necessary and then budget sufficient time and money for collecting and processing enough data to obtain that reliability. Analysis of variance described in Chapter 6 is especially useful for evaluating chronology reliability. If too few samples are available, it may be necessary to pool data from several sources or to resample before the results become statistically significant or before the error is reduced to acceptable limits. As a general rule, this writer prefers to collect too large a sample than to risk collecting too small a sample. This is especially important if it is difficult or expensive to reach the site. If the sample turns out to be excessively large, the longest records with the most ring-width sensitivity can be selected, which will maintain the greatest amount of signal per individual tree and preserve the maximum ratio of signal to noise.

IV. Degrees of Freedom and the Effective Sample Size

A truly random sample of n observations also has n *degrees of freedom*, that is, any one value could vary without affecting the values of the others in the time series. However, if the mean is specified for the random sample, there is one less $(n-1)$ degree of freedom (see Barry and Perry, 1973). This is best seen by considering a sample of two observations. If the mean and one value of this sample are specified, the other value is automatically determined. In a case of 10 observations, if the mean and nine values are given, the 10th value is automatically determined. In these cases the last value is constrained or has no "freedom" because the mean has been specified. For example, if the mean is 1·0 and one observation has a value of 0·9, the other value must be 1·1. If the observation is 0·8, the other must be 1·2, etc. A population with both a given mean and a given standard deviation has $n-2$ degrees of freedom because the standard deviation adds another constraint to the population. In other words, a degree of freedom is lost for each statistic used to define a particular population. A response or transfer function with 10 regression coefficients defining a relationship would use up one degree of freedom for each of the 10 coefficients, one for the mean, and one for the standard deviation.

The number of degrees of freedom are also fewer than the number of observations whenever the values are not randomly distributed through time. That is, one value is dependent on, or to some extent constrained by, the other values. For example, the elements of a series which has been smoothed by a filter or which exhibits autoregression are only partly independent, because a certain amount of their variance is affected by the values of other observations. Consequently, the effective sample size of a time series of standardized ring widths is inversely related to the amount of significant autocorrelation or persistence that is present. All tests of significance must take into account this reduction in sample size. Mitchell et al. (1966) propose that the effective sample size (n') may be approximated by using the first-order autocorrelation as defined in Chapter 6. The calculation is performed as follows:

$$n' = n\frac{1-r_1}{1+r_1} \qquad (7.1)$$

where n is the number of observations and r_1 is the first-order autocorrelation which always is assumed to be positive in the equation. For example, the n' for a series of 100 ring-width width indices with a first-order autocorrelation of 0·5 is

$$n' = 100\frac{1-0\cdot5}{1+0\cdot5} = 100\frac{0\cdot5}{1\cdot5} = 33 \qquad (7.2)$$

This indicates that a 100-year-long ring-width chronology with a first-order autocorrelation of 0·5 instead of having 100 degrees of freedom, has

33 degrees of freedom or about one-third the number of the actual sample size. Any significance testing of this sample must be based upon a sample number of 33 even though there are 100 ring-width indices.

This method of calculating n' from r_1 is only an approximate one and assumes that r_1 is truly representative of the persistence and is relatively independent of other types of non-randomness that are present. It is not uncommon for dendroclimatological investigators to misjudge statistical significance by overlooking the reduction in degrees of freedom which results from persistence in the ring-width measurements.

If a time series with autocorrelation is smoothed by running means or moving averages, additional persistence is effectively introduced and there is further reduction in the effective sample size. For example, a two-year moving average applied to a set of 100 random data points reduces the effective number in the series to one-half, because only every second value in the moving average series is independent. Weighting by using a three-year moving average with weights 1–2–1 has approximately the same effect.

It is interesting to note that some early workers justified smoothing as a means of attaining higher correlations. It is possible that if some of them had realized how much of the increased correlation was due to the technique of smoothing, they might have been less enthusiastic about using such procedures. Smoothing before calibration with climate is no longer a recommended practice, except in cases where the trees themselves appear to be smoothing the effect of a particular year's climate over a large number of subsequent annual rings (LaMarche, 1974a; LaMarche and Stockton, 1974). In such cases the growth model used may include some sort of averaging of climatic data over a period of several years, but if such smoothing is used, the degrees of freedom for tests of significance must be adjusted appropriately.

V. Selecting the Statistical Model

The statistical model for a response function includes a number of weights associated with variables of monthly or seasonally averaged climatic data. Each statistical weight of the response function is usually multiplied by the appropriate climatic data, and the resulting products are summed to obtain the growth estimates. If the response function is a good one, the summed products will be large for years when climatic conditions were favorable and small when climatic conditions were unfavorable. For cases when climatic conditions are near average, the sum of the products will be near the growth average.

Common practice in the past has dictated a reliance on *a priori* models using predictor variables such as monthly precipitation or temperature, the total annual water balance, the number of hours that temperatures are above a critical value, the number of days in the growing season, or some other expression of the growth-controlling conditions that can be derived from available data. Traditionally, a variety of models are tested until one or more are found which produce the best growth estimates. The success of such *a priori* analysis can depend in large part upon the cleverness and insights of the investigator as he constructs the models for analysis. If he has the wrong notions about the growth model, he may never obtain a satisfactory analysis except where he tries so many possibilities that one becomes statistically significant by chance.

This *a priori* approach to modeling has the advantage of being relatively easy to apply since it does not require statistical training and availability of a large computer. Many workers start out with simple *a priori* methods and gradually increase the complexity of their models until they obtain satisfactory growth estimates.

As was pointed out in Chapter 5, there can be a variety of possible linkages between the macroclimate and the growth-controlling factors. It was also shown that trees of varying species and of varying microhabitats can respond differently from one month to the next to the same climatic factors. During certain months growth can be directly correlated with a climatic factor, and during other months it can be inversely correlated with the same factor, depending upon which processes are most limiting to growth at that time. Models A, B, and C diagrammed in Figs. 5.8, 5.9, and 5.10 describe some of the most important biological and physical pathways by which the climatic factors of precipitation and temperature can be linked with plant processes to affect ring widths. The implication is that the more flexible multivariate methods are superior to *a priori* techniques because they let the data themselves determine the coefficients of the response function, and the investigator does not have to anticipate the particular weight given to each variable.

Nevertheless, a great deal of research has involved *a priori* modeling and the averaging of variables over periods of time. Climatic factors, such as precipitation, temperature, sunlight, humidity, and wind are utilized, or transformations are obtained to represent biological conditions such as water stress, evapotranspiration deficit, or number of days with temperatures above a particular threshold. The researcher may try various combinations of climatic data for summed intervals from one-half month to a period of 12 or more months to see which best fits the tree response. The following sections describe a number of examples found in the literature which utilize different variables and apply various degrees of *a priori* modeling.

VI. The Diversity in Variable Selection

The interval of time chosen in *a priori* modeling studies involves a variety of periods throughout the year and a variety of climatic factors. Glock (1950), in a study of arid-site trees, chose the following monthly intervals over which the precipitation was summed:

November–May	May–July
January–May	May–August
January–August	April
March–April	May
March–June	June
March–July	July
May–June	August

Estes (1970), in a study of oak and pine in the central Missouri valley, used a variety of periods within the interval from April through August from which he averaged precipitation. In a study of *Pinus longaeva*, Fritts (1969) totaled precipitation and averaged temperatures for July–November, December–March, and April–June. Tryon and True (1958), in a study on *Quercus* in West Virginia, used precipitation periods as follows:

Current growing season: May–June, May–July, May–August
Preceding season: July–August, July–September

Slåstad (1957), working in Norway, used both precipitation and temperature with annual values calculated for September 1 through August 31. Seasonal values included the periods September 1 through April 30, September 1 through May 31, May 1 through July 31, May 1 through June 30, and June 1 through July 31. Hustich (1945) used temperature, precipitation, and cloudiness for the three-month interval of June through August and for single months within that period. Eklund (1957) utilized half-monthly units from May 1 through August 31 and calculated mean temperatures and precipitation for various intervals within this period. He also calculated the number of days when temperatures were above a chosen threshold and compared these to growth. Schulman (1956) chose precipitation data for 16 intervals from a duration of three months to two years, all representing a continuous time sequence.

Sundry factors have been averaged over a variety of other interesting time intervals, presumably to represent ecologically sound parameters. For example, when working on ring widths of *Pinus halepensis* in southern France, Serre *et al.* (1966) calculated the following variables which were used in subsequent analysis:

(1) The sequence number of the year from 1 to 21.

(2) The number of days after January 1 when the dry summer period started.

(3) The number of days after January 1 when the dry summer period ended

(4) The number of days of frost from November to March.

(5) The last day of frost numbered from January 1.

(6) The total precipitation of the dry period.

(7) The total precipitation of the wet period.

(8) The exposure of the tree.

(9) The exposure of the cores with regard to slope.

(10) The number of the tree.

These 10 variables were used to estimate the natural logarithm of the width of each ring in each radius of each tree. Variables 1, 8, 9, and 10 were required to assess the nonclimatic variability in their sample.

Jonsson (1969) describes a very complex system for computing and representing environmental data which were thought to control growth: The range of a given climatic factor (temperature, light, or soil moisture) was divided into several classes, and the number of days falling into each class was tabulated. Variations among seasons were analyzed as different growth phases defined as a function of the seasonal regime of temperature.

LaMarche and Fritts (1971b), in their study of *Pinus cembra* in the Alps, relate ring-width variations to percentages of advancing glaciers which were known to be associated with growing-season temperatures. Stockton and Fritts (1973) associated the growth of *Picea glauca* located on natural levees along stream channels to fluctuations in levels of a nearby lake, which in turn were related to flooding at the tree sites.

Some of the variables used in analyses are soil–water deficits (Smith and Wilsie, 1961), calculated monthly soil–moisture storage and monthly evapotranspiration deficits (Fritts, 1962b, 1969), evaporation stress and sunshine (Schulman and Bryson, 1965), and moisture surplus (Scott, 1972).

Sometimes the growth data are transformed into logarithms (Serre *et al.*, 1966; Jonsson, 1969). Stockton and Fritts (1971b) used nine equally probable classes to subdivide frequencies of occurrence of growth. Serre (1973) uses frequency of occurrence of arbitrary growth classes. Sometimes smoothing operations are performed beforehand on ring-width data. Glock (1950), Keen (1937), and Schulman (1947, 1956) employed such a scheme, using three-year means in which the central year was weighted twice as much as the end weights. Fahn *et al.* (1963) used 21-year running means.

Although the growth data usually represent the means of many trees, some workers have analyzed ring widths in individual trees. For example, Phipps (1972) listed extremes of growth for each tree and compared the occurrences with extremes in climate. Mitchell (1967) used sophisticated multivariate statistical techniques on chronologies from single trees that were not on climatic stress sites and, as might be expected, he obtained so much variability that results were difficult to interpret.

While many of these studies have used values for climatic factors summed over intervals ranging from half-months to one or more years, a large number of the results have turned out to be inconclusive because the sums do not satisfactorily portray the way the tree responds to climatic factors. High values of factors may be limiting in certain months, while low values may be limiting in other months. Thus, when climatic data are summarized for intervals which include both direct and inverse relationships, the two effects cancel one another and no statistical relationship is evident. In reality, the climatic factor was important but acted differently for different times in the year. In other cases the relationship was nonlinear, or the effects varied in their relative importance throughout the year. Because of many complicating factors of the climate–growth system, dendroclimatic literature is full of conflicting reports as well as genuine disagreements among investigators. (See Glock, 1941, and Agerter and Glock, 1965.)

VII. Testing the Association between Variables

A. *The Sign Test*

There are a number of ways to test whether or not sufficient similarity exists between the actual and estimated data. A simple and nonparametric procedure is to count the number of agreements and disagreements in the two series and to apply a measure of association known as a *sign test*.

One way to employ the test is to tabulate the direction of change in ring widths and climate from one year to the next (the first difference; see Chapter 6) as either an increase ($+$) or a decrease ($-$). No consideration is given to the magnitude of the change. Large differences receive the same weight as small ones. A second way to employ the test is to use the signs of departures from the sample mean. As in the first differences, small departures from the mean are weighted the same as large ones, and the similarity or dissimilarity of these signs is tested. The application involving the direction of change (signs of the first difference) is relatively unaffected by trends and persistence and can be applied to unstandardized ring widths.

All cases in which the changes or departures in growth and climate are the same sign (either + or −) are tabulated as similar. Those cases in which the changes or departures are in opposite directions (+ and −, or − and +) are tabulated as dissimilar. Where no difference in either the tree-ring or climatic data occurs, the case is tabulated as 0 and it is eliminated from the particular comparison. If there is no association between the two variables, about half the occurrences will be similar and half dissimilar. The hypothesis that there is a significant association is accepted if the number of similarities is significantly larger than the number of dissimilarities.

TABLE 7.1 Number of agreements or disagreements in signs[a] required to obtain two significance levels[b]

Total number of cases (N)	Level of significance		Total number of cases (N)	Level of significance	
	0·95	0·99		0·95	0·99
6	0	—	31	9	7
7	0	—	32	9	8
8	0	0	33	10	8
9	1	0	34	10	9
10	1	0	35	11	9
11	1	0	36	11	9
12	2	1	37	12	10
13	2	1	38	12	10
14	2	1	39	12	11
15	3	2	40	13	11
16	3	2	41	13	11
17	4	2	42	14	12
18	4	3	43	14	12
19	4	3	44	15	13
20	5	3	45	15	13
21	5	4	46	15	13
22	5	4	47	16	14
23	6	4	48	16	14
24	6	5	49	17	15
25	7	5	50	17	15
26	7	6	55	19	17
27	7	6	60	21	19
28	8	6	65	24	21
29	8	7	70	26	23
30	9	7	75	28	25

[a] Whichever is smaller. [b] Using two-tailed test, adapted from Tate and Clelland (1957). [c] When $N > 50$, the number required is $(N-1-z\sqrt{N})/2$ where z is 1·96 and 2·58 at the 0·95 and 0·99 levels respectively.

One procedure for testing the significance of such a relationship is shown in Table 7.I (Siegel, 1956; Tate and Clelland, 1957). The number of agreements or disagreements necessary to obtain significance is presented as a function of the total number of cases. It can be seen from the table that in the case of 10 observations all occurrences must fall into one of the two classes for the relationship to be significant at the 0·99 probability level, while all but one occurrence must fall in a particular class to be significant at the 0·95 probability level. For a sample size of 20, one class must include three or fewer cases to be significant at the 0·99 probability level and five or fewer cases to be significant at the 0·95 probability level. In other applications the critical limits may be derived from comparing many cases of improperly matched data sets (Eckstein and Bauch, 1969). The advantages of the sign test are that (1) it is not affected by extremely anomalous data, and (2) it does not require heavy computation. The sign test does not provide a rigorous analysis of significance because it does not take into account the degree of correspondence among individual cases.

The sign test has been used most extensively in European works (Ermich, 1955) where ring widths are converted to logarithms, and decreasing ring width due to tree age is present in the data. Some early North American studies using the signs of the first differences were undertaken by Friesner and Friesner (1941), Friesner (1950), Miller (1950a, b), Lyon (1936, 1943), and McInteer (1947). With the accessibility of computing facilities and programs involving rigorous statistical procedures, analysis by means of the sign test will be used mostly for preliminary probing, early model development, and for independent verification tests. In the last case the sign test is useful in that it measures the direction of the reconstruction either as a first difference or a departure from the calibration mean, without being unduly affected when the climatic reconstruction is extreme or incorrect.

B. Product Means

Another statistic, referred to as *product mean*, takes into account both the sign and magnitude of the departure from the calibration average. It is computed from the product of the actual and estimated yearly departures from the mean value, with the positive and negative products summed separately. If the departure sign is correctly estimated, the product is positive; if the sign is incorrectly estimated, the product is negative. For random guessing the means of the positive and negative products (neglecting their signs) have approximately the same values and for correct reconstruction the mean for the positive products is larger. The difference between the mean of the positive products, m_1, and the mean of the negative products, m_2, (disregarding the sign) is tested with the t statistic as

$$t = \frac{m_1 - m_2}{\sqrt{\dfrac{s_1^2}{n_1} + \dfrac{s_2^2}{n_2}}} \tag{7.3}$$

where s_1^2 and s_2^2 are the corresponding sample variances (which may be unequal), and n_1 and n_2 are the number of items in each sum. If the value of t is higher than the expected value at the appropriate confidence level, then the difference is considered to be statistically significant. (See Parzen, 1967; Freund, 1962; Snedecor, 1956; or other general statistics texts.) A significantly larger mean positive product indicates a tendency for both actual and estimated departures from the average value to be large when the sign is correctly estimated and to be small when the sign is incorrectly estimated. In the latter case approximately average conditions both occurred and were estimated, so the reconstruction can be regarded as essentially correct even though its sign was incorrect.

C. Simple Correlation and Reduction of Error

The correlation coefficient is the most common method for measuring the association between tree rings and climate (Keen, 1937; Weakly, 1943; Hustich and Elfving, 1944; Hustich, 1945; Schulman, 1947, 1956; Holmsgaard, 1955; Eklund, 1957; Slåstad, 1957; Tryon et al., 1957; Tryon and True, 1958; Daubenmire, 1960; Gagnon, 1961; Sirén, 1961; Bray and Struik, 1963; Serre et al., 1966; McGinnies, 1967; Estes, 1970; Parker and Henoch, 1971; Scott, 1972). The computations of this parameter are given in Equation 6.4. The numerator of the equation is the cross product or covariance, a term which takes into account not only the number of cases of agreement and disagreement between variables, but also the relative degree of correspondence. The product emphasizes large deviations from the mean relative to small ones.

Glock (1942) describes what he refers to as the "trend method" of obtaining correlation, which can be used on ring-width data containing growth trends. It incorporates the idea of a sign test of year-to-year differences, taking into account the degree as well as the direction of the correspondence. The first differences are obtained from one value to the next, after which the sums and cross products of the differences are obtained. Although some early workers used this method, it is no better than standard statistics which can be readily tested for significance.

A statistic, referred to as the *Reduction in Error* (R. E.), is described by

Lorenz (1956) and can be applied to independent data to measure association between a series of actual values and their estimates. If it is applied to the dependent data set, the result is equivalent to the square of the correlation coefficient and measures the percent variance calibrated by the relationship.

The computations are as follows

$$RE = 1 - SSR/SSM \qquad (7.4)$$

where SSR is the sum of the squares of the differences between actual data and the statistical estimates (the residuals) and SSM is the sum of the squares of the differences of the actual data from the mean of the dependent data set used for calibration. The theoretical limits for the values of this statistic range from a maximum of $+1$, which indicates perfect agreement to minus infinity. A minus value is treated as zero and indicates no agreement. A major disadvantage of this statistic is that one extremely bad estimate can completely offset the effect of several very good estimates. The R. E. statistic provides a rigorous test of association on independent data and any positive value can be considered as encouraging. Statistical tests for positive values of the R. E. statistic are the same as for the square of the correlation coefficient. The interested reader should consult statistics texts for a more thorough discussion of the correlation coefficient, its interpretations, uses, and limitations (Ezekiel and Fox, 1959; Freund, 1962; Baggaley, 1964).

D. Conditional Probability

A somewhat specialized approach to calibration and the measurement of association between ring widths and climate involves the utilization of contingency tables to assess agreement or disagreement between two data sets. This is accomplished by dividing the tree-ring and climatic data sets into several *equally probable classes* (that is, groups of equal size) and by estimating joint occurrence using chi square (χ^2) and coefficient of contingency statistics. Data from Arizona described by Stockton and Fritts (1971b) offer a good example of this application and are summarized in this section.

The climatic data used in the example are seasonally averaged temperatures and precipitation where each of the four seasons represents three months beginning with winter as December–February, and so forth. If the data are close to a normal distribution, the averages for each year for each season can be placed into three equally probable classes. This is accomplished by calculating the mean and standard deviations of the climatic data and examining a table of the cumulative normal frequency distribution published in most statistical texts to select the value (expressed in standard deviation units) which divides the normal curve into the desired classes. In the above case a distance of 0·43 standard deviation on each side of the mean separates the normal curve into the three desired classes where one-third of the data classed as near normal are within $\pm 0\cdot 43$ standard deviation from the mean, one-third of the data classed as below normal are less than $-0\cdot 43$ standard deviation and one-third classed as above normal are greater than $+0\cdot 43$ standard deviation from the sample mean. Nine climatic classes can be obtained by combining the three precipitation and three temperature classes (Fig. 7.4). The data are treated separately for each of four seasons, so that analyses include both individual seasons and their combinations.

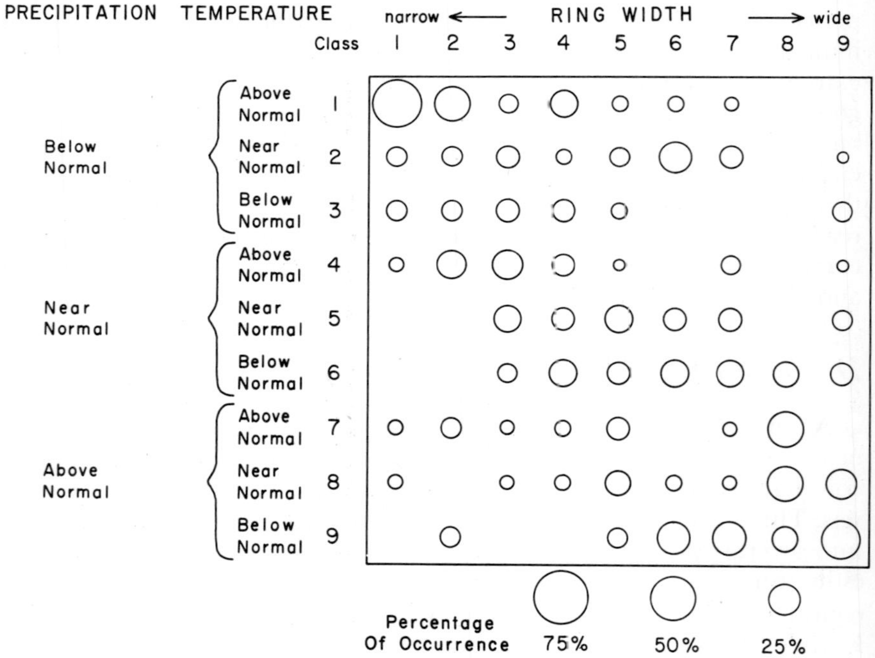

FIG. 7.4. The percentage of occurrences, represented diagrammatically by the size of the circles shown in each cell, between nine combined precipitation and temperature classes shown on the left and nine ring-width classes shown at the top. Data are pooled for autumn, winter, and spring seasons. The larger circles clustered along the diagonal of the table indicate a linkage of below-normal precipitation and above-normal temperature with narrow rings, and above-normal precipitation and below-normal temperatures with wide rings. The circle sizes express the percentages of occurrence of each climate class given the particular ring-width class indicated at the top of each column. (Stockton and Fritts, 1971b)

Normally distributed ring-width indices can be placed into equally probable classes by averaging all data from several sites and transforming them to *standard normal variates* (sometimes called *normalized data*) (z_t) in the following manner:

$$z_t = \frac{x_t - m_x}{s_x} \tag{7.5}$$

where m_x is the mean and s_x the standard deviation of the mean ring-width indices (x_t) in this example for the calibration period 1900–1957 (Equations 6.1, 6.2). These standard variates of ring-width data can then be separated into nine equally probable classes (Table 7.II) as derived from the cumulative normal frequency distribution. The classes range from 1 (very narrow rings) to 9 (very wide rings).

TABLE 7.II Upper class limits used to separate normalized data into 9 equally probable classes[a]

Class Number	Value $\leq$[b]
1	−1·2200
2	−0·7643
3	−0·4306
4	−0·1395
5	0·1395
6	0·4306
7	0·7643
8	1·2200
9	∞

[a]From Stockton and Fritts (1971b). [b]Units are expressed as standard normal variates (Equation 7.5).

The data for each climatic class for each season were matched with the corresponding ring-width classes, and the number of joint occurrences was tabulated for (a) five seasons including the summer prior to and the summer concurrent with the growing season, (b) four seasons including the summer, autumn, winter, and spring prior to the growing season, (c) four seasons including the autumn, winter, and spring prior to growth as well as the summer concurrent with the growth season, and (d) three seasons including the autumn, winter, and spring (Table 7.III). The numbers of occurrences in each class are entered in the various *cells* or boxes of a contingency table, and the percentages or joint occurrences between tree-ring and climatic classes are obtained (Fig. 7.4, Table 7.IV).

Of interest is the hypothesis that the percentage occurrences are more frequent in certain classes and less frequent in others than would be

TABLE 7.III Contingency analysis of ring-width indices and Arizona climate[a]

Seasons	Computed χ^2 [b]	Occurrences	Coefficient of contingency[c]
Previous summer climate	65·8	58	---
Autumn climate	72·9	58	---
Winter climate	100·0	58	0·80
Spring climate	74·2	58	0·75
Concurrent summer climate	6·3	58	---
Five seasons combined	132·9	290	0·56
Four seasons combined using prior summer	132·4	232	0·60
Four seasons combined using concurrent summer	123·1	232	0·59
Three seasons combined without summers	124·7	174	0·64

[a] From Stockton and Fritts (1971b). [b] The theoretical χ^2 at 95 percent confidence limit for 64 degrees of freedom is 73·6. [c] Only seasons with significant chi square values above 73·6 are calculated.

expected by pure chance given no suppositions as to which classes are involved. A test of this hypothesis is accomplished using the occurrences in the contingency table similarly portrayed in Fig. 7.4 and Table 7.IV. A chi square (χ^2), statistic is calculated for joint occurrences of the nine climatic classes with the nine ring-width classes as follows:

$$\chi^2 = \sum^{kr} \frac{(f_{mn} - h_{mn})^2}{h_{mn}} \qquad (7.6)$$

where the term f_{mn} is the observed number of occurrences in a particular cell of the table ($k_m r_n$) representing ring-width class k_m and climatic class r_n. The term h_{mn} is the hypothetical frequency computed as

$$h_{mn} = k_m \frac{r_n}{\Sigma_r} \qquad (7.7)$$

where k_m is the number of occurrences in all nine climatic classes within a particular ring-width class, r_n is the occurrence of all nine ring-width classes within a particular climatic class, and Σ_r is the total of all cells for all classes.

If the computed statistic exceeds the 95% confidence level of the chi-square distribution, it may be concluded that particular climatic classes are significantly linked with particular ring-width classes (Table 7.III), and statements regarding probabilities of occurrence of climatic classes given particular ring-width classes may be obtained from them (Table 7.IV).

7. CALIBRATION

TABLE 7.IV Conditional probability for each climatic class for previous summer, autumn, winter, and spring given a ring-width class[a]

Climatic class	Ring-width class								
	1	2	3	4	5	6	7	8	9
1	0·583	0·292	0·179	0·200	0·100	0·050	0·036	0·125	0·000
2	0·083	0·125	0·179	0·050	0·100	0·200	0·143	0·000	0·025
3	0·083	0·083	0·143	0·100	0·075	0·050	0·036	0·000	0·075
4	0·042	0·250	0·179	0·100	0·075	0·000	0·107	0·000	0·025
5	0·042	0·083	0·179	0·150	0·225	0·150	0·107	0·000	0·075
6	0·042	0·000	0·071	0·200	0·100	0·150	0·179	0·125	0·150
7	0·042	0·083	0·036	0·050	0·100	0·000	0·036	0·250	0·025
8	0·042	0·000	0·036	0·100	0·150	0·100	0·107	0·250	0·200
9	0·042	0·083	0·000	0·050	0·075	0·300	0·250	0·250	0·425

[a] Using the joint occurrence of ring-width and seasonal climatic classes from Stockton and Fritts (1971b).

The coefficient of contingency (C), which measures the degree of linkage or dependence, can be calculated in the following manner:

$$C = \left[\frac{\chi^2}{\chi^2 + \Sigma_r}\right]^{\frac{1}{2}} \tag{7.8}$$

where χ^2 is the chi square computed from Equation 7.6 and Σ_r is the total of all cells for all classes as used in Equation 7.7.

The value of C (Table 7.III) can be calculated only if the chi-square value is significant. The higher the value, the greater the association between the two classes.

The highest coefficient of contingency for a single season as shown in Table 7.III was for ring width and winter climate (0·80). The only other significant single-season relationship was for ring width and spring climate (0·75). Of the calculations using pooled data, the combination of autumn, winter, and spring gave the highest coefficient. The values of the coefficient were lower when summer was combined with autumn, winter, and spring climate. This result is consistent with the statement made in earlier chapters that the trees respond in this semiarid area more consistently with the autumn, winter, and spring climate than with the summer climate.

The occurrences within the individual cells or boxes of the contingency table are converted to percentages by dividing the number of occurrences in each cell by the sum of occurrences for that particular column representing a ring-width class (Fig. 7.4, Table 7.IV). If there had been no relationship between tree growth and climate, the percentages would be similar in all cells of the table, and all circles in Figure 7.4 would be approximately the same size; that is, each ring-width class would have all climatic classes associated with it. However, there was a direct relationship between the two data sets so that the cells with highest frequency of occurrence (the largest circles and largest probabilities) occur along the diagonal line extending from cell 1,1 to cell 9,9. The climate class most likely to be associated with a narrow ring is the dry–warm class while that most likely to be associated with a wide ring is the cool–moist class.

A major disadvantage of the conditional probability technique is that the data of different seasons can be pooled only if the effects of the climatic factors are the same. If they are different the significance of the pooled data is less than for the individual seasons. Two advantages of the technique are (1) the relationship between variables does not have to be linear; that is, the largest values need not be in the cells along the diagonals of the table, and (2) it is possible to make statements about the probability of climatic classes in the past, given the occurrence in a particular ring-width class. The class is read from the normalized values of tree growth (Fig. 7.5) for a particular year in the past as defined by the dashed horizontal lines. The column in Table 7.III for the appropriate ring-width class is entered, and the conditional probabilities for each climatic class given the particular tree-ring class is read from the table.

For example, the ring width for 1872 shown in Fig. 7.5 is narrow and falls in class 1. Entering Table 7.IV under ring-width class 1, the probabilities for an occurrence of climatic classes 1, 2, ...,9 for the climate of summer, autumn, winter, or spring seasons are 0·583, 0·083, ..., 0·042. These data are interpreted in the following manner: Any one of the four seasons for that year has a 0·583 probability of being in climatic class 1 (below normal precipitation and above normal temperature), a 0·083 probability of being in climatic class 2, and a 0·042 probability of being in climatic class 4.

The probabilities may be combined for several classes to make statements about precipitation, or temperature alone, or statements like the following: The probability that climate for 1872 was in either climatic class 1 or 2 is 0·583 + 0·083, or 0·666, and the probability that it was in a class 3 or higher is 1 − 0·666, or 0·334. Also, the probability of occurrence for a particular climatic class during a particular interval of time may be calculated from the table using the growth classes for each year of the interval as follows:

$$Pr(E_n) = \frac{N_n}{\Sigma N} \qquad (7.9)$$

where $Pr(E_n)$ is the probability of occurrence for one of the nine ring-

7. CALIBRATION

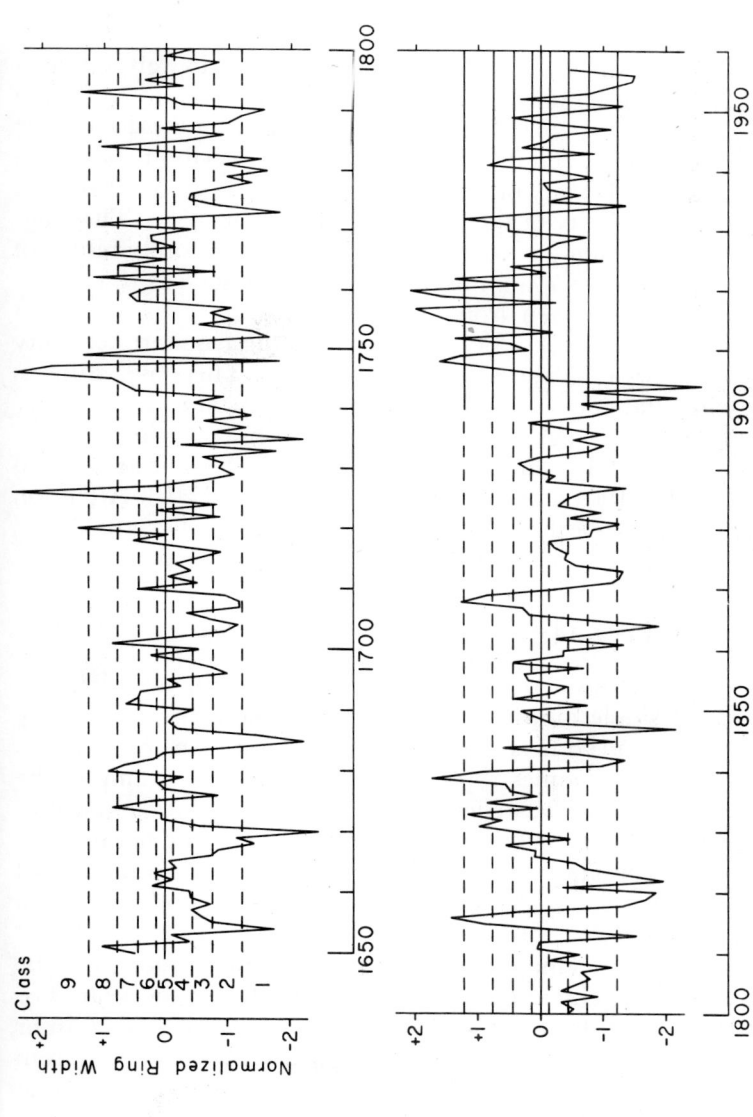

FIG. 7.5. Yearly ring-width values for the state of Arizona are used to make statements of conditional probability concerning climate. The data are the average of four area chronologies expressed as standard normal variates (Equation 7.5). The equally probable ring-width class limits for the calibration period are shown as solid horizontal lines and these are extrapolated for the earlier record as dashed lines. Statements concerning past climate are obtained by identifying the tree-ring class for a year in the earlier record, finding the appropriate column for the growth class in Table 7.III, and reading the conditional probabilities for the climate classes from the rows in the column. (Stockton and Fritts, 1971b)

width classes occurring within the interval which must be a period substantially less than the entire length of record. N_n is the number of yearly occurrences of the ring-width class being considered, and ΣN is the total number of occurrences (years) within the time interval being considered.

Applying the tree-ring data in Fig. 7.5 for a decade of low growth such as 1730–1739 to Equation 7.9 using the data in Table 7.IV, the probabilities for occurrences for class 1, 2, 3, ..., 9 during the decade are calculated to be 0·365, 0·117, 0·103, ..., 0·038. The probability of below-normal precipitation for any of the four seasons during these 10 years is the sum of the probability of the first three classes, or 0·585. The ring-width class for a high-growth decade indicates a probability of occurrence for a season of above normal precipitation such as 1831–1840 to be 0·547. It can be seen from comparing these results and from examining the data in Table 7.IV that in general there is more certainty about drought years than there is about wet years. This observation is consistent with statements made in Chapter 6 about the greater reliability of narrow rings than wide rings from arid-site trees as estimators of climate. For further discussion of the technique refer to Spiegel (1961), Li (1964), and Stockton and Fritts (1971b). Attention will now be directed to techniques for handling limiting effects of factors that vary from one month to the next throughout the year.

VIII. Multivariate Techniques

As shown by the growth models in Chapter 5, ring-width variations are rarely caused by single limiting factors, though one can be markedly more important than others, and a single factor can vary in its importance at different times throughout the year. For example, the relative effect on growth of a particular amount of precipitation in winter is usually different from the effect of the same amount in summer because the two seasons offer different limiting environmental conditions and there are different growth-controlling processes in operation (see Chapters 2, 3, and 4).

Also, the effect of a climatic factor such as precipitation may be conditioned by other climatic factors occurring during the year. For example, temperature determines whether precipitation falls as rain or snow, and temperature governs the rate at which water evaporates from the soil. Therefore, the tree-growth and climatic relationship is best represented as a system with different variables entering into the relationship. In fact, there are so many possibilities for variations in limiting factors, that it is surprising that *a priori* modeling by summing one variable during different periods of time has worked at all. A more logical but complex approach would be to include many variables as growth predictors, to utilize statistical techniques which ascertain the variables that are most important, and to scale the variables in a way to obtain the best results.

A straightforward approach to the problem of analyzing such a multiple variable system is to calculate simple correlations (Equation 6.4) between several variables and ring widths and to select all variables with which growth is significantly correlated as predictors (Hustich and Elfving, 1944; Hustich, 1945; Eklund, 1957; Tryon et al., 1957; Elling, 1966; Serre et al, 1966; Julian and Fritts, 1968; Scott, 1972; and many others). The use of simple correlation has one major limitation, however. Variables of the climatic system are usually highly correlated with each other as well as with growth. Positive intercorrelation within the climatic system can inflate the simple correlations between some noncausal factors and growth, and negative intercorrelation can deflate the simple correlation between important causal factors and growth.

One way to deal with this difference in intercorrelation is to use a statistical technique referred to as *multiple regression* or *multiple correlation* (Hustich and Elfving, 1944; Hustich, 1945; Holmsgaard, 1955; Eklund, 1957; Tryon et al., 1957; Sirén, 1961; Fritts, 1962a, 1962b, 1965, 1969; Schulman and Bryson, 1965; Elling, 1966; Serre et al., 1966; Julian and Fritts, 1968; Scott, 1972).

A. Multiple Regression and Correlation

Multiple regression and correlation employs statistical procedures to resolve variable intercorrelation so that one equation is obtained expressing the relative effect of each variable upon the amount of growth. The statistical model assumes that the variations in some predictands, which in this case are growth, are related in a linear fashion to variations in several predictors, which in this case are the intercorrelated variables of climate. The statistical equation is

$$\hat{y}_t = b_o + b_1 x_{1t} + b_2 x_{2t} \cdots b_m x_{mt} \tag{7.10}$$

where $\hat{y}_t$ is the estimate of the predictand, which varies each year (t) depending upon the values of the predictors; b_1 through b_m are regression *coefficients*, sometimes called partial regression coefficients, associated with each of the m predictor variables x_1 through x_m, and b_o scales the equation to the mean of the predictand. The calculation of the multiple regression equation is a complex statistical procedure derived from the statistical theory of *least squares* which is beyond the scope of this text. The interested reader may consult Ezekiel and Fox (1959), Draper and Smith (1966), Bendat and Piersol (1966), Simpson *et al.* (1960), or some other standard statistical text.

The effect of each variable on growth is scaled by the coefficient and the scaled values are added to obtain the growth estimate. Significance tests assume that the variables are distributed normally. The partial regression coefficients express the relative effect of each variable on the predictand, and the coefficients can be tested using standard errors that are associated with them. Also, a multiple correlation coefficient is obtained which is analogous to the simple correlation in that it is a measure of the linear statistical relationship that exists between the set of predictors and the predictand. The percentage of variance in the dependent variable explained (accounted for or reduced) by the equation is calculated by squaring the multiple correlation and multiplying by 100. The multiple correlation generally increases with the addition of a predictor variable, partly because it may reduce additional variance and partly because an additional degree of freedom is lost.

Multiple regression has been used most often to describe, test, and measure those climatic variables that are related to variations in growth. Only recently has it been applied to climatic reconstruction or growth estimation. For example, Eklund (1957) calculated an index of ring widths for *Picea excelsa* in northern Sweden with reference to the mean growth for 1900–1944. The following regression equation was obtained by selecting only those coefficients that were statistically significant:

$$\hat{y}_t = 99 \cdot 41 + 0 \cdot 9188 x_1 - 3 \cdot 129 x_2 - 2 \cdot 405 x_3 - 0 \cdot 4282 x_4 \quad (7.11)$$

Variable x_1 was the number of days during May 16 through July 31 for year t in which the maximum temperature reached at least 16°C. Variable x_2 was cone yield, expressed in cone point numbers for the year t, and variable x_3 is the cone yield expressed in cone points for the year $t-1$. Variable x_4 is the accumulated daily maximum temperatures for the period July 16 through August 31 of year $t-1$. For simplicity, the subscripts t for the predictor variables are not listed. Precipitation was included in early analyses, but the regression coefficient for precipitation did not significantly differ from zero.

It may be inferred from these results that the widest rings were formed when (1) maximum temperatures for many days during the period from May 16 through July 31 were equal to or greater than 16°C, (2) when the cone yield of both the current season and the previous season were low, and (3) when the accumulated daily maximum temperature for the period July 16 through August 31 of the prior year was low.

Another example is from the work of Serre et al. (1966) in which the following equation for estimating the logarithm of ring width of *Pinus halepensis* was obtained:

$$\hat{y}_t = 3{\cdot}070 - 0{\cdot}5965x_1 - 0{\cdot}01811x_4 + 0{\cdot}00208x_6 - 0{\cdot}00018x_7 \\ - 0{\cdot}23392x_9 + 0{\cdot}06936x_{10} \quad (7.12)$$

where x_1 through x_{10} are the variables listed in section VI of this chapter. The interpretation of these results is listed in the order of the variables entered into the equation as follows: (1) Each successive ring width decreases at an average of 64 microns for each year; (2) the width of the ring decreases by 19 microns for each day of frost during November through March; (3) the width of the ring increases by 30 microns for each 10mm of rain falling during the specified dry season; (4) the width of the ring increases by 5 microns for each 10mm of rain falling during the winter wet period; (5) the rings in the upslope sides of the stem are on average 231 microns wider than those of the downslope sides; and (6) ring widths vary from one tree to the next by an average of 79 microns.

Schulman and Bryson (1965) obtained equations for predicting ring widths of eight deciduous species growing in Wisconsin, the largest sample being from *Quercus rubra*. The equation for estimating ring-width ($\hat{y}_t$) was

$$\hat{y}_t = -0{\cdot}01967x_{18} + 0{\cdot}003151x_{41} + 0{\cdot}002606x_{39} - 0{\cdot}001102x_{11} \\ + 0{\cdot}01199x_{22} \quad (7.13)$$

where x_{18} is evaporative stress for June of year t (expressed in miles per hour times inches of mercury); x_{39} and x_{41} are total precipitation for May and July respectively of year t; x_{11} is mean temperature for May of year $t-1$; and x_{22} is evaporative stress for April of year $t-1$. It may be inferred from the equation that rings in *Quercus rubra* are wide for seasons (1) when evaporative stress in June is low, (2) when total precipitation for both May and July is high, (3) when mean temperature for May of the prior year is low, and (4) when evaporative stress for April of the prior year is high.

The multiple correlation and the regression coefficients may be tested for significance to ascertain whether they could have arisen solely by chance. The appropriate tests which are described in most statistical texts require that the number of degrees of freedom be obtained. The computations for degrees of freedom cannot be made in the usual manner, however, because high autocorrelation in ring-width chronologies reduces their effective sample size. Equation 7.1 is therefore used to obtain n', which is then entered in the following equation along with the number of regression coefficients (k) to calculate the degrees of freedom (d.f.) that apply.

$$d.f. = n' - k - 2 \qquad (7.14)$$

A more efficient method of handling autocorrelation in multiple regression analysis was suggested by Quenouille (1952) and applied to tree-ring work by Fritts (1962a). Variables for prior growth at lags of 1, 2, and 3 years are used as predictor variables along with the variables of climate. If this procedure adequately removes the effects of autocorrelation, the variance of the residual is random with respect to time and the usual tests of significance and degrees of freedom apply. The value for n' in Equation 7.14 then becomes the number of yearly observations, and as above, k is the number of regression coefficients including those for prior growth.

Some workers have used sequential numbers, their squares, and cubes as statistical predictor variables to express systematic variations due to increasing tree age or changes in the site. For example, the value of t is used as a variable along with its square and cube, and the regression coefficients associated with these three variables describe a curve that statistically assesses the variance associated with most trend in growth (Fritts, 1962b; Schulman and Bryson, 1965; Serre et al., 1966). However, it is usually better to remove these systematic variations beforehand by standardization procedures, because the variance represented by trends is usually large and masks the variance related to climatic factors.

Often there is a large number of possible variables from which a smaller number of predictor variables must be selected. Each one that is entered into regression consumes one degree of freedom, so that the number of observations (or the effective sample size) limits the number of variables that can be entered in regression and still be significant. Whenever degrees of freedom (as calculated by Equation 7.14) are less than 10, there is a high likelihood of obtaining large coefficients purely by chance.

For example, a multiple-regression equation with 20 coefficients and 24 independent observations will have two degrees of freedom in the residual variance (Equation 7.14). In such a case, only the largest regression coefficients, those that are greater than 4·3 times the standard error, will exceed the 0·95 level of significance. However, if there had been 31 independent observations, the degrees of freedom would be 9, and all coefficients that are greater than 2·3 times their standard errors would exceed the 0·95 level of significance. [See the t distribution in Ezekiel and Fox (1959), Draper and Smith (1966), or a general statistics text.] It is desirable to retain 20 or more degrees of freedom in the residual variance. Under this condition a record with at least 42 independent observations is necessary to retain 20 degrees of freedom in a multiple regression relationship involving 20 predictor variables; while 30 independent observations are needed to retain 20 degrees of freedom in a relationship involving 8 predictor variables.

Fritts (1962a) proposes that the procedure of stepwise multiple regression be used when it is necessary to select a small number of variables from a large number of possibilities. The stepwise procedure starts with a calculation of simple correlations for all combinations of variables under consideration, including intercorrelations among predictor variables, as well as correlations between predictors and the predictand. The procedure then examines all of the simple correlation coefficients and first selects the variable that is best correlated with the predictand. A regression using that variable to estimate growth is obtained, and all variables excluding the one in the regression are correlated with the residual variance. The one that is best correlated is added to the regression and a new two-variable equation is obtained. Additional variables are added as long as the residual variance is reduced by a significant amount.

Stepwise multiple-regression analysis does lead to certain technical difficulties, since high intercorrelations among the independent variables create instability in the results. Some variables that are entered early in the stepwise process become insignificant when other variables highly correlated with them are entered in regression at a later step. Therefore, the significance of each regression coefficient is tested at each step, and whenever it drops below a level specified beforehand, the insignificant variable is removed and a new regression equation is computed. At each step the regression procedure adjusts for intercorrelations with only the variables already in regression. A large correlation of a variable entered into regression with one that has not entered can prohibit the latter from being selected in the stepwise process, even though it may be well correlated with growth.

Figures 7.6 and 7.7 include examples of several stepwise multiple regression analyses. The former includes the results of four analyses of two stands of *Fagus grandifolia*, one on a well-drained site and one on a poorly-drained site (see soil profiles in Fig. 3.11). The latter includes the results of eight different stepwise multiple regression analyses for two stands of *Quercus alba*, one on a north-facing slope and one on a south-facing slope. In the case of *Quercus* the earlywood and latewood portions of the ring were analyzed rather than total ring width.

The stepwise analysis selected from variables of mean temperature,

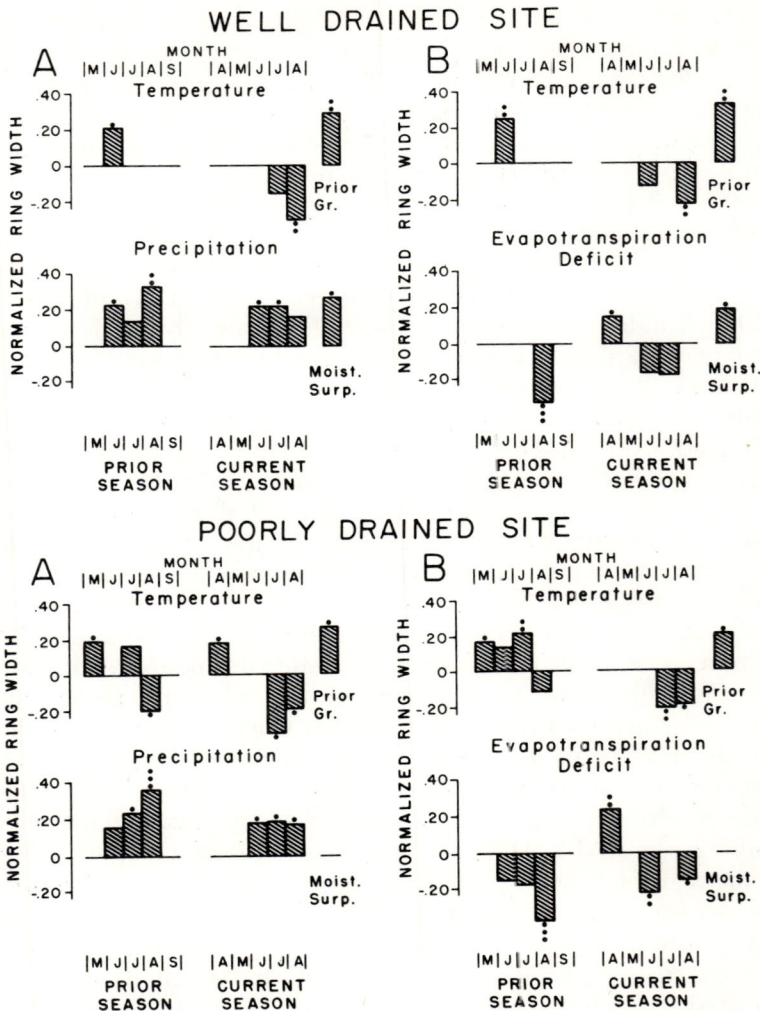

FIG. 7.6. Results of stepwise multiple regression analysis of variations in ring widths for two stands of *Fagus grandifolia* in an Ohio swamp forest. Potential predictor variables available to the stepwise selection process are monthly temperature and precipitation (A), or monthly temperature and evapotranspiration deficits (B) (Thornthwaite and Mather, 1957), for May–September prior to the season of growth and for April–August during the season concurrent with growth. Soil moisture surplus of the winter, spring, and the ring width of the prior year were included as potential predictor variables. The bars represent the standardized regression coefficients for all variables entered into the regression equation with F ratios greater than 1, and the number of dots indicates their significance level: • ≥ 0.95 •• ≥ 0.99, ••• ≥ 0.999. The percent variance reduced ranged from 64% to 72%. (Redrawn from Fritts, 1962b, by M. Huggins)

total monthly precipitation, and evapotranspiration deficits for months May, June, July, August, and September during the year prior to growth and for months April, May, June, July, and August, of the year concurrent with growth. Other variables that were available for selection were moisture surplus and values of prior ring width. Set A shown in each of the figures utilized only precipitation as the moisture variable while set B utilized evapotranspiration deficits. The same temperature variables were available in both cases. The length and direction of the bars portray the magnitude and sign of the regression coefficients for variables that were selected in the stepwise process, and the number of dots indicates the level of significance.

It is notable from these figures that less than half the number of available variables were selected in the stepwise process, and sometimes the particular variables selected varied markedly from one analysis to the next. Nevertheless, the most significant regression coefficients do appear reasonable and are relatively consistent.

For example, most coefficients for precipitation are positive, indicating a direct relationship between precipitation and ring width. Also, precipitation in the middle or latter part of the growing season is most highly correlated with growth. Temperature and evapotranspiration deficits during July and August are most often inversely correlated with ring widths, but earlier in the season and during the prior year coefficients for these variables are both positive and negative, indicating the presence of both direct and inverse relationships.

The significant temperature variables that were entered into the equations differ between analyses using precipitation and analyses using evapotranspiration deficits. Small differences in the intercorrelations among these variables affected the sequence of selection and influenced which variables were included in the final result. Ring widths of *Fagus grandifolia* and earlywood widths of *Quercus alba* are about as highly correlated with the climate of the prior season as with the climate during the season of growth. Latewood of *Quercus alba*, however, is best correlated with moisture and temperature during the season of growth.

All analyses show highly significant first-order autocorrelation; that is, ring width is significantly correlated with the prior ring width, earlywood with the width of the prior season's latewood, and latewood with the width of earlywood produced during the same year.

Sometimes there are a number of ways of representing various limiting factors and it is not certain ahead of time which representation will be the best for statistically predicting growth. For example, temperature may be represented by average, maximum, or minimum daily temperatures; moisture by precipitation or evapotranspiration deficits; air humidity by

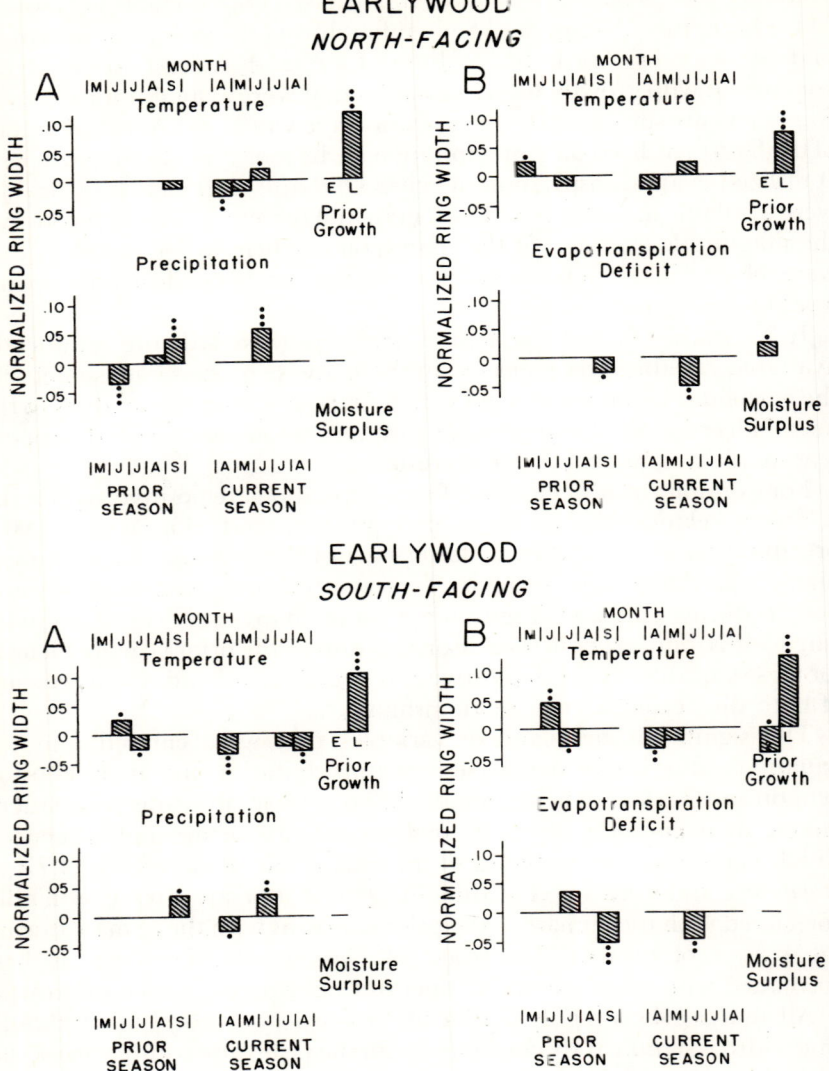

FIG. 7.7. Results of stepwise multiple regression analysis of earlywood and latewood of *Quercus alba* on two opposing slopes in central Illinois. The potential predictor variables are the same as in Fig. 7.6, except that the width of earlywood was included as a predictor variable as well as the width of the earlywood and latewood during the prior year. The analysis and symbols are the same as in Fig. 7.6. The percent variance reduced ranged from 73% to 83% for earlywood and 70% to 81% for latewood. (Redrawn from Fritts, 1962b, by M. Huggins)

LATEWOOD
NORTH-FACING

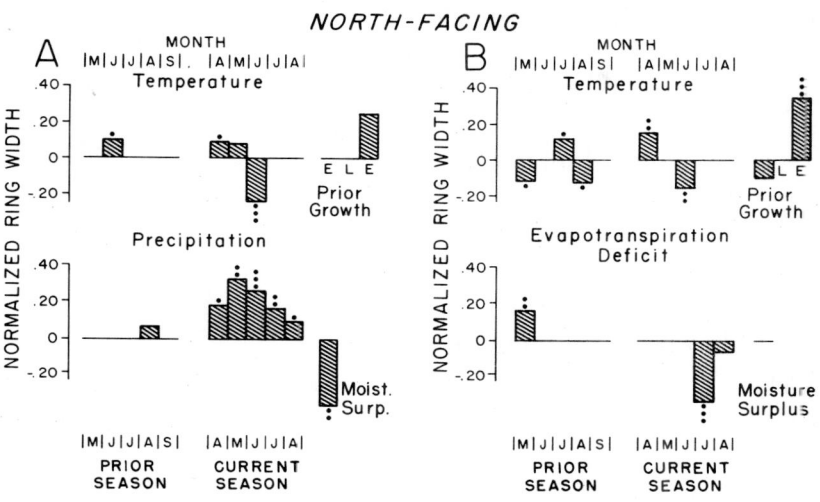

LATEWOOD
SOUTH-FACING

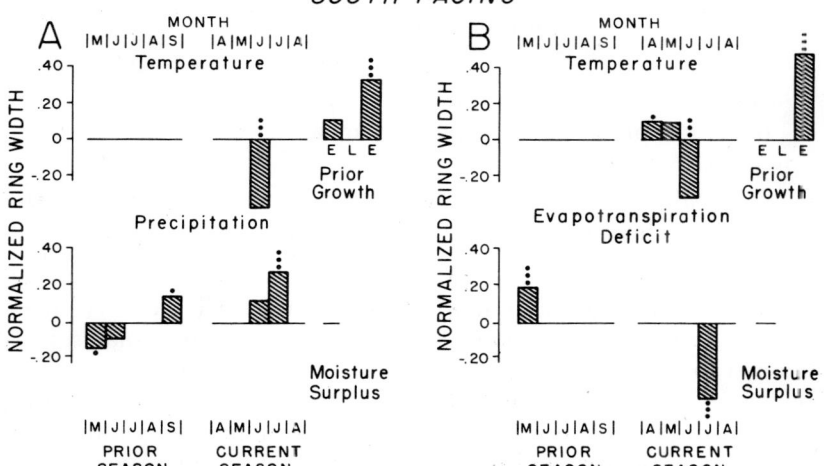

relative humidity or vapor pressure; and solar radiation by Langley units or hours of sunshine. It may also be uncertain as to how many different variables may be analyzed at once or whether relationships are nonlinear or multiplicative, thereby requiring some transformation to be made.

Fritts *et al.* (1956c) ran stepwise regression a number of ways. First, variables of total precipitation and average mean, average maximum, and average minimum temperatures were calculated and grouped into the following eight intervals of time with respect to the particular growing season:

1. June of the previous year
2. July of the previous year
3. August through September of the previous year
4. October through November of the previous year
5. December through February of the current year
6. March through May of the current year
7. June of the current year
8. July of the current year

The variables for regression were selected from the above intervals, and seven different analyses were run using the following combinations of data:

1. Total precipitation
2. Average maximum temperature
3. Average mean temperature
4. Average minimum temperature
5. Total precipitation, average maximum temperature, and the product of these two
6. Total precipitation, average mean temperature, and the product of these two
7. Total precipitation, average minimum temperature, and the product of these two

Eight climatic variables, representing the eight time intervals, were used in analyses 1 through 4, while 24 climatic variables were used in analyses 5 through 7. Stepwise multiple-regression analyses were obtained in the above systematic order, using the tree-ring chronology from each statistical sample as the dependent variable. Multiple regressions were obtained for predicting the ring indices as a function of only the climatic data and then as a function of the climatic data along with variables representing the three preceding ring indices. Thus, 14 separate regression analyses were required to evaluate each tree-ring sample, and in each of these, five to 10 different steps were employed to arrive at the final equation. The final multiple correlations adjusted for degrees of freedom lost (Ezekiel and Fox, 1959), that were obtained with each set of climatic variables (Table 7.V), were multiplied by 100 to obtain percent variance explained by the relationship.

TABLE 7.V The square of multiple correlation coefficients adjusted for sample size for regression analysis of three tree species[a]

Climatic variables used[c]	Sites and Species[b]			
	Wetherill Mesa		Navajo Canyon	
	Juniperus osteosperma 400-year trees	*Pinus edulis* 200-year trees	*Pseudotsuga menziesii*	
			260-year trees	500–800-year trees
Total monthly precipitation	0.78	0.66	0.79	0.75
Average maximum temperatures	0.45	0.64	0.42	0.53
Average mean temperatures	0.39	0.50	0.37	0.47
Average minimum temperatures	0.20	0.38	0.17	0.00
Precipitation and average maximum temperatures	0.87	0.78	0.80	0.79
Precipitation and average mean temperatures	0.87	0.76	0.78	0.82
Precipitation and average minimum temperatures	0.83	0.72	0.85	0.82

[a] Fritts et al. (1965c), see also Table 7.VI. [b] Mesa Verde National Park, Colorado. [c] Including variables for indices of 3 prior rings.

Sometimes the values of variables can be squared or converted to logarithms to allow for nonlinear relationships (Fritts, 1969; also see Fig. 7.3), and cross products between variables can be utilized to allow for multiplicative effects (Table 7.VI). Since the square of the multiple correlation adjusted for sample size increased with the complexity of the analysis, it was concluded by Fritts et al. (1965c) that the percentage of variance did indeed increase for the more complex models and that the complexity was required to handle some of the variable interactions. However, final selection of results depended not only on the significance and variance calibrated by the regression but also on the biological reasonableness of the results (Table 7.VI).

For example, the data in Table 7.V indicate that ring widths of *Juniperus osteosperma* at Wetherill Mesa are more markedly affected by precipitation ($R^2 = 0.78$) than temperature ($R^2 = 0.45$). When both variables are used, significant interaction of these two variables appears during the autumn, winter, and spring months from October to May (Table 7.VI). That is, the product of temperature and precipitation was chosen as a significant predictor. The square of multiple correlation for this case was 0.87. Combinations of precipitation and temperature gave

TABLE 7.VI Climatic conditions which produce narrow rings in three species of trees[a]

Interval summed	Site and species[b]			
	Wetherill Mesa		Navajo Canyon	
	Juniperus osteosperma 400-year trees	*Pinus edulis* 250-year trees	*Pseudotsuga menziesii*	
			260-year trees	500–800-year trees
Prior June, July	no effect	no effect	6 warm, dry[d]	6 warm, dry
Prior Aug., Sept.	no effect	no effect	3 dry[e]	3 dry[e]
Prior Oct., Nov.[c]	3 dry, warm[e]	1 dry[e]	1 dry[e]	2 dry[e]
Prior Dec.–Feb.	1 dry, cool[e]	2 dry[e]	2 dry[e]	5 dry[d]
Current Mar.–May	2 dry, warm[e]	4 dry(cool)[d]	4 dry(cool)[e]	1 dry(cool)[e]
Current June	no effect	no effect	5 dry[d]	4 warm[e]
Current July	no effect	3 warm[e]	no effect	no effect

[a]Fritts *et al.* (1965c), also see Table 7.V. [b]Mesa Verde National Park, Colorado. [c]Numbers 1 to 6 indicate rank of variable importance from high to low. Comma between variables indicates variable interaction with cross products entered into the regression. Parentheses indicate second variable entered but no cross product term involved. [d]Significant $P > 0.95$. [e]Significant $P > 0.99$.

the highest multiple correlations for *Pinus edulis* at Wetherill Mesa and for *Pseudotsuga menziesii* at both Navajo Canyon sites listed in Table 7.VI, but variable interaction was evident for the latter species only during June and July of the summer prior to growth.

Stepwise multiple-regression analysis cannot handle all variable intercorrelations. As was mentioned earlier, certain variables which are important to growth are prevented from entering regression because of their correlation with variables already in regression. This produces differences and inconsistencies which limit the usefulness of stepwise multiple regression for evaluating the effect of intercorrelated climatic variables on ring-width growth.

There are two regression procedures that may be suitable alternatives to the stepwise analysis because they are more efficient for handling the intercorrelations among predictors. One method begins by solving for all variables in regression and then deletes each one in a stepwise fashion. Those that are most insignificant are deleted first. The other approach uses principal components combined with regression. The first alternative has not been used by the writer, since the second seemed to have more advantages. This second method is described in the following sections.

B. Regression after Extracting The Principal Components

Problems due to variable intercorrelation can be circumvented by transforming the predictor variables into a new set of orthogonal or uncorrelated variables called *eigenvectors* or *principal components*. When these new variables are used as predictors in a stepwise multiple-regression analysis, the selection process occurs in a systematic and efficient manner. In addition, a smaller number of transformed variables may be required to obtain the regression relationships, thereby reducing the amount of error as well as the number of degrees of freedom consumed in the stepwise process. Lastly, the transformation ranks the new orthogonal variables in order of their importance allowing the investigator to select the most important variables and to discard the least important ones. Once the regression coefficients for the selected set of orthogonal variables have been calculated, they may be mathematically transformed into a new set of coefficients which correspond to the original correlated set of variables. These new coefficients (sometimes referred to as *weights* or *elements* of the response function) are analogous to the stepwise regression coefficients described in the previous section, except that a coefficient is associated with each original climatic variable, there are usually more degrees of freedom, and there is less error to contend with because of variable orthogonality and because some of the principal components are left out. As a result, the equations derived by means of this method are more reliable than those obtained by the usual stepwise multiple-regression process. Since the procedure is a relatively new and complicated one, the details are described in the following subsections. For background reading of these multivariate techniques see Cooley and Lohnes (1971), Dempster (1969), Morrison (1967), Press (1972), and Seal (1968).

1. Some Matrix Notations

It is necessary and easier for the remaining portions of this volume to describe the computations in terms of matrices and simple algebraic manipulations. The reader already familiar with this terminology may go on to the next subsection.

Matrix notation facilitates the mathematics involved in large grids of data where there are a number of variables and a number of observations of each variable. Probably the most obvious dendroclimatological example of a data matrix is a sequence of measurements taken from a spatial grid of tree-ring or climatic stations. A *matrix* is defined as simply an array of data ordered in columns and rows. Each data matrix can be identified by a letter assigned to it, such as A, B, or C. Portions of three matrices are shown diagrammatically in Fig. 7.8 as blocks divided into rows and columns, which are referred to as *elements*. It is convenient to designate the size of each matrix, that is, the number of rows and columns, by using subscripts. In the examples shown in Fig. 7.8, matrix A

indicated by *a* in each element has *m* rows and *n* columns. This is written in matrix notation as $_mA_n$, where the subscripts refer to the total number of rows and columns of the matrix. The elements of *A* are also subscripted, where the first subscript refers to the number of the row in the matrix and the second to the number of the column in the matrix. The rows might represent different variables used as growth predictors, such as monthly precipitation and temperature, and the columns might represent the individual years for which the precipitation and temperature data are recorded.

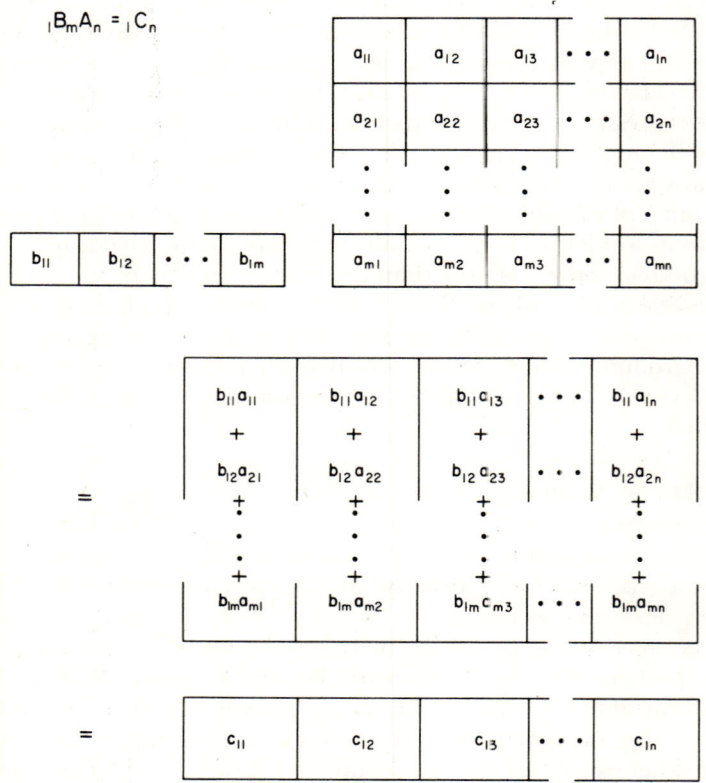

FIG. 7.8. A diagrammatic representation of a matrix multiplication. Matrix *A* is multiplied by *B* to obtain a new matrix *C*. Subscripts designate the number of rows and columns in each matrix. The elements of the matrices are shown as boxes at the top and bottom of the figure. The vertically oriented boxes show the algebraic multiplication of the elements before the columns are summed to obtain the *C* matrix. See text for explanation.

Matrix B in Fig 7.8 is also a matrix but it has 1 row and m columns. This matrix may be written as $_1B_m$. Thus, the elements b_{1i} are arranged in one row with m columns in the figure. This matrix might represent a set of multiple regression coefficients obtained for use with the variable stored in the rows of matrix A to estimate another variable such as ring width. Matrix C, shown at the bottom of the figure, has 1 row and n columns and may be written as $_1C_n$, the elements c_{1j} arranged in one row and n columns.

Only two matrix manipulations are of importance to this discussion. The first is multiplication which can be accomplished between two matrices if the number of columns in the first is the same as the number of rows in the second. In the aforementioned example (Fig. 7.8), matrix B and A can be multiplied since B has m columns and A has m rows. The algebraic representation of this multiplication is shown in the figure and can be written as follows:

$$_1B_m A_n = {_1C_n} \tag{7.15}$$

The subscripts for the columns of B and the rows of A are the same and need not be written twice. Multiplication as shown in Fig. 7.8 involves a particular element in each row of A which is multiplied by the corresponding column element in B: $b_{11} \times a_{11}$, $b_{11} \times a_{12}$, $b_{11} \times a_{13}$ ... $b_{11} \times a_{1n}$; and $b_{12} \times a_{21}$, $b_{12} \times a_{22}$, $b_{12} \times a_{23}$ ... $b_{12} \times a_{2n}$. The products of the elements B and A are then summed in the direction of the columns of A as shown in the lower portion of Fig. 7.8 to become a new matrix, $_1C_n$, with the same number of columns as A and the same number of rows as B (see Equation 7.15).

Using the example cited above, where the elements of $_mA_n$ are observations of climatic data on m variables for n years and the elements of $_1B_m$ are the corresponding regression weights, matrix $_1C_n$ contains the solutions or estimates obtained by applying the regression equation to the data. Thus the multiple regression equation is very conveniently written in matrix notation as in Equation 7.15.

If the same operation were written using summation notation, the elements in matrix C, c_{1j}, are obtained as follows:

$$c_{1j} = \sum_{i=1}^{m} b_{1i} a_{ij}; \quad j = 1, n \tag{7.16}$$

If there had been q rows rather than one row in matrix B, there would have been q rows in matrix C, and the operation would have been repeated as in Fig. 7.8 q times using a different row of B to obtain a different row in C.

The only other matrix operation that is required in the following pages involves obtaining the transpose of a matrix in which the columns and rows are interchanged. The transpose of A is written as A' and the elements of row 1 become the elements of column 1, while the elements of row 2 become the elements of column 2, etc. A transpose is sometimes necessary to arrange the rows and columns properly to make a multiplication. See Searles (1966) for additional discussion of matrix algebra techniques.

2. *Principal Components*

The *principal components,* which are also referred to as *eigenvectors,* are a set of variables which are derived from a data set to be *orthogonal* to (uncorrelated with) each other. In dendroclimatic studies, the principal components may be derived from a set of climatic data or a set of tree-ring data. For example, the climatic data may be mean monthly temperature and a total monthly precipitation. In the example shown below, the data for each month are considered as separate variables, and the variables are arranged in a 28-variable array consisting of 14 months of temperature and the same 14 months of precipitation preceding and concurrent with each growing season from 1924 to 1962. The first variable is the average temperature of June for the year prior to the season of growth and the 14th is the average temperature for July during the period of growth. The 15th variable is precipitation for the prior June, and variable 28 is precipitation for July of the growing season. Although this 14-month interval has been found in prior analyses to include the most important climatic factors thought to affect ring width, other periods of time could be used.

The eigenvectors are representations of the original climatic variables; that is, they include the same information as the original data. There are usually as many eigenvectors as there are variables but the most important eigenvectors, as the name *principal component* implies, account for a large portion of the data variance. The least important eigenvectors express only minor variations in the original data, some of which result from errors, noise, and residual variation imposed by the mathematical constraints in deriving the eigenvectors. It is customary to order these eigenvectors from greatest to least importance (those that express the most to the least variance) and to select a subset which concentrates the climatic data into a smaller number of variables than the original data set.

For the sake of clarity, an actual example is helpful. It will include climatic data from Mesa Verde in southwestern Colorado for 1923–1962. The temperature and precipitation data are designated as matrix $_mF_n$, which is a set of observations for n years on m climatic variables. In the

7. CALIBRATION

example n is 39 years and m is 28 climatic variables, so that matrix F has 39 columns and 28 rows. The lower part of Fig. 7.9 shows the data for two columns (growth-years) of the matrix.

The first step in extracting the eigenvectors of such a data set is to obtain the simple correlation coefficients (Equation 6.4) for each variable with all other variables. If there are m variables, the set of correlations can be defined as an $m - by - m$ matrix, C, in which the element of the first row and first column is the correlation coefficient for variable 1 with itself, and the second element of the first row is the correlation coefficient for variable 1 and variable 2, and the ith row and the jth column is the correlation between the ith and jth variables.

If the data are normalized beforehand (Equation 7.5) to standard units (see lower half of Fig. 7.9) the correlation matrix can be calculated as

$$_mC_m = \left(\frac{1}{n}\right) {_mF_n} F'_m \qquad (7.17)$$

where $_mF_n$ is the set of climatic data described above and the symbol (′) designates the transpose.

The matrix of principal components (eigenvectors) $_mE_m$ is obtained from the correlation matrix $_mC_m$ by a complicated mathematical procedure. The relationship between the correlation matrix and the eigenvectors is

$$_mC_m E_m = {_mE_m} L_m \qquad (7.18)$$

where $_mL_m$ is a set of scaling factors referred to as *eigenvalues*. These scalars represent the importance of each eigenvector and are arranged in order of descending value in the diagonal elements of matrix L, with zeroes in the off-diagonal elements. Such a diagonal matrix is a technical construction necessary to multiply a set of scalars with another matrix. Each diagonal element of the eigenvalue matrix is proportional to the variance reduced by one of the eigenvectors. The percent variance reduced by the eigenvector corresponding to a particular eigenvalue is obtained by dividing that eigenvalue by the sum of all eigenvalues and multiplying by 100.

Each column $_mE_m$ contains an eigenvector consisting of m elements. The eigenvector elements are weights or scalars associated with each of the m variables of matrix F so that the magnitude of each element is proportional to the importance of the corresponding variable in the particular eigenvector. The complete matrix of eigenvectors contains the same information as was in the original correlation matrix. In mathematical terms, the axes have been rotated and the data are expressed in terms of a new set of coordinates, with each eigenvector representing one orthogonal mode of behavior in the original data set. The following illustrates the relationships between the data and the eigenvectors.

The data for the 28 variables during two years at Mesa Verde, one ending in 1936 and the second ending in 1924, are shown at the top of Fig.

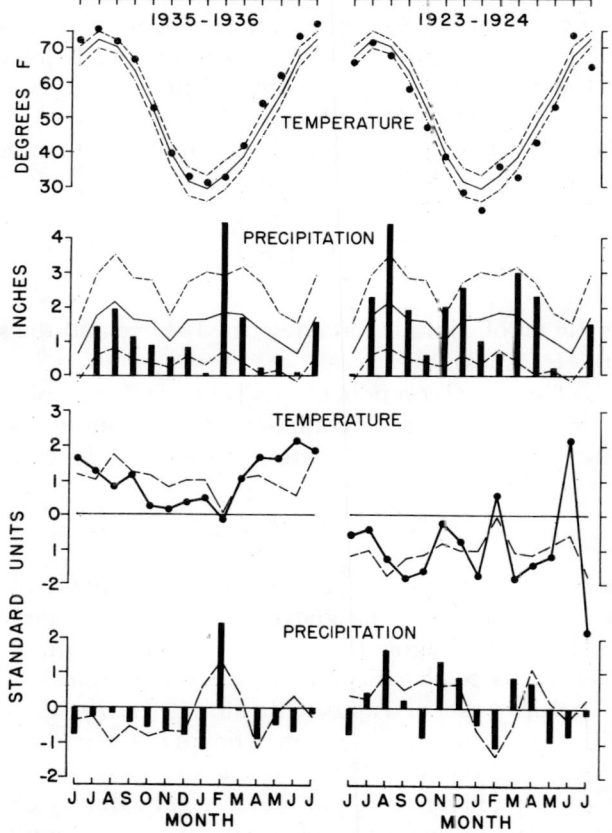

FIG. 7.9. The values for average monthly precipitation and monthly temperature at Mesa Verde for two 14-month periods, one for 1935–1936 and one for 1923–1924, which are arranged as 28 variables for calculating the principal components. The points and bars in the upper half of the figure represent the actual values for temperature and precipitation, respectively. The solid lines indicate the mean for the entire record for 1923–1962; the dashed lines designate distances of plus and minus one standard deviation from the means. The data in the lower half of the figure are normalized and plotted as standard units (z, Equation 7.5). The lines in the lower portion of the figure represent the values for the elements of eigenvector 1 in its positive and negative representations. See text for explanation.

7.9. The values for the mean temperatures and total precipitation are plotted as points and bars respectively. The solid lines behind these data connect the means for each variable for the entire period of record, 1923–62. The dashed lines connect the points representing plus and minus one standard deviation from the mean. The normalized form of the data (see Equation 7.5), expressed as standard units, are shown in the lower half of Fig. 7.9. The values for temperature for 1935–36 were generally above average, and those for precipitation were below average for June through January and March through July. For February the signs of the standard units for both variables were reversed. Comparable data for the year 1923–1924 are shown in Fig. 7.9 on the right.

A plot of the first principal component (eigenvector 1) for the sample data is shown at the top of Fig. 7.10. Since this eigenvector is the first one, it represents the most important pattern or mode of behavior of the original 28 climatic variables. It reduces 13% of the total variance in the original data set of $_mF_n$. Each element of the eigenvector shown in Fig. 7.10 corresponds to one of the 28 variables shown in Fig. 7.9. The magnitudes of the elements for the 14 temperature variables are plotted at the top, and the magnitudes of the elements for the 14 precipitation variables are plotted immediately below them. The pattern formed by the elements of the eigenvectors is of importance—not the signs. That is to say that all monthly weights could have been multiplied by -1 giving the opposite signs, and the elements in Fig. 7.10 could be plotted as the reverse.

The climatic mode expressed by the positive representation of eigenvector 1 represents years in which monthly temperatures (variables 1–14 plotted at the top of Fig. 7.10), with the exception of those in February, are all above average, while those for 10 of the 14 precipitation variables (plotted below those for temperature) are below average. Only values for precipitation of January through March and June are above average in this mode. The climatic mode for the negative representation consists of years when temperatures below average precipitation for the 10 months are above average, and precipitation amounts for the four months are below average.

When the eigenvector matrix (Fig. 7.10) is multiplied by the normalized climatic data shown in Fig. 7.9, a new matrix is obtained called the *amplitudes* or *factor scores* $(_mX_n)$.

$$_mX_n = {_mE'_m}F_n \tag{7.19}$$

The values of the first row of the amplitude matrix corresponding to eigenvector 1 are plotted at the bottom of Fig. 7.10. The years of the growing seasons are substituted for the subscripts 1 to n. The variance of

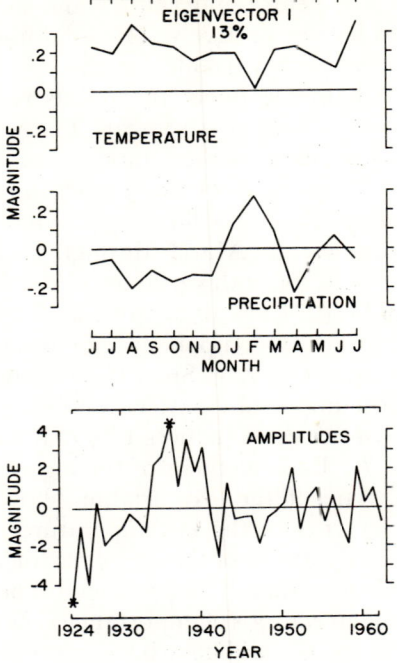

FIG. 7.10. Plot of the magnitudes of the elements of the first and most important eigenvector of Mesa Verde climate, which reduces 13% of the climatic variance, and the corresponding amplitude set. The eigenvector elements for temperature are all the same sign, and the corresponding signs for 10 elements for precipitation have the opposite sign. This arises because temperatures throughout the 14-month period are somewhat positively correlated with each other, but they are negatively correlated with precipitation for 10 out of 14 months. The eigenvector is multiplied with normalized climatic data to obtain the amplitude set. Asterisks mark those elements with the largest positive and negative values indicating the most resemblance of the climatic regime for the year to that particular eigenvector (see Fig. 7.9).

this new time series represents the 13% of the original variance reduced by the first eigenvector.

It may be helpful to mentally carry out one multiplication for one row and column. Refer to Fig. 7.8 and consider the 28 elements of eigenvector 1 shown in Fig. 7.10 as if they are the terms in the boxes of matrix B arranged in one row with 28 columns; and consider the normalized climatic data for 1935–1936, shown in the lower left of Fig. 7.9, as if they are the items of a column in matrix A with 28 rows. The items in the row are individually multiplied by the corresponding item in the column. To help visualize the operation, the corresponding elements of the first

7. CALIBRATION

eigenvector are plotted as dashed lines behind the normalized data in the lower left of Fig. 7.9. It may be noted from the figure that the individual products are generally positive because the data and the elements of the eigenvector both tend to be above average and below average together. When the products (Fig. 7.8) are summed, the result is large and positive so that the element of X, the amplitude for 1935–1936, has a large positive value (see the large positive value with star in Fig. 7.10).

For the year 1923–24 the data are like the negative representation of eigenvector 1 (Fig. 7.9) so that individual products tend to be large and negative and the amplitude for the year (starred in Fig. 7.10) has a large negative value. Eigenvector 1 is plotted in the lower right of Fig. 7.9 with signs of the elements reversed along with the normalized data to facilitate comparison.

Thus the amplitude matrix is a time series which expresses how well the data for each year resemble the particular eigenvector. If the amplitude is large and positive, the climatic data resemble the eigenvector in a direct manner; if it is large in magnitude but is negative, the climatic data resemble the eigenvector in a negative or inverse manner. If the amplitude value is zero, the eigenvector is uncorrelated with the climatic data for that year. Usually a number of eigenvectors are necessary to characterize the climatic regime represented by a particular year, although one or two eigenvectors may be by far the most important.

The upper portion of Fig. 7.11 is a plot of the second most important principal component (eigenvector 2) in this example, which reduces 11% of the variance. The plot of the elements for this eigenvector represents a mode of climate in which the first six months of the 14-month period are either warmer and dryer or cooler and wetter than average, while the eight following months are the opposite; that is, cooler and wetter or warmer and drier than average. As in eigenvector 1, high temperatures usually go with low precipitation and low temperatures with high precipitation. The amplitudes for this eigenvector are plotted at the bottom of the figure and the stars mark those elements with most extreme values.

The climatic data corresponding to the starred amplitudes are shown in Fig. 7.12 and the elements of eigenvector 2 are plotted as dashed lines with the normalized climatic data to facilitate comparison. As noted in Fig. 7.9, the climatic regimes of years with large positive or large negative amplitude again resemble the positive or negative mode of the particular eigenvector, though as before the values for individual eigenvector elements and variables do not match perfectly.

If all 28 eigenvectors are used, the information contained in the complete matrix of amplitudes describes all the original data. In other

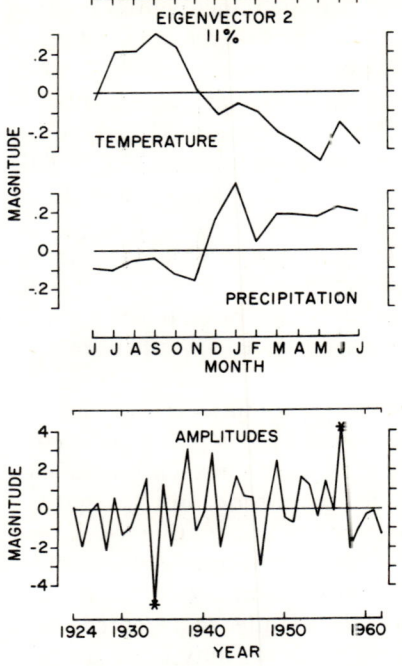

FIG. 7.11. The magnitude of the elements of the second eigenvector of Mesa Verde climate, which reduces 11% of the climatic variance, and the corresponding amplitude set. This eigenvector expresses a mode of climate in which the departures of temperatures for July–November are opposite of the sign for those of temperature during December–July. All elements for precipitation have signs opposite those for temperature, indicating a generally inverse relationship. The eigenvector is multiplied with normalized climatic data to obtain the amplitude set. Asterisks mark those elements with the largest positive and negative values indicating the most resemblence of the climatic regime for the year to that particular eigenvector (see Fig. 7.12).

words, the amplitude matrix contains exactly the same information as the original normalized data set, except that the data are arranged into a set of new variables which are completely uncorrelated with one another and which can be ordered in terms of their importance from those reducing the greatest amount of variance to those reducing the least.

Thus, the extraction of the principal components or eigenvectors actually transforms data representing a number of correlated variables into a new set of uncorrelated variables. The investigator can examine the eigenvalues and discard the least important eigenvectors with the smallest eigenvalues, thus removing unwanted noise and small-scale variability and reducing the total number of variables with which he must deal. This concentrates the information into a small number of

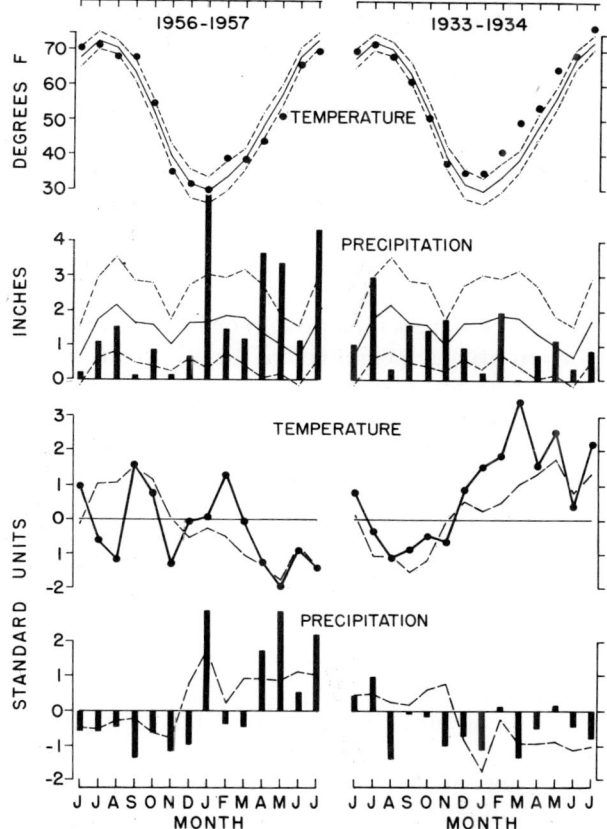

FIG. 7.12. The values for average monthly precipitation and monthly temperature at Mesa Verde for 14 months during 1956–1957 and 1933–1934 which are arranged as the 28 variables used in calculating the principal components. The lines with long dashes in the lower half of the figure represent the values for the elements of eigenvector 2 in its positive and negative representations. See Fig. 7.9 and text for additional explanation.

principal components. Often the first 20 of the 28 possible eigenvectors of climatic data handled in the above manner reduce 90% to 95% of the variance in the original data set. The reduced set of amplitudes is then used as predictor variables in a stepwise multiple regression analysis in place of the original climatic data as described in the next subsection.

3. Response Functions

Since the amplitude matrix represents the original climatic data, the amplitudes may be used to statistically predict a ring-width chronology,

$_1P_n$, using a stepwise multiple regression analysis. The multiple regression equation (7.10) may be rewritten in matrix notation as

$$_1\hat{P}_n = {_1R_p}X_n \qquad (7.20)$$

where $_1\hat{P}_n$ includes the estimated ring-width indices for n years and $_1R_p$ are the significant partial multiple regression coefficients associated with each of the amplitudes of the selected set of p eigenvector amplitudes (all values for insignificant regression coefficients are assigned a value of zero). As for regression using actual climatic data, the particular values assigned to the regression coefficients can be interpreted as expressing the relative importance of each eigenvector amplitude in predicting growth ($_1P_n$). The interpretation of the eigenvector amplitude is considerably more difficult because the real world does not behave like the eigenvectors, as uncorrelated modes of climatic behavior. However, these regression coefficients associated with the eigenvector amplitude can be converted into a new set of coefficients, $_1T_m$, which express the same relationships but in terms of the original variables rather than the eigenvector amplitudes. The equation for the transformation is obtained by substituting Equation 7.19 into 7.20:

$$_1\hat{P}_n = {_1R_p}E_mF_n = {_1T_m}F_n \qquad (7.21)$$

The same statistical estimates of ring-width indices $_1\hat{P}_n$ are obtained by either multiplying normalized climatic data, $_mF_n$, and the response function, $_1T_m$ or by multiplying the amplitudes $_pX_n$ by the regression coefficients, $_1R_p$. Since the elements of the response function, $_1T_m$, are associated with the original climatic variables, not the eigenvector amplitudes, each element can be directly interpreted in terms of an anomaly in the particular variable—monthly temperature or monthly precipitation. Growth is expressed as a response to variations in particular climatic variables.

As was pointed out earlier, the presence of interdependence in ring-width indices measured by the first-, second-, and third-order autocorrelation can affect the tests of significance. Therefore the ring-width indices for three prior years must be included as three additional predictor variables ($_3Z_n$) in the stepwise analysis. The regression coefficients for these three variables are handled in matrix notation as $_1R_3^*$ where the estimations of $_1P_n$ made with this enlarged set of predictors is accomplished by adding $_1R_3^*Z_n$ to the original estimates ($_1\hat{P}_n$).

The same eigenvector amplitudes may be used to derive response functions for several chronologies from a given climatic region, though the regression coefficients will be different for each chronology. The orthogonality of the eigenvectors and their amplitudes allows efficient and orderly use of stepwise multiple-regression analysis. The variables in regression are more likely to remain in the same order of importance as they were when they entered regression, but exceptions to this occur when the autocorrelations in the ring-width indices complicate the relationships.

When estimating $_1P_n$ using the response function $(_1T_m)$, the variance reduced is a direct measure of the amount of climatic information expressed by the chronology. A second estimation of $_1P_n$ can be made using the above response function $(_1T_m)$ and the correction for prior growth $(_1R^*_3Z_n)$ which usually accounts for more variance in $_1P_n$ if autocorrelation in the ring-width chronology is significant. The difference in the variance reduced in the above two calculations provides a measure of the variance attributed to autocorrelation of indices at a lag of one to three years. A check for the residuals is made to ensure that they are randomly distributed with respect to time and that the usual tests for significance are valid.

The confidence limits for each element of the response function, $_1T_m$, are calculated from the respective standard errors (see Draper and Smith, 1966). These standard errors are in turn obtained from the standard errors of the elements of $_1R_p$ as defined in Equation 7.20. The required transformation is

$$_mS_m = {_mE_p}U_pU_pE'_m \qquad (7.22)$$

where $_pU_p$ is a *diagonal matrix* of the standard errors of the elements of $_1R_p$, and $_mS_m$ is a symmetric matrix, the diagonal elements of which are the square of the standard errors of the elements of $_1T_m$. This operation simply distributes the square of the errors to each standard error in proportion to the squares of the elements of the eigenvector.

The desired confidence limits are obtained by multiplying each diagonal element of $_mS_m$ by the appropriate F value with v_1/v_2 degrees of freedom and then taking the square root. Here v_1 has a value of 1 and v_2 is the effective sample size less the number of nonzero regression coefficients in matrices $_1R_p$ and $_1R^*_3$ and less two more degrees of freedom (Equation 7.14). The residuals of the regression are almost always sequentially independent (having no significant autocorrelation), so the effective sample size is the same as the number of years of observations.

Figure 7.13 includes several response functions obtained for a chronology representing the mean of six sites along the eastern slope of the Rocky Mountains (Julian and Fritts, 1968; Fritts et al., 1971). Stepwise multiple regression was run using the amplitudes of climatic eigenvectors to estimate growth. Each successive step utilizes an increased number of eigenvector amplitudes. The regression weights at five different steps were multiplied by the eigenvectors (as in Equation 7.21) to obtain five different response functions ($_1T_m$). The first response function included only one regression coefficient associated with the amplitude set which is best correlated with the ring-width chronology. In the example it was the amplitude of the first and most important eigenvector. The regression coefficient is multiplied by the weights of eigenvector 1 to obtain a response function that reduces 36% of the growth variance (Fig. 7.13). This response function can be interpreted as representing the condition of yearly climate which leads to above average growth, characterized by below average temperature for any month and above average precipitation for any month.

The vertical lines in Fig. 7.13 indicate the 0·95 confidence intervals used to test variable significance. In the above example all but two variables (elements of the response function) have a 0·95 or higher probability of being significant from zero.

The second response function shown in Fig. 7.13 is the result of stepwise multiple regression analysis of the same climatic amplitudes using the three most significant regression coefficients. The three coefficients were multiplied by the appropriate three eigenvectors, and the three products were summed for each of the 28 elements of the response function $_1T_m$. This response function assumes a more complicated shape and the variance reduced is 67·3% of the total growth variance. It can be interpreted by the significant elements to indicate that temperatures in May are the only temperature variables that are directly related to growth, while temperatures in seven of the 14 months are significant and inversely related to growth. Variables for precipitation are significant and directly related to growth in nine of the 14 months. Precipitation for the July prior to growth is the only precipitation variable that is inversely related to growth at this step.

The stepwise multiple-regression procedure was continued, and the response functions which were obtained for steps with 7, 12, and 20 amplitude variables are included in Fig. 7.13. Since the last response function reduces 97·7% of the ring-width index variance, the elements of the last response function may be considered to most resemble the actual manner in which the trees on the particular sites are responding to fluctuations in climate.

The response function with 20 variables can be interpreted as showing significant inverse correlation of growth with temperatures in July of the prior summer, November, April, and July during the year of growth. The elements for September, March, and May suggest the presence of a direct relationship between growth and temperatures during these three months. The results for precipitation indicate an inverse and significant relationship for January, and a direct and significant relationship for the prior June, October, February, April, May, and June. The prior growth

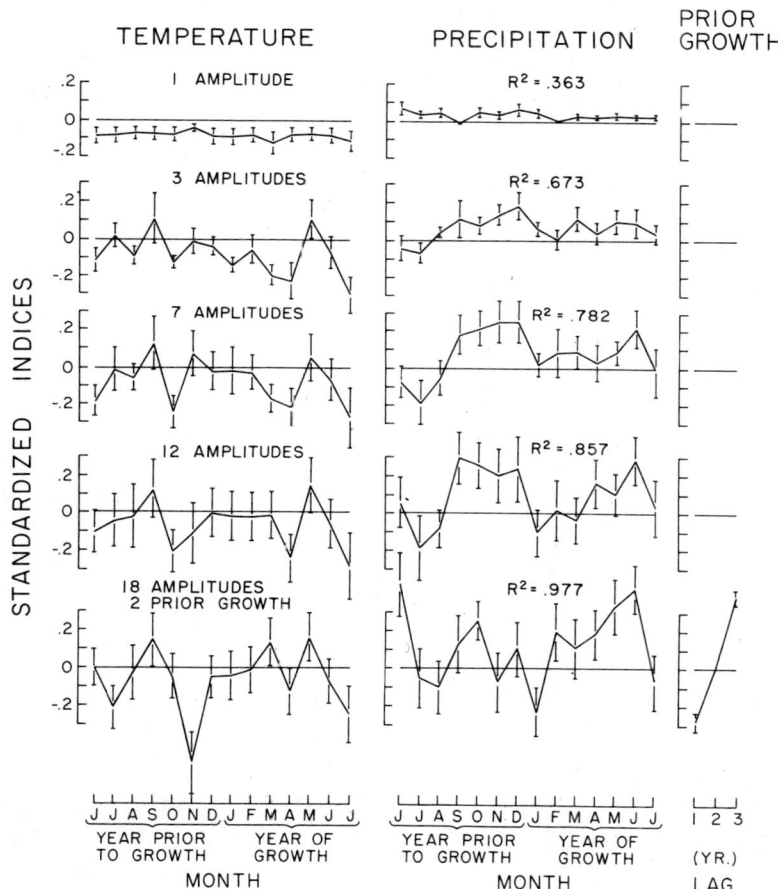

FIG. 7.13. Response functions obtained from a stepwise regression analysis using amplitudes of eigenvectors and prior growth to estimate a ring-width chronology representing six *Pinus ponderosa* sites along the lower slopes of the Rocky Mountains, Colorado. Steps with 1, 3, 7, 12, and 20 predictor variables are shown. The regression coefficients for amplitudes, $_1R_t$, are converted to response functions, $_1T_m$, (Equation 7.21), along with the approximate $\pm$ 0.95 confidence limits (Equation 7.22). When response functions are complex, as in this example, a linear combination of many eigenvectors is needed to obtain the best-fitting relationship. Prior growth was entered into regression after the step with 12 variables. The percent variance reduced can be calculated by multiplying the R^2 by 100.

at a lag of one year is inversely correlated with the ring width, although at a lag of three years a direct correlation exists.

The changing shapes of the response functions in Fig. 7.13 illustrate how the regressions on increasing numbers of eigenvector amplitudes can be used to describe the manner in which climatic factors affect growth. At first the amplitudes that are most highly correlated with the tree-ring chronology are entered into regression, and the linear combination of the eigenvectors express the general pattern of growth response to climate. As more and more steps are added, the finer details of the response of growth to climate begin to take shape. The method implies that the more an amplitude series resembles the ring-width chronology, the more the corresponding eigenvector resembles the manner in which the trees respond to climate.

The stepwise multiple-regression analysis is terminated when the F ratio for the variable entering regression is $<1\cdot0$. In this type of analysis the F ratio for entering variables is used not as a test of significance, but rather as a measure of the ratio between the variance reduced and the error variance. When the variance due to the addition of a variable is less than the error of variance (F is $<1\cdot0$) there is nothing gained by including it as a predictor in the regression equation. But as long as the variance added by regression is greater than the error variance, some improvement can be made. If a large number of confidence intervals begin to intersect the zero line of the response function indicating they are insignificant, however, it may be desirable to choose a response function for an earlier step. Also, a higher F level for termination of analysis can be chosen if the response function is to be used for some sort of statistical estimation.

The results of stepwise multiple regression on the amplitudes of the eigenvectors are different from stepwise multiple regression on the actual climatic variables. For example, Fig. 7.14 includes the results of a stepwise multiple-regression analysis using the same climatic and tree-growth data as in Fig. 7.13 except that the actual climatic data rather than amplitudes of eigenvectors are used as predictor variables. The plotted bars represent the regression coefficients for the climatic variables, R^2 is the square of the multiple correlation coefficient, and the response functions for comparable steps to those in Fig. 7.13 are shown without the confidence intervals.

At step 1, the regression coefficient for June precipitation is entered which reduced 39·8% of the variance; while in the response function for step 1, 26 of the 28 elements of the response function are significant—reducing 36·3% of the variance. At step 3, temperature of July, precipitation of October, and precipitation of June are entered as significant, reducing 67.2% of the variance. The response function exhibits a complicated pattern with 18 elements out of 28 significant, reducing 67·3% of the variance.

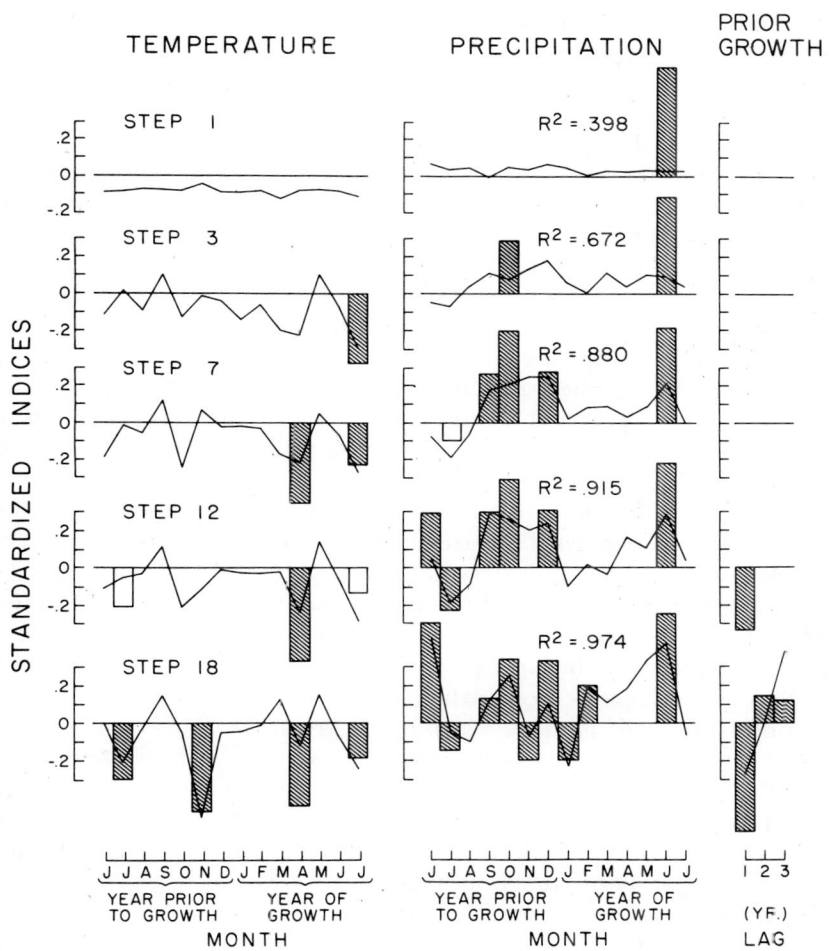

FIG. 7.14. The regression coefficients for stepwise multiple regression on the same climatic and tree-growth data utilized in Fig. 7.13. The comparable response functions as shown in Fig. 7.13 are drawn at each step to facilitate comparison. The shading indicates those partial regression coefficients that are significant at the 95% level of confidence. The percent variance reduced can be calculated in the same manner as in Fig. 7.13.

At step 7, the regression analysis entered seven variables, six of which are significant, reducing 88·0% of the variance. The comparable step in the response function included 13 significant elements. At step 12 there are only 10 regression coefficients entered, because a coefficient for November precipitation, entered at step 8, is deleted at step 12, probably as a result of its intercorrelation with other variables entered into the regression. Prior growth for a lag of one year is entered as a significant predictor variable. Only eight of the coefficients are significant and the equation reduces 91·5% of the variance. The response function for the comparable step has 10 elements which are significant.

Step 18 was the final step as the F ratio at higher order steps dropped below values of 1. A total of 16 variables are entered and are significant, including partial regression coefficients for prior growth at all three lags and reducing 97·4% of the variance. The last step for the response function utilized 20 amplitude variables and reduced 97·7% of the tree-growth variance; and 14 of the elements plus two coefficients for prior growth are significant.

It is evident from these results that the regression coefficients at the various steps often, but not always, coincide with the larger elements of the response function. However, there are important differences. For example, at step 18 the significant regression coefficients are associated with four variables of temperature and are negative, indicating the predominance of a general inverse effect of temperature on growth. Yet the response function indicates that the weights for temperatures are significant and positive for three additional months. The regression coefficients for precipitation stress several apparent relationships from month to month, but the large positive weights for precipitation in April and May are not represented by the coefficients of the stepwise analysis of the untransformed climatic variables. Apparently, intercorrelation among predictor variables prevents variables for April and May precipitation from entering the equation, although the variables may have been limiting to growth. Three prior growth variables are entered by the stepwise analysis while only two variables for prior growth are utilized in the response function analysis.

In general, there are usually more significant items in the response function results than in the straightforward stepwise multiple regression analysis. The response functions also have a weight for all variables which is sometimes helpful in visualizing the growth response. Lastly, as in the case of spring precipitation for step 18, the results from the response function seem more reasonable and consistent than the results from the straightforward regression analysis.

4. Transfer Functions

Transfer functions are obtained in a manner similar to response functions, the primary difference being that transfer functions are obtained by performing regression analysis using the ring-width chronologies as predictors and the climatic data as the predictands. Thus, the data $_mF_n$ in Equation 7.17 would be ring-width chronologies which are obtained from sites differing in geographic location, in species, or in microenvironments.

The principal components, $_mE_m$ (eigenvectors of ring-widths, Equation 7.18), express modes of growth behavior (Fig. 7.15). The examples shown include the seven most important eigenvectors derived from chronologies at the 26 sites described by Fritts (1965). The first eigenvector accounts for 27% of the tree-growth variance and is characterized by a marked resemblance among the chronologies of the Colorado River Basin and surrounding areas. This pattern can be observed as a positive growth anomaly during the decades 1616–1625 and as a negative growth anomaly during 1896–1905 (see Fig. 1.13). The second principal component (Fig. 7.15) represents anomalies in which higher-than-average growth occurs over northwestern United States, while an area of lower-than-normal growth is centered over New Mexico. The growth for the decade 1521–1530 resembles this pattern in a positive sense, and the growth for 1626–1635 and 1931–1940 resembles the pattern in an inverse sense (see Fig. 1.13).

The remaining five patterns each reduce a decreasing amount of variance and express smaller and smaller scale variations in growth. The anomalies expressed by the fourth principal component can be observed in the growth anomalies for the decade 1806–1815; those of the sixth resemble in an inverse sense the growth anomalies during 1776–1785 (see Fig. 1.13). The first seven principal components representing the largest scale growth anomalies reduce a total of 67% of the growth variance. As in the case of eigenvectors of climate, the higher-order principal components may be discarded to eliminate the small-scale variations, since the smallest ones are likely to represent errors, noise, the least important features of growth, or growth anomalies that are not related to macroclimatic variations.

In order to appreciate what the spatial eigenvectors of growth can tell us, consider the following theoretical situations. If all the chronologies of a region represent exactly the same information, only one principal component, with elements of the eigenvector weighting all chronologies equally, would reduce all the variance. On the other hand, if all growth chronologies are independent of every other chronology, theoretically there would be the same number of eigenvectors as chronologies and they

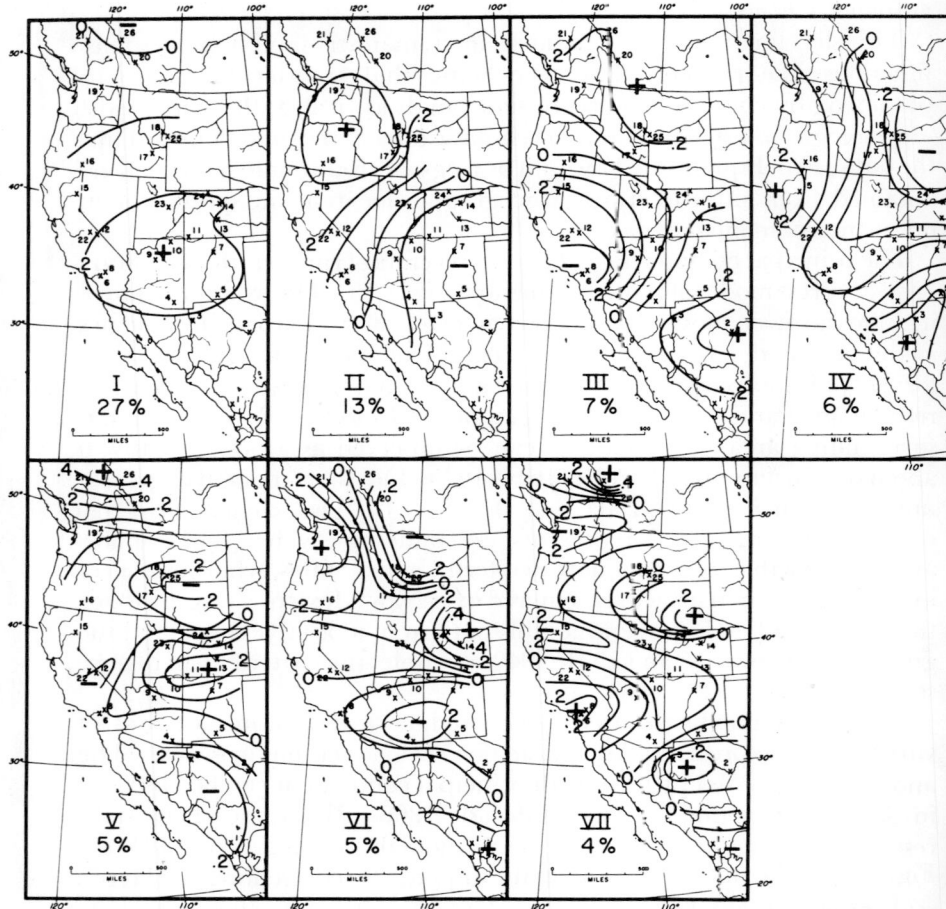

FIG. 7.15. The first seven eigenvectors of tree growth from 26 sites throughout western North America (Fritts, 1965). The percentage growth variance reduced by each eigenvector is shown.

would all reduce about the same variance. However, the real world is somewhere between these two extremes. If there is a large amount of redundancy in the several variables, most of the variance is reduced by a small number of eigenvectors. As more and more of the data sets behave independently, more and more eigenvectors are required to reduce the variance.

The growth eigenvectors and the normalized growth data for each year are multiplied (Equation 7.19) to obtain the amplitudes which express the amount of similarity of growth patterns during each year to each eigenvector pattern. These amplitudes can then be used as predictors of climate in a multiple regression equation (Equation 7.20) as in the response function analysis. Such regressions on the amplitudes of growth eigenvectors utilize the spatial variability of the chronologies to statistically reconstruct the anomalies in the macroclimate that had originally produced the corresponding anomalies in the growth.

The similarity between spatial anomalies of growth and climate were noted by LaMarche and Fritts (1971a). They examined the most important eigenvectors for the ring-width indices from western North America and found that they resembled eigenvectors of precipitation for the same region described by Sellers (1968). Fig. 7.16 includes selected principal components of monthly precipitation that occurred during 1931–1966 which matched the first four principal components of growth for A.D. 1700–1930. The first principal component exhibits large-scale anomalies over the Colorado Plateau for both precipitation and tree growth. The second component expresses opposite anomalies between northwestern United States and New Mexico and southwestern Texas. The third component for precipitation in October and for tree growth exhibits high values over the east slope of the Rocky Mountains and low values over California and the extreme Southwest. The third component for March precipitation and the fourth component for tree growth exhibit positive areas over the Pacific Southwest, the northern Rocky Mountains, and the northern Great Plains, while negative areas can be noted over the southwestern Plains and the extreme Northwest.

One way to obtain the actual transfer function is to calculate a multiple regression equation ($_1R_p$, Equation 7.20) which uses the eigenvector amplitudes of tree growth as predictor variables to estimate a climatic variable in a particular region or site. Since there is sometimes a lag of growth for one or more years behind the occurrence of certain climatic variables, the statistical calibration model may include growth predictors which lag one or more years behind the climatic occurrence. Also, autocorrelation may exist in the tree growth data, and it may be necessary in some cases to include growth for prior years so that the constraint due to prior growth is separated from the signal of climate (see Figs. 1.10 and 1.11).

In general, transfer functions are derived in the same manner as response functions, only the variables used for predictor and predictand sets vary. However, the model for transfer functions can be simple, utilizing only a small number of predictor variables, or it can be extremely complex. While the complex models can potentially reduce a large amount of the variance in the predictand for the dependent period, the simpler models using only the large-scale principal components may

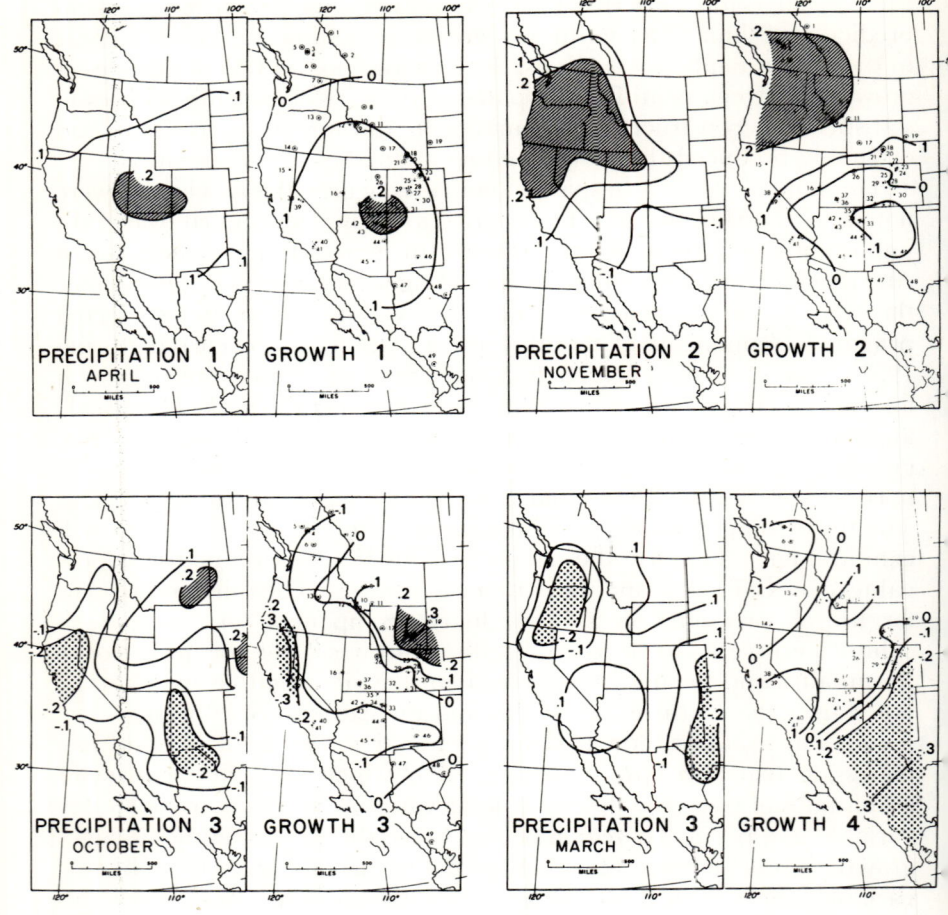

FIG. 7.16. Similar patterns in the spatial anomalies of precipitation and growth occur over western North America. The precipitation data (Sellers, 1968) were obtained by principal component analysis of monthly precipitation data for selected climatic regions of the United States from 1931–1966. The growth data were obtained by analysis of 49 tree-ring stations over western North America for 1700–1930. Examples are selected from the first three principal components of monthly precipitation and are matched with similar principal components of tree growth. (Redrawn from LaMarche and Fritts, 1971a)

hold up better, as seems to be the case when they are tested against independent predictand sets.

Now that the working tools of dendroclimatic analysis have been described, attention will be directed to analyzing the relationships and reconstructing past climate. The first portion of Chapter 8 deals with examples of response function analyses. Examples of various models used to obtain transfer functions and their applications will be described more fully in the latter portion of Chapter 8 and in Chapter 9 of this book.

Chapter 8

Interpretation of Climatic Calibrations, Reconstruction, and Verification

I	Introduction	376
II	Response Functions.	377
	A. Examples of response functions	377
	B. Analyzing variations among response functions	382
	C. Frequencies of positive and negative function elements. . . .	391
	D. Interpreting response functions	393
III	Strengths and Weaknesses of Response Function Analysis	400
IV	Significance of Response Function Capability	401
V	Assessing Effects on Growth of Varying Climate	402
VI	Climatic Reconstruction and Verification.	405
VII	Inferences from Chronologies With Different Growth Responses . .	407
VIII	Reconstruction Using Multivariate Transfer Functions . . .	412
	A. Climatic variations in arid regions affecting streamflow . . .	412
	B. Climatic variations and hydrologic conditions in subpolar regions .	415
	C. Climatic variations and studies of sea surface temperatures . .	421
	D. Climatic variations affecting biological systems	425
	E. Variations in temperature and precipitation.	428

I. Introduction

A variety of techniques have been described for calibrating ring-width variations with variations in climate and for reconstructing past variations in climate. The simplest calibrations usually involve few statistics and rely upon biological insight for recognition and documentation of some type of relationship. Climatic data are converted to growth estimates and compared to the actual growth measurements. Ring-width variation is then utilized to infer past variations in climate.

Most dendroclimatic studies include some quantitative analysis. At the end of Chapter 7 two recent innovations in quantitative methods were described. The two which will be further discussed here are (1) the multivariate response function, which is used to describe the ring-width response to variations in climate, and (2) the multivariate transfer function, which transforms values of ring width into estimates of climate.

The first portion of this chapter describes the details of six different response functions and then summarizes an analysis of 127 North American response functions. Differences are examined in terms of seasonal variations in the environments and their effect on the growth-controlling processes. The last portion draws upon recent studies to illustrate the variety of approaches to dendroclimatology, ranging from the use of simple logical inference to the use of rigorous calibration, reconstruction, and verification. This chapter focuses on single station and local climatic reconstructions, while the last chapter describes regional and synoptic studies employing spatial analysis.

II. Response Functions

A. Examples of Response Functions

Response functions like those described in Chapter 7 have been calculated for a large number of North American tree sites (Fritts, 1974) and more recently for a number of European study sites. All response functions described here include 14 weights (coefficients or elements) associated with variables of monthly mean temperature and 14 weights with variables of total monthly precipitation. Three additional weights are included to allow assessment of growth on the relationships during prior years. Other climatic factors and monthly intervals can be used instead if the data are available and if there is a reason to believe the factors are important in the growth–climatic relationship.

The three plots in Fig. 8.1 represent response functions for semiarid conifers showing weights for temperature on the left, weights for precipitation on the center right, and weights for prior growth on the extreme right. The vertical bars which delimit the 95% confidence limits are used to test for significance.

The uppermost plot for Quartz Mountain *Pinus ponderosa* illustrates a weakly linked relationship, since most elements of the response function have confidence limits intersecting zero and therefore must be considered insignificant. A total of 46% of the variance is accounted for by climate and 39% by prior growth so that climate and prior growth together explain a total of 85% of the ring-width variance.

Although no temperature weight is significant, those for summer are generally negative and those for winter are positive. Weights corresponding to precipitation are more often positive, but only the one

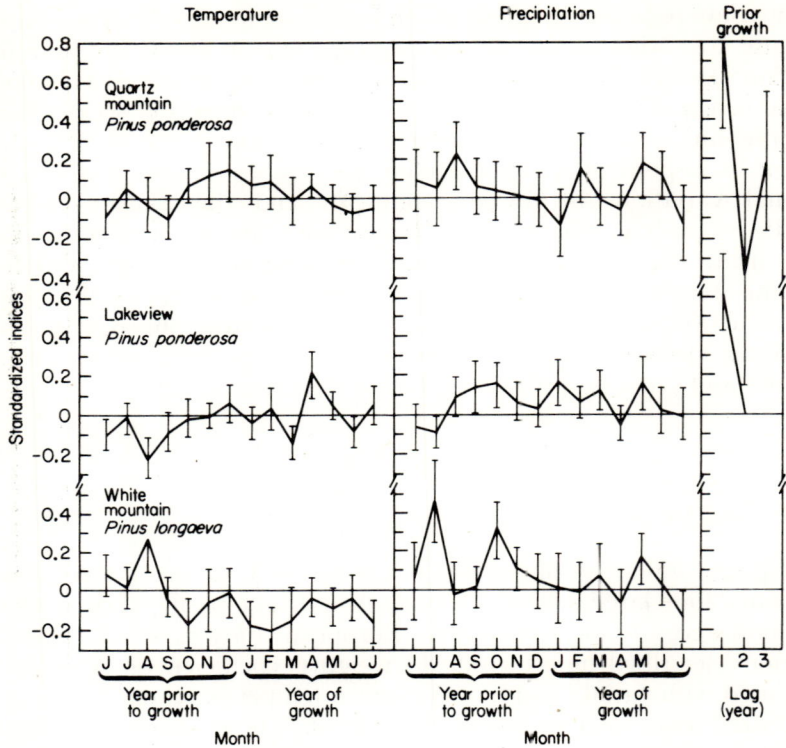

FIG. 8.1. Three response functions for ring-width indices of *Pinus ponderosa* on Quartz Mountain, Washington, and Lakeview, Oregon, and *Pinus longaeva* in the White Mountains of California. Each response function includes 14 weights for average monthly temperatures and 14 weights for total monthly precipitation from June of the year prior to the season of growth through the July concurrent with growth. Three additional weights are associated with prior ring-width index at lags of 1, 2, and 3 years. A positive weight indicates a direct relationship of the climatic variable or prior growth variable to ring width, and a negative weight indicates an inverse relationship. The vertical lines designate the 0·95 confidence level.

for August is significant. The weight for prior growth at a lag of one year is positive and significant, indicating an important direct relationship, but at lags of two and three years the respective weights are negative and positive, both being insignificant.

The excessively wide spread of the confidence limits indicates that any one element of this response function provides information of questionable reliability. However, the predominance of water stress can be inferred from the large number of negative weights for temperatures during summer, and low temperature limitations can be inferred from the large number of positive weights for winter temperature. In addition, the positive weights for precipitation suggests that moisture is limiting, especially in late summer of the preceding year and May through June of the growing year, while the positive weight for prior growth at a lag of one year indicates the presence of first-order autocorrelation.

The plot for Lakeview *Pinus ponderosa* (Fig. 8.1) is more typical of arid sites as it represents a stronger climate–growth linkage. It is similar to the response function for the Quartz Mountain site in that climate accounts for about the same amount of ring-width variance. A total of 48% is accounted for by climate and 39% by prior growth, giving a total of 87%. It differs, however, in that more weights for climate are significant. This occurs because fewer eigenvectors are used in deriving the response function, so that the error terms for fewer eigenvectors are included in the confidence limits.

Weights for variables of temperature are negative and significant for the prior June, August, and March, and the current June. Only the weight for April is significantly positive. For precipitation, the only significant negative weight is for the July prior to growth, while significant positive weights for precipitation are those for September, October, January, March, and May. The weight for growth at a lag of one year is the only prior growth variable that was entered into the response function equation, and its positive sign indicates the presence of first-order positive autocorrelation.

The four negative weights for temperatures imply that water stress is limiting for at least four months during the year. However, the positive weight for temperature in April indicates that higher than average temperatures at that time are favorable to growth, perhaps as they influence the breaking of dormancy and the resumption of physiological activity in the tree. The weights for precipitation indicate that above-average precipitation during September, October, January, March, and May are beneficial to growth, undoubtedly as they affect the water balance of the trees. However, lower than average precipitation in the prior July is significant, perhaps because it interacts with the growth which is occurring at that time. In short, annual rings in the Lakeview trees are shown to be wide when the autumn, winter, and spring are wetter than average, and when late summer of the prior season and early spring are cooler than average, changing to warmer than average by April.

The plot in the lower part of Fig. 8.1 is typical for arid-site *Pinus longaeva* from the White Mountains of California. Climate accounts for 63% of the growth variance, and the effect of prior growth is not significant. The weights for temperature of the prior August are significantly positive, and those for October, January, February, and the current July are significantly negative. The weights for precipitation are generally positive, with those for the prior July, October, and May being significant. The weight for precipitation during the current July is negative and significant.

It may be inferred from this response function that rings are wider than average in White Mountain *Pinus longaeva* when temperatures are (1) higher than average in the prior summer favoring prior growth conditions, and (2) lower than average in late autumn, winter, and spring maintaining low respiration rates within the trees. Wide rings also result when precipitation is (1) higher than average in the prior July and October and the current May, which helps to maintain a favorable water balance, and (2) is lower than average in the July concurrent with the growing season, a result with no clear-cut explanation.

The plots in Fig. 8.2 are response functions for *Quercus alba* from eastern North America and for two conifer species in Western Europe. In the first example, climate accounts for 59% of the ring-width variance and prior growth accounts for 2%. The significant positive temperature weights are for the prior July, November, and February, the latter two being months when the tree is leafless and apparently dormant. In June the temperature relationship is inverse and significant. The significant weights for precipitation all are positive and include the months of the prior August, May, June, and July, all times when the tree is physiologically active. The weight for prior ring width is insignificant.

It can be inferred from this response function that rings are widest when (1) precipitation is higher than average and (2) temperatures are lower than average during the growing season, both of which are factors that help maintain a favorable water balance in the trees. Above-average temperatures and moisture prior to the growing season can also favor growth since they promote rapid physiological activity within the trees.

The variance explained by climate for *Pinus halepensis* near Marseille, France, is 42%, and that explained by prior growth is 26%, making a total of 68%. The negative weights for temperatures in October, February, and April are significant. All weights for precipitation are positive for the period from August through May except for October. The predominant effect of winter moisture on growth is likely an effect of high precipitation on soil moisture recharge and winter photosynthesis. Temperature is apparently less important to growth, although the inverse relationships are probably due to water stress, and the direct relationships due to respiration and preconditioning phenomena occurring in October, February, and April.

The variance explained by climate in the third response function (Fig.

8.2) for *Pinus sylvestris* in northern Sweden accounts for 44%, while prior growth accounts for 20%. The weights for temperature in the prior June and the concurrent July are positive and highly significant, and this is inferred to indicate the major control of growing season temperatures on ring width. The positive temperature weights for September and April are smaller but also significant; those for the prior July and August are negative and significant. As might be expected for the northern tree line, fewer weights for precipitation are significant. Those for June of the year prior to growth and for March and April are positive, while those for May and July are inverse. Ring width at a lag of one year is the only prior

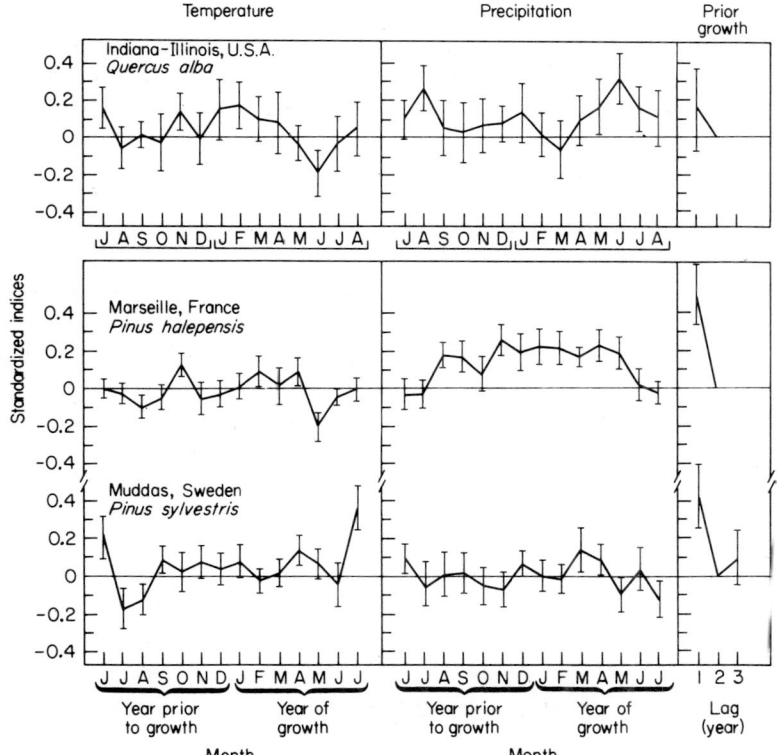

FIG. 8.2. Response functions for ring-width indices of *Quercus alba* in Indiana and Illinois (Ashby and Fritts, 1972), *Pinus halepensis* in Marseille, France (courtesy of Françoise Serre), and *Pinus sylvestris* from northern Sweden (courtesy of B. Jonsson and T. P. Harlan). The 14-month interval for the first differs from the latter two by one month. See Fig. 8.1 and text for further explanation.

growth variable that is significant and its weight is positive. It is inferred from positive response to temperature and inverse response to precipitation that on this site Model Part C (Fig. 5.10) is dominant.

The varying signs of the six response functions in Figs 8.1 and 8.2 indicate the presence of both direct and inverse relationships; the confidence limits indicate varying significance; and the presence or absence of prior growth variables indicate varying autocorrelation relationships. Such relationships with prior growth can be visualized as climatic phenomena, for they incorporate effects of climate on growth in prior years which are transmitted to a later growing season through the autocorrelation phenomenon.

Because of the statistical interaction of prior growth values with the prior summer climate, the weights for prior summer climate must be interpreted with caution if ring width at a lag of one year has been entered as a part of the statistical relationship. The regression portions out the growth and climatic effects as it attempts statistically to reconcile their intercorrelations. If climate and growth are highly correlated with one another, their individual weights in the response function may be diminished or be of opposite sign.

The Muddas site response function (Fig. 8.2) provides a good example of this kind of interaction. July temperature and concurrent ring width are directly correlated. From this fact it is concluded that temperature of the prior July and prior growth are also directly correlated. However, the weight for the response function for July temperatures in the prior summer is negative, while the weight for prior growth is positive. The equation has separated the effects of the two directly correlated variables statistically holding all but one variable constant. Thus, when prior growth is held constant the statistical effect of temperature with prior growth held constant is inverse. If prior growth had not been used as a variable, the strong positive linkage between prior temperature, prior growth, and ring width would have prevailed, and the weight for temperature of the prior July would likely have been positive and significant.

B. *Analyzing Variations among Response Functions*

Response functions have been obtained for a large number of coniferous species and sites in western North America. The smoothed response functions shown in Fig. 5.12 are the means for five different species or species complexes. An analysis of these same data including response functions for 127 individual ring-width chronologies (Fritts, 1974), is presented below. Most of the response functions were calculated

8. INTERPRETATION OF CLIMATIC CALIBRATIONS

using monthly mean temperatures and total monthly precipitation for homogeneous climatic regions within state boundaries. In a few cases climatic data for single stations were utilized, but the results for regional climatic data usually surpassed those for single-station data in explaining a greater percent of ring-width variance.

The weights or elements in each response function (excluding variables for prior growth) were transformed to the same variance by dividing each element of the response function by the standard deviation of all 28 elements, then multiplying by a constant. A multivariate procedure referred to as cluster analysis as described by Tryon and Bailey (1970)

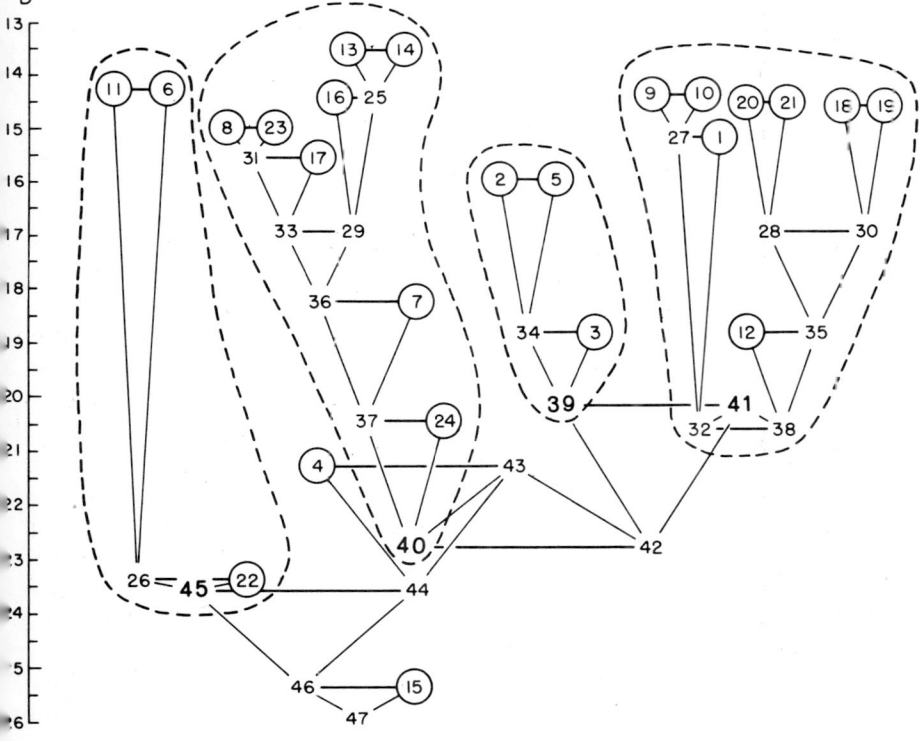

FIG. 8.3. A diagrammatic representation of the relationships among the clustered groups of response functions. The circled numbers represent the basic clusters of three or more response functions. Each cluster is paired with another which most resembles it as indicated by solid horizontal lines joining numbers, and the pairs are plotted as a function of their similarity. The most similar have low values of D and are plotted near the top of the figure. The dashed lines enclose four distinct sets of response functions as described in the text. (From Fritts, 1974)

was used to sort the response functions and to classify them into meaningful groups and subgroups. The cluster analysis was necessary because the differences among the 127 response functions were often subtle and difficult to sort and classify by visual inspection. The dashed lines shown in Fig. 8.3 encircle four categories of response functions which are merged into four large clusters shown near the bottom of each encircled set as numbers 41, 39, 40, and 45. Five relatively independent features of the response functions were identified as differentiating factors in the process of clustering (Fritts, 1974), and these factors were used to sort and group the 127 response functions into their respective classes.

Figure 8.3 is a diagram representing the clustering response function hierarchy, where the ends of the branches terminate in 24 distinct groups, each of which includes three or more similar response functions. These 24 groups are represented by the circled numbers 1 to 24 and are referred to in the following discussion as the basic clusters. Each is paired with the cluster it most resembles and the pairs are indicated by a horizontal line drawn between them (Fig. 8.3). The vertical position of each pair in the figure expresses the dissimilarity as measured by the scale marked D on the left margin of the figure. Pairs at the bottom are the least similar while those at the top are most similar. The most similar pair is identified first in the clustering process and the two clusters are merged into a single cluster group. The next most similar pair is then identified and these clusters merged. Pairing and merging continues until finally all the clusters are merged into a single group, shown as number 47 at the bottom of the figure. The cluster groups to be merged become less and less similar as more response functions are included.

The elements of the mean response function for all 127 sites represented by cluster 47 are shown in Fig. 8.4. The average weight for each of the 14 elements for temperature is shown in the upper portion of the plot; the average for each of the 14 elements for precipitation is shown in the lower portion. The average for prior growth at each of the three lags is also shown, even though these elements were not utilized in the clustering procedure. This mean response function for all sites has mostly positive weights for precipitation, representing the generally positive association of arid-site tree growth with precipitation. The means for elements of the response function associated with temperature are nearer to zero than those for precipitation, because the temperature elements for individual response functions exhibit both positive and negative signs. The average elements of temperature that depart most markedly from zero and are most often negative, are those for the months of October, March, May, and June.

This averaged response function confirms the generally stated

8. INTERPRETATION OF CLIMATIC CALIBRATIONS

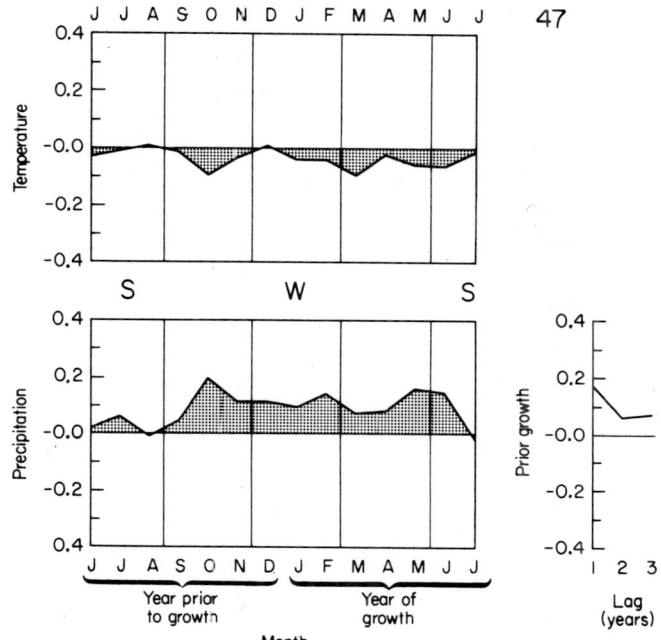

FIG. 8.4. The mean values for the 31 elements of all 127 response functions representing the 28 climatic variables and the three variables for prior growth (cluster 47, Fig. 8.3). The magnitude of the averaged elements and the sign indicate the average importance and the average direction of the relationship that exists. Vertical lines divide the seasons: S, W, and S designate prior summer, prior winter, and the summer concurrent with growth. (From Fritts, 1974)

conclusion of many of the early studies that tree rings from arid-site conifers are narrowest when precipitation during the year is low and when temperatures during certain months of the year are high with respect to normal conditions at each site.

It is apparent from the study that certain clusters exhibit a relatively high degree of resemblance and others do not. The largest differences revealed by clustering appear to result from varying site factors, such as exposure, elevation, slope, and geographic location. Species differences are present (see Fig. 5.12), but interestingly enough they are not as marked as the differences due to site characteristics.

The mean response function for any one of the four encircled cluster sets is different from the mean response function for the other sets (Fig. 8.5). These differences may be interpreted as the most important differences among the 127 response functions. The average response

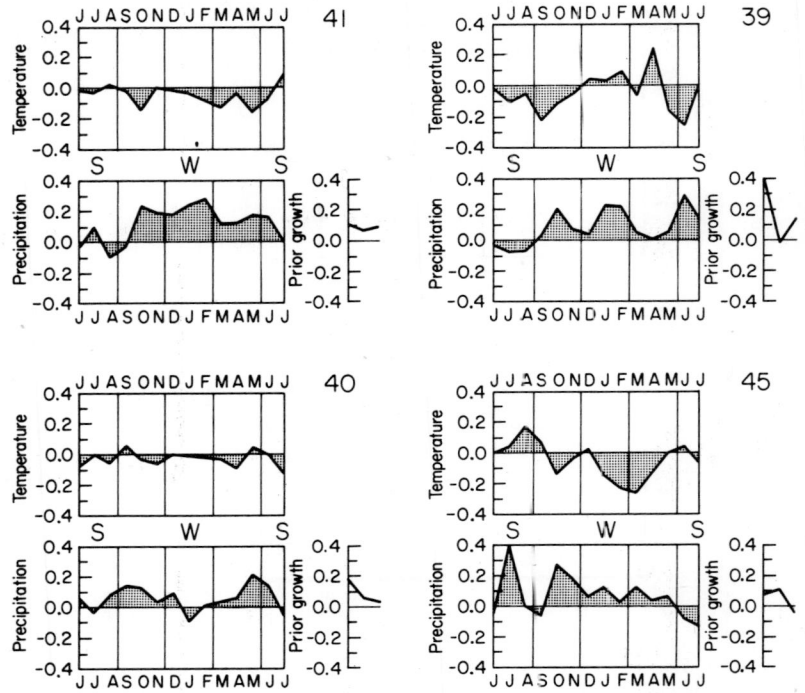

FIG. 8.5. The mean values for the 31 elements of the response functions for the four encircled groups of clusters shown in Fig. 8.3 as expressed by clusters 41, 39, 40, and 45. The magnitude of the elements and sign indicate the importance and direction of relationships that exist. Vertical lines divide seasons as in Fig. 8.4.

function of cluster set 41 (top left, Fig. 8.5) resembles that of cluster 47 (the mean of all response functions shown in Fig. 8.4). In these response functions there is generally a direct relation between growth and precipitation, especially during the latter portion of autumn, winter, spring, and the first month of summer, June. Temperature on the average is inversely related to growth for the six months of October, February, March, April, May, and June.

The branches associated with cluster set 41 shown in Fig. 8.3 terminate with eight basic clusters. One group, indicated by number circles 19, 18, 21, 20 and 12, is related to cluster set 41 through an intermediate cluster, 38, on the right. Another group, 9, 10, and 1, is related to cluster set 41 through an intermediate cluster, 32, on the left. The differences as well as the similarities within the set of basic clusters can be expressed by averaging the mean weight of response functions in each basic cluster,

8. INTERPRETATION OF CLIMATIC CALIBRATIONS

then plotting all mean weights which were significantly greater or less than zero (in this case at the 85% probability level) (Fig. 8.6). The upper plots on the left of this figure include the statistically significant mean elements of the response functions in basic clusters 19, 18, 21, 20, and 12. The plots for 9, 10, and 1 are shown on the right.

All response functions in the first two groups have generally large positive values for the elements associated with precipitation for autumn, winter, and spring as is shown in the mean response function for cluster set 41 (Fig. 8.5), though certain variations among the response functions of the basic clusters are evident. Many of the elements for variables of temperature are negative, representing a predominance of inverse relationships for temperature in autumn, winter, and spring (Fig. 8.6 top). In a few cases, particularly significant elements are positive, indicating a direct relationship between temperature and growth in certain months.

Basic clusters 19, 18, 21, 20, and 12 include response functions where the elements for temperature and precipitation during the prior summer and for those at the end of the current summer (July) are frequently correlated positively with growth. However, for basic clusters 9, 10, and 1, the significant weights for summer temperatures are either negative or insignificant, while those for midspring (April) are either positive or insignificant.

Thirty-two percent of all response functions were included in the clusters associated with 41. The trees are on arid, exposed, and well-drained sites located on the mountains and mesas of Arizona, New Mexico, Colorado, and Utah (Fritts, 1974). The chronologies are most typical of arid-site trees as described by Schulman (1956) and include 73% of the *Pinus edulis* stands sampled in this particular study. *Pinus edulis* is an arid-site species restricted to low elevations and latitudes in southwestern North America (see Fig. 4.16). Also included are 43% of the sites with *Pseudotsuga menziesii* and 26% with *Pinus ponderosa*.

Three closely related basic clusters of response functions are associated with cluster set 39 (Fig. 8.3). The means for the elements of the response functions included in 39 are shown in the upper right portion of Fig. 8.5. The elements for precipitation are large and positive for certain months in the prior autumn and winter, and the summer concurrent with growth. The elements associated with temperature are, on the average, negative for the prior summer, autumn, and late spring, and early summer concurrent with growth, while the means associated with temperatures of April and the winter months are positive. The means of the significant elements of the response functions in the three basic clusters associated with 39 (basic clusters 5, 2, and 3) are shown in the left

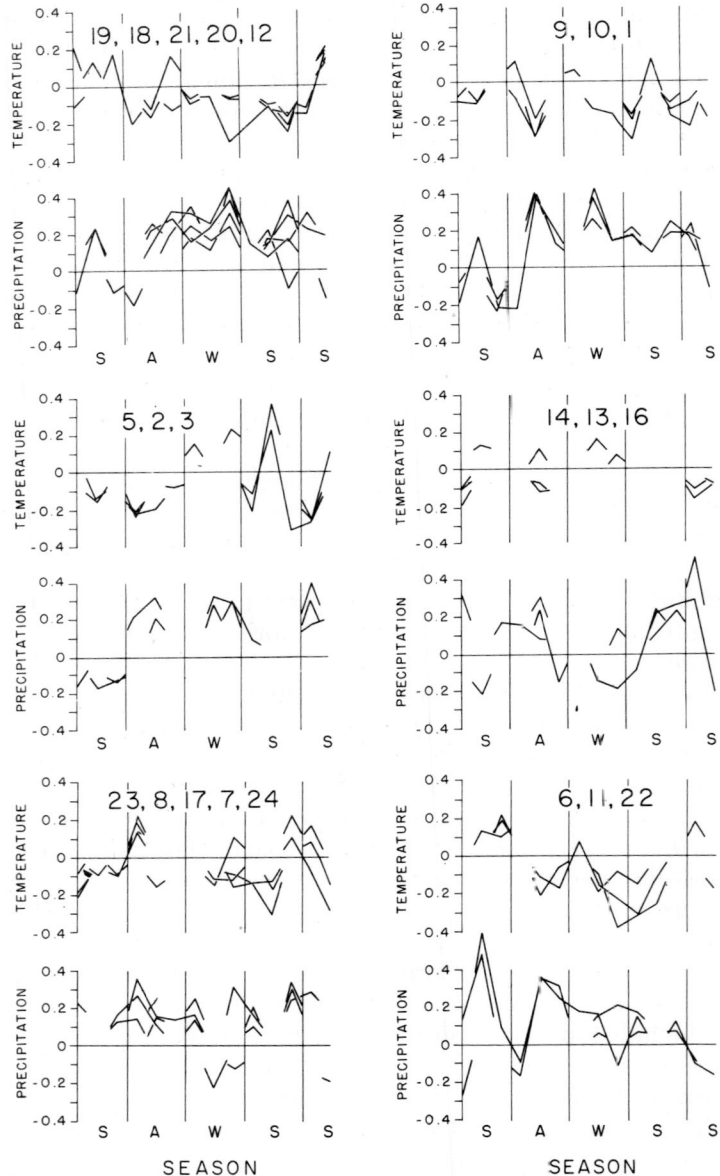

FIG. 8.6. The means of significant elements in the response functions included in each basic cluster are plotted as a function of the variable they represent. Plots of similar clusters are superimposed to show the kinds of variability that occur among the basic clusters. Numbers at the top of each set designate which of the basic clusters from Fig. 8.3 are represented. Variables are plotted as in Fig. 8.5.

middle portion of Fig. 8.6. Basic clusters 5 and 2 both emphasize direct relationships of growth with April temperatures, while for 3, the significant direct relationship is for temperatures in February.

The principal feature of this group of clusters is the direct response of growth to temperatures prior to the beginning of cambial activity. Basic clusters 5 and 2 represent response functions for 22% of the *Pinus ponderosa* and *P. jeffreyi* sites. Basic cluster 3 includes response functions for a site of *Pseudotsuga menziesii* in Montana, *Pinus edulis* in Arizona, and *P. flexilis* in Wyoming. All are on steep west-facing slopes.

The response to temperatures in another basic cluster (15) shown near the bottom of Fig. 8.3 resembles basic cluster 3 in that there is a marked direct relationship with February temperature, but it differs in that March shows a significant direct temperature effect. Therefore, basic cluster 15 was not classed with any of the four large sets and appears in the diagram in Fig. 8.3 nearest the mean cluster for the entire set (47).

Cluster set 40 (Fig. 8.5) portrays the mean for the third set of response functions encircled by dashed lines in Fig. 8.3. The means for elements associated with precipitation (especially for winter) are not as large as for those in 41 and 39. This indicates a lower importance of winter precipitation as a limiting factor on growth but still shows a similar dependency on warm-season moisture. The mean values of the weights for temperatures are small, with some indicating direct and some indicating inverse effects.

The significant elements of each basic cluster, 14, 13, 16, 23, 8, 17, 7, and 24, are shown in the middle and lower portions of Fig. 8.6. Many of the elements associated with temperature for basic clusters 14, 13, and 16 do not appear in the plot because they are statistically insignificant. In two cases where winter response is significant, a direct relationship between temperature and growth is apparent. During the warmer seasons, the elements for temperature are more often negative, indicating that growth and temperature relationships at that time of year are more commonly inverse than direct. It is difficult to generalize here because there are considerable differences among the elements for temperature in the response functions representing basic clusters 13, 14, and 16.

The elements in basic clusters 23, 8, 17, 7, and 24, as shown at the lower left of Fig. 8.6, often express a direct relationship between growth and temperature during early autumn, late spring, and summer of the growing season. Within this cluster, elements for precipitation during winter months are in some instances positive, but in other instances negative.

The response functions of set 40, in general, express a direct relationship with precipitation in autumn, late spring, and early summer, but differ markedly in the signs and values for temperature elements. In a few cases there is an inverse or low correlation of growth with winter precipitation and a direct correlation of growth with temperatures in early autumn or late spring. The set includes 40% of the *Pseudotsuga menziesii* sites, 43% of the *Pinus ponderosa* sites, and 20% of the *Pinus edulis* sites used in the cluster analysis.

Elevations of the sites are often high and exposures are usually other than south-facing. The habitats are apparently relatively cool and moist where low temperatures in the cooler seasons would tend to be severe and hence lead to physiological inactivity or to direct temperature and growth relationships. Climatic conditions occurring during the growing season are emphasized more than those during midwinter and this suggests that models A and C, described in Chapter 5, are the most important for describing these particular climate–growth relationships.

The averages for elements of the response functions of the basic clusters associated with cluster set 45 (Fig. 8.3) are shown at the lower right portion of Fig. 8.5, and the differences among the three basic clusters are shown on the lower right portion of Fig. 8.6. Large positive values for elements of the response functions can be noted for the summer and autumn seasons prior to growth. The weights associated with precipitation in the prior July, October, and November, and for temperatures in the prior July and August are large and positive. However, the weights for elements of temperature are large and negative for autumn and late winter through early spring. Those for precipitation in June and July, the period of the growing season, are also negative.

This type of response to climate expresses a negative feedback relationship whereby a summer of extreme climate appears to favor the current growth in one direction and the following year's growth in the reverse direction. Such a negative feedback counters tendencies for positive autocorrelation, and the mean for the element associated with prior growth at a one-year lag is smaller than in the other basic clusters. However, the effects on growth of the climatic variations during winter and spring are similar to those of other groups in that precipitation is directly related and temperature is inversely related. Unlike many other response functions, the weights for temperatures during winter and early spring are larger than those for precipitation, indicating that temperature is more important than precipitation during the winter period.

These clusters include response functions for high-elevation arid sites, mostly for *Pinus longaeva* (also see Fig. 8.1). It can be noted from Fig. 8.3 that basic cluster 22 differs markedly from the mean for the other two basic clusters in this set. This particular basic cluster includes response functions for extremely high-elevation sites near the upper tree line for *Pinus longaeva* and one for *Abies concolor*. The weights in the response functions for these sites differ in that growth is directly correlated with temperatures during the concurrent summer as well as during the summer prior to growth. This response, which is associated with cold high-elevation sites of arid regions, is consistent with the response of conifers at high northern latitudes, in that the temperatures of the growing season are important and directly related to ring width.

LaMarche and Stockton (1974) computed response functions for eight high-elevation sites of *Pinus longaeva* and *Pinus aristata* at the upper tree line. The values of the elements they obtained for temperature during the growing season were also large and positive. The elements for precipitation in late summer of the prior year, autumn, and spring were more often positive than those for precipitation in winter, except in the case of some of the more arid sites. The weights for winter temperatures were positive as often as they were negative. LaMarche and Stockton's results confirm further the generalization that ring widths of low latitude trees at the upper tree line are often directly correlated with temperature and sometimes directly correlated with precipitation. The results differ from those of Fritts (1974) reported above, in that growth at their high-elevation sites was not as strongly coupled with winter and early spring temperatures as occurred in the more arid habitats.

C. Frequencies of Positive and Negative Response Function Elements

Certain significant features are evident from a frequency analysis of the signs of the response function elements. The occurrences of significant positive and negative elements of the 24 basic clusters were examined, weighted for the number of response functions they represented, and tabulated by season of occurrence. Table 8.I includes the percentage of response functions in which significant positive and negative weights were exhibited in each of the five seasonal periods (Fritts, 1974).

These data show that the relative importance of direct and inverse relationships differs markedly between variables for temperature and precipitation and from one season to the next. The elements associated with precipitation are more commonly positive, while those associated with temperatures are more commonly negative. During the winter period the weights for temperatures were significantly positive in 33%, significantly negative in 57%, and not significant in 10% of the cases. During the warmer seasons of the year the weights for temperature were

TABLE 8.I Percent response functions that exhibit a significant positive or negative weight for at least one month within a particular season.[a]

Season	Climatic Variable[b]			
	Temperature		Precipitation	
	Positive (%)	Negative (%)	Positive (%)	Negative (%)
Prior Summer	35	57	49	40
Autumn	31	69	87	28
Winter	33	57	90	18
Spring	22	73	93	8
June–July	36	71	46	31

[a]From Fritts, 1974. [b]If coefficients with both positive and negative signs occur for a given season (but for different months), the occurrences are tallied in both the positive and negative columns.

significantly negative from 57% to 73% of the time. The weights for elements of precipitation during the winter and spring were significantly positive 90% and 93% of the time respectively. However, in June and July concurrent with growth the weights for precipitation were significantly positive only 46% of the time.

The variability in the sign of the weights for precipitation at the time of growth in June and July may be attributed to three causes. (1) The ring may be partially formed by that time so that climate may have a limited effect, (2) only two months (June and July) comprise this season as compared to three months for the other seasons, so the probability for a significant weight is two-thirds that for the other seasons, (3) the precipitation for July is generally more weakly correlated or inversely correlated with prior climatic conditions, only partly for climatic reasons. July is positioned at the end of the 14-month period. As a result, a smaller portion of the temperature and precipitation variance for this month is reduced by the first and most important principal components than is reduced by the same principal components for variables well within the 14-month interval. With less variance reduced by the principal components, the elements for July at the end of the 14-month period are on the average lower than the elements for months well within the 14-month interval.

The last condition mentioned above may also reduce the magnitude of the elements corresponding to the June marking the beginning of the 14-month interval. As mentioned earlier, the weights for factors of the prior summer climate may also interact with prior growth, and together they share a role in the predictand variance they reduce. Hence, the response function weights for climate of the prior season are lower and sometimes the opposite sign than if prior growth had not been entered as a ring-width predictor.

D. Interpreting Response Functions

Although response functions cannot prove cause-and-effect relationships, they should at least be explainable in the light of current knowledge of tree growth. In addition, they may indicate previously unrecognized relationships. This section will examine the elements of response functions for possible systematic and meaningful patterns associated with particular species, site conditions, or geographical regions. Details of the physiological processes and microenvironmental factors are covered in Chapters 2 through 5, and the details of the data for the particular sites are described by Fritts (1974).

It is important to keep in mind that each element of a growth response function represents the net effect of whatever correlation exists between environment and growth. The results of the response function are constrained by the year-to-year fluctuation of the climatic variables above and below their mean values for the particular season. If the mean of the tree's environment and the range of variability are near the optimum for physiological processes, the natural variations in climate may not be sufficient to limit the processes, and there is likely to be no correlation and no significant response function elements. If the mean and range of variability in the tree's environment are near the limits of one or more plant processes, there is likely to be a marked effect on these processes and subsequent correlation between climate and growth. Further, if two processes act in opposite directions, they may cancel one another's effect and result in a lack of correlation between climate and growth.

For example, during a given month, temperature at a site may range between 10°C and 15°C, which is optimal for many processes. No process would be limited by temperatures varying within this range, and the element of the response function for temperature would be insignificant. If temperatures at the site actually range from 0°C to some value such as 10°C, the high temperatures may be near the optimum for certain processes such as net photosynthesis, but those near 0°C may be limiting. Significant correlation with climate may result, and the corresponding elements in the response function would be large, positive, and significant. If, however, the site is very cool (such as on a north-facing slope), leaf temperatures on the average may be below 0°C during the winter so that chlorophyll in the leaves is deactivated. The brief periods of warm temperatures (above 0°C), unlike the very cold periods, may be unfavorable to growth because they result in high respiration and high consumption of food. In such a case, due to the brief warm periods, the relationship between temperature and processes affecting growth would be inverse, and the element of the response function would be negative.

In other situations, temperature in winter may range between 0°C and above when temperature is positively correlated with important processes and between 0°C and below when temperature is inversely correlated with important processes. These two opposing situations can cancel one another's effects, in which case the corresponding elements would be near zero and insignificant. On the other hand, during the same time period high precipitation may occur more frequently in meteorological situations when temperatures are near 0°C than may occur when temperatures are at either the warm or cold extreme. If such were the case, precipitation rather than temperature could correlate with the growth-related processes and the response function would exhibit a significant element, when in fact temperature, not precipitation, actually is responsible for the physiological effect. While such noncausal associations may not be a particular problem to a dendroclimatologist who is interested primarily in the information on climate contained in the chronology of widths, they do cause difficulties when one attempts to speculate on the physiological basis for the result.

In addition to the statistical associations described above, a meaningful interpretation of response functions must also be based upon a knowledge of how the various plant processes are limited by environmental factors and how the microenvironmental factors are linked to the monthly temperature and precipitation. Since linkages may be different for different seasons, and since the factors limiting growth processes may vary from one month to the next, it is also important to evaluate the elements of the response function in terms of the different seasons and to consider very carefully the transitional periods between seasons when the microclimates are changing from one limiting condition to the next. Furthermore, the energy balance, the water balance, and the climatic regime may vary markedly for different exposures, latitudes, and elevations (Figs. 4.16 and 5.2). It is well also to consider the distribution of the tree species in question because trees on sites near the limits of the species may be particularly susceptible to climatic variations.

Many kinds of relationships can complicate the interpretation of growth response functions but observations on features of environments, sites, and species, and an understanding of plant processes, can be utilized to eliminate many of the problems and allow for acceptable confidence in explaining the meaning of the result. Inferences can always be made on the basis of the best available information, and hypotheses can be made and tested as new samples are obtained. A certain amount of verification is achieved as one is able to predict the type of responses function on the basis of site and species information along with the statistical characteristics of the chronology that results.

For example, the response functions for *Pinus ponderosa* on sites near its arid and low elevation limits in northern Arizona show a more marked correlation of growth with winter precipitation than with summer precipitation, while response functions for the same species on sites at greater distances from the species' arid lower limits show a greater correlation of growth with summer precipitation than with winter precipitation. This observation can serve as a hypothesis about the importance of winter and summer moisture on arid versus moist sites or warm versus cold sites. The hypothesis may then be tested by classifying available response functions for the same and other species on the basis of nearness of the site to the arid limits and looking for differences in the mean response function of the grouped sets. Since many response functions are found to be consistent with this hypothesis, the hypothesis is accepted as a valid one. Of course, physiological evidence can provide the ultimate verification, and it may lead to a more detailed and accurate explanation of the dominant physiological cause-and-effect relationships.

Observations of differences in the response functions associated with latitude, exposure, elevation, soil type, and other factors have led to many inferences which are testable by classification of response functions and examination of the classes for significant differences in the results. The following discussion is a summary of the best inferences drawn from such classified response functions and other information. Some of the generalizations will undoubtedly be modified and improved in the future, some will remain unchanged, and others will be abandoned as they are found to be inconsistent with new research results. The items described here reflect certain inadvertent biases in that a major portion of research at the present time involves arid-site conifers from North America. Information gathered by workers in other regions has been examined for its consistency, but there are certainly many species, site factors, and situations that have not been examined adequately.

The following discussion treats the macroclimatic factors of monthly mean temperature and total monthly precipitation separately, with each divided into those relationships that are direct and those that are inverse. This classification is further divided into subgroups consisting of varying seasons, sites, and species.

1. *Temperature*

(a) Direct Relationship. Direct relationships between ring width and

summer temperatures are often found in response functions for trees near the upper elevational or latitudinal limits of the species, especially for those on north-facing slopes or in other unusually cool microclimates. This undoubtedly occurs because low temperatures limit respiration, photosynthesis, and other biochemical processes essential for rapid growth (Chapters 2 and 4). Low temperatures can further lead to reduced ring width by delaying the beginning of the growth period and by terminating growth before the end of the normal growing season.

Some direct correlations with temperatures in winter may result from interacting conditions within a tree site. High temperatures in winter can reduce snow cover, promote infiltration of moisture into the soil, cause soil temperatures to be unusually high, and result in high net photosynthesis. Near sea level in northwestern United States where cloudy weather is common, a direct correlation of ring-width growth in conifers with winter temperature may result from the association of cool temperatures with cloudy weather, low light intensities, and reduced net photosynthesis (Chapter 4).

(*b*) *Inverse Relationships.* Ring width of conifers on semiarid sites is often inversely correlated with variations in monthly temperatures. The most obvious explanation for this inverse relationship is that temperature is the driving force for diffusion of water, and anomalously high temperatures lead to rapid evapotranspiration and accentuated conditions of water stress (Chapter 3). Such effects of temperature are most apparent (1) in the summer when high temperatures are common and low temperatures are not likely to limit growth-controlling processes, (2) at low elevations where air temperatures are most likely to be high for many months during the year (Fig. 4.16), (3) on south-facing slopes, where there is abundant radiant energy and plant temperatures are high, (4) in trees exposed to wind which causes plant temperatures to approach those of the air and steepens the leaf–air moisture gradients, (5) in trees on shallow well-drained soils or with restricted root distributions where little moisture is available, and (6) in individual tree sites nearest the warm and arid limits of the particular species distribution.

An inverse relationship may be generated by the direct effect of temperatures on respiration, causing available foods to be utilized more rapidly and net photosynthesis and storage of food to decrease (Chapters 4 and 5). Also, high air temperatures may force plant temperatures to exceed the optimum levels for processes of growth, assimilation, and hormone synthesis (Chapters 2 and 4). Inverse correlations between temperature and growth can also result when root temperatures are likely to be low during winter periods of little snow or in the late autumn and early spring. High air temperatures during times when the roots are cold or frozen can lead to severe internal water stress and injury to the leaves and tissues exposed to the air (Chapter 4).

Higher than normal temperatures in the autumn can delay winter hardening, making the tissues more susceptible to frost injury and adversely influencing growth in the following spring (Chapter 4). Higher than normal temperatures late in summer or early in the autumn can prolong growth of certain tissues, which consume food reserves so that less is available for growth during the next year (Chapter 2). However, late in the growing season water is also likely to be deficient, in which case an inverse relationship between temperatures and growth might predominate (Chapter 2).

There are other interactions among plant processes which could lead to either direct or inverse correlation with temperature. Extremely high or low temperatures can reduce apical and needle growth affecting the production and movement of growth-regulating substances and the sizes and number of cells produced by the cambium (Chapters 2 and 4). At times of flooding when low amounts of oxygen in the soil may limit processes in the roots, high temperatures can enhance the further depletion of oxygen in the soil as well as induce water stress (Chapters 3 and 4). A review of Chapters 2, 3, and 4 will suggest other ways in which temperature can directly or indirectly affect ring width.

2. Precipitation

(a) *Direct Relationships.* Variations in monthly precipitation are most commonly directly correlated with ring widths in trees on semiarid sites, the most obvious explanation being the association between low precipitation, low soil moisture, and internal water stress, which limits physiological processes affecting growth (Chapter 3). The more precipitation there is, the more moisture is available in the soil, and the longer is the period of time before water stress becomes the most limiting factor to growth.

Such a model has been used frequently in tree-ring studies, as it is relatively apparent from experimental results. However, the response functions and field investigations on arid sites suggest that water stress very often limits processes other than cambial activity and creates internal conditions that at some later time affect growth (Model B, Fig. 5.9). This sequence is seen most clearly in trees nearest the arid limits, on well-drained sites, at low elevations, and on southerly exposures where the response functions often show significant and direct relationships between growth and precipitation during almost any month of the year. Some of the most important processes involved in this direct relationship between precipitation and cambial activity are net photosynthesis, assimilation, apical growth, leaf development, root growth, and both the production and translocation of growth regulators.

Precipitation relationships can vary during the nongrowing season, depending upon whether moisture typically falls as rain or snow. If soil moisture is low and soil temperatures are above 0°C, either rain or snow can add moisture to the soil. If the soil is frozen, additional precipitation falling in the form of snow cannot penetrate the soil and may not have any significant effect on available moisture. Snow early in the season can insulate the soil and contribute to warm soil temperatures and rapid water absorption by the roots during the winter period, while an absence of early snow can lead to low soil temperatures and the freezing of roots. In addition, precipitation falling as snow may be subject to wind, blowing it away from certain exposures and depositing it as drifts in others. Variations in wind direction or intensity from one year to the next can complicate further the growth–precipitation correlations as they affect drifting snow.

(*b*) *Inverse Relationships.* Inverse correlation between precipitation and growth is less common than direct correlation especially in arid sites, and when it does occur it often exists because precipitation is correlated with other more limiting factors. For example, in sites where soil drainage is poor, excessive moisture may reduce soil oxygen and inhibit the growth of roots. An unusually deep snow and an increase in the length of time a particular site is covered by snow may extend the length of the photosynthetically inactive period and delay the beginning of growth (Model C, Fig. 5.10), thus leading to an inverse relationship between precipitation and ring width. Another related inverse effect of precipitation on growth can occur in autumn when unusually dry weather can lead to water stress and increase the winter hardening, which prevents destruction of plant tissues by severe winter cold (Chapter 4). Other inverse correlations with precipitation may result from the favorable effects of soil moisture on flowering, fruiting, and other phenomena that compete with cambium for the reserves of food (Chapter 5). Heavy snow, ice, hail, and lightning that are associated with high precipitation can damage the plant and occasionally lead to a reduction in growth.

3. Interactions between Climatic Variables

Precipitation may have an influence on the elements of the response via its correlation with cloud cover, air temperatures, and other factors influencing processes in the trees. At sites which are frequently limited by low temperatures, precipitation and cloud cover may considerably reduce the direct radiation and cause additional cooling of the plant

8. INTERPRETATION OF CLIMATIC CALIBRATIONS

(Chapter 5). If no other factor is more limiting, the association of cloud cover with low temperatures may become more limiting to growth than either factor alone. The response function may show an inverse precipitation effect as well as a direct temperature effect, although the former may not occur frequently enough to give rise to a significant weight.

Such inverse relationships between precipitation and temperature may also lead to unexpected results. Cloud cover, increased humidity, and cooler temperatures associated with rainy periods in summer can have a favorable effect on net photosynthesis and growth in arid-site trees because of the corresponding reduction in the heat load and improved water balance (Chapters 3 and 5). This could conceivably produce a direct correlation between growth and precipitation of a region due to the frequent occurrence of afternoon clouds, even though the amounts of rainfall falling on the tree sites might be insufficient to affect growth.

Clouds in winter may prevent excessive cooling at night and maintain plant temperatures near those of the daytime (Chapter 5). At high elevations winters with little precipitation may be accompanied by warm days and very cold nights, which would contribute to high respiration in the day without any compensating photosynthesis, and would result in depletion of food reserves available for spring growth (Chapter 4).

Interactions can occur among prior growth and the concurrent climatic factors if drought is followed by more favorable conditions in late summer and early autumn (Chapter 2). The growth processes which ordinarily consume foods are checked by the drought, and all photosynthates go into food reserves which are then available for the next year's growth. Also, the moisture of the prior growing season may affect the number, size, and kind of buds produced, the amounts of leaf and root growth, and the degree of winter hardening, all of which in turn determine the functional tissues that influence the amount of growth (Chapter 5). The net effect of drought may be a reduction in present growth, but an increase in future growth if other factors are not limiting. The opposite type of interaction has already been mentioned for *Pinus longaeva* from arid sites where high precipitation during the growing season leads to low growth in that year, but high growth in the next year. This may be due to diversion of food substances to the growth of needles and roots, which in turn enable the tree in the following season to better utilize its environment for growth.

Interactions may arise from antecedent climatic conditions which affect the tree through preconditioning. An extremely cold winter can cause the tree to become completely dormant (all processes essentially inactive) for several months. Variations in precipitation and soil moisture are then less likely to affect processes because of their inactivity. In the spring, as temperatures rise and the trees become more active, an abundant soil moisture reserve is more likely to have a beneficial effect. Also if the annual precipitation is high enough to fully recharge the soil during the winter, late winter precipitation may run off and have no physiological effect unless they influence variables such as light and temperature, which become limiting to processes affecting growth. The amount of precipitation and temperature may also be correlated with frequency of various types of destructive processes which indirectly lead to a change in the forest habitat and the variations in growth. For example, climate can be a contributing factor to the occurrence of fire, insect infestation (Fig. 5.6), ice damage, landslides, and flooding which in turn can lead to variations in growth.

III. Strengths and Weaknesses of Response Function Analysis

It can be seen from the above discussion that response function analysis provides an exceptionally precise tool for identifying tree-growth and climatic relationships. The fact that variables of monthly temperature and precipitation can account for as much variance as they do attests to the validity of the method. For example, in 80% of the response functions mentioned by Fritts et al., (1971), climatic variables reduced more than 50% of the ring-width variance. The median percentage of variance due to climate is approximately 60% to 65%, and for 20% of the response functions climatic variation accounted for more than 80% of the chronology variance. If one regards the variance due to prior growth as an effect of climate in prior years, an even larger percentage of total variance is due to climate. Considering the fact that response functions represent a linear expression of very complex biological systems responding to the microenvironments, and that the microclimates are only partially coupled to the macroclimate of a large climatic region, the percentage of variance due to climate appears to be a substantial amount.

However, the response function technique does have a number of weaknesses. The most important one is that elements in response functions need not be causally related to ring-width growth. Any factors that are correlated with temperature and precipitation inadvertently affect the response function relationships. A second limitation is that strong nonlinear interactions among variables may not be represented adequately and may not be fully taken into account.

It is possible that additional flexibility in the statistical model for the response functions could be achieved by using curvilinear regression or other appropriate transformations (Jonsson, 1969) or by adding cross products as factors to allow for multiplicative effects. Such changes could lead to significant loss of degrees of freedom and perhaps impair the ability to analyze individual months and sites. A certain number of degrees of freedom could be added by increasing the length of the climatic record with which the ring-width chronology is analyzed. However, the quality of many early climatic records is questionable and, in the case of region-wide data, the record is likely to include fewer reporting stations so that it represents a less reliable record of variations in the macroclimate. Thus, it is not at all promising that the response function analysis can be improved markedly by adding more observations and variables to handle nonlinear effects.

Another limitation is that the response function in its present form does not allow for differentiation between climatic factors that have an immediate effect on growth (high frequencies) and those that may have a prolonged affect lasting several years to several decades (low frequencies). Such differences in tree growth and environmental relationships which involve frequency variations may be handled by filtering the data before analysis or by use of power and cross-power spectrum techniques (Chapter 6).

IV. Significance of Response Function Capability

As tree-ring analysis is applied to a wider variety of sites and climatic conditions, even more diversity in response functions is expected to result. Although the presence of such diversity appears to substantiate claims made by early critics that the tree growth–climatic relationships are exceedingly complex, it does not follow that they are too complex to allow for climatic analysis. Actually the complexity is an asset, for the greater the diversity in the response to climate, the greater the information on climate.

For example, the response functions for two chronologies from trees on neighboring sites may be similar in all respects except for the weights of one climatic factor and season, such as precipitation for winter. Differences between the chronologies could occur only when winter precipitation was either higher or lower than the average. In such a case the two chronologies could be subtracted one from the other to estimate the variation in winter precipitation. To be more specific assume that:

Trees on a hypothetical site A respond to precipitation over all seasons of the year;
Trees on a hypothetical site B respond only to precipitation of the autumn, winter, and spring;
Trees on site C respond only to precipitation of autumn and spring; and
Trees on site D respond only to precipitation of spring.

Then the following can be obtained:

Annual precipitation is proportional to indices of A;
Summer precipitation is proportional to the indices of A minus the indices of B;
Winter precipitation is proportional to the indices of B minus those of C;
Autumn precipitation is proportional to indices of C minus indices of D; and
Spring precipitation is proportional to the indices of D.

Thus, it is at least theoretically possible to obtain estimates of climate for different seasons if the response functions indicate sufficient diversity in the chronologies used to estimate climate. In order to obtain as much diverse information as possible, a number of sites and species in an area should be sampled with adequate replication and each summarized separately. Each ring-width chronology may then be calibrated directly with available climatic data, and the calibrations used to ascertain what information the associated chronologies have recorded regarding past climate. Those chronologies with the most diverse set of response functions can be selected and a transfer function calculated to convert the variations in the selected chronologies to variations in climate.

The development of this multivariate procedure has substantially changed the data-collection requirements of dendroclimatological research. In the past the emphasis was placed on selecting trees from extremely arid or cold sites which responded similarly to variations in either drought or temperature. With such selection the ring-width chronologies were all inferred to represent the appropriate climatic element. Now one can somewhat relax the emphasis on collection from the extremely limiting environments, since the multivariate procedures offer convenient, objective, and rapid methods of analyzing the diversity of the response. However, the selected sites must be sufficiently stressful so that some climatic factors are limiting and there is sufficient variability in ring structure to crossdate and calibrate with climate.

V. Assessing Effects on Growth of Varying Climate

As was mentioned earlier, each element or weight of a response function is proportional to the relative effect of the calibrated climatic variable on ring width. Responses functions shown thus far are in normal form; that is, the units are expressed in units of the standard deviations of each variable. When each element of a response function is divided by the appropriate standard deviation of the climatic variable it represents, the elements are transformed to units expressing the effect of inches or millimeters of precipitation and degrees of temperature on ring width (Fritts et al., 1971). It is then possible to substitute into the equation particular departure values of each climatic variable for each month and calculate the absolute effect of climate upon ring width as measured at the climatic station. If specific departure values are applied to all weights, including substitutions of the derived growth departure for prior years, some interactions among variables and prior growth are taken into account.

Figure 8.7 is an example of the use of response functions for a number of relatively arid-site trees from the western United States to assess the effects on tree growth of hypothesized changes in global climate which could result from the operation of a fleet of supersonic transport aircraft (Cooper *et al.*, 1974). A reduction in temperature of 2°C (3°C for sites north of latitude 40°N) for each of the 14 variables for monthly temperature was multiplied with each element of the response function, which had itself been divided by the appropriate standard deviation. Calculations also included precipitation changes of $+5\%$ and -5% (10% for sites north of latitude 40°N). The calculations were carried out for five successive years, and estimated anomalies in growth were multiplied by the elements of prior growth for lags of up to three years. The result represents an estimate of the percentage change in ring-width growth that would be expected from the proposed climatic anomaly operating over a period of five years (Fig. 8.7). This of course does not take into account community changes which may alter competition and affect growth.

Growth increased in nine out of the 12 cases shown in Fig. 8.7, a result attributable to the lowering of temperatures, which reduces both water stress and respiration (Models A and B, Figs. 5.8 and 5.9) and hence augments the available water and food supplies. In the three sites where growth was reduced, it was hypothesized that the lowering of temperatures caused a reduction in physiological processes that directly govern ring-width (Model C, Fig. 5.10).

The effect of both a 5% decrease and 5% increase in precipitation (solid line and dashed line) indicates that the specified changes in precipitation would have a much smaller effect than temperature on growth. A decrease in precipitation consistently reduces growth. The differences in the shapes of the curves are due to varying effects of prior growth.

These results differed from computations for other forest sites using simulation models of productivity developed for certain forest biomes (Cooper *et al.*, 1974). The simulation of some models was based on measurements of actual plant processes on cool and moist sites. Consequently, those models predicted a reduction in growth due to the lowering of temperatures, while the response functions derived from more arid-site trees indicated that the detrimental effects of water stress due to high temperatures would be most limiting to growth.

o

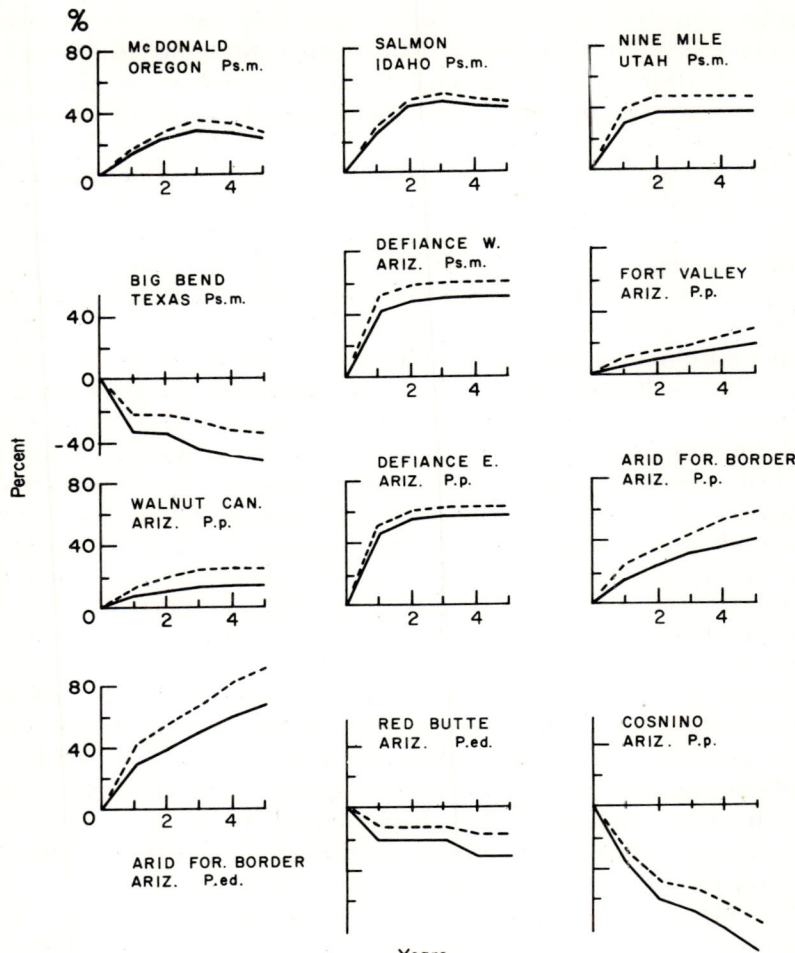

FIG. 8.7. The calculated ring-width response to the most extreme climatic changes suggested to result from operation of a fleet of supersonic transport aircraft, shown for 12 sites. Calculations included a 2°C decrease in temperatures and a 5% increase (dashed lines) or decrease (solid lines) in precipitation for each month throughout the year, which were multiplied with the appropriate element of the response function for sites south of 40 N. (For sites north of 40°N, the specified change in temperature was 3°C and in precipitation ± 10%.) The calculations were carried out for a succession of five years and the estimated growth applied to the elements of the response function for prior growth at lags up to three years. Ordinate values are in terms of percentage of present normal ring width at each site. Response functions and site data are published from Fritts (1974). Ps.m., P.p., and P.ed. indicate species *Pseudotsuga menziesii, Pinus ponderosa,* and *Pinus edulis* respectively.

The response function analysis described by Fritts (1974) demonstrates the existence of a wide variety of climatic factors that can affect ring width. This suggests that more physiological studies on extreme sites may reveal new growth–environmental relationships not encountered in more optimal environments.

Since the response functions can be used to ascertain the effects of climatic variations on growth over a wide area and a variety of sites, they may be used to test the appropriateness of extrapolating information gathered from areas intensively studied for ecosystem analysis to other locations, especially to those representing stress sites near the natural limits of the forest.

VI. Climatic Reconstruction and Verification

The primary goal of dendroclimatology is the reconstruction of past climate. A typical approach to reconstruction has been to identify what climatic parameters correspond to particular variations in ring width, to examine the ring-width record for anomalous variations in growth, and then to deduce the appropriate variations in the climatic parameters from the anomalies in growth (Schulman, 1956; Sirén, 1963).

Figure 8.8 is an example of such a study in which past variations in growth at Mesa Verde, Colorado, were used to infer variations in past climate (Fritts, *et al.*, 1965c). The study was made to test a hypothesis which asserted that a so-called "great drought" apparent from tree rings for A.D. 1276–1289 was of sufficient severity to force abandonment of the cliff dwellings at about the same time.

Ring widths in living trees of *Pseudotsuga menziesii* were examined and were found to be directly related to variations in precipitation and, to a limited extent, inversely related to variations in temperature for a 13-month period extending from June of one year through June of the next (see Table 7.VI). The climate for particular monthly intervals that were significantly related to growth are listed as follows in order of decreasing importance: (1) precipitation and temperature for March through May, (2) precipitation for October and November, (3) precipitation for August and September, (4) temperature for the June concurrent with growth, (5) precipitation for December through February, and (6) temperature and precipitation for June and July of the year prior to growth (Fritts *et al.*, 1965c). Fritts (1974) confirmed these generalities by calculating the response function for the same data and sites.

There was also a direct relationship between yearly ring-width index values and the index for the ring formed three years earlier. The standardized ring-width chronology was corrected for this third-order autocorrelation by subtracting the normalized value for prior growth after it had been multiplied by the appropriate regression coefficient and the result denormalized. A five-year moving average of the corrected indices was plotted for every even-numbered year (Fig. 8.8). The five-year average was chosen to bring out long-term growth variations

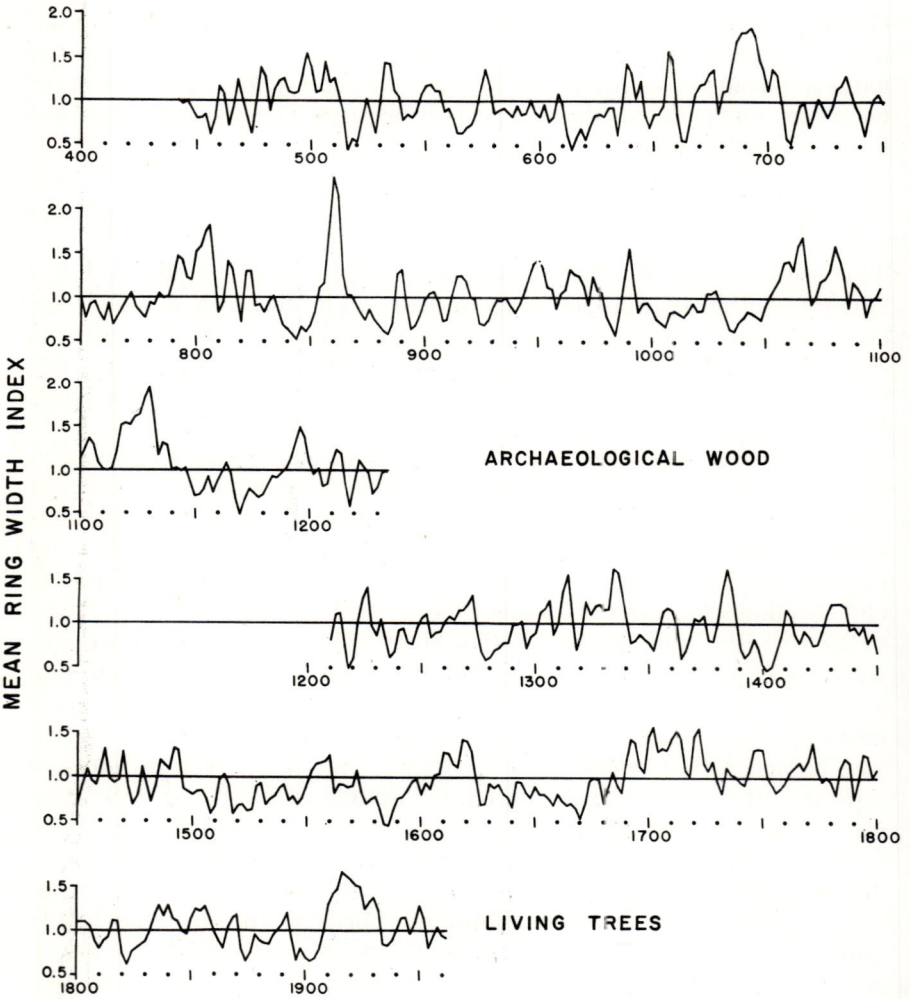

FIG. 8.8. Five-year running means of ring-width indices from *Pseudotsuga menziesii* at Mesa Verde, Colorado, corrected for autocorrelation and plotted on every even year from A.D. 442 through 1962. (Modified from Fritts *et al.*, 1965c)

expressing climatic conditions thought to be detrimental to agriculture and food reserves of the mesa inhabitants. An archaeological ring-width chronology obtained from the dwelling timbers was similarly treated to extend the record back in time.

It was inferred from the calibration that intervals of low growth were also intervals of low precipitation and high temperature, emphasizing the months in order of importance listed above. Fritts *et al.* (1965c) noted the interval A.D. 1273–1285 was characterized by low growth and could be inferred to be dry. However, they also noted that the departures in growth from a mean value of 1 for that particular time appeared to be surpassed between A.D. 512 and 1673 by 10 other intervals characterized by low growth. Therefore, it appeared to these workers that the so-called "great drought" was not as severe as droughts at other times and was probably only one of several factors in a chain of events that led to the disappearance of the prehistoric people from Mesa Verde at that particular time. The only verifications of these reconstructions are similar variations in ring-width chronologies of other sites and other species throughout Mesa Verde and in neighboring areas.

It was also noted that no period of low growth appears to have occurred after 1673 that was as low as the growth between 1276 and 1289. The growth of 1895 to 1906 was low, but the interval was shorter and the indices were larger than the corresponding values during the "great drought" of 1276–1289. Therefore, drought was inferred to be less severe after 1673, which is consistent with other notions about the onset of the cool moist period (sometimes called the Little Ice Age) that began in the 1300's and brought cool moist conditions in southwestern United States early in the 17th century (Fritts, 1965). Ring-width indices for the early 1900's at Mesa Verde suggest the climate early in the 20th century was unusually wet and cool, although in the last few decades the ring-width indices indicate that more average conditions have prevailed.

VII. Inferences from Chronologies with Different Growth Responses

Information obtained from a regression analysis or from a response function applied to a single chronology provides only a generalized estimate of variations in past climate since the response is integrated over one or more years. It is difficult to determine the season of occurrence of a particular climatic anomaly associated with variations in growth and whether the anomaly resulted from precipitation deficits, excessively high temperatures, or both. However, it was suggested in Section IV that the response and transfer function analyses can provide a solution to this problem, if ring-width chronologies are available from two or more sites or species which respond in different ways to the same conditions of climate. The following study of LaMarche (1974a) applies this principle by using two ring-width chronologies in the same species and area which respond differently to anomalies in temperature and precipitation in the respective tree sites.

The two ring-width chronologies were developed from *Pinus longaeva* in eastern California. One was an arid-site chronology from the forest border derived by Edmund Schulman and utilized by Fritts (1965). The other was a 5405-year ring-width chronology which was developed at the upper elevational limits of the species by LaMarche and Harlan (1973). Response functions of ring-width variation indicate that these trees were responding in a direct manner to variations in temperature for the spring, summer, and autumn seasons. Moisture was less often limiting, though narrow rings were sometimes associated with extremely dry years. LaMarche (1974a) also found that the ring-width record in the high-elevation chronology was consistent with trends in temperature during the spring, summer, and autumn at both neighboring and distant localities where weather records were available.

However, response function analysis indicated ring-width variations in drought-sensitive *Pinus longaeva* in the same area but at lower elevation were directly related to precipitation and inversely related to temperature (Fritts, 1969, 1974) (Fig. 8.1). Not only do the chronologies at the upper and lower elevational limits differ in their response to particular climatic factors, but the record from the high-elevation chronology was inferred by LaMarche (1974a) to average the temperatures over a period of 15 or more years. As a result, the ring-width chronology for the high-elevation trees expressed more low-frequency variance than the chronology from the more arid-site trees at lower elevations. The first-order autocorrelation in the former ranged between 0·6 and 0·8, but in the latter it ranged between 0·0 and 0·3. The large differences in the amounts of variance at different frequencies in the two time series, along with the difference in lag of from 1 to 15 years between the climatic effect and the tree-growth results, prohibited a straightforward calibration using transfer function analysis (see Chapter 6). Therefore, LaMarche elected to simply plot 20-year means of ring widths from the high- and low-elevation trees and to infer the appropriate climatic anomaly from the observed departures in growth.

The mean ring-width departures for the two sites and the inferred climatic anomalies are shown in Fig. 8.9. Specific departures in past growth which sometimes persisted for more than 300 years were inferred to be long periods of anomalous temperature and precipitation, which

8. INTERPRETATION OF CLIMATIC CALIBRATIONS

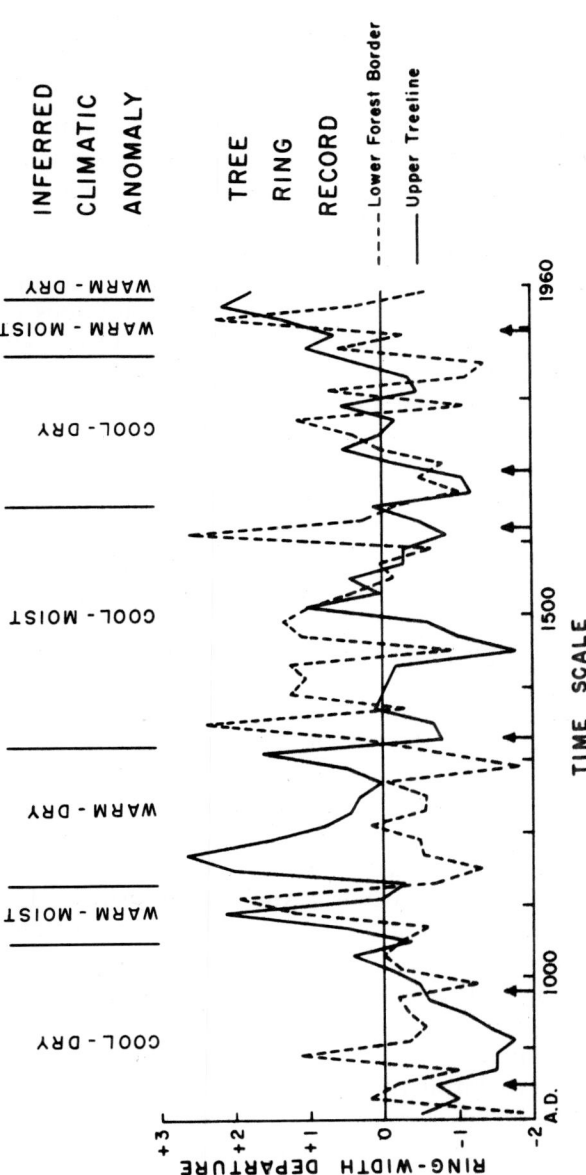

FIG. 8.9. The 20-year average growth, expressed in standard normal values, in *Pinus longaeva* on lower forest border and upper tree line sites of the White Mountains of California, and the precipitation and temperature anomalies inferred from the departures in ring width. Arrows show dates of glacial moraines in the nearby mountains. (From LaMarche, V. C., Jr. (1974). *Science* **183**(4129), 1043–1048. © 1974 by the American Association for the Advancement of Science)

were punctuated by shorter episodes showing climatic variations of a different kind.

LaMarche further linked the inferred temperature–precipitation anomalies to features of the general circulation. This was accomplished by examining the existing meteorological records for the area, identifying the months and years when the anomalies in moisture or temperature had been especially marked, and examining maps of the corresponding heights of the 700 mb pressure surface (about the same as the atmospheric pressure at 3 km) over western North America (Fig. 8.10).

It was noted that the most common anomalies, especially during spring and autumn in recent decades, have been warm–dry and cool–moist situations (Maps C and A, Fig. 8.10). The warm–dry conditions were associated with a low pressure *trough* over the North Pacific Ocean and a high pressure *ridge* over the Rocky Mountains which brought warm dry continental air into California. Cool–moist and cool–dry anomalies (the latter were less frequent in the modern period) were associated with an upper-level trough which caused cold moist air to flow from the ocean or cold dry air to flow southward off the North American continent (see Maps A and B, Fig. 8.10). The warm–moist conditions occurred in summer and involved anomalous extension of what is referred to by meteorologists as the upper level Bermuda High, bringing moisture and warmth from the east into California during July.

LaMarche inferred from the past ring-width record that not only were temperatures often lower in the past at the tree sites but that they were the result of a more frequent occurrence of an upper level trough over western North America than occurred during the modern period. Such an inference assumes the principle of uniformitarianism in that the climatic anomalies in the past are assumed to have been similar to those of the present, except in frequencies of occurrence.

The inferences of LaMarche seemed to be verified by their consistency with the past circulation over western North America suggested by Lamb (1969a), who used different kinds of evidence for his reconstructions. It was noted by LaMarche, however, that while the general circulation over western Europe and elsewhere in western North America was similar to that over California, there were periods of time when the records of climatic variations over the two continents differed.

The 100-year mean growth at the upper tree line plotted for each century since 3000 B.C. is shown in Fig. 1.15 in Chapter 1 (Gates and Mintz, 1975), along with data on expansion and contraction of glaciers primarily in western North America and northern Europe as described by Denton and Karlén (1973). Expansion of glaciers which occur in the same general regions of the trees are clearly associated with periods of low

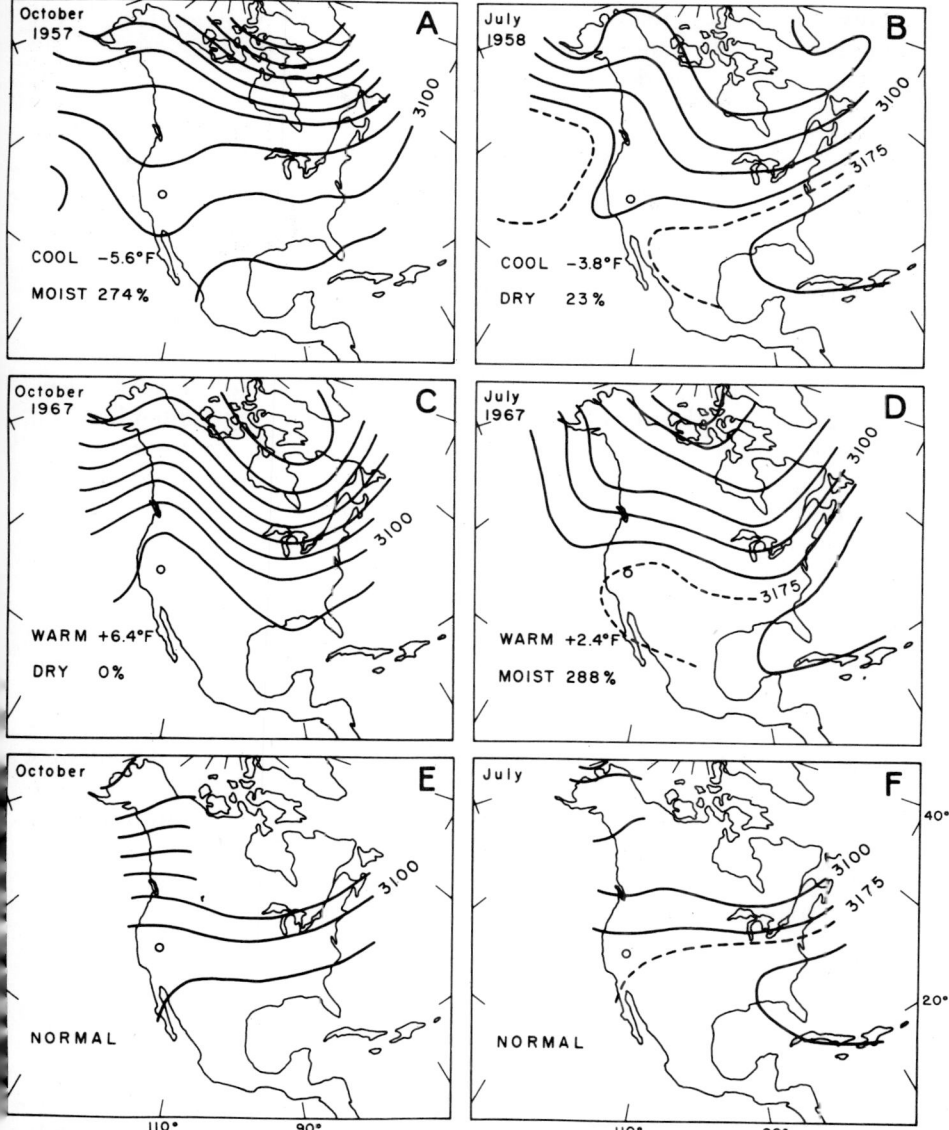

FIG. 8.10. Examples of 700 mb pressure height (about the same as the atmospheric pressure at 3 km) maps for selected months associated with large precipitation and temperature anomalies in the White Mountains of California. Different frequencies of these or similar circulation patterns in the past were offered as explanation of the climatic anomalies inferred from intersite comparisons of ring widths (Fig. 8.9). (From LaMarche, V. C., Jr., 1974, *Science* **183**(4129):1043–1048. © 1974 by the American Association for the Advancement of Science)

growth, which are inferred to be cool. The glacier data, though not dated on as fine a time scale, serve as a kind of independent verification of the climatic anomalies reconstructed from the tree rings.

VIII. Reconstruction Using Multivariate Transfer Functions

A. *Climatic Variations in Arid Regions affecting Streamflow*

Stockton (1971a 1975) and Stockton and Fritts (1971a) applied the techniques of dendroclimatology to the reconstruction of past streamflow from river basins in semiarid western North America. A model is used which assumes that the amounts of annual streamflow for the water year, the twelve months from October through the following September, is a function of precipitation, temperature, and evapotranspiration as they vary throughout the river basin and from season to season. Since ring widths of trees in the watershed vary in response to the same climatic parameters, direct calibration of ring widths with variations in streamflow is obtained, using modern hydrological measurements to establish the correlation between the two. The calibration is then used to estimate the streamflow for the years in the past when no hydrologic records existed, using long tree-ring chronologies from the river basin.

Methods were first tested on two watersheds of different size, one in northcentral Arizona and one in westcentral New Mexico. Several tree-ring samples were obtained from *Pseudotsuga menziesii* on different sites within each watershed. For such studies, careful statistical analyses were made, and response functions for temperature and precipitation data were obtained for streamflow as well as for ring width.

For example, in one watershed both streamflow and tree growth were found to be directly correlated with precipitation during all months of the water year. However, the response functions for tree growth often emphasized the moisture of late autumn and winter, while the response functions for streamflow emphasized precipitation in April and July. Also, high temperature extremes during the months of December, April, and May apparently promoted high streamflow but were limiting to tree growth in the studied site. This is explained by the effect of higher than normal temperatures during the three months mentioned above which favor snow melt and streamflow but adversely affect the physiological conditions in the trees that in turn cause lower growth. Differences were also noted among the response functions for the trees in different sites, and it was thought that a multivariate transfer function was appropriate to reconstruct streamflow.

A power spectrum analysis (see Chapter 6) of differences in data in regard to frequency revealed that there was more low-frequency variance in the tree-ring data than in the records of streamflow. This discrepancy, which was attributed to differences in the statistical generating processes of the two types of data, suggests that the best reconstruction of streamflow may require tree growth lagged for a number of years.

Statistical models listed in Table 8.II describe the ways tree rings were matched with streamflow for seven calibration attempts and includes the percentage of streamflow variance accounted for in each case. The establishment of the coefficients associated with the predictors of each equation involved the following procedure: (1) extraction of the eigenvectors of tree growth (Equation 7.18), (2) regression analysis using the eigenvector amplitudes of growth and sometimes actual values of prior streamflow to statistically estimate streamflow, and (3) conversion of the coefficients for the amplitudes to the transfer function (Equation 7.21), the elements of which become weights identified with the coordinates of the original tree growth (Stockton, 1975).

For hydrological reasons Models 1, 2, and 3 (Table 8.II) were constructed to utilize values of both prior streamflow and tree growth to predict streamflow. One difficulty with such a model is that only estimates of the prior streamflow values, not the actual data, are available for years before the calibration period. This introduces an additional error factor into the prediction equation. However, the reconstructions using Models 4 through 7, which did not possess the additional error,

TABLE 8.II Various models using different sets of lags for reconstructing annual runoff of the upper San Francisco River from ring-width chronologies in the watershed and the percent variance reduced[a]

Predictand	Variables of model Ring-width predictors	Prior runoff predictors	% Variance reduced
1. Runoff (f_t)[b]	$x_{t-3}, x_{t-2}, x_{t-1}, x_t$[c]	with $f_{t-3} + f_{t-2} + f_{t-1}$[d]	71.5
2. Runoff (f_t)	$x_{t-2}, x_{t-1}, x_t, x_{t+1}$	with $f_{t-2} + f_{t-1}$	64.0
3. Runoff (f_t)	$x_{t-1}, x_t, x_{t+1}, x_{t+2}$	with f_{t-1}	72.2
4. Runoff (f_t)	$x_{t-3}, x_{t-2}, x_{t-1}, x_t$		60.5
5. Runoff (f_t)	$x_{t-2}, x_{t-1}, x_t, x_{t+1}$		55.1
6. Runoff (f_t)	$x_{t-1}, x_t, x_{t+1}, x_{t+2}$		72.2
7. Runoff (f_t)	$x_t, x_{t+1}, x_{t+2}, x_{t+3}$		66.0

[a] From Stockton (1971) and Stockton (1975). [b] Runoff for water year t. [c] Ring-width indices for 3 years prior to year t through year t. [d] Runoff for water year 3 years prior to year t through the year prior to t.

were shown by power spectrum analysis to exhibit a different time-series structure than the actual streamflow records. Thus it was concluded by Stockton (1975) that Model 1 is superior to Models 3 and 6, even though the reconstructions using the latter two models have an additional source of error. In each case of reconstruction, the streamflow record from a nearby gauging station was used to verify the reconstructions.

Stockton (1975) applied this same modeling and calibration procedure to reconstructing streamflow from the Colorado River Basin, which covers 283,605 km². Seventeen tree sites within or near the watershed were used, and the streamflow record for Lee Ferry, Arizona, adjusted for diversion of water from 1896 to 1961, is calibrated with the growth record of the 17 tree sites. The same models shown in Table 8.II were used, and Model 5 was shown to yield the best reconstruction, with a calibrated streamflow variance of 82% for the dependent data set (see Fig. 7.2). The transfer function from the calibration was applied using the 17 ring-width chronologies to reconstruct streamflow as far back in time as all ring-width records are complete. These reconstructions have already been presented in the first chapter (Fig. 1.16).

The inclusion of the growth indices of the two prior years (x_{t-1} and x_{t-2}) (See Model 5, Table 8.II) appears to adjust for differences between the generating processes of the trees and the generating processes of streamflow that produce the respective time series (see Chapter 6). The indices for one year following (x_{t+1}) account for a lag of one year between the climatic input and the resulting streamflow, as well as for lags of tree growth behind the occurrence of climate.

Stockton (1975) compares the power spectra of (1) the actual streamflow record, 1896–1961, (2) the reconstructed streamflow record 1896–1961, and (3) the entire reconstructed streamflow record 1564–1961. He concludes that the frequency distribution of the reconstructions resemble those of the actual streamflow sufficiently to accept model 5 as the best.

The reconstructed long-term mean for the flow of the Colorado River for 1569–1961 was 13·0 million acre feet, which was over 2 million acre feet less than the mean of the gauged record for 1896–1961 of 15·03 million acre feet. The estimated variance for the longer reconstructed record, which was corrected for variance lost in regression, was 17·01. This was less than the corresponding figure of 18·42 for the variance of the actual streamflow during period 1896–1961.

Later work of Stockton and Jacoby (in press) and Stockton (1976) examined differences in ways the hydrological data were adjusted and attempted calibrations using different time periods. Since they obtained estimated long-term means as high as 14·0 million acre feet, they argued that a reasonable long-term estimate of the mean annual flow for the Colorado River is 13·5 million acre feet which is an estimate with a ±0·5 million acre feet error.

Stockton points out that these statistics, representing the longer reconstructed period, will improve estimates of expected water yield because they are derived from a larger sample. Furthermore, he notes that the reconstructions (Fig. 1.16) indicate that droughts were more common in the past than during the period of record after 1896. Also, since 1569, the wetness of the period 1905–1930 was matched in extent only by one other period, 1601–1621. Stockton recommends that the corrected statistics using the longer record from tree rings should be used to manage Lake Powell, a newly created reservoir on the Colorado River, because they represent a record six times as long as the actual gauged record and they have a higher probability of being closer to the actual flow of future years.

Stockton and Jacoby (in press) verified their reconstructions by means of independent streamflow data estimated from an upstream station for the interval 1896–1914. The trends in the two data sets were generally in agreement except for a few years of high flow. They conclude that the overall reconstructions from tree rings appear unbiased in that the number of times they exceed the values estimated from the upstream flow are the same as the number of times they are less.

B. *Climatic Variations and Hydrologic Conditions in Subpolar Regions*

Tree-ring analyses of hydrologic situations are possible in subpolar as well as arid regions. Stockton and Fritts (1973) used the rings from *Picea glauca* growing on the natural levees of the Peace–Athabasca Delta system along the western border of Alberta, Canada, to reconstruct past variations in the levels of Lake Athabasca.

A dam obstructing river flow in the headwaters of the Peace River appeared to cause a permanent drop in the water levels of the lake and the nearby delta system, both of which are downstream from the dam site. High water levels in the past had resulted in flooding and mineral enrichment of soils within the delta region. With the apparent drop in maximum water levels and a decrease in flooding, scientists feared that the ecology of the delta might be permanently altered. While measurements back to 1935 recorded relatively high lake levels and associated flooding, fragmentary evidence existed suggesting that these records may not have been representative of the water levels and flooding history occurring over a longer period of time. Reconstruction of a longer record was thought necessary to make the best decisions on future management of the delta system.

Although the trees on the levees in the area were near neither their arid nor cold limits, where climatic factors would limit growth, it was thought that local conditions of soil moisture and flooding associated with the levels in the waterways may have been recorded by the growing trees. Replicated cores were extracted from each of 10 or more trees from six sites distributed within the delta system (Fig. 8.11).

At first only one site was sampled (site 1), and the rings were examined for variability in width and for crossdating to ascertain whether a

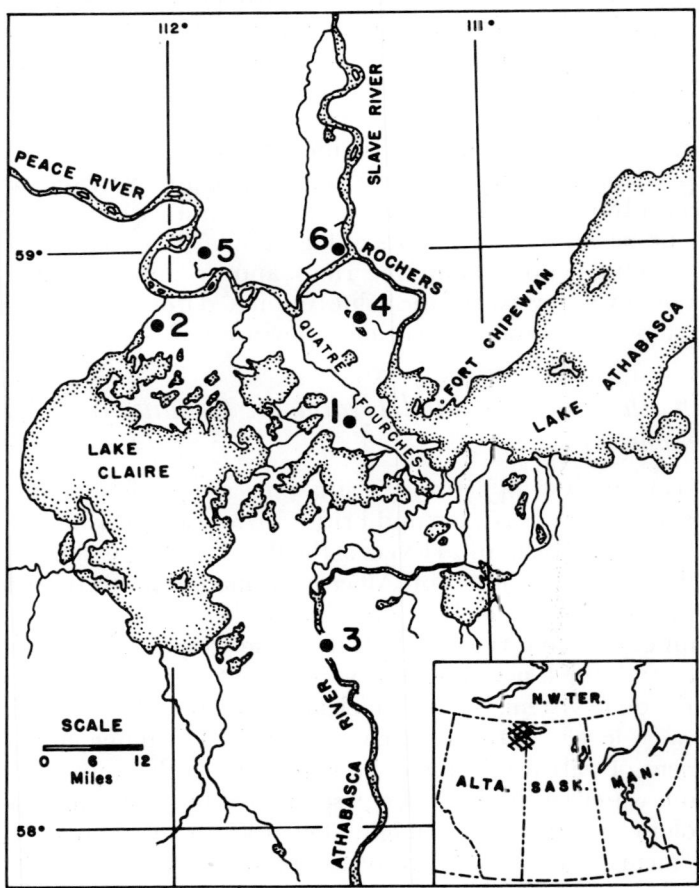

FIG. 8.11. Location of six sites of *Picea glauca* sampled throughout the Peace–Athabasca Delta system to reconstruct past levels of Lake Athabasca. (From Stockton and Fritts, 1973, courtesy of the American Water Resources Association)

regional factor such as flooding was affecting growth. Both variability and crossdating were evident, and 54% of the ring-width variance was common to all sampled trees of the site. The percentage of variance remaining in the group chronology was higher than had been found for many trees on mesic sites to the south, but it was somewhat lower than that for trees in arid forests. Also, it appeared that the ring-width chronology was correlated to some extent with the available lake level records.

With these promising results, Stockton returned to the delta region and sampled five additional sites (Fig. 8.11). Some sites were chosen to represent different drainage classes and others were chosen because of their strategic location as potential monitors of the water levels in the inlets and outlets to the delta system.

It was noted that *Picea glauca* grew most vigorously on the well-drained sites at the top of levees. When a stream meander is cut off, water levels become higher, soil aeration apparently declines, there appears to be a deterioration of *Picea glauca* growth, and species typical of wetter habitats begin to invade the forest. A model was created in which it was assumed that growth at each site during the year of flooding, t, and during the following year, $t+1$, may be correlated with the water level in the lake. The further assumption was made that growth at each site may be either directly or inversely related to the water levels, depending upon the drainage characteristics at the soil, the proximity to a waterway, and the timing of the water level fluctuations within each year.

The ring widths in each of the sites were crossdated and standardized as in Equations 6.7 and 6.8, and the principal components (eigenvectors) were extracted from the six ring-width chronologies. The most important eigenvector reduced 61·2% of the variance (Fig. 8.12), and all of its elements corresponding to the six sites have large negative weights. The second eigenvector reduced 13·6% of the variance, exhibiting positive weights for sites 2, 5, and 6, which are the three sites most closely associated with the Peace River, and negative weights for the other sites, which are more closely related to the Athabasca River (Figs 8.11 and 8.12). The third eigenvector reduced 10% of the variance and included a positive weight for site 3 and a negative weight for site 4, with weights near zero for the other four sites. This analysis indicates a 61% similarity in variance of all six tree sites (eigenvector 1), but suggests that two important differences exist among the tree sites associated with 13·6% and 10·0% of the tree growth variance, respectively.

The dependent variables chosen for this analysis were the mean water levels in Lake Athabasca for three 10-day intervals which provide an index to the amount and timing of flooding in the delta system. The first is

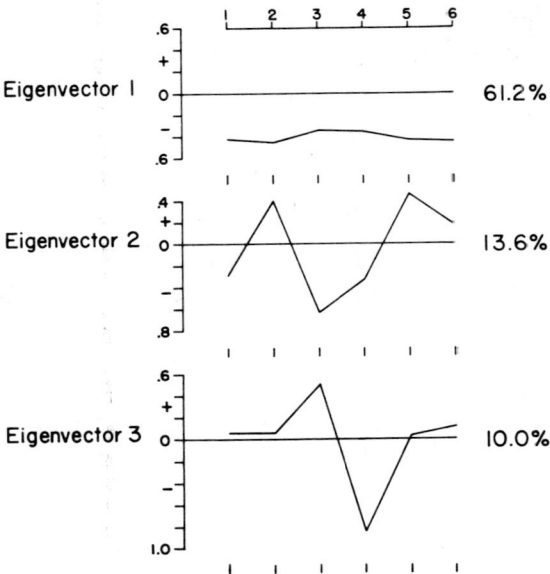

FIG. 8.12. The first three eigenvectors of the six ring-width chronologies from the Peace–Athabasca Delta area sites located in Fig. 8.11. (From Stockton and Fritts, 1973, courtesy of the American Water Resources Association)

the measured height of water during May 21–30 at about the time growth begins, the last, September 21–30, represents conditions after growth had ceased, and the middle period is July 11–20. The statistical model used is:

$$\hat{Y}_t = b_0 + b_1 X_{1,t} + b_2 X_{1,t+1} + b_3 X_{2,t} + b_4 X_{2,t+1} + \cdots$$
$$b_{11} X_{6,t} + b_{12} X_{6,t+1} \qquad (8.1)$$

where $\hat{Y}$ is the predicted lake level during one of the three 10-day periods at year t; $X_{i,t}$ and $X_{i,t+1}$ are the mean ring-width indices for site 1 for years t and $t+1$; and b_0 is a constant adjusting for differences in variable means. Actually, canonical regression was used (Glahn, 1968), but all canonical weights were retained. The results are the same as if three multiple regressions had been run for each of the three 10-day periods, with 12 predictor variables representing ring width.

The transfer function weights are plotted in Fig. 8.13. Site 1 is poorly drained and the weights indicate that the most pronounced effect of higher than average water level is a reduction in growth during year $t+1$. Sites 2 and 3 include trees on better-drained soils, so the weights are more often positive as high rather than low water levels are associated with high growth. At site 2, higher than average water levels in May have a pronounced and immediate effect on growth, but if the high water occurs later in the season, the effect appears in the next year's growth $t+1$. Site 4 is poorly drained and the largest response to high water is inverse, apparently due to poor aeration, and the response occurs in the first-year growth.

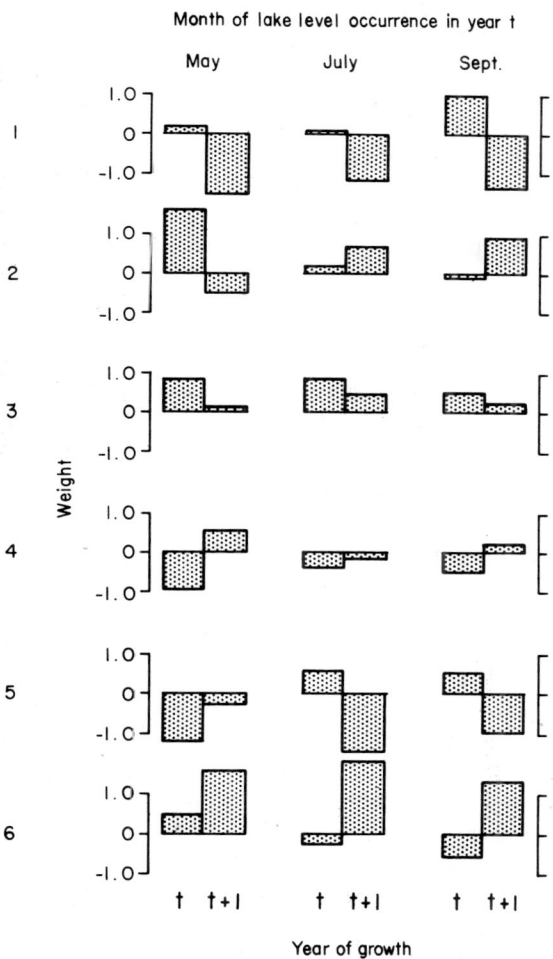

FIG. 8.13. Plots of all the regression coefficients of the transfer function used to reconstruct levels of Lake Athabasca for three different 10-day intervals using the variations in ring widths from six different tree sites. Positive values designate direct relationships between the growth and lake level and negative values designate inverse relationships. (Redrawn from Stockton and Fritts, 1973)

The elements of the transfer functions for sites 5 and 6 have opposite signs, apparently for hydrologic reasons. The sites are located along the Peace River (Fig. 8.11), where the former site is approximately 25 miles (40 km) above a major outlet (Rochers) from Lake Athabasca, while the latter is at the confluence of the Peace River and the outlet. High growth at the two sites is associated with high water levels in the nearby river. However, when the trees at the confluence (site 6) exhibit higher growth than site 5 upstream, a proportionately higher than average flow is indicated from the outlet than from the Peace River, reconstructing high water levels in the lake. When growth is higher for trees at site 5 than at site 6, the relative flow in the Peace River is greater than that of the outlet, reconstructing levels which were probably lower in the lake.

The lake levels that are reconstructed from tree rings for years 1810 through 1967 have been included in Chapter 1 (Fig. 1.17). Stockton and Fritts infer from the extended lake-level record that the variance of the May 21–30 lake levels for the 1935–67 interval is only one-third the variance for the longer record; the variance of lake levels for July 11–20 during the 1935–67 interval is one-half that of the longer record; and the variance for September 21–30 is similar to that of the longer record. The mean levels for the longest available record 1810–1967 do not appear to be different from the 1935–1967 measurements. The tree rings indicate that the long-term natural variability of early season levels of Lake Athabasca is greater than one would expect from the 1935–1967 measurements. Extremely wet and extremely dry years occurred more frequently during the 1810–1934 interval than during the more recent period.

Management of the Peace–Athabasca Delta system now incorporates the construction of a temporary earthen dam across one of the outlets every few years, artificially creating a high lake level and an associated flooding of the neighboring terrain until the impounded high water washes out the dam. This management procedure has resulted because it is believed that the occasional high water shown by analysis of tree rings is all that is necessary to maintain the delta system in its present ecological state.

Quantitative verification was not possible because of the shortness of the calibration record. However, the reconstructions were verified to a limited extent by historical references, direct observations, and botanical evidence. A dendrochronological method of verification using the methods of Alestalo (1971) might have been to return to the area after the reconstructions were made and to date leaning trees on the wave-cut terraces to see if the dates correspond to the high-water years indicated by the above dendrohydrological analysis (Alestalo, 1971).

C. Climatic Variations and Studies of Sea Surface Temperatures

Surface water temperatures in the North Pacific Ocean influence the storms that enter various regions over the west coast of North America, making possible the calibration of tree growth (as representative of climate) with sea surface temperatures. Mild conditions occur in the extreme west of North America and cold dry conditions occur in the East when the central North Pacific is anomalously warm and the eastern North Pacific and northern Gulf of Alaska are anomalously cool. Wet winters develop when the portion of the North Pacific between Hawaii and California is anomalously warm (Namias, 1969, 1971, 1973).

Since anomalous variations in ring widths of trees along the Pacific Coast are a result of anomalous variations in precipitation and temperature at those sites and since these factors result from anomalies in the movement of storms over the North Pacific, variations in ring widths from both coastal and inland sites can be modeled as sources of information on past anomalous variations in temperatures of the ocean surface. Douglas (1973) tested the possibility by analyzing five standardized ring-width chronologies from sites in southern California along with features of sea surface temperatures at three coastal stations in California (Port Hueneme, Balboa, and La Jolla). He found that the monthly data in sea surface temperatures could be divided into three natural seasons, where winter is November through February, spring is March through June, and summer is July through October. Autumn was not defined, because it was thought to represent only a short period bridging the warm temperatures of summer and the cold temperatures of winter.

Douglas examined several statistical models relating tree growth and climate and concluded that the best reconstructions were obtained for sea surface temperatures by using the following equation.

$$\hat{Y}_t = b_0 + b_1 X_{1,t-1} + b_2 X_{1,t} + b_3 X_{1,t+1} + b_4 X_{1,t+2} + \cdots$$
$$b_{17} X_{5,t-1} + b_{18} X_{5,t} + b_{19} X_{5,t+1} + b_{20} X_{5,t+2} \qquad (8.2)$$

where $\hat{Y}_t$ is the estimated sea surface temperature for a station and season in year t; the b's are regression coefficients associated with tree-ring indices from sites 1 through 5 for years $t-1$, t, $t+1$, and $t+2$; and b_0 is a constant. A stepwise multiple regression was used, and only those coefficients with an F ratio greater than 2 were entered into the regression equation.

The results of the calibration using these 20 possible predictors from five sites are shown as model 1 in Table 8.III, and they include the

TABLE 8.III Several statistics for equations using 3 kinds of models for reconstructing sea surface temperatures at 3 coastal locations in California for 3 seasons[a]

Model number	1			2			3		
Number of predictors	20			12			10		
Kind of predictor	Indices for 5 sites[b]			3 Eigenvector amplitudes lagged[c]			10 Eigenvector amplitudes of indices lagged[d]		
Station and season	$\%\ s_x^2$[e]	F ratio	No. of variables	$\%\ s_x^2$[e]	F ratio	No. of variables	$\%\ s_x^2$[e]	F ratio	No. of variables
Port Hueneme									
Winter	45	6.3	4	27	6.1	2	42	5.6	4
Spring	61	7.7	6	26	3.8	3	52	6.4	4
Summer	53	5.4	6	33	5.3	3	14	4.4	1
Balboa									
Winter	24	9.2	1	23	4.0	2	47	6.0	3
Spring	52	9.9	3	67	6.5	7	70	7.3	7
Summer	63	7.5	5	56	7.9	4	34	4.3	3
La Jolla									
Winter	23	4.1	3	11	2.6	2	37	3.5	4
Spring	41	5.4	5	15	3.6	2	67	7.3	6
Summer	46	6.7	5	46	6.7	5	40	4.1	4

[a] From work of Douglas (1973), and unpublished results. [b] Indices from 5 sites were lagged for 4 years from $t-1$ to $t+2$, making a total of 20 possible predictor variables (Equation 8.2). [c] The 3 most important eigenvector amplitudes calculated for the 5 chronologies for year t and the amplitudes lagged as in Equation 8.2. [d] The 10 most important eigenvector amplitudes of data used in model 1. [e] Percent calibrated variance.

percent variance calibrated, the F ratio for the total regression, and the number of tree-growth variables selected in the stepwise process. The reconstructions for this model which calibrated the largest percentage of the variance are for sea surface temperature of the spring and summer seasons. In four out of six cases the percent calibrated variance exceeds 50%.

The ring-width indices were substituted into the appropriate regression equation and estimates of sea surface temperatures were obtained for A.D. 1611–1964 (Fig. 8.14). Douglas attempted two additional reconstructions of sea surface temperatures using eigenvector amplitudes of tree growth. In one analysis (Model 2, Table 8.III) eigenvectors were extracted for the five chronologies of tree growth. The first three eigenvectors which reduce 87% of the variance in tree growth were selected, the amplitudes of the eigenvectors computed, and the amplitudes for year $t-1, t, t+1$, and $t+2$ used as predictors of sea surface temperatures (3 eigenvectors and 4 lags making a total of 12 predictors). In the second analysis (Model 3, Table 8.III) the data for the five tree sites and for years $t-1, t, t+1$, and $t+2$ were subjected to eigenvector analysis (5 trees at 4 lags making a total of 20 variables). The first 10 amplitudes of eigenvectors were selected to statistically estimate sea surface temperatures since they reduced 89% of the tree growth variance. No marked improvement in reconstruction was observed, and in most cases the calibrated variance was less and the F ratios lower than when the actual ring-width values were used as predictors.

Douglas attempted several verifications of the multiple regression reconstructions (Table 8.III, Model 1), including a comparison of some early records at San Diego for 1854–1872 with reconstructed temperatures at nearby La Jolla. He found that the winter reconstructions are significantly correlated with these early records, but unfortunately no significant agreement between spring and summer temperatures and those of the early records was observed.

Other independent records suggest that positive anomalies in sea surface temperatures occurred within the California current in 1869, 1871, 1877, 1885, and 1889, while negative anomalies occurred in 1864, 1882, and 1886. The sign of the anomalies is correctly reconstructed by tree rings in seven of the eight years (not in 1871).

Unusual occurrences of numerous warm water organisms off the California coast were reported in the middle of the 19th century, a period that Douglas reconstructs to be particularly warm. Douglas has also compared his reconstructed sea surface temperatures with deposits of Pacific hake fish scales taken from cores of the Santa Barbara Basin. He notes that decades with large year-to-year variability in sea surface

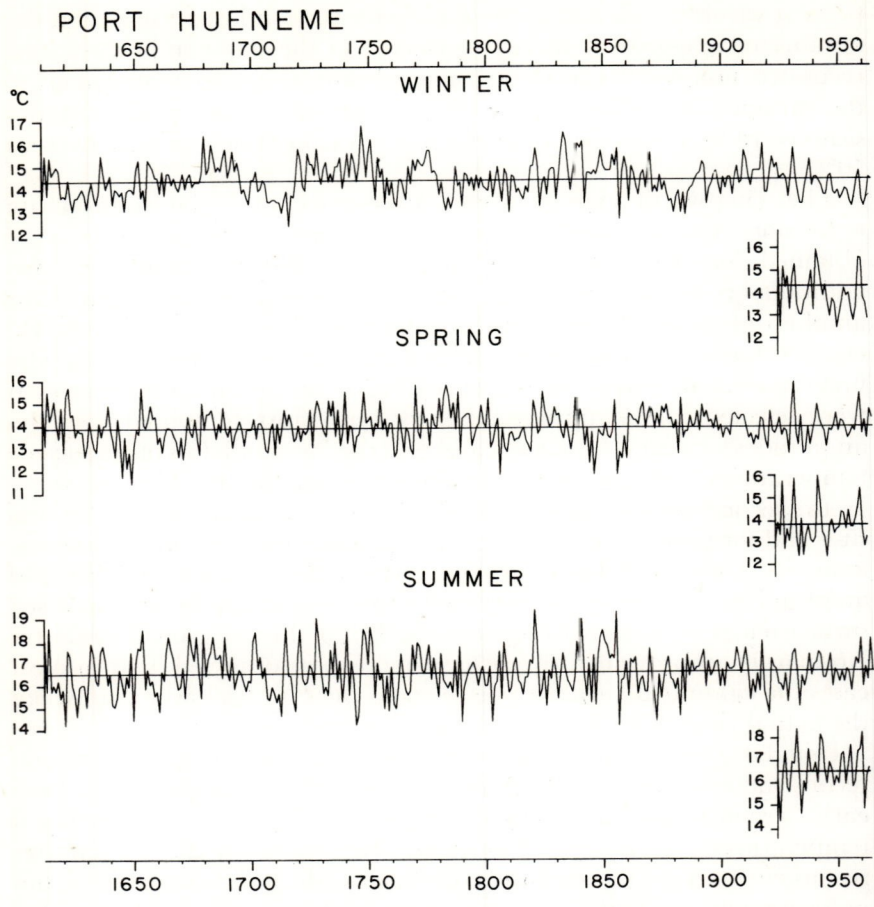

FIG. 8.14. The reconstructed sea surface temperatures for Port Hueneme as estimated from five ring-width chronologies in Southern California. The transfer function consisted of the significant multiple regression coefficients selected from 20 possible ring-width predictors. Predictor variables were the indices for years $t-1, t, t+1$, and $t+2$, where t is the year in which the sea surface temperatures were estimated (see Table 8.III). The actual sea surface temperatures used for calibration are plotted below and to the right of the estimates. (From Douglas, 1973)

temperatures are also decades with large counts of hake scales. The correlation between number of fish scales and the standard deviation of the reconstructed sea surface temperatures for each decade proved to be significant ($r = 0.35$). These results suggest that large fluctuations in sea surface temperatures may have resulted in the death of large numbers of fish whose scales were deposited in the ocean sediments.

Douglas (1973) summarizes the paleoclimatic significance of his results by pointing out that decades of warm sea surface temperatures occurred in the 1610's, 1680's, 1770's, 1830's and 1850's. This condition is hypothesized to be the result of weak atmospheric pressure gradients along the California coast. He suggests that a contrasting temperature distribution prevailed in the decades of the 1620's, 1640's, 1670's, 1760's. and 1880's, during which cold conditions may have resulted from a strong pressure gradient parallel to the California coast.

Douglas attributes these conditions to possible variations in the strength and positioning of the North Pacific high pressure system in summer and in the location of an upper-level low pressure trough off the coast of North America, particularly in late autumn.

Comparing the results using eigenvector amplitudes as predictors with those using actual growth indices (Table 8.III), it can be seen that regression on amplitudes in this case is inferior to the use of the ring-width indices themselves. This appears to be true for cases in which the numbers of variables are not excessive and where correlation among the predictors is not extremely high. However, when there are many highly correlated variables and the numbers of variables approach the numbers of observations, principal component analysis is more appropriate. The number of variables can be reduced to a reasonable size, and the smaller-scale variances representing noise can be eliminated by excluding those principal components which reduce little variance.

D. Climatic Variations Affecting Biological Systems

Since it was shown that ring-width variations can be linked with variations in sea surface temperatures and thereby with hake fish population, it was thought that other sea organisms might be affected by sea surface temperatures and also subject to population estimation and reconstruction backward in time. Clark et al., (1975), investigated this possibility by using tree rings and yearly catch statistics of *Thunnus alalunga* (albacore tuna), a fish which is known to be affected by sea surface temperatures. The percentage of annual catch north of San Francisco, California, during the years 1938–1961, were the only available quantitative measurement of the *Thunnus alalunga* population.

The predictors used were the eigenvector amplitudes of ring width for 49 sites in western North America which are similar to those shown in Fig. 7.15 and described in detail in Chapter 9. A stepwise multiple regression analysis was used to obtain the following equation:

$$\hat{Y} = 35 \cdot 12 + 5 \cdot 76 A_2 + 5 \cdot 13 A_8 + 8 \cdot 56 A'_4 - 6 \cdot 17 A'_9 + 11 \cdot 54 A'_{10} \qquad (8.3)$$

where $\hat{Y}$ is the estimated percentage of albacore tuna and A and A' are eigenvector amplitudes of tree growth corresponding to the catch year and the following year, respectively. The subscripts refer to the eigenvector number. The five variables of the ring-width amplitudes for the dependent period from 1930 to 1961 accounted for 83% of the fish-catch variance, a statistically significant result.

Figure 8.15 shows the annual estimated percentage of albacore catch reconstructed by tree rings, assuming a constant level of fishing technology where the shaded area in the figure represents the dependent data set. The elements of the transfer function associated with the tree growth during the year of the catch (a) and during the year following the catch (b) are calculated and plotted (Fig. 8.16) to check the reasonableness of the calibration relationship. Each map represents the departure in growth associated with higher catch percentages and anomalously high sea surface temperatures north of San Francisco.

Since the tree-ring data were from arid-site trees, high growth could be assumed to indicate anomalously cool, cloudy weather with above normal precipitation, and low growth could be assumed to reflect warm, sunny, and dry conditions. During the fishing season (Fig. 8.16a) tree growth associated with high catch was low in northwestern United States, indicating that fewer than normal storms were entering this area. This would indicate the weather over the ocean also was unusually sunny and mild—warm conditions with high solar radiation which would warm the waters and favor the movement of fish into the region. These warm conditions would then be expected to result in increased evaporation during the following autumn and winter which would lead to increased

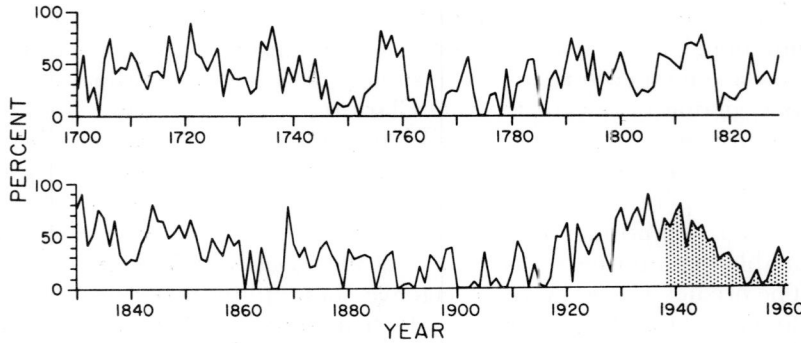

FIG. 8.15. Percent of the total catch of *Thunnus alalunga* (albacore tuna) taken north of San Francisco estimated from the calibration equation applied to ring-width variations for A.D. 1700 to 1961 from western North America. Shading indicates the dependent calibration period. (From Clark *et al.*, 1975, courtesy of *Nature*)

cyclonic activity, increased precipitation along the coast north of San Francisco, and, ultimately, above-average growth in the Northwest during the following spring (Fig. 8.16b). With deflection of storms into the Northwest, fewer storms would be expected to pass through the Great Basin where tree growth was below average. Other growth anomalies were also consistent with the weather systems that might be expected preceding and following a summer of warmer than normal sea surface temperatures and high fish catch north of San Francisco.

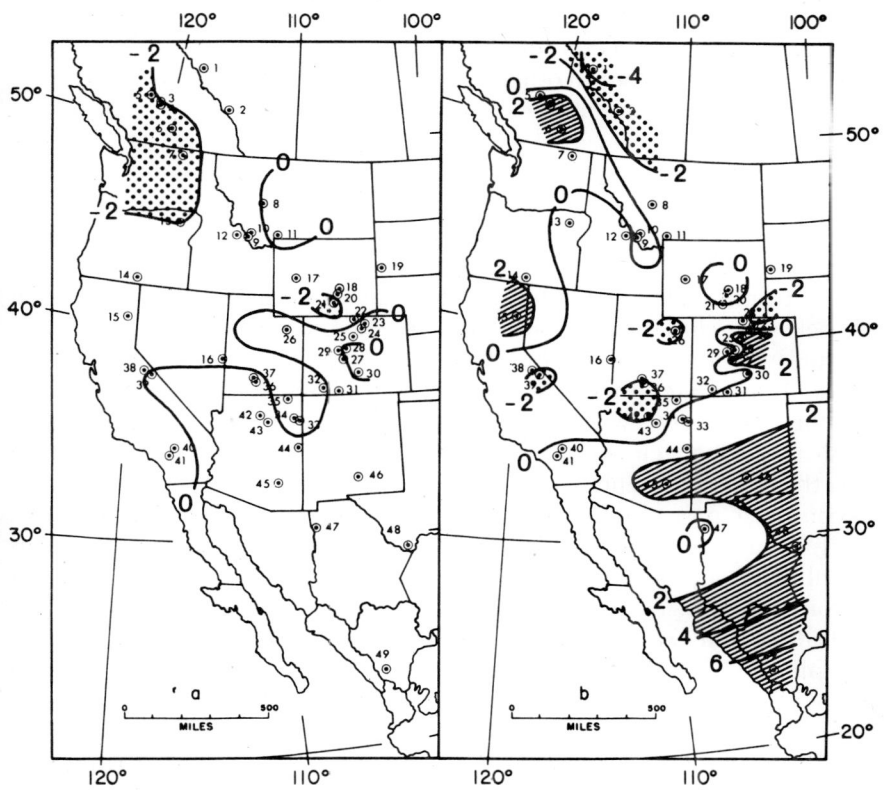

FIG. 8.16. Tree-growth anomaly patterns associated with above normal catch percentages of *Thunnus alalunga* (albacore tuna) by the fishing fleet north of San Francisco. Map *a* shows the growth anomalies during the year of the catch and Map *b* shows the growth anomalies for the year following the catch. Both are obtained as linear combinations of the eigenvectors and then weighted as indicated in Equation 8.3. (From Clark *et al.*, 1975, courtesy of *Nature*)

Qualitative data on fish catch for the independent period prior to 1938 were found to be consistent with the percentages reconstructed from tree rings. These results substantiated that there was a meaningful relationship linking the albacore tuna and ring-width variations. The authors suggested that this catch reconstruction indicates the presence of long-term changes in the ocean system (Fig. 8.15) which are not evident in the shorter record of catch statistics available since 1938.

Clark et al. (1975) conclude that the success of this study suggests that tree-ring variations can be used to objectively reconstruct other past climatically caused biotic variations. They believe the study provides useful information about the climatic controls of *Thunnus alalunga* (albacore tuna) distribution and that similar studies can help in anticipating variations in other biotic populations that could result from future climatic variations.

E. Variations in Temperature and Precipitation

An experiment is described in this section which tests the applicability of eigenvector patterns of tree growth over western North America to the reconstruction of monthly temperature and total monthly precipitation at selected climatic stations back in time prior to record-keeping. Monthly mean temperature and monthly precipitation were obtained for Boise, Idaho; Sacramento, California; and Phoenix, Arizona. Data from all climatic stations with long and homogeneous records were calibrated with the eigenvector amplitudes of tree-growth patterns over western North America. The regression equations obtained were used to reconstruct past climate at the three stations from the amplitudes of spatial variations in tree rings. A portion of the climatic record was withheld from calibration and used as independent data to test the reconstructions.

Calibration was accomplished in the following manner: (1) Amplitudes of the first 15 eigenvectors of tree growth for 49 sites throughout western North America at lags of t and $t+1$ were selected as possible predictors of climatic variables for the years 1911–1960. (The amplitude data are for eigenvectors similar to those in Figs 7.15 and 7.16 and are described in more detail in Chapter 9). The climatic predictands included monthly temperature and precipitation data for June through December of the year $t-1$ and January through July of year t so as to correspond to the 14-month year used for the response function analysis (Fritts, 1974). However, in this particular test they were the predictands, not the predictors. (2) Stepwise multiple regression analysis was used to obtain the appropriate and significant regression weights for statistically estimating each monthly climatic variable using the 15 amplitudes for year t and for year $t+1$. (3) The resulting regression equations were then applied to independent tree-ring amplitude data to reconstruct the monthly climatic data for the independent period, extending from the beginning of the record through the year 1910. The actual data from the independent period were then compared to the reconstructed data using the reduction of error statistic (Equation 7.4).

In general, the multiple regression equations for reconstructing the individual monthly data did not hold up when applied to data for the independent period, even though the regression equations for the dependent period 1911–1960 were highly significant. Out of a total of 84 possible regression equations (14 months of precipitation and 14 months of temperature at three stations), the reductions of error using independent monthly data were zero or negative 72 out of 84 cases. The tree-ring chronologies which are integrated records of climate over an entire year simply did not have sufficient information in their differences to reconstruct the monthly climatic data.

Therefore it was hypothesized that the tree rings might reconstruct averages for the 14-month year even though they cannot reconstruct data for the individual months. To test this hypothesis the monthly reconstructions for temperature and precipitation obtained in the above test were averaged (totaled) and compared to the independent climatic data which were also averaged (totaled) for the 14 months. Data for both June and July were used twice to allow computation of the 14-month interval. The reduction of error statistic was recomputed and a sign count made of the independent data set (see Chapter 7) to test whether the predicted departure (from the dependent mean) was correctly reconstructed (Table 8.IV).

TABLE 8.IV Verification statistics of climate reconstructions using tree-ring data to estimate mean temperature and precipitation for a 14-month period[a]

Station	Variable	Reduction of error	Sign test — number of Agreements	Disagreements
Boise	Temperature	0·000	21	19
Boise	Precipitation	+0·120[b]	28	14[b]
Sacramento	Temperature	+0·292[c]	43	14[c]
Sacramento	Precipitation	0·000	25	35
Phoenix	Temperature	+0·336[c]	24	9[b]
Phoenix	Precipitation	+0·332[c]	17	4[c]

[a]Estimates for individual months were summed for 14 months and compared with the sums of actual measurements of climate prior to 1911. [b]Significant at the 0·95 level. [c]Significant at the 0·99 level.

It is seen from the table that the relationships are significant for both temperature and precipitation at Phoenix, for temperatures at Sacramento, and for precipitation at Boise. The lack of predictability of temperature at Boise may be explained as either the result of localized climatic data, which can be poorly correlated with the climate of the larger region which affected the spatial patterns of tree growth, or the result of a paucity of highly sensitive tree-ring chronologies in the area. The same may have been the case for precipitation at Sacramento, where a large number of zero values for summer months limit the variance usable for calibration and where there are no chronologies near the climatic station. However, four of the six analyses did pass the 0·95 probability level of significance for independent data, and it is concluded that even single station data at certain localities can be reconstructed from tree rings if the data are averaged over several months.

It was hypothesized that reconstruction of climatic data for single stations, or for several stations representing a small region, is more likely to be accomplished using tree-ring chronologies located near the climatic station. The following is a preliminary report on a test of this hypothesis made by Shatz (personal communication), utilizing 21 long and homogeneous precipitation records that were paired with at least one yearly ring-width chronology at distances from the climatic station ranging from 2 to 60 miles (3 to 97 km), with an average distance of 17 miles (27 km).

Monthly precipitation for each year was totaled for the March–June interval. Data for the years 1909–1960 were available from all 21 climatic stations, so 32 out of these 52 years were selected at random to be used for calibration with the 21 paired ring-width chronologies. The same 32 years of information were used from all data sets and the remaining 20 years of data serve as independent checks. In addition, a number of the climatic records include valid information prior to 1909 and these were also utilized as additional independent data. All climatic and tree-ring data were normalized using the means and standard deviations for the 32-year calibration set.

Stepwise multiple regression analysis was employed using the 32 randomly selected years, as shown in the following model:

$$\hat{y}_t = b_1 x_{t-1} + b_2 x_t + b_3 x_{t+1} + b_4 x_{t+2} + b_5 x_{t+3} \tag{8.4}$$

where $\hat{y}_t$ is the normalized estimate of the total precipitation for the four-month period for year t, and b_1 to b_5 are significant regression coefficients associated with normalized ring-width indices for one year prior to year t

8. INTERPRETATION OF CLIMATIC CALIBRATIONS

Pair number	1	2	3	4	5	6	7	8	9	10
Species	P.e.[d]	P.e.	Ps.m.[e]	Ps.m.	P.e.	Ps.m.	P.p.[f]	P.f.[g]	Ps.m.	Ps.m.
Distance in km between pairs	97	11	43	32	40	11	24	3	13	40
Rank of tree-ring statistics	5	11	2	7	3	6	15	12	4	1
Subscripts of significant coefficients	2,5	3	2	3	2	2	2,4	2	2,3	3
Percent variance calibrated	34	13	36	17	18	42	22	23	51	27
Independent test 1909–1960										
signs +	13	8	12	11	15	15	13	9	13	11
signs −	7	12	8	9	5[b]	5[b]	7	11	7	9
positive product mean	0·65	0·31	2·66	1·44	0·73	2·94[b]	1·21[b]	1·26	2·11[b]	1·36
negative product mean	0·80	0·31	1·70	1·05	0·78	0·81	0·15	1·26	0·79	0·62
reduction of error	−0·29	−0·12	−0·09	0·02	0·14	0·40[c]	0·37[c]	−0·17	0·25[b]	0·04
Independent test pre-1909										
signs +	9	3	11		18	3	5		22	2
signs −	0[c]	1	3		6	0	4		6[c]	1
positive product mean	1·24		4·60[c]		0·97		0·92[c]		3·72	
negative product mean	0·00		0·39		0·45		0·25		1·75	
reduction of error	0·70[c]		0·34[b]		0·34[c]		0·26		0·15[b]	
Verified by independent data	✓		✓		✓	✓	✓		✓	

[a]Preliminary analysis by Shatz (unpublished). [b]Significant 0·95 level. [c]Significant 0·99 level. [d]*Pinus edulis*. [e]*Pseudotsuga menziesii*. [f]*Pinus ponderosa*. [g]*Pinus flexilis*.

through three years following year t. The values for all insignificant regression coefficients in Equation 8.4 are set at zero.

A total of 10 out of the 21 pairs exhibited significant calibration statistics with the total precipitation for these four months (Table 8.V), with percentages of calibrated variance ranging from 13% to 51%. Additional information provided in the table includes the particular variables entered into the regression with significant partial regression coefficients, and the rank of the tree-ring chronology based upon first-order autocorrelation and mean sensitivity statistics. The first in the rank had the highest mean sensitivity value and the lowest autocorrelation coefficient.

A comparison of the data from the 11 chronologies that did not calibrate showed no consistent relationship between the degree of calibration and either the distance between tree and climatic stations or their elevations. However, 8 of 10 of the chronologies which did calibrate were ranked 10 or higher, confirming that the statistical quality of tree-ring chronologies, rather than their closeness to climatic stations, is important to establishing significant climatic relationships. For some indeterminable reason, perhaps due to the site itself, the calibration for the highest ranking chronology (Table 8.V, no. 10) did not hold up in the verification.

A total of 6 out of the 10 chronologies which exhibited significant calibration were verifiable with independent data (Table 8.V). In all six cases of verification, the regression coefficient for growth of year t was significant, while in three of the four cases that did not verify growth of year $t+1$ was the variable selected for calibration by the use of the stepwise regression procedure. Apparently there was insufficient linkage at a lag of one year to hold up when tested against the independent data set.

Since six of the significant calibration relationships were also shown to have significant verification statistics, it can be concluded that precipitation for a four-month interval in the spring can be reconstructed to some extent from variations in tree rings from nearby sites. However, these reconstructions using a chronology from only one site exhibit considerable statistical noise, since the highest percent variance reduced for the dependent set was only 51%.

There are several possible explanations for the low calibration percentage:

(1) Some of the climatic records may have been faulty and included information irrelevant to the climate at the tree sites.

(2) The tree-ring chronology may not have included enough variance related to the climatic parameter with which it was calibrated, a situation

which arises when (a) the nonclimatic noise in the chronology is too large, or (b) the trees simply do not respond to the particular climatic information with which they were calibrated.

(3) A large part of the tree-ring chronology variance may have been related to climatic factors not included in the calibration, and this uncalibrated variance acted as noise.

(4) The trees may have responded to precipitation directly during certain months in the calibrated season and inversely during other months in the calibrated season, so that there was no net correlation of growth with the average precipitation for the four month interval.

It is apparent from the results described in this chapter that while it is often possible to successfully reconstruct seasonably averaged data such as streamflow, lake levels, and climatically linked population data, attempts to reconstruct precipitation on a station-by-station basis involves a large amount of noise and the climatic reconstructions are likely to include considerable error in the statistical estimates. One method of minimizing this noise is to average a large number of chronologies and climatic data stations and relate the spatial patterns in one to the spatial patterns of the other. Such an approach is now feasible by using the averaging capabilities of principal components. The following chapter describes these rather recent developments.

Chapter 9

Reconstructing Spatial Variations in Climate

I	Introduction	434
II	The General Nature of Dendroclimatographic Analysis	437
III	The Statistical Model	438
IV	A Feasibility Study	439
V	Recalibration	450
VI	Climatological Studies	455
	A. Identification of pressure types	455
	B. Characterization of types	457
	C. Reconstruction of pressure types	468
VII	Summarization of Reconstructions for Winter Using the Pressure Types	470
VIII	Verification of Reconstructions for Winter	476
	A. Verification of types using the dependent data set	477
	B. Verification of types using the independent data set	479
	C. Consistency among different data sets	482
IX	Summarization of Reconstructions for Summer Using the Pressure Types	488
X	Verification of Reconstructions for Summer Using Independent Tree-ring Data	491
XI	Verification Using Journals, Historical Data, and Various Proxy Records of Climate	499
XII	Applications to Climatological Problems	500
XIII	Present and Future Prospects of Dendroclimatology	503

I. Introduction

The atmosphere of the earth is a fluid system being continually mixed on scales of motion ranging from large planetary whirls to minute molecular motions. Day-to-day changes in the weather are generally caused by the movement of intermediate-scale low pressure and high pressure systems (cyclones and anticyclones) which migrate over a few thousand kilometers of the earth's surface. These intermediate-scale systems can produce weather conditions lasting from a fraction of a day to a number of days over large geographic regions, and when changes in weather occur they are often associated with the approach and passage of one or more of these pressure systems.

Air masses which are separated from one another by boundaries called *fronts* are intermixed by these systems. The temperatures and moisture of different air masses vary so that the passage of a front is often marked by abrupt changes in air temperature and moisture as well as wind direction. These intermediate-scale cyclones and anticyclones may be modified locally by associated smaller-scale systems such as hurricanes, tornadoes, and thunderstorms. Such localized phenomena, at times connected with frontal passage, can create apparent discontinuity and variability in the circulation system when viewed from a point on the earth's surface.

The *cyclones* and *anticyclones* are themselves part of still larger planetary circulations, which are most obvious at altitudes of several kilometers above the earth's surface. Such large-scale systems may move very slowly or persist for months and even for years over a region, steering the cyclones and anticyclones into certain areas and away from other areas over the surface of the earth.

These large-scale systems can affect the weather over the entire globe, causing anomalies of one kind to occur in particular sectors of the systems which are correlated with anomalies (often of different kinds) in other sectors. For example, the sub-Saharan drought of 1968 through 1973 is apparently associated with higher-than-normal pressure in the planetary systems over western Europe, and lower-than-normal pressure south of the Azores and in east central Africa (Winstanley, 1973; Namias, 1974). It may also be associated with anomalous weather occurring elsewhere around the globe.

Both the intermediate- and large-scale circulation systems produce a certain amount of spatial coherence in climate. If weather records for a given set of years are taken from neighboring stations, they usually resemble one another and are directly correlated. Stations separated by greater distances, such as those on opposite sides of continents and mountain systems are usually less well correlated, although some correlation may still exist. Depending upon their relationship to the planetary systems that steer the cyclones and anticyclones over the specific regions, the correlations may be either direct or inverse. For example, winters may be dry in Pacific southwestern North America during years in which many winter storms are passing along the Pacific Northwest, and similarly, years of cold winters in eastern United States may be years of warm winters in the West. As a result of such long-distance correlations, there is a certain amount of information on climatic variability that is not random over the globe.

The storage and exchange of energy among various components of the earth–atmosphere system can dampen the variability of the circulation systems and create a certain amount of nonrandomness in climatic records. For example, the rates of cooling in autumn and of warming in spring are governed by the rates of energy released and stored at the land and water surfaces. An anomalously warm or cold ocean may be responsible for a persisting climate over nearby continents (Namias, 1974; Gates and Mintz, 1975). The presence of a widespread early snowfall can affect the albedo of the earth and hasten the onset of winter or a late snowfall can delay the beginning of the warmer spring climate (Kukla and Kukla, 1974). Large ice masses, such as those that occurred 17,000 years ago, can be associated with extremely long-term climatic changes even in regions where the ice was not present (Gates and Mintz, 1975).

Persistence in the various parameters of weather and climate can lead to substantial autocorrelation in the climatic record (Jones, 1975) and to intercorrelations among climatic variables in different regions (Walker and Bliss, 1932; Bjerknes, 1962, 1966; Namias, 1953, 1969; Mitchell *et al.*, 1966; Lamb and Johnson, 1959). Kutzbach (1970) describes several major centers of action in the Northern Hemisphere and characterizes their spatial and temporal variations. Blasing (1975) describes several large-scale modes of pressure variation over the North Pacific and North American continent and discusses some innovative procedures for analyzing them.

The term *synoptic climatology* (Barry and Perry, 1973) is applied to studies of large-scale variations in climate in which the general circulation of the atmosphere must be considered over both time and space, while the term *climatography* is applied to the mapping of climate. Such studies often include analysis of averages, variances, or anomalies of various kinds, expressed as departures from long-term means. They may include differences in means from one period to the next, frequencies of occurrences of various climatic types or events, correlations among variables distributed over space, and other representations of climate as it varies through both time and space. The term *dendroclimatography* which has been introduced in Chapter 1, is the name of the relatively new field in which the spatial variations in tree rings are employed to reconstruct and map the spatial variations in past climate. Figures 1.13 and 1.14 include examples of early studies in which tree-growth maps for western North America are used to infer regional variations in climate. The following pages summarize some of the most important developments in this type of research, some of which utilize techniques described in earlier chapters, and others which employ new techniques designed specifically for mapping analysis.

This chapter will include a general discussion of dendroclimatographic data and examples of statistical models that are useful. A particular feasibility study is then described which documents the potential of the field, raises some climatological problems, and suggests certain improvements which are incorporated in subsequent reanalysis work. The most pertinent results from climatological analyses and reconstructions growing out of the feasibility study are summarized. In particular, a new procedure for identifying and describing variations in climate is discussed and used to interpret the climate reconstructed from the spatial variations in tree growth. Subsequent sections describe the details of the reconstructions, their verification, and other applications to analysis of spatial variations in climate. The chapter concludes with a discussion of present and future prospects for dendroclimatic reconstruction and related tree-ring work.

II. The General Nature of Dendroclimatographic Analysis

It has been established that ring widths are especially suitable for analysis of past climate because the growth layers can be dated accurately. The methods described below require absolute dating so that all observations in the spatial arrays of both climate and tree rings can be placed in their proper time sequence.

Inferences of year-to-year climate from growth rings are complicated, however, by the fact that a portion of the tree-growth response can lag behind the occurrence of a climatic event (see Figs. 1.10 and 5.9), and the climate of various seasons can affect growth differently. An understanding of the growth response for specific species and sites is therefore necessary for constructing appropriate models for calibration. Results from different models are tested in order to select the best statistical relationship. This is then applied to yearly tree-ring data to obtain yearly climatic estimates.

A climatographic model was suggested by the work of LaMarche and Fritts (1971a) in which eigenvectors of spatial variations in tree growth were found to resemble the eigenvectors of spatial variations in precipitation over the same area and period of time (Fig. 7.16). This suggested that a spatial pattern of climatic anomaly may produce a spatial pattern of tree-growth anomaly, and that the eigenvector amplitudes of climate might be calibrated with the eigenvector amplitudes of tree growth (Equation 7.20).

The climate–growth models used to test this possibility are similar to those in Table 8.II, the only difference being that the data vary through space as well as time. The models assume that modes of variation in precipitation, temperature, or other climatic parameters, measured over a grid of weather stations, are linearly related to corresponding modes of tree growth. Maps of tree growth are then transformed to corresponding maps of climate.

III. The Statistical Model

The objective is to obtain a transfer function $_pT_m$ which can be applied to m tree-ring chronologies associated with n years ($_mF_n$) to obtain estimates at p climate stations for n years of record ($_p\hat{P}_n$). This can be accomplished by solving Equation 7.21 for each of p predictand data points

$$_p\hat{P}_n = {_pT_m}{_mF_n} \tag{9.1}$$

but such solutions involve considerable computations, and excessive numbers of degrees of freedom are used up.

A technique referred to as *canonical analysis* is applied to obtain an optimal solution as it combines redundant data from neighboring stations into orthogonal covarying components (Glahn, 1968; Blasing, unpublished). Two orthogonal matrices, $_pA_g$ and $_kB_g$, are derived which can be multiplied by matrices $_nX_p$ and $_nY_k$ respectively to obtain two orthogonal sets of amplitude variables, $_nW_g$ and $_nQ_g$.

$$_nW_g = {_nX_p}{_pA_g} \tag{9.2}$$

$$_nQ_g = {_nY_k}{_kB_g} \tag{9.3}$$

where g is equal to the smaller number of p or k. Each W variable is orthogonal to all other variables in the $_nW_g$ set, and each Q variable is orthogonal to all other variables in the $_nQ_g$ set. Furthermore, each variable in $_nW_g$ is correlated with only one variable in Q, and is orthogonal to all other Q variables in the $_nQ_g$ set. This is expressed mathematically as

$$\left(\frac{1}{n}\right)_g {_gW'_n}{_nQ_g} = {_g\Lambda_g} \tag{9.4}$$

where $_gW'_n$ is the transpose of $_nW_g$ and $_g\Lambda_g$ is the diagonal matrix of *canonical correlations* (with zeros in the off-diagonal elements) between each W variable and the corresponding Q variable with which it is matched. The g (diagonal) elements of $_g\Lambda_g$ are arranged in order from the highest to lowest canonical correlation coefficient.

According to Blasing (unpublished), if $g = k$ (that is $k \leq p$), then the equation to estimate Q from W is

$$_n\hat{Q}_g = {_nW_g\Lambda_g} \qquad (9.5)$$

Applying Equations 9.2 and 9.3, Equation 9.5 can be expanded:

$$_n\widehat{Y_gB_g} = {_nX_pA_g\Lambda_g} \qquad (9.6)$$

or, multiplying both sides by $_gB_g^{-1}$

$$_n\hat{Y}_g = {_nX_pA_g\Lambda_gB_g^{-1}} = {_nX_p T_g} \qquad (9.7)$$

where the transfer function $_pT_g$ is composed of a set of *canonical regression coefficients* that transform values of $_nX_p$ into estimates of $_nY_g$.

As in the case with eigenvectors, small-scale variance may be eliminated by selecting only the first h canonical variates, so that only the large-scale features of growth are used to reconstruct the large-scale features of climate. If all g canonical sets are used, the elements of $_pT_g$ are the same as the usual multiple regression coefficients with p elements of X statistically predicting k elements of Y. Such multiple regression provides the best fitting relationships for the calibration period, but more degrees of freedom are consumed and they may not hold up on independent data as well as the reduced set of canonical variates.

Thus, canonical analysis is used: (1) to obtain covarying modes of behavior between the two variable sets, (2) to eliminate the small-scale modes which have the lowest covariance between sets, (3) to maximize the number of degrees of freedom in the final analysis, and (4) to gain efficiency and stability by solving for a reduced number of variables.

IV. A Feasibility Study

The tree-ring data used for the first dendroclimatographic reconstruction test described in the chapter consisted of 49 chronologies from western North America (Stokes *et al.*, 1973). LaMarche selected this tree-ring set from a collection of several hundred sites for analysis of possible sunspot and tree-growth relationships (LaMarche and Fritts, 1972). It includes a variety of topographical situations and species, ranging from central British Columbia, Canada, to Durango, Mexico, and from California to South Dakota.

The climatic data used in this study are monthly values of surface pressure for the Northern Hemisphere adjusted to mean sea level and tabulated for 1900 through 1962, excluding the years for 1939-1944 (these missing data were later obtained or estimated). Preliminary examination had shown substantial correlation of tree growth with pressure data at sizable distances from the tree sites. Thus, a grid was chosen to include data for 100 points located at every 10 degrees from 25° to 65° N latitude and 165°E to 5°W longitude.

Since the ring-width index represents the integrated effect of climate for several seasons in the growth year, the seasonally averaged climatic data were used for all four seasons. This introduces a change from the more conventional meteorological approach which uses January data to represent winter and July data to represent summer. Each season is defined as follows:

Summer	July, August, September
Autumn	October, November, December
Winter	January, February, March
Spring	April, May, June

The growth for a particular year was matched with pressure data for the prior summer, autumn, and winter seasons, as well as with that of the spring in which growth began. Thus the variables of pressure include 100 grid points for four seasons which were to be estimated from 49 ring-width chronologies.

The size of the problem actually exceeded the capacity of the computer and would have consumed all available degrees of freedom, so the following simplifications were made. All data for each of the 400 pressure variables and the 49 tree-ring variables were normalized (Equation 7.5). The pressure data for each season were subjected to principal component analysis, and the first eight and most important eigenvectors, reducing approximately 80% of the variance, were selected for further study. The amplitudes of the eigenvectors (Equation 7.19) were obtained for each season, providing a total of 32 variables of pressure (eight for each of four seasons). We shall refer to these as $_nY_k$.

Similarly, the first seven eigenvectors of tree growth and their amplitudes (LaMarche and Fritts, 1971a) were used and they reduced 57% of the variance. A more recent analysis of the same data is shown in Figs 9.1 and 9.2, which utilizes the amplitudes of the first 10 eigenvectors to reduce 63% of the tree-growth variance. We shall refer to these as $_nX_p$.

The eigenvector variables for pressure were orthogonal only for the eight variables within each season, and therefore further analysis was required to reduce them to completely orthogonal components. The first part of canonical analysis was run as in Equations 9.2, 9.3, and 9.4. Matrices $_nW_g$ and $_nQ_g$ were obtained, where the first are the seven canonical amplitudes of the 49 ring-width chronologies (Equation 9.2) and the second are the seven canonical amplitudes of pressure for the four seasons (Equation 9.3). Matrix $_pA_g$ (Equation 9.2) was multiplied by the eigenvectors of tree growth, $_mE_p$, to obtain matrix $_mZ_g$ as follows:

9. RECONSTRUCTING SPATIAL VARIATIONS IN CLIMATE 441

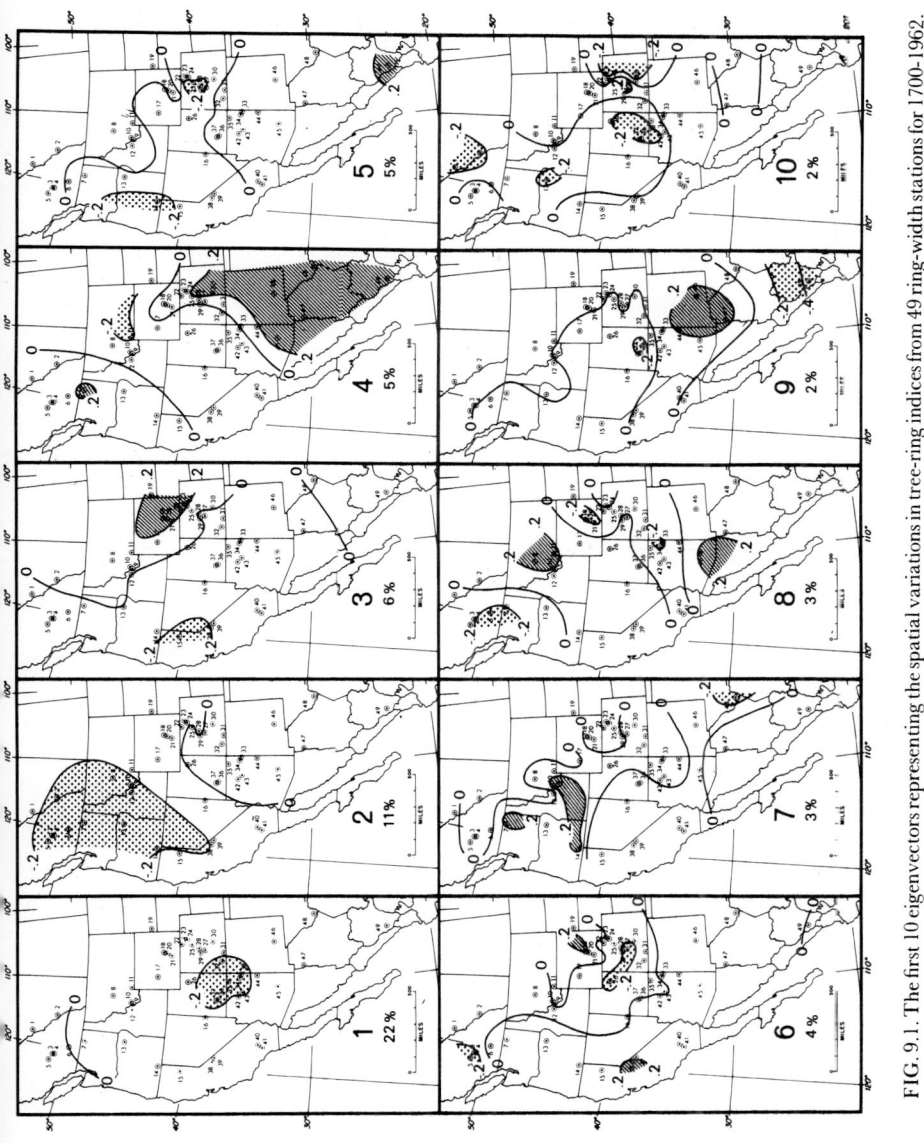

FIG. 9.1. The first 10 eigenvectors representing the spatial variations in tree-ring indices from 49 ring-width stations for 1700-1962. The eigenvectors are ranked in the order of their importance, that is, in order of the variance they reduce which is indicated by the percentage figure in lower left of each map. (Drawing by M. Huggins.)

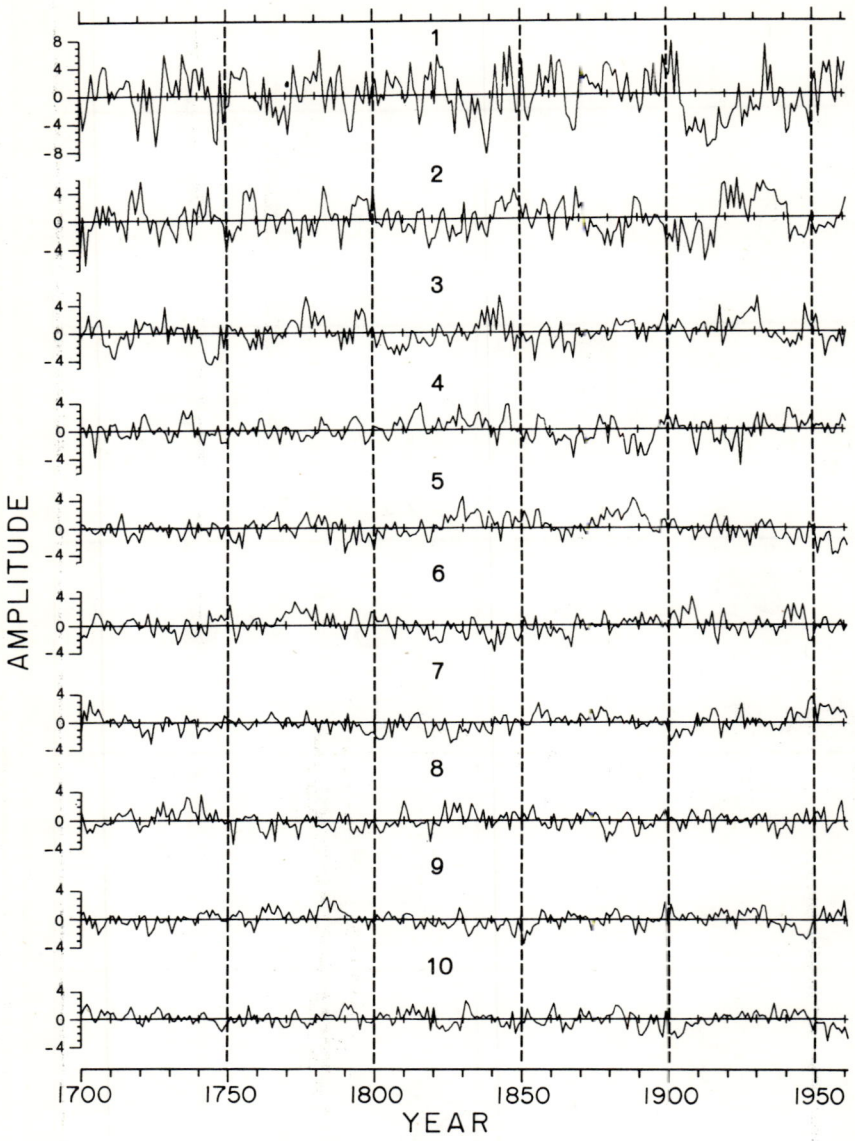

FIG. 9.2. The amplitudes of the first 10 eigenvectors shown in Fig. 9.1 plotted in standard deviations (normalized units) to show their variations through time. Note the decrease in the variance associated with the increasing rank.

9. RECONSTRUCTING SPATIAL VARIATIONS IN CLIMATE

$$_m Z_g = {_m}E_p A_g \tag{9.8}$$

This matrix includes the eigenvectors of tree growth canonically weighted to give maximum relationship with variations in pressure. The values for the elements are ordered according to the canonical correlation coefficients and mapped in Fig. 9.3, while the values for amplitudes $_1 W_g$ are plotted in Fig. 9.4.

The transfer function that was obtained in this particular study differed from Equation 9.7, since the appropriate computer program had not yet been developed. The amplitude series $_n W_g$ were used as predictors of the normalized pressure for 400 seven-variable multiple regression analyses. Hence,

$$_p \hat{P}_n = {_p}R_g W''_n \tag{9.9}$$

where $_p R_g$ is the complete set of 2800 regression coefficients relating the seven canonically weighted amplitudes of growth to the pressure at 100 grid points for four seasons. The regression coefficients were applied to transform $_g W''_n$ for the years 1700-1899 into the estimates of $_p P_n$ the latter representing past anomalies in pressure for 1700-1899. The estimates of $_p P_n$ were converted to millibar pressure anomalies by multiplying each item of $_p \hat{P}_n$ by the appropriate standard deviation for the season and grid point.

Figure 9.5 includes examples of the mean pressure reconstructions and the corresponding mean growth anomalies for two selected decades. Although the differences between the two growth anomalies are not large, there are substantial differences in the pressure estimates. Analyses of these pressures, which are described later in the chapter, indicate that the mid-North American climate during the summers of 1731-1740 was dry and warm, but during winter there were unusually frequent outbreaks of cold Arctic air. During 1771-1780 the summers were anomalously cool and moist but the winters were milder than normal.

The anomalies of pressure for each season were averaged by pentads and mapped. The maps showing the most marked pressure anomalies for winter are included in Fig. 9.6a, b, and c. These maps clearly show periods of anomalous pressure centered over the Gulf of Alaska which are associated with anomalies of opposite sign over Hudson's Bay or the North Atlantic. During most of the 19th century (Figs 9.6b and c), higher than normal pressures persisted in the Gulf of Alaska, indicating a weak Aleutian Low and cold air in the Arctic. However, in the 18th century, surface pressures were lower for the North Pacific, indicating a stronger

Aleutian Low and somewhat warmer temperatures in North America, a pattern that was common during the 20th century.

The percent pressure variance reconstructed at all grid points for each season during the calibration period averaged 23·7% for summer, 20·6% for autumn, 24·0% for winter, and 18·5% for spring. The best reconstructions were for the pressure over the subtropics of both oceans, over the north Pacific Ocean, and over the North American continent

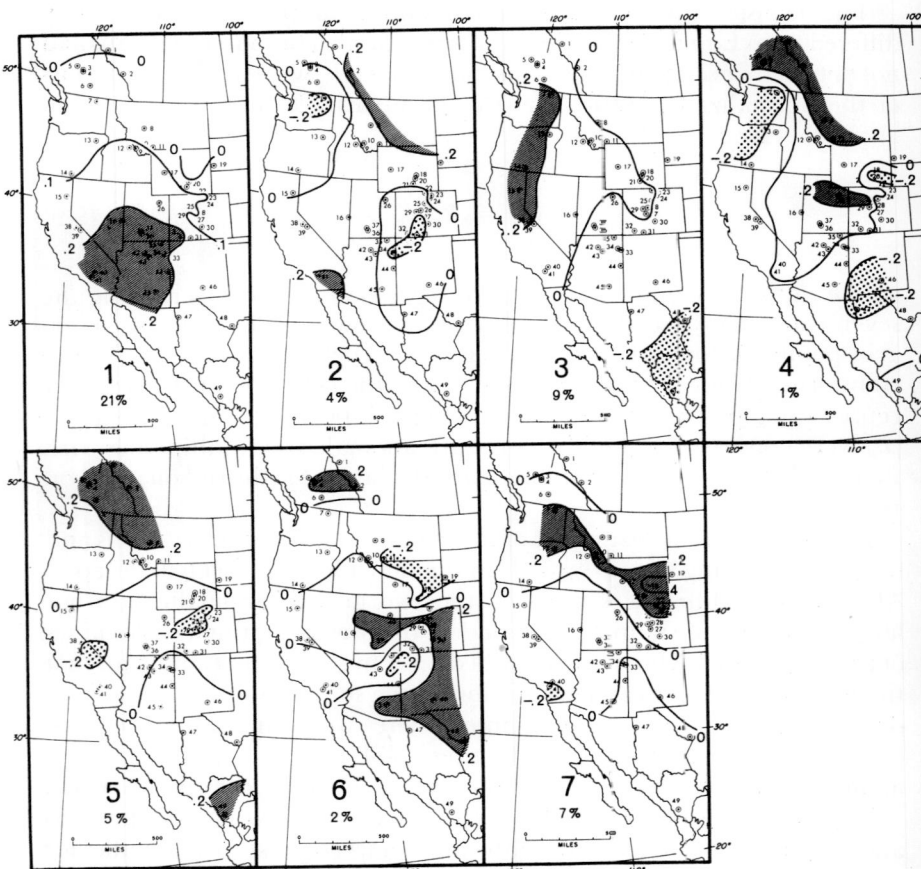

FIG. 9.3. Plots of the eigenvector patterns of growth canonically weighted to give maximum correlation with anomalies in surface pressure ($_m E_p A_g$, Equation 9.8). The maps are ranked from high to low correlation with the pressure anomalies, and the percentage in the lower left of each map is the variance in the growth that was reduced by each weighted set. Note that the weights which are ranked in the figure according to the correlation with the pressure anomalies are no longer in the order of the variance in growth that was reduced. (Fritts et al., 1971)

where the percentages of variance reconstructed ranged from 30% to 50%. The percent variance reconstructed for the North Atlantic was so small that the reconstructions in that area were judged to be insignificant (see Fritts *et al.*, 1971).

It was concluded from this feasibility study that the multivariate technique was very promising but that the task of reconstruction might be improved by (1) changing the pressure grid to eliminate the area of poor reconstruction in the Atlantic Ocean and to include more area in the Western Pacific and the Asian mainland, (2) defining the seasons differently, (3) including tree growth in the model for at least one year before and for two years after the occurrence of climate, and (4)

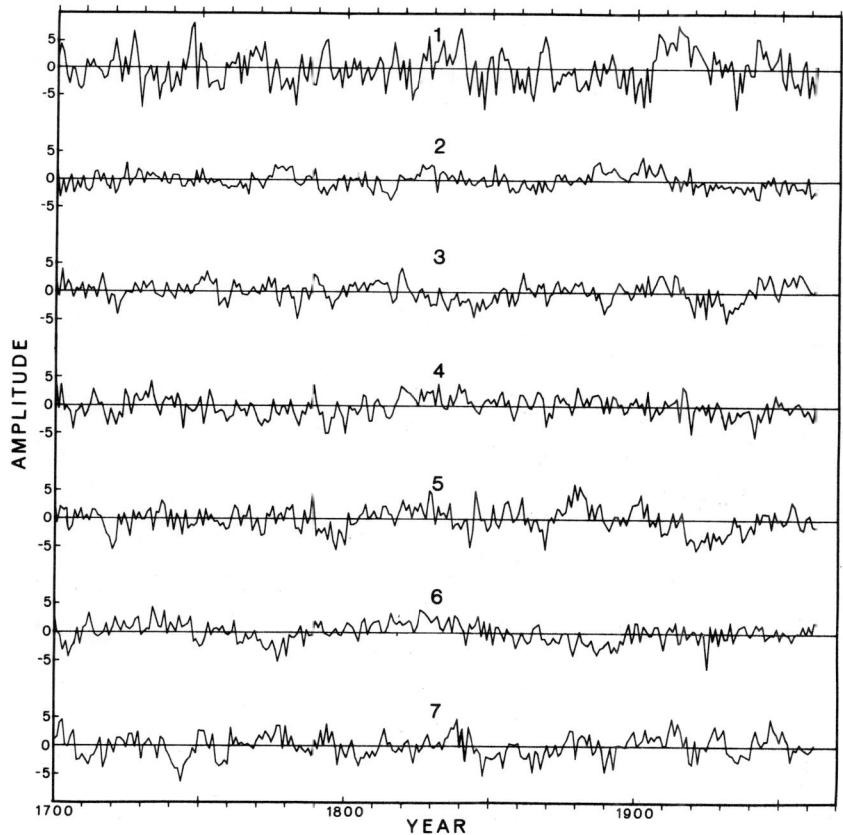

FIG. 9.4. Plots of the amplitudes of the seven eigenvectors of tree growth ($_tW_g$) canonically weighted to give maximum correlation with surface pressure. Plots are ordered from high to low correlations with pressure. (Fritts *et al.*, 1971)

446   TREE RINGS AND CLIMATE

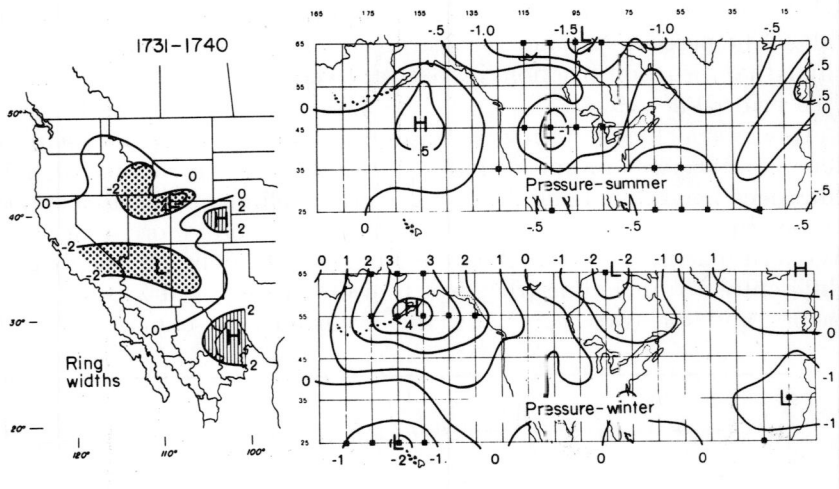

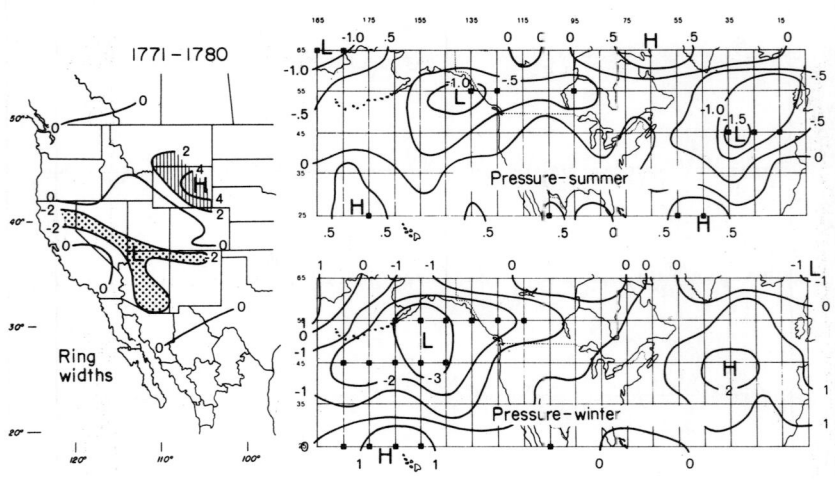

FIG. 9.5. The mean anomalies in growth (left) for two decades of the past are substituted into the transfer function to obtain reconstructed anomalies in surface pressure (right) for summer and winter seasons. Square dots indicate those reconstructed data points that exceed two standard errors of estimate and may be inferred to be significant. (Drawing by M. Huggins)

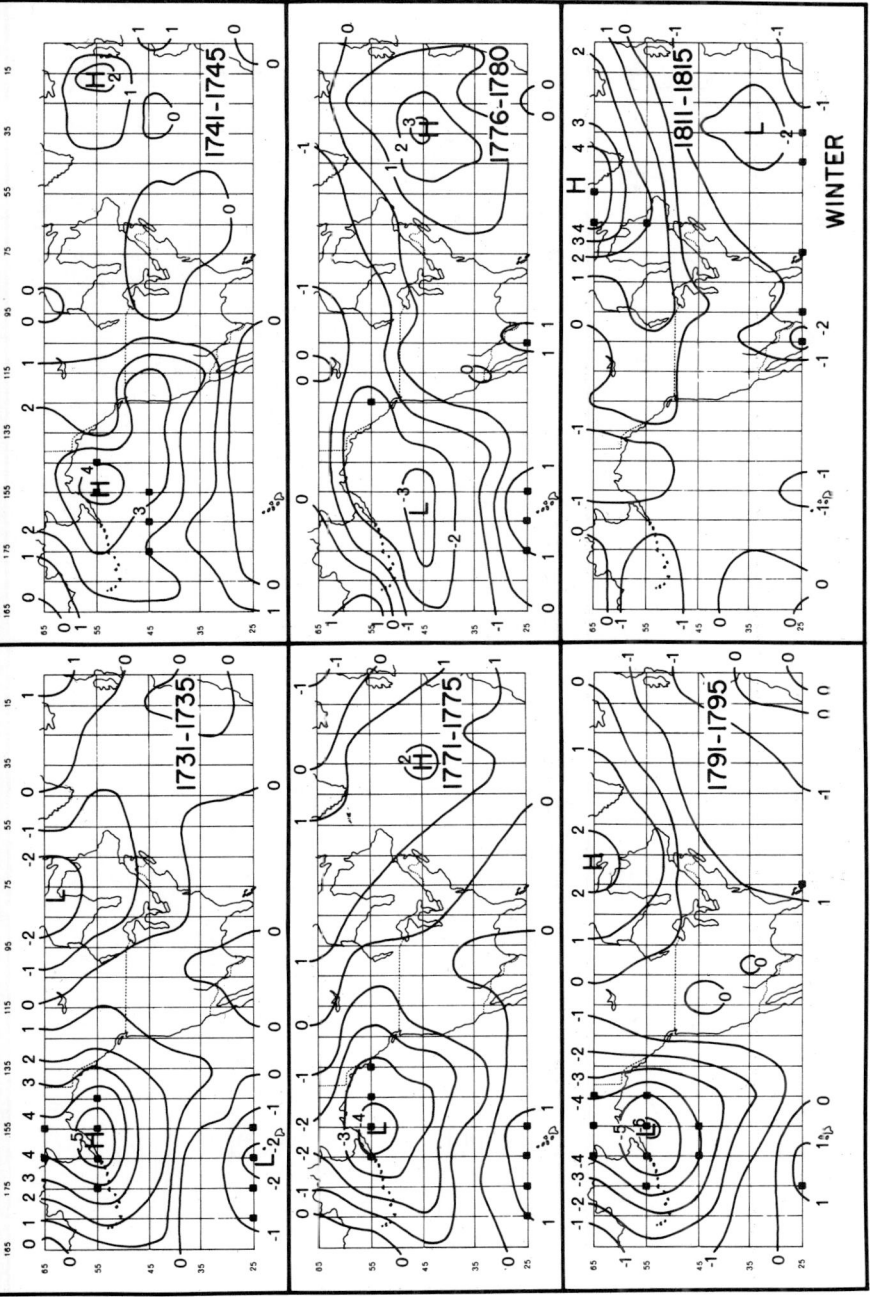

FIG. 9.6. Reconstructed pressure anomalies for January–March expressed in millibars for selected pentads during the interval (a) 1731–1815, (b) 1816–1845, and (c) 1846–1945 as derived from anomalous ring-width indices. Square dots indicate the five-year mean anomaly at the grid point was greater than twice the standard error. Only pentads with five or more pressure estimates exceeding two standard errors are shown. (Fritts et al., 1971)

(a)

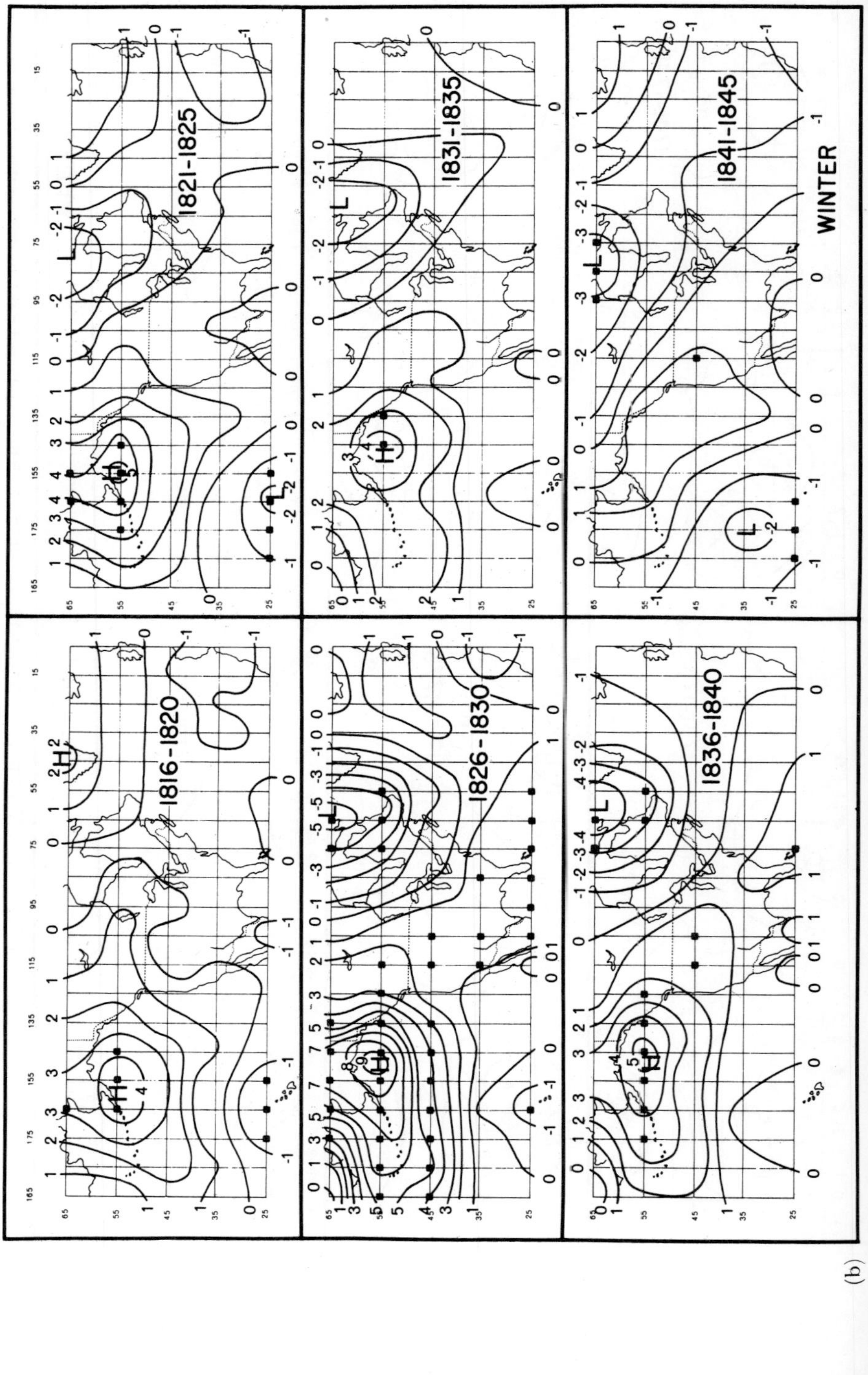

448 TREE RINGS AND CLIMATE

9. RECONSTRUCTING SPATIAL VARIATIONS IN CLIMATE

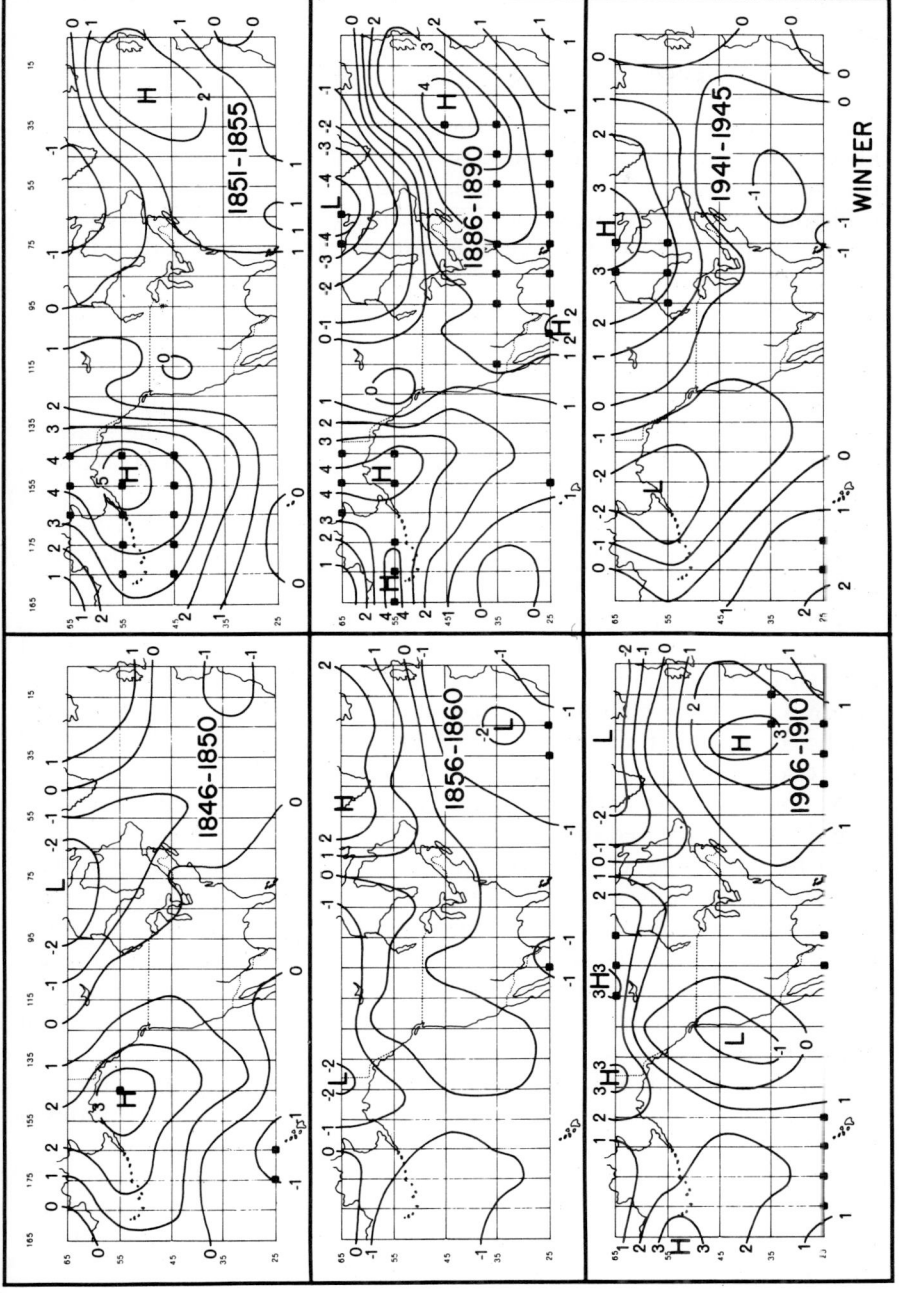

(c)

analyzing each season independently. The following sections describe several reanalyses.

V. Recalibration

The data and models utilized in the feasibility study were altered in an attempt to improve the climatic reconstructions. The same sources of pressure data were used, but the pressure grid was changed to include 96 grid points from 20°N latitude to 70°N latitude and from 80°W longitude across the Pacific Ocean to 100°E longitude. The grid included data at every 20° longitude for 60° and 70°N and every 10° longitude at latitudes 20°, 30°, 40°, and 50°N. The east Asian coast was included partly to allow use of its long climatic record as independent checks, and partly to include a larger area upwind of the tree rings.

The pressure data for the revised grid were examined using principal component analysis and other techniques to ascertain the natural climatological seasons for the particular grid that was selected (Blasing, 1975). December, January, and February were found to be the most similar winter months and were chosen to represent that season. June was found to differ significantly from July and August, so the latter two months were used as summer.

The recalibrations were obtained using the eigenvector amplitudes from the correlation matrix of the same 49 ring-width chronologies used in the feasibility analysis (Equations 7.18 and 7.19, Figs 9.1 and 9.2). The eigenvectors differed only in that the entire 1700-1962 period was analyzed rather than the 1700-1930 interval, and additions and corrections had been made on a few of the chronologies.

Not only was the statistical model changed but the pressure and tree-ring data were treated differently: (1) The mean for each pressure data point from 1899-1966 was subtracted from each seasonal value, but no division by the standard deviation was made. Thus the actual pressure variance was preserved. (2) The eigenvectors of pressure were extracted by season for the entire calibration period 1899-1966 using the covariance rather than the correlation matrix. This allowed a greater weight to be placed on areas where the pressure variations were greatest. (3) Different numbers of eigenvectors were used to represent the pressure and tree-growth variance, depending upon the model to be tested. (4) The tree-ring climatic models involved various lags by utilizing the amplitudes of the tree-ring data for year t and lagging them by one or more years to determine which combination of lags best fit the relationship. As a result, the amplitudes between lags were often nonorthogonal. (5) Canonical regression was used, as in Equation 9.7, to obtain the coefficients for reconstructing amplitudes of pressure from amplitudes of tree growth.

Only the canonical sets which calibrated more than 1% of the variance in the pressure amplitudes were selected. The canonical regressions for reconstructing the amplitudes of pressure were multiplied by the appropriate eigenvectors of pressure (Equation 7.21) to obtain the pressure transfer functions. The amplitudes of tree growth for each year were then multiplied by this transfer function to reconstruct the yearly anomalies in pressure.

The data used to assess lagging relationships including amplitudes for the ring widths in year $t-1$, t, and $t+1$ up to year $t+k$ ($_nX_p$, Equation 9.7) may be written as follows:

$$(_nX_p)_{t-1},\ (_nX_p)_t,\ (_nX_p)_{t+1},\ (_nX_p)_{t+k} \qquad (9.10)$$

However, all the amplitudes could not be used because they would have exceeded the available degrees of freedom. Several calibrations were attempted which involved varying numbers of pressure amplitudes, varying numbers of tree growth amplitudes, varying lagged relationships, and varying the season (Table 9.1). The square of the multiple correlation, which was used to measure the calibrated amplitude variance, is also partially affected by the degrees of freedom that are used up in the calibration. The percent variance due to the loss of degrees of freedom can be estimated from the square of the correlation expected for random numbers (R^2 random) as follow:

$$R^2\ \mathrm{random} = \frac{k}{n-1} \qquad (9.11)$$

where k is the number of predictor variables and n is the effective sample size, which in this case is assumed to be equal to the number of years used for calibration. The percent amplitude variance calibrated is subtracted from the R^2 random (Table 9.1) and the square root taken to estimate d, a measure of the statistical fit above that expected for random numbers. This value is analogous to, and may be tested as, the correlation coefficient.

The following inferences were drawn from the various calibration statistics (Table 9.1):

(1) More significant variance was calibrated (that is the values of d were generally larger) for those models using 10 or fewer eigenvector amplitudes of pressure. It appears that high-order amplitudes of pressure are too small scale or have too little variance to be useful in these climatic reconstructions.

(2) Little significant variance was reduced by models incorporating greater lags than those for years t and $t+1$ as shown in Rows 4, 5, and 6 in the table.

TABLE 9.I Selected statistics from calibrations using amplitudes of tree rings (X) to reconstruct amplitudes of pressure (Y)

	Variables[a]				Season Reconstructed									
					Winter Dec–Feb			Early Spring Mar–Apr		Late Spring May–June		Summer July–Aug		Autumn Sept–Nov	
Line	Y	X	df[b]	R^2 Random[c]	R^2 Amp[d]	d[e]		R^2 Amp[d]	d[e]	R^2 Amp[d]	d[e]	R^2 Amp[d]	d[e]	R^2 Amp[d]	d[e]
1	20	20_t	43	0.318	0.360	0.205						0.354	0.190		
2	20	20_{t+1}	42	0.323	0.347	0.155						0.350	0.164		
3	20	$15_t, 15_{t+1}$	32	0.484	0.532	0.219						0.526	0.205		
4	20	$10_{t-1}, 10_t, 10_{t+1}$	32	0.484	0.507	0.152						0.546	0.249		
5	20	$10_t, 10_{t+1}, 10_{t+2}$	31	0.492	0.519	0.164						0.527	0.158		
6	20	$8_{t-1}, 8_t, 8_{t+1}, 8_{t+2}$	29	0.525	0.565	0.200						0.555	0.173		
7	10	20_t	43[f]	0.318[f]				0.393	0.274	0.380	0.249	0.388	0.265	0.457	0.390[h]
8	10	$10_t, 10_{t+1}$	42[f]	0.323[f]				0.392	0.263	0.449	0.355[g]	0.424	0.318[g]	0.464	0.369[g]
9	10	10_t	53[f]	0.159[f]				0.214	0.235	0.248	0.298[g]	0.217	0.241	0.274	0.336[g]
10	10	10_{t+1}	52	0.161	0.292							0.280	0.345[g]		
11	6	10_t	53	0.159		0.365[h]									

[a]Subscript t represents the calendar year of the predictand, except for autumn when it represents the following year. [b]Degrees of freedom. [c]R squared for random numbers. [d]R squared for reconstruction of pressure amplitudes. [e]Square root of the difference between R squared amplitudes and R squared random numbers. The highest and most significant value for a particular season is printed in bold type. [f]One less degree of freedom and higher R squared for random numbers used for autumn. [g]Significant ($p \geq 0.95$). [h]Significant ($p \geq 0.99$).

(3) In the case of autumn, winter, and early spring, the best models were those including no lag and associated with growth commencing in the concurrent spring season. This result is reasonable in that there is ample time for climate during these three seasons to modify conditions which in turn influence important plant processes. Since no value of d for early spring was significant, the spring seasons were classified differently in subsequent analysis.

(4) The best model for late spring (May–June) includes growth in years t as well as $t+1$. Since the growing season usually starts in May, it is reasonable that a portion of the climatic influence is transferred to the next year's growing season.

(5) The best model for summer includes only the amplitudes for year $t+1$, but the model using years t and $t+1$ (Table 9.I, Line 8) is also significant. Apparently growth slows down sufficiently during this season so that climate for year t has less influence on the width for year t than it does for year $t+1$.

The percent variance calibrated at each pressure grid point was calculated by subtracting the residual variance from the total variance at each point, dividing the result by the latter and multiplying by 100 to obtain a percentage figure. The results for all grid points are mapped in Fig. 9.7 for three different winter and summer models. The areas of significant calibrated variance for winter and summer are different, but many similarities among the models for a given season are apparent.

For example, the areas of significant calibrated variance for winter include Alaska and the Canadian Arctic, the extreme North Pacific (including an area south of Korea), Mexico and the United States Southwest, the central North Pacific along latitude 20° west of Hawaii, and for some models, in mainland China.

The areas of significant calibrated variance for summer include the Canadian and east Siberian Arctic, western United States, a large area in the central North Pacific, and the Asian mainland along 100°E longitude (the western boundary of the pressure grid).

It was concluded that pressure can be reconstructed for a substantial portion of the grid, although the area of significant reconstruction is less than half the area used for calibration. It is hoped that pressure reconstructions may be improved in the areas of poor calibration by utilizing tree rings from additional sites, especially those from Canada, Alaska, and the Asian continent.

An inability to reconstruct pressure in a given region does not imply that other climatic variables cannot be reconstructed. For example, precipitation anomalies in certain regions may be associated with increased frequencies of cyclones and anticyclones and increased numbers of frontal passages but not with anomalies in the mean seasonal pressure in that particular region. In such cases, temperature or precipitation may be reconstructed even though there is little possibility of reconstructing the pressure variance.

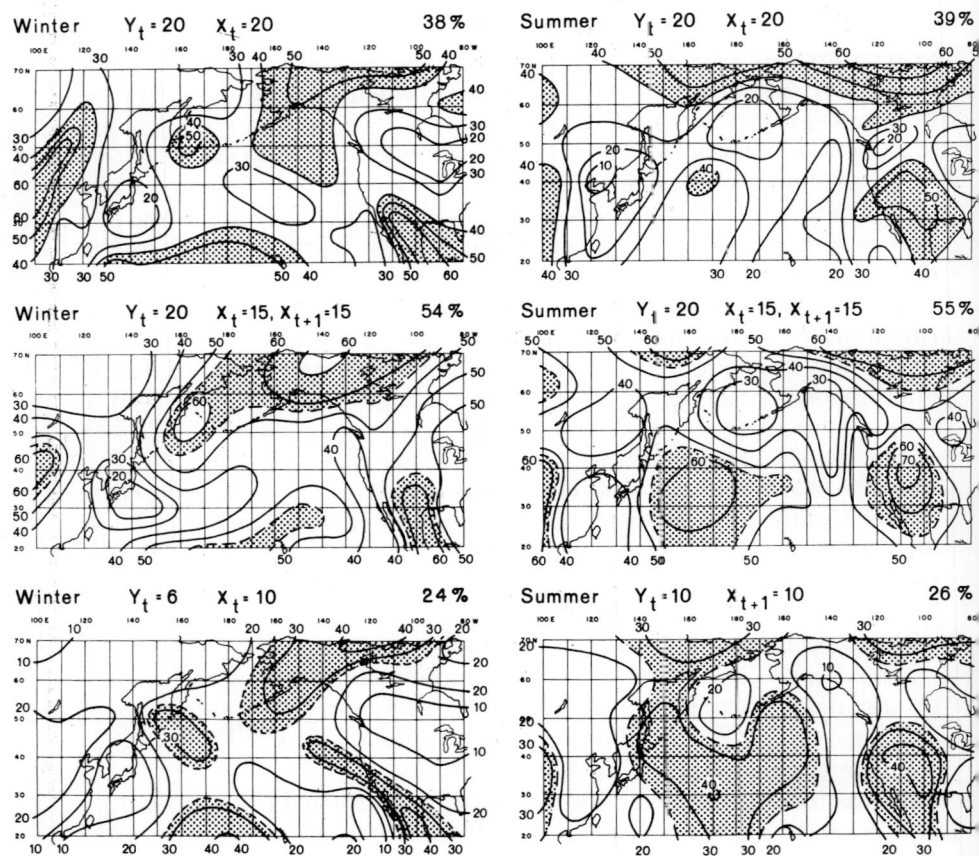

FIG. 9.7. The percent variance calibrated over the spatial grid of pressure using six different models included in Table 9.I. Shading designates the areas where the percent variance exceeds the approximate 0·95 confidence limits and can be considered to be significant. The variable numbers and the total percent of pressure variance reduced are shown above each map. (Drawing by M. Huggins)

With the exception of early spring, the models with the highest values of d (values in bold type in Table 9.I) are significant. Those for winter and summer were selected for the following analyses.

VI. Climatological Studies

When the feasibility study was undertaken to reconstruct spatial variations in pressure from tree growth, it was thought that the anomalies in pressure would be readily interpretable in terms of anomalies in storm tracks, air flow, temperature, and precipitation. However, it soon became apparent that heretofore no summary had been made of the seasonal climate over the selected grid of pressure, and there was little reliable basis for interpreting the new reconstructions. Some climatologists had looked in depth at only the months of January and July, while others dealt only with average conditions over a particular region. Still other investigations were limited to analyses of particular years and seasons or to consideration of only the last few decades, during which measurements were available from the upper atmosphere as well as the surface.

Therefore, Blasing (1975) undertook the task of studying the 20th century climate in terms of the surface pressure variance over the newly selected grid. He developed a technique which was used to classify and identify anomalous patterns in the seasonal pressures. Each of these anomaly patterns, which were called *pressure types* (Fig. 9.8) were in turn associated with anomaly patterns of cyclone occurrences, temperature, and precipitation in the United States, and tree-ring widths in western North America. The most pertinent results to dendroclimatographic analysis are described and summarized in the following sections.

A. *Identification of Pressure Types*

Eigenvector techniques have mathematical properties which limit their usefulness in describing climate. The eigenvector patterns, unlike the natural pressure patterns, are constrained to be mutually orthogonal. Also the eigenvectors are constrained to express both positive and negative modes of behavior like mirror images of the same pattern, yet the actual modes of climate rarely behave in exactly opposite fashions. Years with anomalously high pressures in certain regions are not always balanced by years with anomalously low pressures in the same regions. As a result, features observable in climatic data require linear combinations of two or more eigenvectors for adequate characterization (see Julian, 1970, and Blasing, 1975).

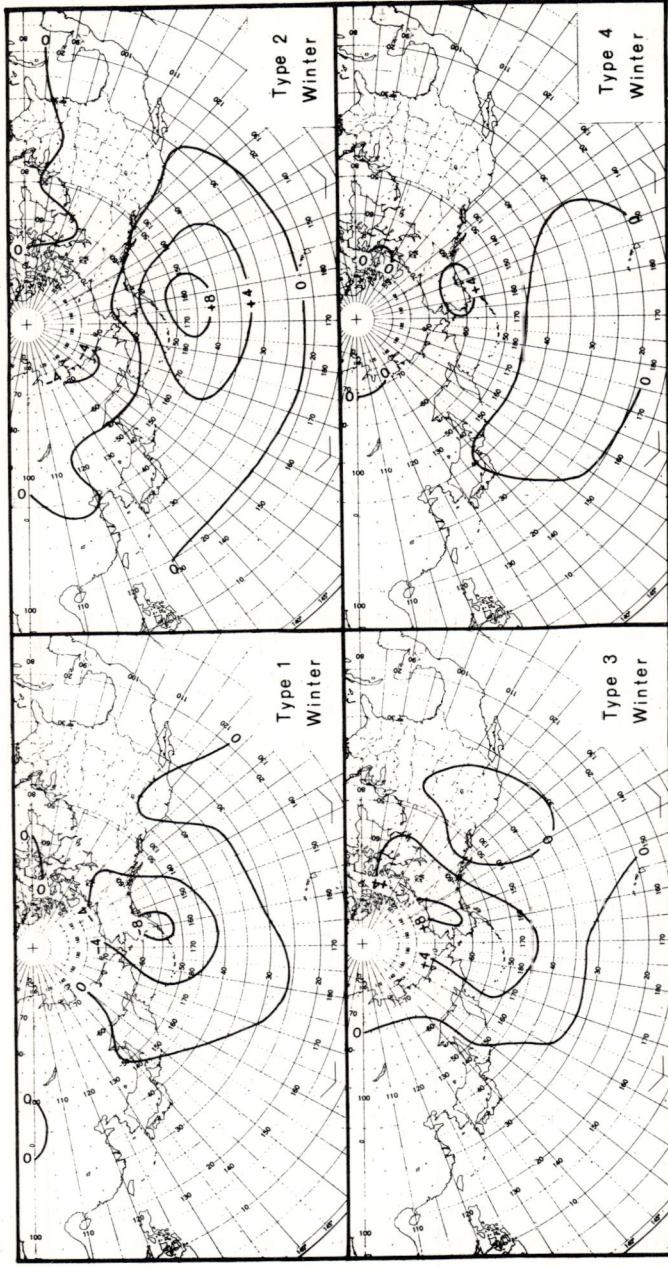

FIG. 9.8. The spatial correlations anomaly types of pressure for winters from 1899 through 1966. The number of the type is assigned in order of its importance during the 20th century. (Blasing, 1975)

The typing procedure developed and programmed by Blasing (1975) circumvents the above constraints of eigenvector analysis. It finds modes of seasonal climatic behavior (or growth or any other selected variable) that are recognizable in the data for certain years included in that analysis, but are not necessarily associated with an anomaly of opposite sign. Therefore the types, unlike eigenvectors, are characterized only in their positive representation.

The statistical procedure, which was first described by Lund (1963), computes the correlations among patterns in a field of variables at a given time with patterns at other times. In this particular analysis, the pressure departure patterns of each year from 1899-1966 are correlated with all other pressure departure patterns for the 68 years used in the analysis. The resulting correlation coefficients are then searched to identify that pattern in pressure which is highly correlated with the greatest number of other pressure patterns.

The procedure Blasing developed differs from that described by Lund in that it averages data for the selected year with the five other best correlated departure patterns (Blasing, 1975). The averaging process filters out some of the noise due to individual yearly variations, and it preserves the essential features that give rise to the correlations.

The mean pressure anomaly of the six best correlated years becomes the first spatial correlation anomaly type, which is then correlated with the remaining yearly departures. All anomaly patterns that are correlated with the six-year mean (that is, which are equal to or greater than a prespecified correlation value) are classified as that type and are removed from the data set. The procedure is repeated on the remaining data until the correlation coefficients among yearly anomaly patterns fall below a minimum acceptable value, at which time the analysis stops and the remaining years not associated with any type are left unclassified. Four spatial correlation anomaly types of pressure were identified for winter (Fig. 9.8) and five were identified for summer (Blasing, 1975).

B. *Characterization of Types*

The data on cyclones, temperature, precipitation, and tree growth were each averaged for the five years which were best correlated with each summer and winter pressure type, the averaged normalized ring-width departures were mapped (Blasing, 1975). The resultant plots (see Fig. 9.9) represent the anomaly patterns in each variable corresponding to each pressure type. Since these data are used later in the chapter to describe past climate, the most important results are summarized in the following sections.

1. *Winter*

The four pressure anomaly types for winter were numbered in the order of their importance, plotted on maps, and isobars drawn as in Fig. 9.8. Type 1 exhibits a negative anomaly in pressure centered over the Alaskan Peninsula. Type 2 consists of a positive pressure anomaly south of the Alaska Peninsula and a negative pressure anomaly over nothern Siberia. Type 3 exhibits a positive pressure anomaly centered over Alaska, with a region of slight negative anomaly along the west coast of North America. Type 4 includes an area of positive pressure anomaly along the south coast of Alaska and a large area of weak negative anomaly over the North Pacific Ocean.

Characteristics of Type 1 winters are shown in Fig. 9.9. The most characteristic winters of this type were those ending in 1931, 1953, 1945, 1912, and 1927 listed in order of decreasing correlation with the mean type. An above-normal number of cyclone-days occurs in the area of negative pressure anomaly over the Alaska Peninsula. High cyclone frequencies also occur in certain areas of Asia, in northwest Canada, in areas north of 30°N from 140°E through southern California to Texas, and in the western portion of the Mississippi Valley.

Climatologically, the latter two areas represent cyclone tracks along which the storms intensify when warm moist air from the Gulf of Mexico is drawn into them, increasing precipitation in the central Plains and along the east slope of the Rocky Mountains (Fig. 9.9). Counterclockwise circulation around the area of negative pressure anomaly results in anomalous northward or northeasterly flow of relatively warm air from the Pacific Ocean into western Canada and northwestern United States, causing temperatures there to be above normal. With fewer storms, precipitation along the west coast of the United States is below average.

Blasing (1975) states that "the upper air pattern for Type 1 winters involves a strong Aleutian Low and a tendency for a broad ridge [high pressure] over the western United States." When such winters prevail, there is also a smaller than normal number of polar outbreaks of cold air in the United States and above normal precipitation along the south coast of Alaska because of the greater number of storms there.

The tree-growth pattern associated with Type 1 winter indicates that low growth occurs in Canada and the Great Basin, while in the extreme Southwest and southern Rocky Mountains growth is high, probably due to the abundant winter moisture brought into the area by the northerly flow. The high growth over Wyoming, southern Idaho, and southern Montana probably reflects mild winters and the upslope flow of moist air from the south that may occur there.

Type 2 winter is best exemplified by the years 1937, 1949, 1932, 1911,

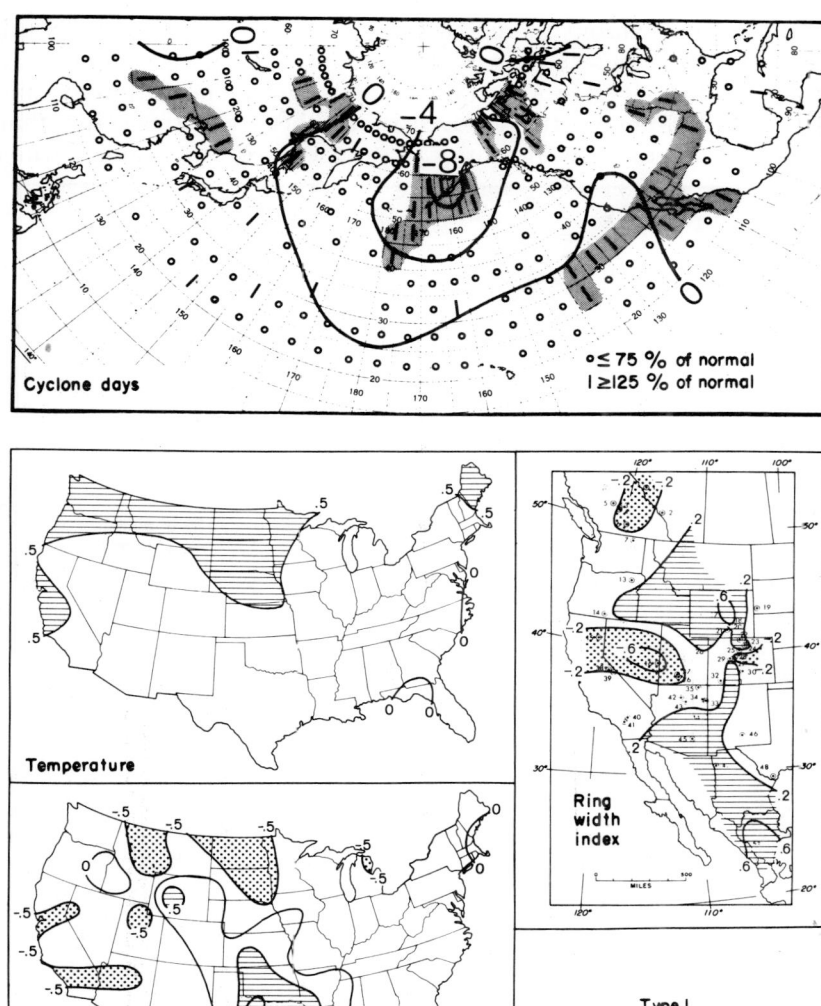

FIG. 9.9. Anomalies of climate and ring-width index occurring in the five years characterizing Type 1 winters. Cyclone days are expressed as percentage of normal for each five-degree gridsquare. Shading indicates large areas of anomalously high cyclone frequencies. Isobars of pressure are superimposed on the cyclone-day anomaly map. Temperature, precipitation, and ring-width index anomaly values are in standard deviation units for each parameter at each station or site. (Blasing, 1975)

460   TREE RINGS AND CLIMATE

and 1952, and it exhibits anomalously high pressures in the central North Pacific, where frequencies of cyclones are low (Fig. 9.8). A higher than normal number of cyclones enter the United States along the Oregon and northern California coast and move east–northeast bringing colder than normal temperatures to the entire western United States and warmer

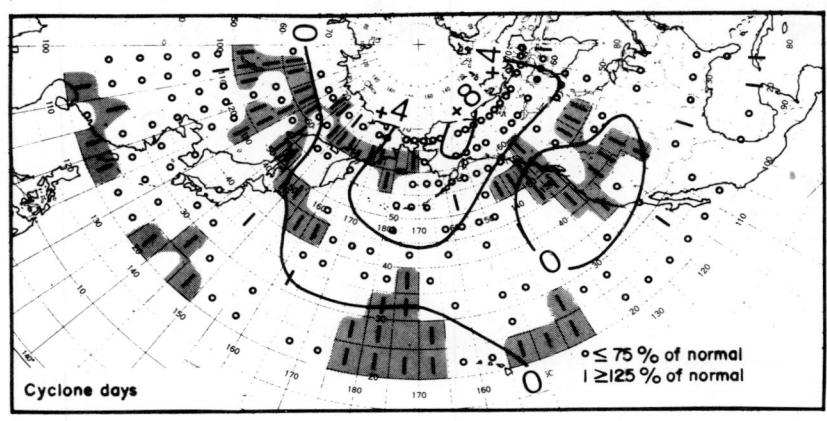

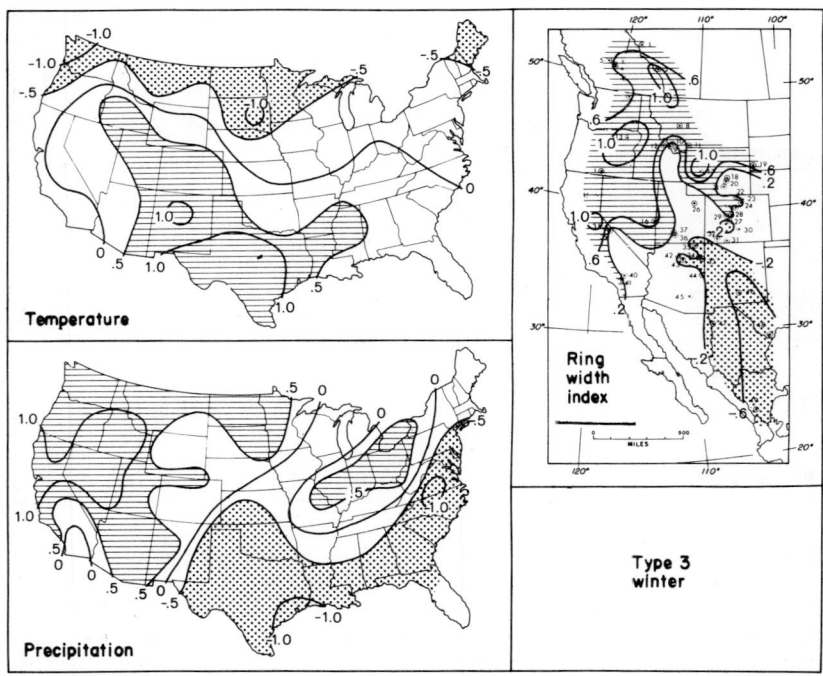

FIG. 9.10. Same as Fig. 9.9, but for Type 3 winters. (Blasing, 1975)

than normal temperatures to the eastern United States. Precipitation is much above normal in the Ohio and lower Mississippi valleys, somewhat above normal in the northern Great Plains and Great Basin, and below normal in the extreme Pacific northwest. The tree rings are wider than normal over the Colorado plateau of the Southwest and lower than normal over northern California, Oregon, Idaho, and Washington. Illustrations and more details on this and other winter types are presented by Blasing (1975) and Blasing and Fritts (in press).

The anomalies associated with Type 3 winters are shown in Fig. 9.10. The years best correlated with the type are 1907, 1916, 1956, 1904, and 1909. The area of positive pressure anomaly over the extreme North Pacific, Alaska, and northwestern Canada is associated with a smaller than normal percentage of cyclone days there. A larger than normal number of cyclones enters the continent farther south along the Canadian and United States west coast. Blasing (1975) suggests that there is an upper air ridge over the North Pacific Ocean, a well-defined trough (an area of lower pressure) off the west coast of the United States, and a weak ridge over west central United States. Lows entering the west coast of the United States often head northeastward in the direction of the upper air flow, although it is not uncommon for them to enter British Columbia and follow a southeastward course. The number of storms are near normal for central and eastern North America. Anticyclones following the cyclones across North America often remain farther north than normal, leaving the central and southern United States with warm air. Cyclones are also common in Asia for latitudes between 50°N and 65°N.

During Type 3 winters the northern United States is anomalously cool because of the southward flow of air from the Arctic, but temperatures elsewhere in the United States are above normal. Precipitation in western United States is high due to the frequent passage of storms. The warm, moist air from the Gulf of Mexico meets the cold air farther north than normal causing abnormally high precipitation in the Ohio River Basin and western Appalachian Mountain areas. Lower than normal precipitation in the southern United States is due to the low percentage of cyclones there.

In all respects except the growth pattern, Type 3 winters are opposite those of Type 1. While tree growth for Type 3 years is high in northwestern North America where precipitation is also high, it is low in New Mexico and extreme western Texas where temperatures are higher than normal and precipitation is low.

Type 4 winter is best exemplified by the years 1962, 1963, 1957, 1965, and 1930, with anomalously high pressure centered over the southern coast of Alaska and slightly below-normal pressures over the central

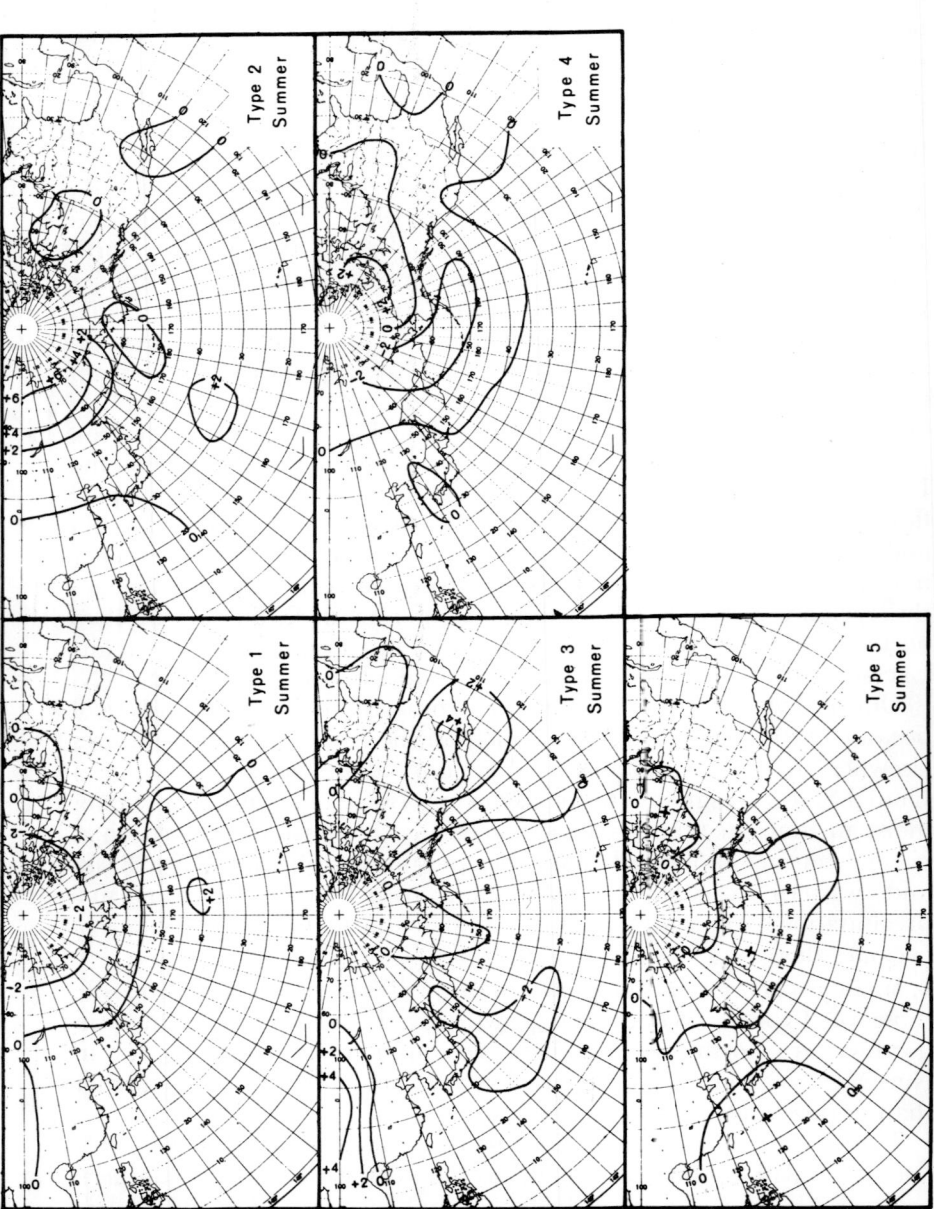

FIG. 9.11. The spatial correlation anomaly types of pressure for summer from 1899 through 1966. (Same as Fig. 9.8.)

North Pacific (Fig. 9.8). Cyclones are more frequent than normal along the 45° parallel in the central North Pacific. They are also more frequent over the Baja California and Hudson's Bay areas. Temperatures are much below normal in the entire Northwest and the northern Plains—especially over the upper Mississippi Valley—and precipitation is below normal in the central and southern Plains, the northeast, and west coast regions of the United States. High growth occurs in northern British Columbia, in the northern Sierra Nevada, the Great Basin and central Rocky Mountains, and in southern Arizona; and low growth occurs in the Cascades and northern Rocky Mountains, southern California, northern Arizona, and New Mexico.

2. *Summer*

The five pressure anomaly types for summer as described by Blasing (1975) are included in Fig. 9.11. Type 1 exhibits higher than normal pressures in the central North Pacific and lower than normal pressures in the Arctic. Type 2 exhibits a higher than normal presure anomaly in northeastern Siberia and in the western North Pacific. Type 3 is characterized by anomalously high pressure along the western borders of the grid in central Asia, over the Pacific Coast and the Great Basin of the United States, and over the Bering Sea. Lower than normal pressures occur over the western North Pacific Ocean and over eastern North America. Type 4 has many features in common with a typical winter anomaly pattern showing anomalously low pressure over the Bering Sea and the Gulf of Alaska, and higher than normal pressure anomalies over the Arctic Ocean, northern Alaska, and northwestern Canada. Type 5 exhibits a weak positive pressure anomaly over northeastern Asia, southern Alaska, and the northern parts of the North Pacific. Positive anomalies are found in central Canada and in the China Sea.

Figure 9.12 includes the climatic data associated with Type 1 summers for 1955, 1937, 1949, 1935, and 1933. Anomalously high pressure over the Pacific Ocean indicates a northwesterly displacement of the north Pacific High. As a result, storms are displaced northward and mean summer pressures are low along the Arctic Ocean. Cyclones are more common in the Caribbean Sea, across Mexico, in areas of western United States, and along the Washington and Oregon coast. More cyclones than normal also occur in the central North Pacific to the southwest of the high pressure area.

The West Coast and Great Basin area, where Type 1 summer cyclones occur more frequently than average, is a region where such activity is actually quite rare. The cyclone days, though above normal in number, represent weak, short-lived systems associated with a northerly displacement of a quasi-stationary low that develops over the Southwest during the summer. Warm, dry air is advected into western United States bringing drought and higher than normal temperatures except for areas along the Mexican border of California and Arizona where rainfall is above average.

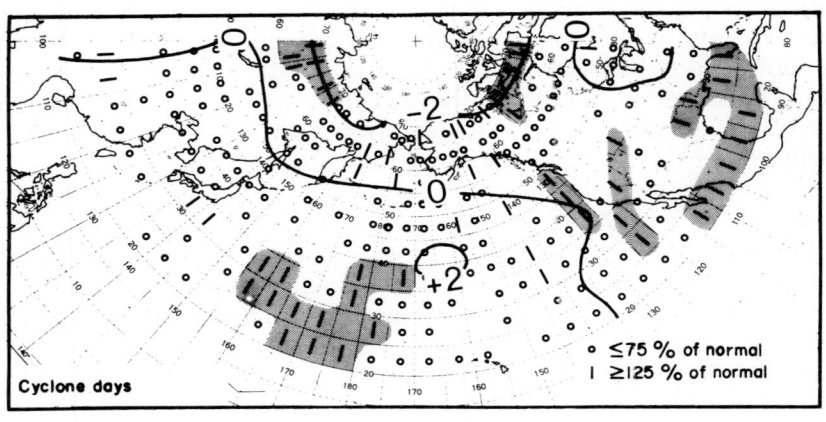

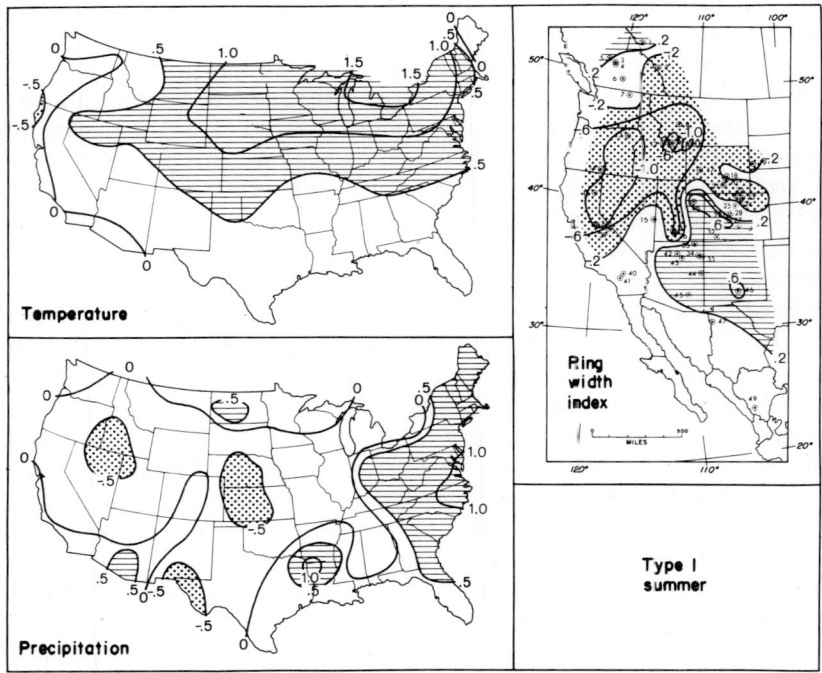

FIG. 9.12. Same as Fig. 9.9 except for Type 1 summers. (Blasing, 1975)

Blasing (1975) suggests that the Atlantic Subtropical High is displaced northward bringing above-normal temperatures to the eastern United States, especially in the Great Lakes. The association of this condition with hurricanes in Type 1 summers accounts for the higher than normal precipitation in the east and Gulf states, as well as the large number of cyclone days in the Caribbean and central Mexico.

The circulation patterns associated with Type 1 summers resemble those described by Namias (1960) which occur in conjunction with drought in the Great Plains of the United States. During such summers there is a tendency for a drought-producing anticyclone to appear over the United States. However, if it is small in size, the eastern coast is typically under the influence of a moist tongue of air from the Bermuda High to the east over the Atlantic. Some good examples of Type 1 circulation occurred during the dust bowl years of the 1930's, while the best example of this pattern occurred in the summer of 1955 bringing extreme drought to the Plains and moisture-laden hurricanes Connie and Diane to the East.

The higher than normal precipitation anomaly along the boundary of the north central United States is associated with excessively severe storm activity, often accompanied by hail and heavy local precipitation. Blasing (1975) suggests that these features are triggered by the southern end of cold fronts which are located in Canada. Above-normal precipitation also occurs in the portion of the southern Alaskan coast called the panhandle. Tree growth occurring in Type 1 summers is much below normal in northwestern United States but above normal in northern British Columbia and southwestern United States.

Type 2 summer is best exemplified by the years 1922, 1921, 1920, 1917, and 1919, with higher than normal pressures over northern Siberia and the western North Pacific, and lower than normal pressures and greater frequency of cyclones southwest of the Alaskan mainland, over central Canada, and west of Baja California. Temperatures are above average over the northern Rocky Mountains and the Great Basin, and lower than normal over southern Arizona, New Mexico, and coastal areas of the United States Pacific Northwest. Precipitation is generally below normal along the western Canadian–United States border, and above normal over Arizona. The growth anomalies are lower than normal in Canada, the Pacific Northwest, Idaho, and central Wyoming, and high throughout the entire Southwest. Illustrations and more details on this and other summer types are presented by Blasing (1975).

Type 3 summer is best exemplified by 1927, 1928, 1930, 1929, and 1926, and it represents an extreme north–south (meridinal) circulation with marked high pressures along the western boundary of the pressure grid (100°E longitude) and over the southwestern United States (Fig. 9.11). Pressure is anomalously low over the western North Pacific. Temperatures are above normal throughout the Pacific Northwest of the United States and southern Arizona and generally below normal throughout the southern Rocky Mountains, much of the central Plains and Midwest, and especially over the Great Lakes. Precipitation is much above normal throughout the central Rocky Mountains and high plains from southern New Mexico to southern Montana.

The tree-growth anomalies for Type 3 summers exhibit low growth in the Canadian Rocky Mountains and North Pacific states, southern Idaho, Nevada, northern California, and southern Arizona, but high growth throughout western Montana, Wyoming, Colorado, New Mexico, northern Arizona, and north central Mexico.

Type 4 summer is best exemplified by the years 1909, 1908, 1903, 1950, and 1964, and it is characterized by lower than normal pressures and an increased number of storms from northeastern Siberia to the Gulf of Alaska, including more than normal storms over the ocean west of the North American continent. Pressures over western North America are somewhat lower than normal but higher than normal over northwestern Canada. A higher than normal number of cyclones move inland in a southeasterly direction along the western Canadian coast, bringing cool temperatures to large portions of the United States. Precipitation is above normal over the western United States/Canadian boundary, Arizona, southern California, southern Nevada, and large areas in the Mississippi Valley and Great Lakes. Tree growth for this cool, relatively moist type is anomalously high except in Colorado, New Mexico, and the Colorado Plateau of Arizona and Utah.

Summer Type 5 resembles Type 1 in that it is associated with drought (Fig. 9.13). The five best examples of Type 5 are 1963, 1957, 1959, 1936, and 1954, characterized by higher than normal temperatures throughout the central United States and by a drought area extending from the Plains to the eastern seaboard and the Northeastern United States.

The Pacific Subtropical High is displaced north of its normal position, and areas of weakly positive pressure anomaly and below-normal cyclone days occur in the extreme North Pacific, southwestern Pacific, the south coast of Alaska, and central Canada. Upper air features include two anomalously deep troughs of low pressure which occur off both the east and west coasts of North America associated with high-pressure ridging over the central portions of the United States. The upper-level jet stream in the Central Pacific is displaced to the north and at times becomes split, the southern current plunging south bringing low pressure and cold air to the Pacific Northwest, and the northerly current flowing over the Arctic

9. RECONSTRUCTING SPATIAL VARIATIONS IN CLIMATE

and then returning southward in the trough along the east coast bringing a southward advection of cold, dry air (Fig. 9.13).

Type 5 summer differs from Type 1 in that the upper level troughs are more apparent, the Atlantic Subtropical High is not as likely to be

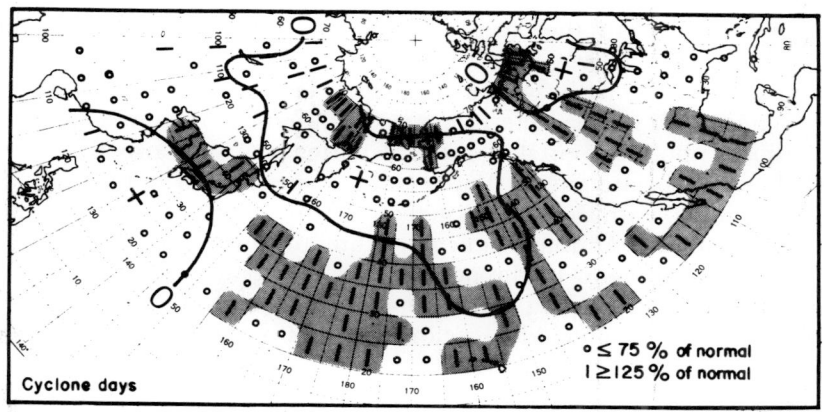

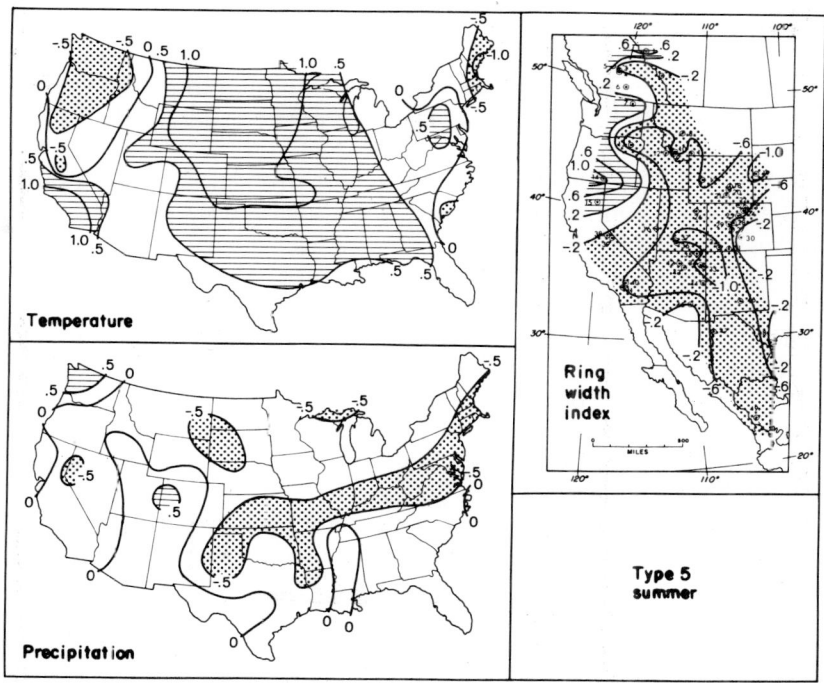

FIG. 9.13. Same as Fig. 9.9 except for Type 5 summers. (Blasing, 1975)

displaced northward, and drier air is advected into the eastern portions of the continent. Under a Type 5 regime, drought conditions occur over the eastern half and central portions of the United States associated with the descending air entering the eastern trough.

Tree growth for Type 5 is below normal for large areas of western North America, with the exception of northern California, Oregon, and Washington where growth is above normal.

It is apparent from the work of Blasing (1975) that each pressure type is associated with a characteristic set of climatic conditions, synoptic situations, upper air circulations and patterns of tree growth. If it is possible to associate one or more of these types that are present in the modern period with the reconstructed pressure anomalies for past years, it would then be possible to infer the climatic conditions that must be associated with them. In the next section attention will be directed to reconstructing and typing the anomalies in past atmospheric pressure. Subsequent sections include the climatic interpretation and verification.

C. Reconstruction of Pressure Types

It was noted in Section V that the best calibrations use models with 10 or fewer eigenvector amplitudes of both tree growth and pressure (Table 9.1). The question arose as to what is the minimum number of variables necessary to reconstruct a particular phenomenon such as the pressure types.

Blasing (1975) addressed himself to this problem by developing a minimum-sized model that would reconstruct only the large-scale patterns represented by the four winter pressure types. It was reasoned for the model development that the best reconstructions of the particular types could be obtained by using only those eigenvector amplitudes that occurred consistently in the five type years (see subsection B). Therefore the values for the eigenvector amplitudes of pressure were tabulated for each year that was best correlated with the type (Table 9.II). An examination was made of the absolute value of the eigenvector amplitude, the mean amplitude for the five years, and the consistency in amplitude sign.

For example, the amplitudes of eigenvector 1 are large and their signs consistent for all years within Types 1, 2, and 3 (Table 9.II). The amplitudes of eigenvector 2 for pressure are large and signs are consistent for those years in pressure Types 2, 3, and 4, although the signs of the amplitudes are less consistent. Blasing noted that the amplitudes for eigenvectors 1-6 were moderately consistent for at least one type but were generally inconsistent for all higher-order pressure eigenvectors. He concluded that the pressure types were sufficiently well represented by the first six eigenvectors to justify their exclusive use in reconstructing them. Another set of statistics, V^2 (Sellers, 1968), which measures the percent variance reduced by each amplitude during each year, were examined and these data substantiated the decision to use only the first six.

TABLE 9.II Amplitude values of the first 10 pressure eigenvectors associated with years characterizing winter pressure types[a]

Type	Year	1	2	3	4	5	6	7	8	9	10
1	1931	33·12	15·73	3·14	−9·73	1·99	−0·16	2·53	−3·48	−7·50	−0·63
	1953	20·08	−7·20	7·82	−8·09	7·31	0·98	2·07	5·50	−1·80	1·75
	1945	18·90	1·47	8·41	−3·37	−1·52	−0·63	−0·87	0·08	−6·17	−0·05
	1912	10·60	−4·58	−1·37	−4·87	2·06	−3·29	1·44	−4·12	−3·72	0·49
	1927	13·66	−5·10	−0·96	−2·89	−1·71	−0·12	3·78	−0·64	−2·36	−3·90
2	1937	−29·16	−25·06	−5·34	12·17	−1·36	7·72	−2·44	−1·47	0·93	−3·18
	1949	−22·66	−18·88	−6·85	13·61	−8·49	−1·79	−2·75	1·72	−2·38	−0·06
	1932	−16·26	−19·76	0·66	4·84	10·68	4·07	4·21	1·48	1·62	−3·05
	1911	−21·80	−6·24	−1·26	−3·44	5·08	0·50	−3·26	−2·44	−3·68	−2·21
	1952	−9·84	−9·37	−3·28	0·97	1·64	5·78	0·43	4·36	2·99	0·01
3	1907	−32·02	28·84	−8·55	−11·06	−4·51	−9·44	−4·20	−0·06	−0·89	2·04
	1916	−28·80	10·36	−4·46	−6·69	−4·82	1·56	2·60	0·32	1·27	1·69
	1956	−19·99	12·71	−4·51	1·23	4·29	21·65	6·75	0·35	−2·89	−1·35
	1904	−26·07	10·67	−0·12	−2·64	3·17	−12·41	4·16	5·83	6·22	2·12
	1909	−15·76	7·95	−16·05	−13·89	−6·55	−3·16	−1·89	8·82	3·41	−6·95
4	1962	−9·82	11·39	3·34	16·94	−1·21	−1·94	0·59	3·64	−6·31	0·22
	1963	15·30	26·54	1·97	21·38	0·49	−3·04	0·82	0·54	−4·53	1·74
	1957	−15·80	11·02	23·69	4·82	7·86	0·38	−6·10	−0·22	−2·41	−0·54
	1965	−2·42	9·76	7·53	4·51	2·98	7·43	1·66	7·64	6·01	0·29
	1930	−3·84	4·87	4·68	8·04	7·03	−1·31	−2·86	−7·02	5·23	−4·99
Total % Variance Reduced by Eigenvector[b]		34·7	16·8	10·6	8·1	5·6	4·7	2·9	2·4	1·7	1·5

[a]Blasing, 1975 [b]For all years from 1900–1966.

The amplitudes of the tree-growth eigenvectors were examined in the same way. The first seven amplitudes of tree growth had both consistent signs and high mean values for all five years associated with at least one of the four types. There were several type years for eigenvectors 8 through 10 in which both the magnitudes of the amplitudes were 2·0 or greater (the average expected for random numbers) and the V^2 were sufficiently large. No amplitude for eigenvectors above 10 reduced enough variance or appeared systematically related to any type, so only ring-width eigenvectors 1 through 10 were used in the calibration and reconstruction. A similar analysis for summer led to the selection of the first 10 eigenvectors of pressure and the first 10 eigenvectors of tree growth.

The square of the multiple correlation for predicting the amplitude variance for these two minimum-sized models for winter and summer (Row 11, Table 9.I) are 0·292 and 0·280, but when these data were converted to total pressure, variance percentages declined to 24% and 26% (Fig. 9.7). The areas of significant calibrated variance for the minimum-sized models were not substantially different from those for the larger models, except that there was less significant variance over the Asian continent. On the other hand, the area of statistically significant reconstruction was greater for the summer minimum-sized model than for the larger models even though the total percent variance calibrated was less. Therefore, the minimum-sized models were inferred to be superior to the larger ones, especially for reconstructing the pressure types. Except when stated otherwise, the results from these minimum-sized models are used in all subsequent analyses.

VII. Summarization of Reconstructions for Winter Using the Pressure Types

Once a transfer function is obtained (Equation 9.7) it can be applied to the tree-ring data for each year to obtain the reconstructions. For the following discussion the reconstructions are departures of pressure (expressed in millibars) for a particular year and season over 96 grid points. Isobars are drawn to obtain pressure anomaly maps for half of the Northern Hemisphere. The reconstructions are obtained for each model and season back to 1700, totalling 263 maps. Figures 9.14 and 9.15 include examples of four such seasonal maps.

The selected years were chosen because historical records for North America indicated interesting anomalies in climate, and the reconstructions could be examined for consistency with the available facts. The year 1815-16 was chosen because the volcano Tambora erupted late in 1815, ejecting clouds of dust into the upper atmosphere. No outstanding anomaly was recorded for the winter, but the following summer was exceptionally cold in New England and became known as "the year without a summer" (Ludlum, 1966). Temperatures were reported much below normal, snow fell in June in New York, Vermont, New Hampshire, and Maine, and frosts were reported for July and August as far south as Pennsylvania and New Jersey, with drought extending into early autumn. The disastrous year was culminated with a killing frost in northern New England before the end of September (Ludlum, 1966).

The reconstruction for winter (Fig. 9.14) indicated a Type 1 pattern which is shown in Fig. 9.9 to bring generally warm, mild weather that would not be likely to evoke any recorded comment. However, the reconstruction for summer resembled Types 2 and 5, both of which are accompanied by anomalous cold and drought in the New England states. (See section VI for details on circulation of Type 5 summers.)

Figure 9.15 includes two winters known to have marked contrasts in climates. The winter (December, January, and February) of 1866 was

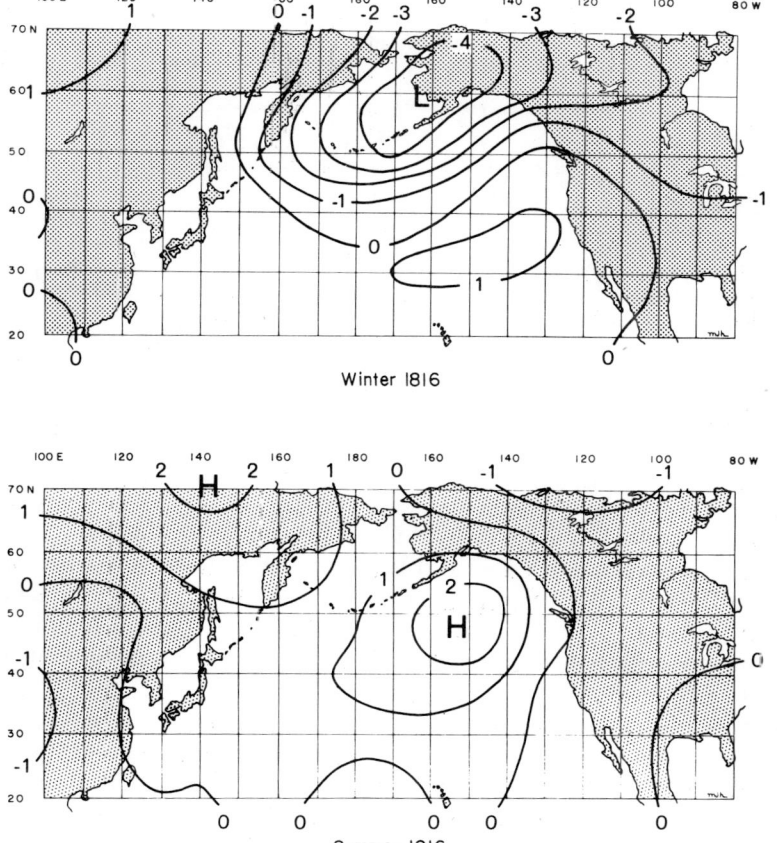

FIG. 9.14. Reconstructions of anomalies in pressure derived from tree rings for an interesting year in the 19th century. (Drawing by M. Huggins)

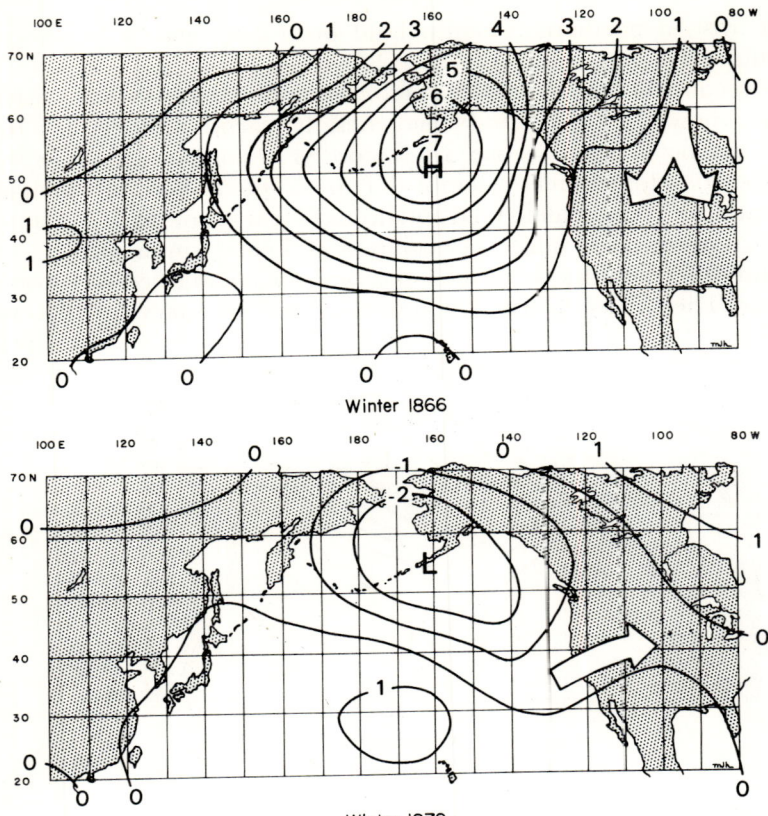

FIG. 9.15. Reconstructions of anomalies in pressure for winter derived from tree rings for two years known to be years of contrasting climate. The arrows show anomalies in atmospheric circulation inferred from the plots. (Drawing by M. Huggins)

reported by Stockman (1904) and Ludlum (1968) to be severely cold in the north central Great Plains, averaging 9·6°F above zero (−12·4°C) in central Minnesota. Temperatures were 3·3°F (1·8°C) below average for Fort Levenworth, Kansas, and 6·6°F (3·7°C) below average for Chicago, Illinois.

The reconstruction for the winter of 1866 (Fig. 9.15) exhibits an extremely strong Type 3 pressure pattern with anomalously high pressures over Alaska (Fig. 9.10) with an inferred anomalous flow of air from the Arctic as indicated by the arrows. This is consistent with the reported low temperatures in the north central states.

The winter of 1878, on the other hand, was unusually warm. Temperatures were as high as 14·0°F (7·8°C) above average at Fort Snelling, 10·0°F (5·6°C) above average at Fort Levenworth, and 11·4°F

(6·3°C) above average at Chicago. This winter as reported in the *Monthly Weather Review* for December, January, and February 1877-1878 was characterized by an abnormally strong Aleutian Low with numerous storms entering western Canada and northwestern United States. Heavy rains were reported in Texas and the Gulf Coast states, while there was a deficiency of moisture in the interior of the country. The reconstruction for winter of 1878 is a well-developed Type 1 pattern, which is consistent with the *Monthly Weather Review* report. The strengthened Aleutian Low apparently caused an anomalous influx of moist and warm air from the central North Pacific and the Gulf of Mexico (see arrow in Fig. 9.15) with anomalously high temperatures throughout the United States and high rainfall in Texas as reported.

Because of the immense amount of data and the difficulty in summarizing these reconstructions one year at a time, it was necessary to find a rapid way of characterizing the climatic patterns. One method used in the preliminary analysis (Fig. 9.6) involves averaging the data by decades or pentads and plotting mean maps. However, it was noted that the climatic variations did not confine themselves to arbitrarily assigned intervals of time. Contrasting anomalies within the interval can cancel one another so that the resulting mean does not represent the climate correctly.

An alternative method is to associate each reconstructed pressure anomaly with the best correlated 20th century type as described by Blasing (1975) and as summarized in the prior section. The frequency of occurrence of each type could then be studied through time, and associated anomalies in temperature and precipitation inferred from the types.

The occurrences of the types were identified in Fig. 9.16 by correlating the spatial patterns of the reconstructions with each of the pressure types (Figs 9.8 and 9.11). The correlation coefficients were examined, and each seasonal reconstruction classed as the type with which it was most highly correlated. When no coefficient was significant ($r \geq 0\cdot3$), the type with the highest correlation was assigned, but the case was designated as insignificant. The full-sized bars plotted in Fig. 9.17 represent those cases in which the correlation was significant, while the shorter bars represent those cases in which the correlation was insignificant (Blasing and Fritts, in press).

In addition, the actual pressure data for the 20th century were correlated with each pressure type and tabulated in the same fashion as the reconstructions. The types were rearranged in the order of their importance during the period of reconstruction (18th and 19th centuries) and their occurrences plotted in Fig. 9.17. For ease in interpretation, the

474 TREE RINGS AND CLIMATE

predominant climatic anomaly within the United States that is associated with each type is indicated to the right of the plotted data for the 20th century.

It can be inferred from Fig. 9.17 that both the distribution through time and the frequencies of occurrence of the types have varied considerably in the past two centuries from those of our present century. Type 1 winters, which exhibit a strengthened Aleutian Low, higher than normal temperatures, and anomalously low precipitation in the far West

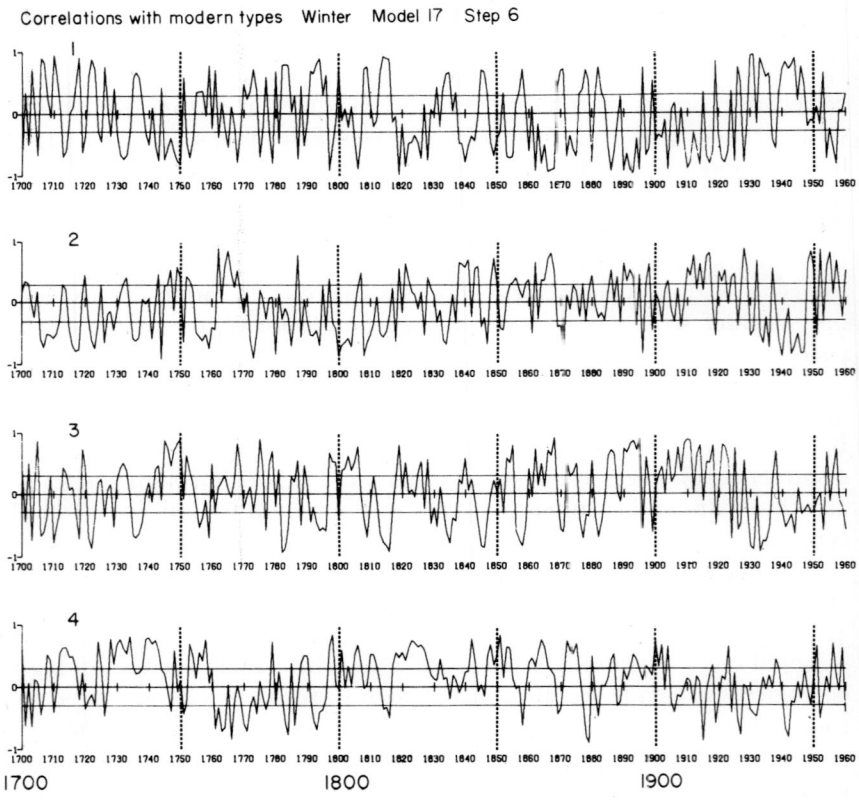

FIG. 9.16. Plots of the correlation coefficients between the pressure anomalies reconstructed for each year from 1700 through 1960 and each winter type (see Fig. 9.8) where the number of observations for each correlation coefficient are the values of the anomalies at the 96 grid points. The line at a correlation of $+0.3$ marks the approximate 0.95 confidence level. All years with correlations above this line can be identified as belonging at least in part to that particular type. The years with negative correlations with a type, even though significant, have no concrete interpretation except that the pressure anomaly reconstructions for the year were definitely not the particular type.

(Fig. 9.9) have occurred most frequently all three centuries. They occurred most often in the 18th century reconstructions (37%) and least often in the 19th century (28%), as compared to the present century (34%).

Type 3 winters (Fig. 9.10), which exhibit positive pressure anomalies centered over Alaska and are generally warm and wet in western United States but cold in the extreme north, appear only slightly more important

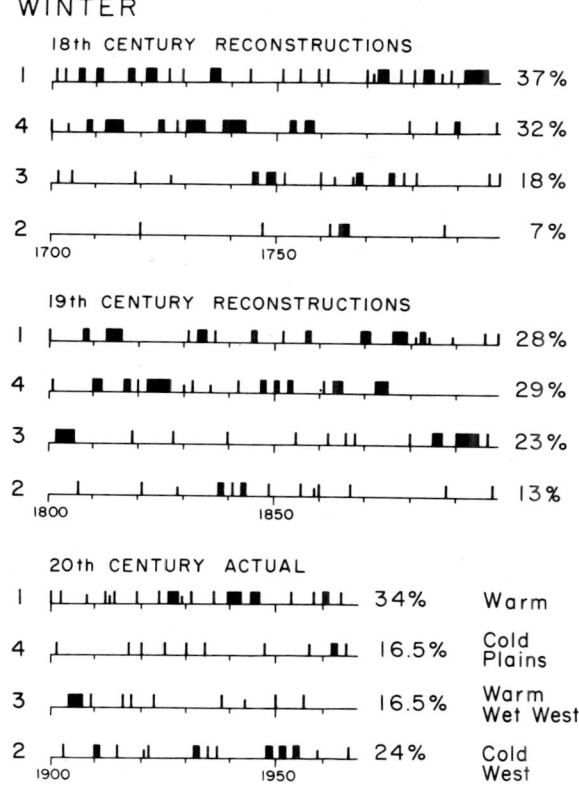

FIG. 9.17. The occurrences of the four winter pressure types as reconstructed from tree rings for A.D. 1700 through 1899 and as actually occurred in the pressure data from 1900 through 1966. Every year is identified as the type with which it is best correlated. The short bars designate those cases for which the correlation was less than 0·3 and therefore insignificant. The type numbers represent the order of their importance during the 20th century, but the order used in the figure represents their relative importance during the 18th and 19th centuries. The percentage occurrences of significant correlation are shown on the right, and the predominant climatic anomaly over the United States associated with each type is indicated as abbreviated descriptions next to the plots on the lower right. (Blasing and Fritts, in press)

in the two earlier centuries than in the current one. They occur 18% and 23% of the time compared to 16·5% for the 20th century.

Type 4 winters, with higher than normal pressure over the south coast of Alaska, lower than normal pressures over the central North Pacific, and anomalously cold outbreaks of arctic air into the northern Plains, occurred 32% of the time in the 18th century and 28% of the time in the 19th century, but only 16·5% of the time in the current century. This cold winter type was clearly more frequent in the past and there is a suggestion in recent records that it may be increasing in frequency at present (Blasing, 1975).

Type 2 winters, which exhibit a strong positive pressure anomaly in the north Central Pacific, a weak negative anomaly to the north and east (Fig. 9.8), and colder than normal winters in the far western United States, occurred 7% of the time in the 18th century, 13% of the time in the 19th century, and 24% of the time in the 20th century.

Figures 9.16 and 9.17 also show distinct climatic variations within the three centuries. For example, the first half of the 18th century can be characterized as varying between winters of general warmth and moderate dryness (Type 1) and winters of extreme cold in the high Plains (Type 4). Beginning in 1745, Type 3 winters were more common, with warmth and moisture in the West, cold in the extreme northern U.S. (Fig. 9.10), and the extremely cold winters in the Plains (Type 4) occurring less frequently. With the possible exception of the 1780's, these more mild conditions continued until 1810.

The cold Type 4 winters occurred with greater frequency between 1810 and 1875, interspersed with periods of warmth (Type 1). In the latter part of the 1840's and the 1850's, Type 2 (cold West) occurred somewhat more frequently. Starting with 1885, Type 3 dominated the circulation until near the end of the century. The persistent occurrence of Type 3 was broken by the occurrence of other types at the turn of the century, but beginning in 1905 Type 3 again occurred more frequently until near the end of the first decade. All four types occurred between 1910 and 1938, followed by a high frequency of occurrence of Types 1 and 2 until late in the 1950's. Since that time, Type 4 increased in frequency. For more details of these particular climatic reconstructions see Blasing (1975) and Blasing and Fritts (in press).

VIII. Verification of Reconstructions for Winter

The typing scheme not only provides a convenient means for characterizing the climatic reconstructions, but enables the comparison of the reconstructions with the climatic anomalies that actually occurred.

This section describes three ways in which the reconstructions were compared with actual climatic data to establish verification. (1) The reconstructed types for the winters of 1900-1961 (the dependent data set) were compared to the types derived from the actual pressure data. (2) The temperature and precipitation anomalies associated with the reconstructed pressure for years prior to 1899 were compared with the temperature and precipitation anomalies for the type years of the 20th century. (3) Calibrations were made for the relationships between tree-ring anomalies and temperature and the reconstructions of these data prior to the calibration period were compared not only to independent temperature data but also to the patterns of temperature expected during the years of the 18th and 19th centuries reconstructed to be of each pressure type (Fig. 9.17).

A. Verification of Types Using the Dependent Data Set

Although the percent variance calibrated is highly significant, an additional test was made by comparing the type patterns in the reconstructed pressure field with those in the actual pressure field. The correlation coefficients of both the actual and reconstructed data with each type were examined and all significant cases, that is those equal to or greater than 0·3, were identified and classified as belonging to that type. If the correlation coefficient for more than one type was significant, the case was identified as belonging to more than one type. After both the actual and the reconstructed pressure fields were so classified, a contingency analysis was made in the following manner: A reconstruction was considered correct if both the reconstruction and actual data were significantly correlated with the same type. It was also assumed that the tree response to climate can sometimes lag one year behind the occurrences of actual climate, so that situations with a significant correlation lagging one year behind climate were tallied as correct along with those that were significant at no lag.

The number of years actually typed using pressure data from 1899 to 1960 are shown in Table 9.III (line 1) along with the number of years reconstructed as each type (line 2). The first question tested was: Given a year in which the actual pressures resembled a type, how many of the reconstructions were of the same type? Line 3 in the table includes these cases, while line 4 includes additional correct reconstructions allowing for a lag in the tree-ring responses one year behind the actual occurrence of the type.

The second question tested was: Given a reconstruction of a type, how many cases in the actual data were of that type? The no-lag associations

TABLE 9.III The number of joint occurrences between the actually occurring pressure types and the types occurring in the reconstructions for winters 1899–1960[a]

Line	Case	Type[b]				Any Type[c]
		1	2	3	4[d]	
1.	Years actually typed	21	19	18	11	56
2	Years reconstructed as the type	21	24	22	11	57
3.	Correctly reconstructed with no lag	13	12	10	3	34
4.	Additional correctly reconstructed given the actual occurrence with a one-year lag in tree growth	2	4	4	2	10
5.	Additional correctly reconstructed given the reconstructed occurrence preceded by the actual occurrence	3	4	4	3	12
6.	Total lines 3 + 4	15	16	14	5	44
7.	Total lines 3 + 5	16	16	14	6	46
8.	Percentage correctly reconstructed given the actual occurrence (line 6/line 1)	71	84	78	45	79
9.	Percentage actual occurrence given the reconstruction (line 7/line 2)	76	67	64	55	81
10.	Percentage actual frequency during the 20th Century (line 1/no. yrs)	34	31	29	18	90
11.	Percentage reconstructed frequency during the 20th Century (line 2/no. yrs.)	34	39	35	18	92

[a]Reconstructions using minimum model for winter, Table 9.II, line 11. [b]Includes all cases where $r \geq 0.3$. [c]Cases where more than one type was identified in a particular year are tabulated only once in this column. [d]The best correlated years representing Type 4 occurred in winters of 1962, 1963, and 1966, which is after the period with continuous tree-ring data. Thus reconstructions and verification of Type 4 for the three years of highest actual correlation were not tested.

for this occurrence are the same as those in line 3 (Table 9.III). However, the lagged associations differ from line 4 in that the cases of reconstructed types—not the actual occurrences—are examined, that is cases where a particular reconstruction is preceded one year by an actual pressure pattern of the same type. Line 6 in Table 9.III is the total number of successes given the actual type, and line 7 is the total number of successes given the reconstructed type. These data are then expressed as percentages in lines 8 and 9. In lines 10 and 11 the numbers of types in the actual and reconstructed data shown in lines 1 and 2 are expressed as percentages of all years examined for the 20th century.

The frequency of occurrence of types during the calibration period was nearly the same for the reconstructions as for the actual data (57 out of a total of 61 years). The results in line 8 in the table indicate that a total of 77% of the years were reconstructed as the correct type, with the actual occurrences of Type 2 being correctly reconstructed 84% of the time and occurrences of Type 4 (the most poorly correlated type) correctly reconstructed only 45% of the time. The analysis period did not include the three characteristic years of Type 4, 1962, 1963, and 1965. It is clear that given a reconstruction of a particular type in the dependent data set, the probability that the climate was actually correlated with that type is 0·79. However, the reconstructions for Type 4 are more likely to be in error than those of the other three types.

B. *Verification of Types using the Independent Data Set*

A more rigorous test of verification can be achieved by using climatic data for years outside the dependent set. While little reliable data on pressure exists prior to 1899 for North America and the North Pacific, there is more reliable information on temperature and precipitation available for a number of weather stations. It was reasoned that if the pressure reconstructions for the 19th century were correct, then those years reconstructed to be of a given type should exhibit precipitation and temperature anomalies resembling those in the 20th century belonging to that type.

An examination of the correlation coefficients between the pressure reconstructions and the four winter pressure types (Fig. 9.16) revealed that 18 out of the 25 cases for winters for the interval 1874-1898 are highly correlated with a type (only those cases where r was equal to or greater than 0·5 are used). Four winters resemble Type 1; one resembles Type 2; 11 resemble Type 3; and two resemble Type 4 (Blasing, 1975 and Blasing and Fritts, in press). A test was made as follows. Temperature and precipitation data for the United States, available in the World Weather Records (Clayton, 1927), during the 1874-1898 period were averaged, normalized, and plotted on maps for all years of a given type. Figures 9.18 and 9.19 include the results for years of Types 1 and 3 respectively, which were the types most often reconstructed. The anomalies were compared and tested against the anomalies associated with the 20th century occurrences (Figs 9.9 and 9.10).

The four winters of the 19th century identified as Type 1 were remarkably in agreement with those prevailing in the 20th century

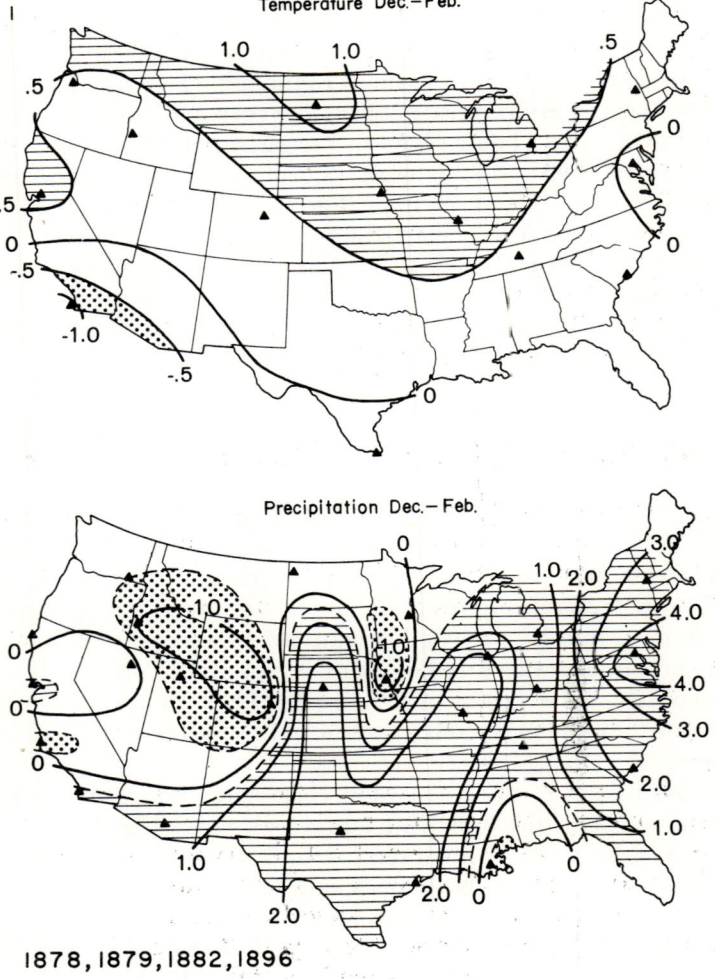

FIG. 9.18. The average anomaly of winter temperature and precipitation (expressed in standard deviation units) that occurred in the four years identified as being Type 1 in the independent period 1875-1898. The triangles indicate the location of the stations used as data points. The similarity of the climatic anomaly with those in Fig. 9.9 verify that the inferred climatic anomalies deduced from Fig. 9.9 are correct. (Blasing, 1975)

during Type 1 years. Both exhibited anomalies of high temperature throughout northern United States, low precipitation in the west, and a tongue of high precipitation associated with the northward flow of air from the Gulf of Mexico into the midcontinent. Only the heavy precipitation in the eastern third of the country is in disagreement. This

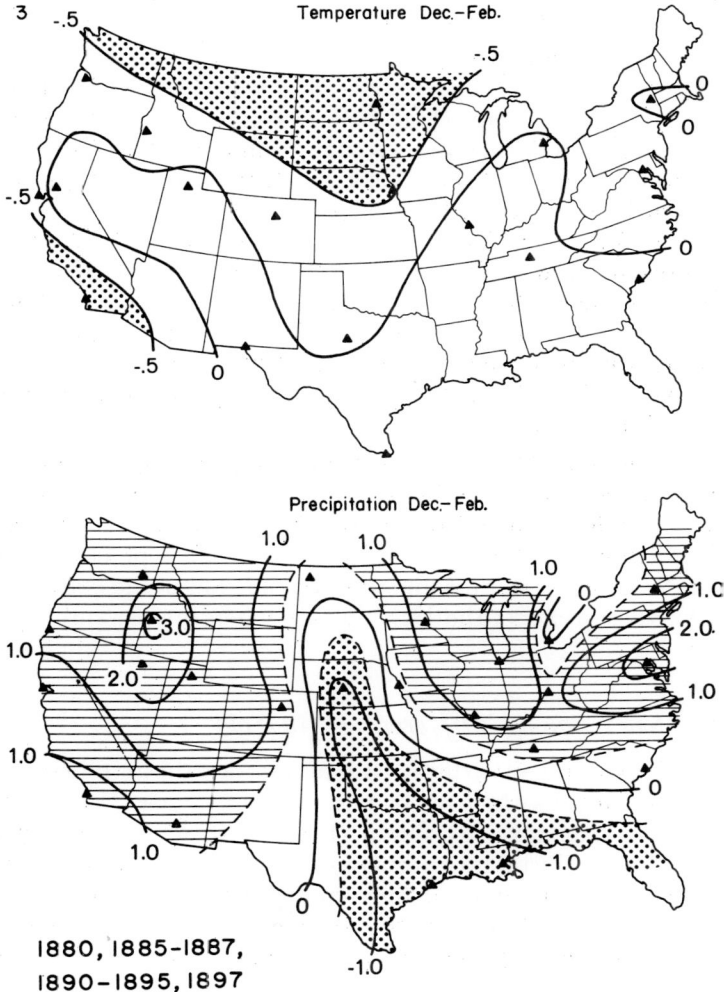

FIG. 9.19. The same as Fig. 9.18 except that it is for 11 years of Type 3 climate. The similarity of the climatic anomaly with those in Fig. 9.10 verify that the inferred climatic anomalies deduced from Fig. 9.10 are correct. (Blasing, 1975)

disagreement is not considered to be important because, (1) heavy precipitation can occur in the eastern United States during certain Type 1 winters, as in the winter of 1944-45, (2) the east coast of the United States is beyond the reconstructed pressure grid, and (3) predictability of pressure for the portion of eastern United States included in the calibrated grid, using the western tree-ring data set, is known to be low (Fritts et al., 1971) (also see Fig. 9.7).

The temperature and precipitation anomalies associated with the 11 years reconstructed to be Type 3 winters (Fig. 9.19) are also in accord with the 20th century maps (Fig. 9.10). Temperatures in the northern states are lower than normal in both maps, while they are above normal from Nevada through southern Texas, the Mississippi Valley, and the Southeastern United States. Both maps show a tongue of anomalously cold air extending southward into the Plains. Similar high precipitation anomalies occur throughout the west and in the Ohio River Valley, and precipitation is low in the southern Plains and the Gulf Coast. The only important difference between the two maps is that low precipitation extends farther north for the 11 years in the 19th century. As was the case for Type 1 years, the differences in precipitation anomalies along the east coast are unimportant.

Though represented by few years, results for the other two winter types are just as encouraging (see Blasing, 1975, and Blasing and Fritts, in press). It is therefore concluded that the major anomalies in temperature and precipitation which were associated with each type in the modern record, at least for the middle and western portions of the continent, are verified for those years of the 19th century when the pressure types are reconstructed.

C. Consistency among Different Data Sets

The third approach to verification is to calibrate the amplitudes of tree growth with spatial variations of temperature at selected stations, to apply the tree-ring data to obtain the reconstructions, to test these reconstructions against data withheld for independent verification, and to check the reconstructions for their consistency with those obtained in the pressure analysis. Only the most successful analysis is included to illustrate the design of the experiment and some of the analysis limitations.

1. Calibration and Verification on Independent Temperature Data

An array of 18 North American climatic stations with long and apparently homogeneous records of temperature were selected. All stations had complete data back to 1894, and the longest record extended back to 1870.

Eigenvectors were extracted from the spatial variations in winter temperature for 1910-1959 and the amplitudes of the first three eigenvectors, which together reduced 81·6% of the variance, were calibrated with the amplitudes of the first 10 eigenvectors of tree growth (Figs 9.1 and 9.2). The model resembles the one shown in line 11, Table 9.I.

Table 9.IV lists the 18 stations that were used, the reduction of error statistic (RE; Equation 7.4), the counts of signs, and the results of the product mean analysis (see Chapter 7). The overall calibrated temperature variance for the dependent period, 1910-1962, is listed in the

TABLE 9.IV Calibration and independent verification statistics for reconstructing December–February temperatures for 18 climatic stations[a]

Station Name	Statistics[c]					
	RE[b]		Sign	Count	Product	Mean
	Dependent	Independent	+	−	+	−
Abilene, Texas	0·086	0·026	13	10	1·6	1·0
Albuquerque, New Mexico	0·153[d]	−0·217	10	6	1·2	2·1
Boise, Idaho	0·287[e]	−0·002	21	19	6·1[e]	2·0
Brownsville, Texas	0·146[d]	0·214[e]	24	15	3·7[e]	1·1
Denver, Colorado	0·243[e]	0·020	25	11[d]	2·0	2·2
Dodge City, Kansas	0·190[e]	0·086	20	14	4·0	2·5
Edmonton, Alberta, Canada	0·218[e]	0·394[e]	17	4[e]	13·4[d]	4·1
El Paso, Texas	0·115[d]	−0·008	13	16	1·0[d]	0·4
Helena, Montana	0·241[e]	0·106	19	9	8·4[e]	2·5
Miles City, Montana	0·248[e]	−0·056	14	3[d]	8·0	9·7
Phoenix, Arizona	0·160[e]	−0·127	18	14	1·2	1·2
Red Bluff, California	0·186[e]	−0·300	20	11	1·1	1·1
Riverside, California	0·173[e]	−0·543	15	12	1·8	1·9
Salt Lake City, Utah	0·228[e]	−0·035	23	11	5·0[d]	2·4
Spokane, Washington	0·279[e]	0·039	19	8	6·0[e]	2·0
Winnemucca, Nevada	0·311[e]	−0·172	15	9	8·2[d]	4·6
North Platte, Nebraska	0·223[e]	0·109[d]	21	13	5·4	3·2
Roseburg, Oregon	0·273[e]	−0·463	20	11	1·1	1·2

[a]Amplitudes of 49 arid tree-ring chronologies used as predictors. [b]Reduction of error statistic is the same as the square of the multiple correlation for dependent period. [c]See text for explanation of statistics. [d]Significant P ≥ 0·95. [e]Significant p ≥ 0·99.

table as the RE statistic. It averages 0·209, but the individual values ranged from a low of 0·086 for Abilene, Texas, to a high of 0·311 for Winnemucca, Nevada. A total of 17 out of 18 of these calibrations were statistically significant.

The calibration equation was applied to the tree-ring data to estimate temperatures from 1700 to 1909 at each of the 18 climatic stations. Figure 9.20 includes one of the best and one of the poorest examples of verification. The reconstructed temperatures were then compared to the available independent temperature data for each of the stations prior to 1910 using the above mentioned verification statistics (Table 9.IV).

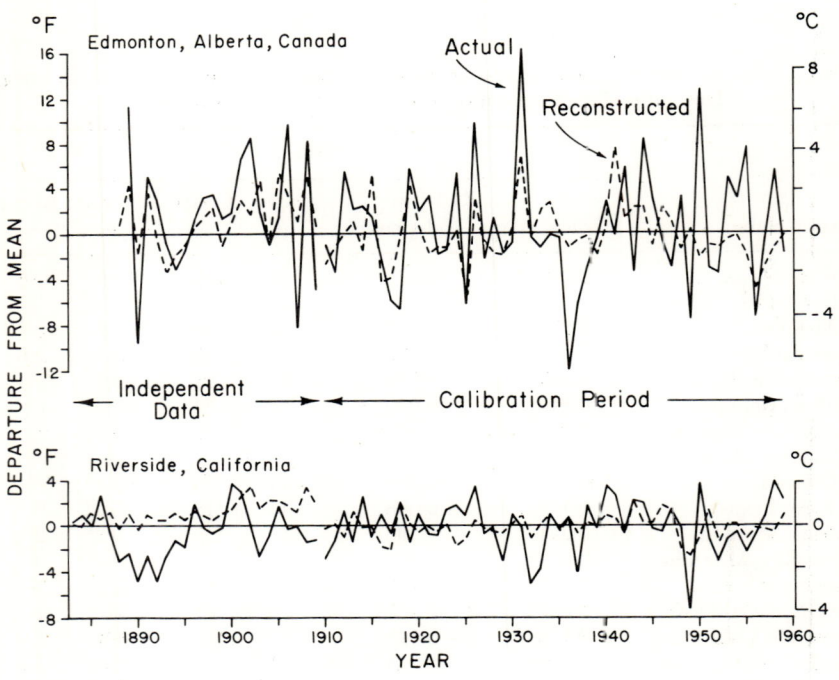

FIG. 9.20. The actual and reconstructed winter temperature for two calibrated stations included in Table 9.IV. The reduction of error statistic (RE) for the calibration period indicates relatively good agreement (0·218 for Edmonton and 0·173 for Riverside). The same statistic for the independent data indicates significant verification for Edmonton (RE = 0·394) and absence of verification for Riverside (RE = −0·543). The reconstructions for the latter station are rejected. (Drawing by M. Huggins)

The reduction of error for the independent data is positive at eight stations and exceeds the 95% confidence level for three of them. The sign of the winter temperature departures for the 18 stations is correctly reconstructed 327 times and incorrectly reconstructed 196 times, which is highly significant. The sign is correctly reconstructed more often than not for all but one station, but the sign counts for only three of these stations are significant. The positive and negative product means pooled for all stations are 4·4 and 2·1 respectively, and the positive product means are significantly greater than the negative product means for eight stations. The locations of the eight stations which exhibited the best reconstruction statistics are indicated as solid triangles in Fig. 9.21.

To test for the possibility that large-scale reconstructions were more accurate in certain years than others, the signs were examined year by year for all cases when six or more stations reported data (Table 9.V). During 11 of the 34 years examined, there were more positive than negative signs than could have occurred by chance. For 22 out of 34 years the values of the positive product means were two or more times those of the negative products. A total of 15 out of those cases with 10 or more observations were significant.

It is concluded from these data that verification statistics indicate significant agreement among the general patterns in climate for approximately one-half of the years available for testing and for 11 out of the 18 stations. However, the reconstructions for all 18 stations were utilized in the subsequent comparisons to allow analysis of spatial variations in climate.

2. *Comparisons of the Mean Values of Temperature Reconstructions Stratified by Pressure Types*

Since the same tree-ring data are used in the reconstruction of pressure and temperature and these two climatic data sets are physically related to one another, it is hypothesized that the reconstructions for the independent data periods should be mutually consistent only if the statistical relationships of both climatic variables with tree rings are real and significant. A test for the presence of a real linkage is made by stratifying the temperature reconstructions for winters from 1700 through 1899 according to the pressure types in the reconstructions (Fig. 9.17). The mean anomalies for reconstructed temperature in the years of each type for the 18th and 19th centuries (Fig. 9.21) could then be checked for consistency with the temperature anomalies in the 20th century that were known to be associated with the type years (Figs. 9.9 and 9.10). Years with Type 1 winters are reconstructed to be anomalously warm, especially in the north (Fig. 9.21), a pattern which is consistent with the

TABLE 9.V Independent verification statistics for reconstructing December–February temperatures tabulated by year for all cases with 6 or more reporting stations[a]

	Statistics			
	Sign Count		Product[d]	Mean
Year	+	−	+	−
1876	1	5	0·5	1·7
1877	5	1	3·0	5·5
1878	6	1	3·5	0·7
1879	8	1[b]	1·4	0·7
1880	6	3	1·9	0·9
1881	7	3	5·7[b]	0·1
1882	6	5	1·5	3·7
1883	13	0[c]	4·2[c]	0·0
1884	10	3	1·0[b]	0·0
1885	8	5	5·7[b]	1·4
1886	11	3	2·0	0·7
1887	10	5	3·9[c]	0·3
1888	13	2[c]	2·5[c]	0·3
1889	6	10	8·8	1·2
1890	14	2[c]	3·4[c]	0·1
1891	5	11	5·9	1·9
1892	3	13	0·8	1·3
1893	6	11	3·3	2·0
1894	2	16	4·0	2·0
1895	3	15	0·9	1·6
1896	15	3[c]	4·0[c]	0·7
1897	14	4[b]	3·0[b]	0·4
1898	12	6	2·2[b]	0·5
1899	10	8	3·6[b]	1·6
1900	15	3[c]	3·1[c]	0·1
1901	15	3[c]	8·2[c]	0·7
1902	16	2[c]	7·4[c]	0 5
1903	7	11	3·8	5 3
1904	13	5	2·9	1·5
1905	13	5	4·9	10·2
1906	15	3[c]	7·0	5·1
1907	10	8	3·8	4·1
1908	14	4[b]	13·9[c]	2·3
1909	11	7	4·7	2·7

[a]See Table 9.IV and text for explanation.
[b]Significant positive agreement ($P \geq 0.95$).
[c]Significant positive agreement ($P \geq 0.99$). [d]Not tested for cases with less than 10 observations.

modern period (Fig. 9.9). The same pattern was again verified by actual temperature data in the late 19th century (Fig. 9.18). The general departure pattern reconstructed for temperature in Type 3 years (Fig. 9.21) also resembles the patterns for the 20th century (Fig. 9.10) and for the 19th century verifications (Fig. 9.19), though some differences in the placement of the zero lines are apparent.

The mean temperature anomalies reconstructed for Type 2 and Type 4 years in the 18th and 19th centuries (Fig. 9.21) are also in accord with the associated temperature anomalies in the 20th century (Blasing, 1975, Blasing and Fritts, in press). Therefore, it is concluded that the results from both the temperature and pressure calibrations are consistent with each other, although the pressure reconstructions appear more reliable than those for temperatures, as the individual reconstructions of temperature exhibit sizable errors with unreliable estimates for particular stations and years.

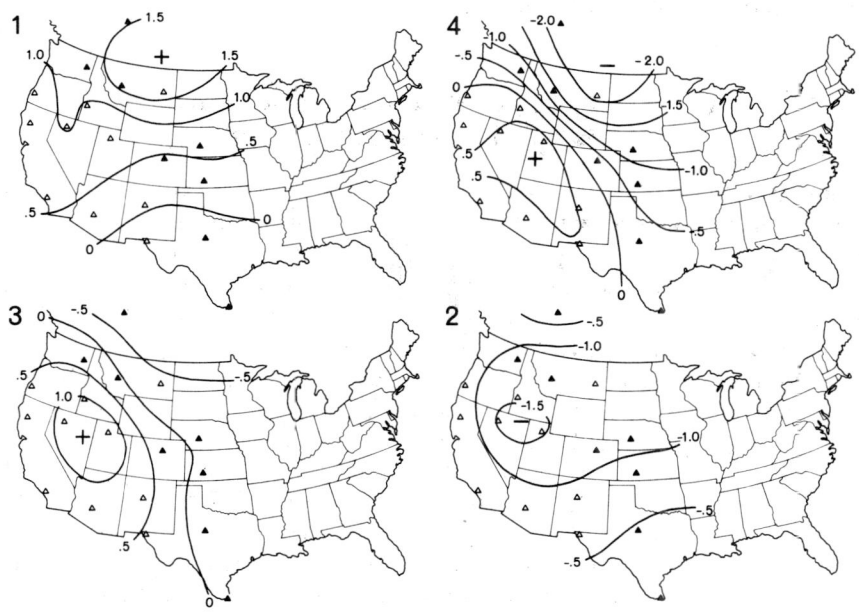

FIG. 9.21. The mean temperatures reconstructed for the winters 1700-1909 and stratified according to the types for 1700-1899, as in Fig. 9.17. The mean values are expressed in standard deviation units and the triangles show the locations of the 18 climate stations. The filled triangles are those stations with the best statistics and hence the most reliable reconstructions. (Drawing by M. Huggins)

Verification analyses using independent climatic data not used in calibration can provide objective and conclusive evidence that past climatic variations are accurately reconstructed. The best verification results are obtained when spatial variations in climate for a season are reconstructed from spatial variations in ring width. This occurs because the trees are natural integrators of climatic variations occurring over a number of months throughout the year. When principal component analysis is applied to many different tree-ring chronologies, the large-scale variations in the eigenvectors of growth are shown to be closely related to the large-scale variations in macroclimate.

These results may be formulated into the following working hypotheses: Climatic reconstructions for tree-ring data can be improved by (1) using data already integrated over space or over intervals of time representing at least several months, (2) averaging data sets via principal component or canonical analyses, or (3) averaging the reconstructions themselves by climatic type, by region, or for meaningful intervals of time as determined by typing analysis.

IX. Summarization of Reconstructions for Summer using the Pressure Types

Climatographic reconstructions of pressure, temperature, and precipitation have now been described for winter months, and the same techniques have been applied to climatic data for other seasons of the year. This section describes the reconstruction of pressure for summer and applies the summer pressure types to characterize the variations that have occurred during the last three centuries.

The reconstruction model for summer utilized growth at a one-year lag, so that the estimates from a particular set of tree rings are actually for the climate occurring in the preceding year. The reconstructions for each year are in turn correlated with five summer types to obtain a plot similar to the one shown in Figure 9.16 (Blasing, 1975; also see Figs 9.11, 9.12, and 9.13). The summer type best correlated with each reconstruction in the 18th and 19th centuries was assigned to that year and plotted in Fig. 9.22. The actual data for the 20th century are treated in the same way and are plotted in the figure along with a word description of the predominant anomaly in climate over the United States.

As for winter, the types were rearranged in the order of their relative importance for the 18th and 19th centuries. When the type occurrences in the reconstructions during the 20th century were compared to those for the actual pressure, it was noted that 60 out of 61 of the years were typed

9. RECONSTRUCTING SPATIAL VARIATIONS IN CLIMATE

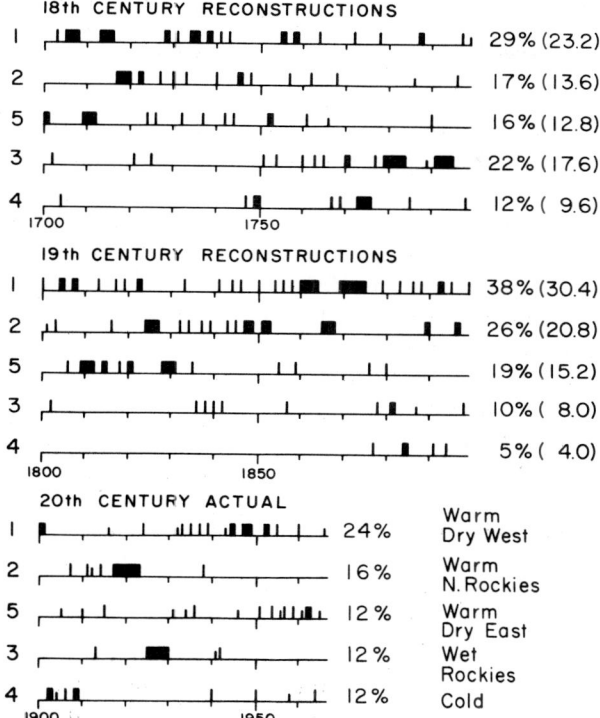

FIG. 9.22. The occurrences of the five summer pressure types as reconstructed from tree rings for 1700 through 1899 and as occurred in the pressure data from 1900 through 1966. The reconstruction percentages exceed the actual by 20%, the parentheses enclose the appropriate adjustment (see also text). See legend to Fig. 9.17. (Drawing by M. Huggins)

($r \geq 0.3$), while only 48 actual occurrences were typed for the same period of time. It was concluded that the reconstructions were overestimating the types and that the frequencies of types should be adjusted by multiplying the percentage by 0.8 ($48/60 = 0.8$). The brackets in Fig. 9.22 indicate this adjustment.

It is apparent from the figure that the adjusted frequencies for the reconstructed summer climates of the 18th and 19th century varied from those in the 20th century. Summer Type 1, which was associated with dry and warm conditions in the central and western United States (Fig. 9.12), was reconstructed 23.2% of the time in the 18th century and 30.4% of the time in the 19th century, as compared to 24% of the time during the 20th century. Type 2, which was associated with anomalous warmth in the Rocky Mountains, was estimated to have occurred 13·6% and 20·8% of the time in the 18th and 19th centuries, but occurred 16% of the time in the present century.

Summer Type 5, which is like Type 1 in that it is dry and warm in the central United States but differs from it in that drought is present in the eastern U.S., was estimated to have occurred 12·8% of the time in the 18th century and 15·2% in the 19th century, as compared to 12% during the 20th century.

Type 3, which is characterized by wet conditions in the Rocky Mountains, occurred most frequently in the 18th century (17·6% of the time) and least frequently in the 19th century (8%) as compared to 12% of the time in the current century. On the other hand, Type 4, which is characterized as a cold, winter-like circulation in summer, was estimated to have occurred 9·6% in the 18th century and 4·0% in the 19th century as compared to 12% in the 20th century.

Examination of the distribution in these occurrences of types within each century (Fig. 9.22) reveals interesting changes in frequency through time. For example, Types 1, 2, and 5 are estimated to have occurred most frequently throughout the first half of the 18th century. Type 3 with wetness in the Rocky Mountains and Type 4 with cold summers are estimated to have occurred more frequently in the latter half of the century. During the 19th century Type 1 is estimated to have occurred at a frequency greater than one out of three years. Type 5 is estimated to have been frequent during the first four decades, and Type 2 is estimated to have occurred during the 1820's through 1860's as well as near the end of the century. During the 20th century the types were often clustered in distinct periods of time. Type 4 predominates during the first decade, which is followed by a high occurrence of Type 2 until 1925 and then by Type 3 until 1931. The drought Types 1 and 5 became most important in the 1930's through 1960's, after which other types occurred more frequently. A comparison of Type 1 and 5 occurrences among the three centuries indicates that the frequency of drought in the 18th century was the same as in the current one but was more frequent in the 19th century by approximately 10%.

As was the case for winter, the pressure anomaly types provide a unique and precise means of summarizing the climate of the past without mapping and analyzing each yearly reconstruction. A variety of periods dominated by particular atmospheric circulation regimes can be identified and interpreted in terms of the modern climate, without obscuring other types of variation that may be mixed with them. The data on yearly occurrence of types are also available in Figs 9.17 and 9.22 for comparison with other information on past climate.

The information provided in the figures deals with half-hemisphere-wide generalizations about climate. As such, the yearly reconstructions may or may not be in agreement with data at any one location, creating apparent inconsistencies of detail. As data from more reporting climatic stations and more tree-ring stations can be incorporated in the reconstructions, the results of the estimated and actual occurrences should be increasingly in agreement.

X. Verification of Reconstructions for Summer using Independent Tree-Ring Data

A number of verifications of the reconstructions for summer have been attempted with varying degrees of success. The reconstructions themselves were more difficult to create and verify than those for winter because (1) the summer season that was used included two rather than three months, (2) the variance of pressure in summer is less than that of the winter, (3) the climatic variations of summer are not as well correlated over space as those of winter, and (4) the summer climate can affect growth more during the following year than during the current one, which leads to an uncertainty as to the placement of reconstructions in the time sequence. In addition, many response functions described in Chapter 8 exhibit smaller weights for the climate of July and August than for the climate of the three winter months, indicating that the trees do not respond as markedly to variations in summer climate.

For the purposes of this volume, only the summer verification methods are presented which differ from those already described for winter. One verification test for summer used independent tree-ring data from Alaska and Northwest Canada back to A.D. 1800 (Blasing and Fritts, 1975). This information was well dated, well replicated, and from northern tree line sites. Therefore, the Arctic tree-ring data offered a unique opportunity for an independent test of summer circulation in the North American Arctic which was outside the area covered by the 49 arid-site chronologies.

For purposes of the analysis, the narrow rings from the Arctic sites were all assumed to be indicators of anomalously cloudy, wet, cold climate, while the wide rings were assumed to be indicators of anomalously sunny, dry, mild climate. The reconstructed pressure anomaly was interpreted as a weather map in terms of the expected anomaly in storm tracks and the expected direction in wind which controlled the temperatures at the tree sites (see Blasing and Fritts, 1975). That is, the anomalous wind direction is inferred to be parallel to the isobars in the anomaly maps, with low pressure to the left and high pressure to the right.

Three characteristics of the Arctic ring-width chronologies complicated the results. (1) Frost can sometimes damage the tree and cause abrupt and persistent changes in growth; (2) the ring-width response appears to lag one to 15 or more years behind the original causal conditions of climate (LaMarche, 1974a; LaMarche and Stockton, 1974); and (3) the period used for pressure reconstruction, July and August, did not coincide exactly with the spring and summer period to which Arctic trees respond. The first two difficulties caused a lag in the growth response behind climate which filters out some of the year-to-year variations in climate, while the last created a certain amount of inconsistency in the results (Blasing and Fritts, 1975).

The test involved two checks. In the first, the mean pattern in Arctic tree growth associated with each pressure type during the 20th century was compared to those for the 19th century for years reconstructed to be the same type (Fig. 9.22). It was hoped that some consistency in growth pattern would be evident, in spite of the fact that the tree growth sometimes lagged behind climate and that some pressure types may not affect the climate of the North American Arctic. In the second test, the growth patterns in the Arctic were themselves typed and then the mean anomaly in reconstructed pressure for the years of each growth type were examined for their consistency with the patterns in the growth types.

The first test was inconclusive because pressure types such as 1, 2, and 5 exhibit weak gradients in the Alaskan area, or, as in the case of Type 3, the most characteristic years were too weakly represented to allow comparison of the two results during the period in common among the data sets, 1800-1939. The second test, which was more successful, is described in the following paragraphs.

The standardized yearly ring-width chronologies from the Arctic were normalized for 1800-1939 and anomalies of growth were typed using the procedure of Blasing (1975). The resulting growth type occurrences were not as randomly distributed through time as those of the pressure types but were confined to distinct periods, apparently because of persistence and autocorrelation in the growth sequences. However, during periods of transition two or more growth types are interspersed with one another.

The upper portion of Fig. 9.23a includes maps of the average normalized patterns for the first four growth types (a) and is labeled according to the first and last year associated with each one. Figure 9.23b includes the mean reconstructed pressure anomaly associated with the particular years of growth. The reconstructed pressure anomalies for the summer model (Table 9.I, line 3) were used first, but later the results from the minimum-sized model (Table 9.I, line 10) were applied and found to be consistent with those described in the following pages. The first growth type is complicated by lagging effects of extreme climatic conditions occurring late in the 18th century, so the discussion will start with the second type. Growth Type 2, which occurred in years from 1812 through 1828, exhibits anomalously low values throughout Alaska and adjoining areas in Canada, except for higher than average growth near the Alaskan coast. Growth Types 3 and 4 show similar anomalies but with a larger area of higher than average growth for a number of chronologies in south Alaska and adjoining areas in Canada. It may be inferred from these patterns that during 1812-1851 conditions in the north were unfavorable to growth and thus were cold and stormy, while further south where growth was higher, conditions were warmer than average and less stormy.

During 1812-1851 (Fig. 9.23b) the North Pacific High was reconstructed to be displaced north of the positions it occupied during the calibration period for 1900-1962, and the summer storm track was also inferred to be displaced to the north. Storms were inferred to be most common in the vicinity of the Arctic coast along a path which turns southward in western Canada, at least for the intervals dominated by growth Types 2 and 3, which occurred from 1812 to 1836. This is consistent with the growth, as storms would bring cloudy weather to the north slope of Alaska, advect cold air southward and eastward from the Arctic Ocean, and bring low temperatures to the areas of low growth in the northern Arctic. The southward flowing air would become warm as it moved over central Alaska and this, in conjunction with the northward displacement of the North Pacific high pressure system, would bring sunny weather and above-normal growth along the southern coast.

Growth Type 1 in Fig. 9.23a is markedly different from growth Types 2 and 3 while the associated pressure anomalies are almost identical. One explanation for the discrepancy was that the climate changed in the last decade of the previous century and the years representing growth Type 1 were transitional to a new mode of climate. An examination of type occurrences shown in Fig. 9.22 confirmed this possibility, as pressure Types 3 and 4 dominated the reconstructions from 1770 to 1795 but occurred rarely in the following century. The anomalies associated with pressure Type 3 were somewhat higher than normal over the Bering Strait and Northwest Canada, and lower than normal over central and eastern Alaska. This is a markedly different anomaly from the pressure reconstructed for 1804-1811 shown in Fig. 9.23 and one which would be more consistent with growth Type 1. In order to test this possibility further, the available growth data associated with pressure Type 3 for the years 1790-1795 were averaged (Blasing and Fritts, 1975). The pattern was remarkably in agreement with growth Type 1 which followed during

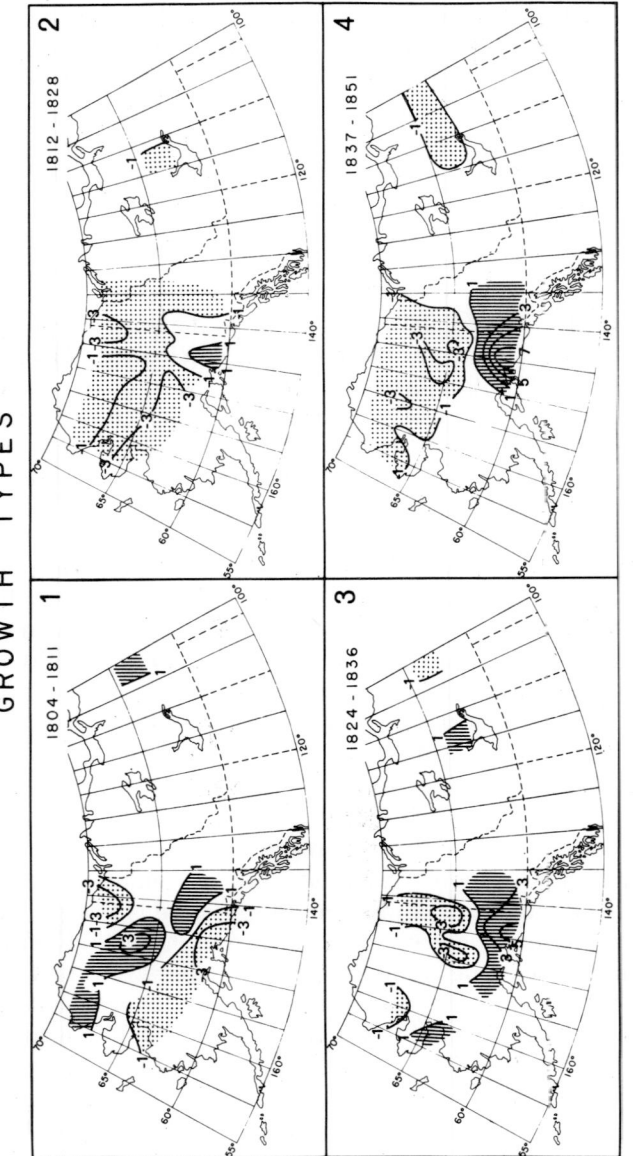

FIG. 9.23a

FIG. 9.23. The anomaly pattern for (a) the four Arctic tree-growth types occurring in the first half of the 19th century, and (b) the reconstructed pressure anomalies averaged for the years of that growth type from the application of the transfer function to the 49 chronologies from arid North American tree sites. The Arctic tree growth is expressed in standard deviation units as departures from the mean for the period 1800-1939. The years shown in the figure are the first and last of those included in the particular growth type. (Blasing and Fritts, 1975)

9. RECONSTRUCTING SPATIAL VARIATIONS IN CLIMATE

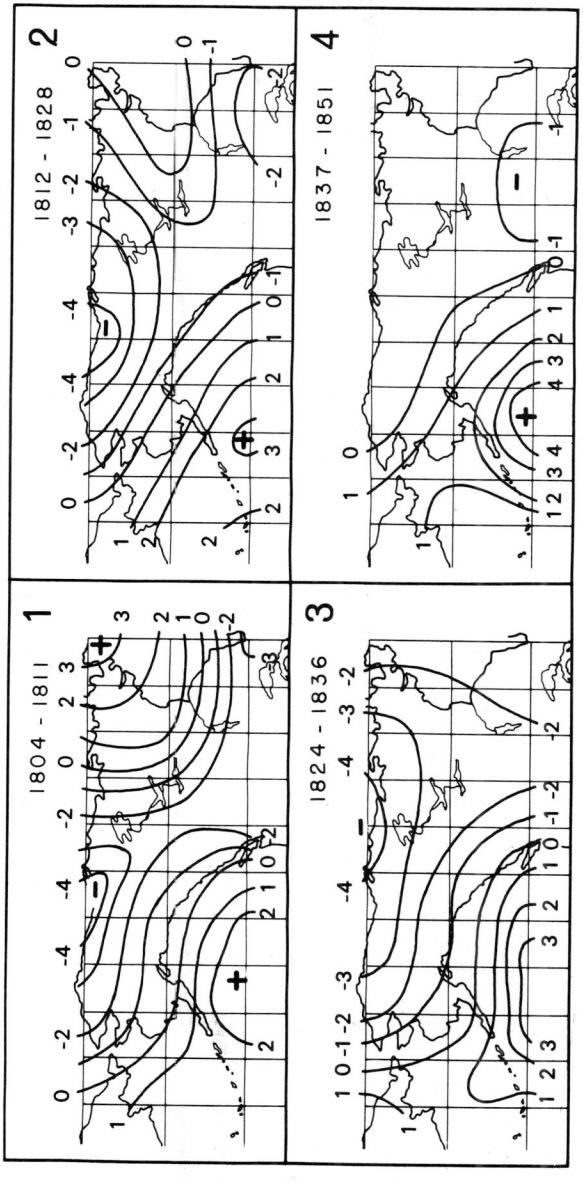

FIG. 9.23b

the 1804-1811 interval. Therefore growth Type 1 was inferred to be transitional representing a lag in growth response behind a marked change in climate.

It was therefore concluded that the growth pattern in the Arctic trees for 1804-1811 represented in part a lagging response behind a marked pattern in climate during the latter part of the 18th century and in part a transitional growth response to a new climate pattern existing for the 1804-1811 period. By 1812 (16 years after Type 3 summers were reconstructed to have occurred) the growth pattern reached a new equilibrium with the current climate, a growth pattern which persisted for approximately 40 years. However, the pressure anomalies associated with growth Type 4 (Fig. 9.23a) suggested that there was a return to normal pressure throughout the continental North American Arctic, so that this growth type might also be a transitional one.

The growth anomaly for Type 5 (Fig. 9.24a) occurred during 1845-1863 with high growth throughout Alaska, especially in the area near Anchorage. The associated pressure anomaly shows a reintensified pressure gradient resembling the prior periods with Types 1-3 with one apparent but important difference. The area of anomalously low pressure extended across the Bering Strait to Siberia so that the wind over western Alaska was from a westerly direction rather than from the northwest, a condition which might be expected to bring warmer air into the western and northern parts of the state. The influx of warmer air across the Alaskan interior is also consistent with the continuation of anomalously high growth in the area along the south coast, as occurred in growth Types 3 and 4.

Growth during the last 40 years of the 19th century (Types 6 and 7) continued high in northwest Alaska, but it decreased to the northeast and south. The anomaly in pressure reconstructed for years of growth Type 6 suggests continued westerly flow from the north Pacific Ocean, with some southward flow from the Arctic Ocean into the Mackenzie Delta where growth was average or below normal.

For the last two decades of the 19th century, which were dominated by growth Type 7, the positive anomalies of growth for northern Alaska become less marked and the inferred climate more like that of today, while negative anomalies in growth become more apparent to the south. The mean pressure anomaly for the years of Type 7 is reconstructed to be close to normal over most of Alaska. However, there appeared to be more storms and a stronger component of onshore air flow from the south than previously, which brings lower summer temperatures along the coast and warmer temperatures inland.

After the turn of the century, 1901-1914, the anomalously high

9. RECONSTRUCTING SPATIAL VARIATIONS IN CLIMATE

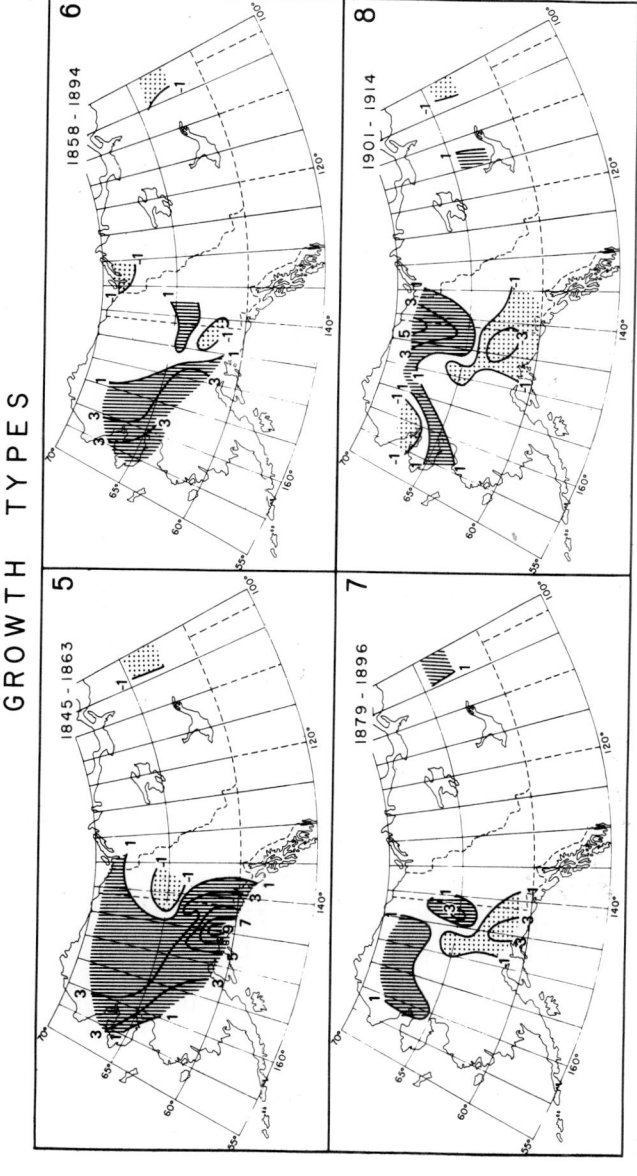

FIG. 9.24a

FIG. 9.24a and b. Same as in Fig. 9.23 but for the latter half of the 19th and early 20th century, which include the period dominated by four additional growth types. (Blasing and Fritts, 1975)

498 TREE RINGS AND CLIMATE

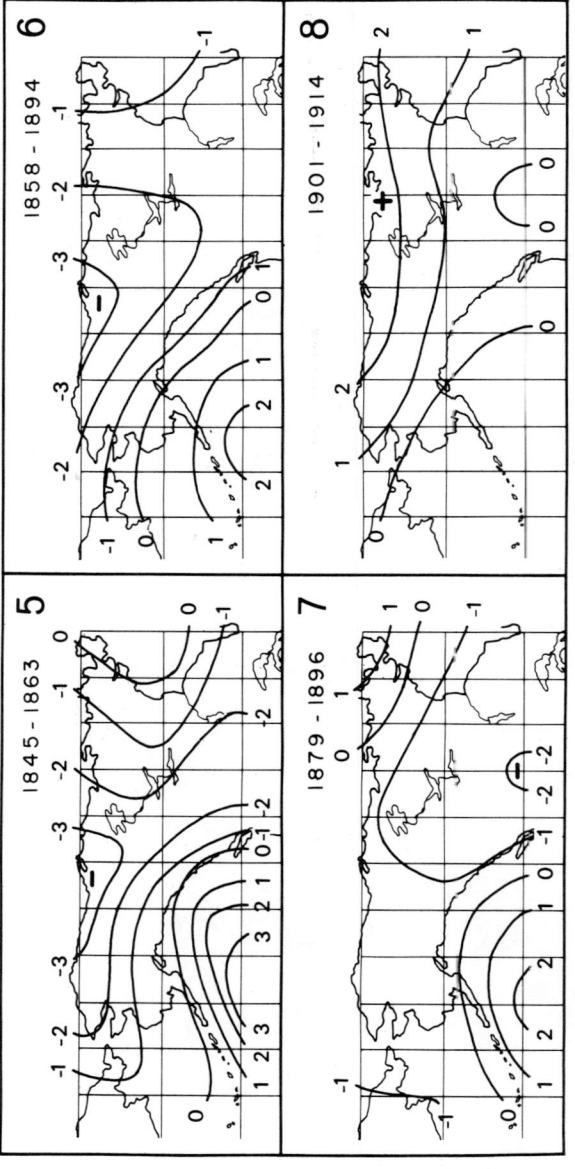

FIG. 9.24b

pressure in the Gulf of Alaska disappeared. This occurred in conjunction with continued lower than average growth in the south, colder temperatures, more storms, and increased cold air flow from the south and southeast off the Gulf of Alaska. Since the northern areas of Alaska and Canada are under a regime of anomalous southerly winds flowing over the land area, the air is warmed and temperatures become higher and more conducive to high growth. One exception to this general warming pattern in the North is the reduced growth in the northern Seward Peninsula, where coastal conditions may have counteracted any ameliorating effect of the more southerly winds.

While the above analysis is speculative, it does show how tree growth itself can be used as direct evidence of climatic variation and applied to verify or document climatic reconstructions. The Arctic tree-ring data appear to be consistent with the pressure anomalies reconstructed for the same area and period of time. Attention shall now be directed to other types of data that may be used to verify, validate, or complement dendroclimatographic reconstructions.

XI. Verification using Journals, Historical Data, and various Proxy Records of Climate

Official meteorological records and tree rings are not the only data sources available for climatic reconstruction and verification. A variety of other information is available from historical materials, journals, harvest logbooks, records of freezing and thawing of lakes, phenological observations on flowering and fruiting of trees, and a variety of chronicles. Much valuable information has been gathered from such references to climatic events, including, for example, the freezing dates of Lake Suwa and the blooming dates of cherry trees in Japan (Arakawa, 1954, 1956); sea ice data (Bergthorsson, 1962); the records of pioneers, explorers, and early forts (Lawson, 1972, 1974; Wahl and Lawson, 1970); vintage dates and other historical events in Europe (Ladurie, 1971); data from ships' logs (Lamb, 1972); and lake level and other hydrologic and geologic records (Antevs, 1952; Fritts and Cathey, 1971). Such records must be scanned critically, evaluated, and compiled by region and date to facilitate their comparison with climatic reconstructions. At present few of these data have been applied to verification of tree-ring reconstructions, and only one example from Blasing and Fritts (1975) is cited here.

Daily temperature data were obtained from Kutzbach and Wendland (personal communication) for the Russian fort at Illoolook, Unalaska Island, near modern Dutch Harbor for 1829-1833. The maximum and minimum temperatures for July and August were averaged and compared to the anomalous wind inferred from the pressure field over the island as reconstructed from the arid-site tree rings. When the anomalous wind was estimated from the west–northwest, which occurred in 1829 and 1831, the observed temperatures at Illoolook were several degrees lower than during years when the anomaly was estimated from the southwest or west–southwest (Table 9.VI). For these years the reconstructed pressure anomalies were verified.

TABLE 9.VI Comparison of inferred anomalous wind direction from reconstructed pressure anomalies and mean July–August temperatures[a] for 5 years at Illoolook, Unalaska

Year	Anomalous direction of the inferred geostrophic wind	Mean summer temperature
1829	west–northwest	53·5°F
1830	southwest	58·6°F
1831	west–northwest	51·4°F
1832	west–southwest	58·1°F
1833	west–southwest	58·2°F

[a]Data courtesy of Kutzbach and Wendland, University of Wisconsin.

More and more proxy data on climate are becoming available through the work of many scientists—data which not only serve as verification but also offer additional climatic information. Some of these data are not dated to the exact year or season of occurrence. Precisely dated, continuous tree-ring evidence could provide the necessary time control, structure, and continuity to weave them into a coherent year-by-year history. In addition, the spatial patterns derived from tree rings could resolve conflicting evidence. For example, temperatures reconstructed to be anomalously high for one area and anomalously low for another would be consistent if they were on opposite sides of an anomalous low pressure trough. Northerly flow to the east of the trough would produce warm temperatures, while southerly flow to the west would produce cool temperatures.

XII. Applications to Climatological Problems

The precisely dated and objectively derived tree-ring reconstructions are expected to contribute significantly to man's understanding of climate. As mentioned above, these climatic reconstructions are unique in that they provide a continuous time sequence, and they can be used to place other less well-dated information into its proper time period and location within a coherent framework. In addition, the reconstructions are objectively derived, not subjectively inferred, so that they provide a precision not found in most paleoclimatic data.

The temporal accuracy of the reconstructions can be of special value to climatological analysis. For example, dynamic modeling of climate (Gates and Mintz, 1975) will eventually require accurate and well-dated maps of past variations for making comparisons (Sheppard, 1966). Various hypotheses about climatic linkages to physical phenomena such as atmosphere, tidal forces, variations in solar energy output or sunspots, variations in atmospheric turbidity resulting from volcanic eruptions, and so forth, can be tested against the long climatic record derived from tree-ring analysis. In addition, hypotheses about man's possible role in changing climate may be tested by searching for analogies in the reconstructed record.

Although reliable climatic forecasts are not presently feasible, some estimates of possible future climates would be useful for planning purposes. For example, it is argued by some workers (Bryson, 1974; Lamb, 1966) that the climate has been so anomalous in recent times that the climatic statistics derived from the recent record are unrealistic and therefore do not adequately portray the expectations for the future. Thus, the tree-ring chronologies spanning a longer period of time may provide a more realistic record, and the statistics of the reconstructions can be compared to those for the shorter climatic record to identify the extent and nature of the bias in the shorter record and to adjust its stastistics accordingly.

The calibrated tree-ring record from the selected 49 sites and the pressure types reconstructed from them shown in Figs 9.17 and 9.22 can be applied to this problem. For purposes of illustration, assume that the frequencies for the 20th century pressure types in the two figures represent the known record of climate from western North America and the North Pacific. The frequencies of 20th century pressure types along with those for the reconstructions in the 18th and 19th centuries are summarized in Fig. 9.25 for both winter and summer. In the case of summer, the dashed lines indicate the actual reconstructions and the solid line the frequencies after adjustment as discussed in Section IX.

The hatched bar on the right is a mean value of the frequencies obtained by pooling the 200 years of estimates and the 62 years of actual pressure occurrences. Some marked anomalies of the 20th century are in evidence. Winter Type 4, which produces cold weather in the Plains, occurred 10% less frequently, and winter Type 2 with a cold western United States occurred 11% more frequently than would be indicated if the longer 262-year mean were used. Any decisions concerning these patterns of climate, based exclusively upon the 20th century data, would be in error, and should be modified using the longer-term estimates. Data from the 20th century concerning pressure Types 1 and 3 are evidently more in line, as they are within 2% of the long-term estimates.

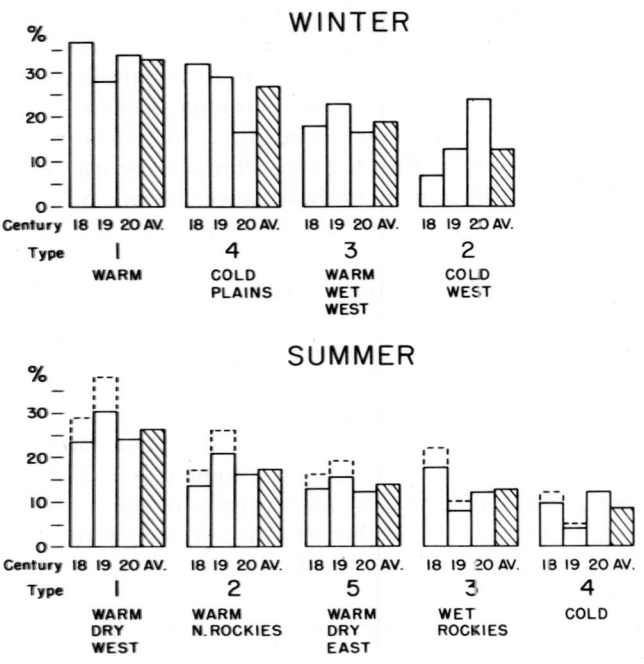

FIG. 9.25. The percent frequencies of reconstruction and occurrence for the pressure types during the individual last three centuries as compared to the averages for the 262 years spanned by the actual data and the reconstructions together. Since the 20th century data are anomalous, the longer-term averages may be better estimates of the expectations for future modes of climate.

The data for summer (Fig. 9.25) indicate that the frequencies during the 20th century are closer to the long-term estimates. However, each of the drought and warm types, Types 1, 2, and 5, exhibits a 1% to 2% lower anomaly and amount to a 5% lower drought frequency. Type 4, the cold summer type, occurred 4% more frequently.

These kinds of analyses can be used to identify the bias in the modern record, which in this example is toward warmer winters in the midcontinent, colder winters in the West, less frequent drought, and cooler temperatures in summer over the middle and eastern portions of the continent.

XIII. Present and Future Prospects of Dendroclimatology

This volume has attempted to describe the most important concepts and principles of the emerging discipline of dendroclimatology. Growth of trees has been shown to fluctuate in many ways, depending upon physical, physiological, and age variations of plant tissues as well as genetic, soil, and microclimatological variations among trees. These variations can affect growth only by limiting the rates of plant processes in one or more organs of the tree.

The details of the interrelationships are exceedingly complex, but the ring characteristics and the essential nature of the climatic relationships can be statistically measured, modeled, and the net effects of climate on growth expressed in response function form. Transfer functions can be objectively derived and applied to past tree-ring information to obtain estimates of past variations in climate.

The climatic reconstructions described require that all rings be precisely dated, that the environmental data with which they are calibrated be accurately placed in time, and that there be environmental data independent of that used in the calibration to serve as checks of the reconstructed climate.

It has also been shown that calibration need not be restricted to the specific microenvironments where tree samples were taken. In fact, it is often better to use regional averages of climate rather than specific site data for calibration, because the macroclimatic variations averaged for a region are not only highly correlated with the microclimates of the different tree sites but they are often more representative of large-scale changes in climate. In addition, the large-scale variations are more closely linked to the major variations in world climate.

The accuracy of the climatic reconstructions at any given location or time period is expected to improve as more diverse tree-ring data sets are obtained from a wider variety of species, sites, regions, and continents. In general, the greater the number of trees and sites used for a spatial sample, the greater the diversity in the tree-ring responses to climate. Similarly, the greater the areal coverage of the tree sites, the more information that can be extracted from the ring-width data on past world climate. It is no longer necessary to confine sampling to extremely arid or cold sites, and it is no longer desirable to select samples responding to the same climatic factor. The multivariate statistics perform their own selecting and scaling of anomaly variance in a way which obtains the maximum information on the calibrated variables of climate.

Data from the growth rings other than their widths are expected to add new dimensions to climatic analysis. Such data may include the character of cells and intracellular structures, variations in wood density throughout the ring, microchemical variations from one ring to the next, fluctuations in isotopes of carbon, hydrogen, and oxygen, or possibly the residual magnetic field which is measurable from specimens of wood. In addition to the tree-ring reconstructions, other well-dated information on climate can be used, such as that available from oxygen isotope analysis of ice (Dansgaard et al., 1971), the annually layered sediments in lakes and ocean basins (Swain, 1973), and layering found in certain species of corals (Buddemeier, 1974), as well as very long climatic records themselves. As long as such data can be dated to the exact year, as are tree rings, they can be entered into the multivariate equation along with the tree-ring chronologies, and together they can be used to establish past hemispheric or even worldwide climatic variations.

Usable tree-ring information appears to be available from temperate climates of both the Northern and Southern Hemispheres. The investigator must examine the appropriate species and sites for the older climatically affected trees and look for the contorted stem forms indicating general stress and old age. While at present there appears to be little success in dating ring structures in the Tropics, there are many possible species and promising habitats in that area that have not been examined; information from the Tropics is too valuable for any potential sources to be arbitrarily ruled out. Of course, the ultimate test of any sample is the crossdating and the resultant amount of calibration possible with variations in climatic factors.

Not all climatic data are equally useful in calibration, for there may be large errors and inconsistencies in poorly collected and prepared data. Such errors will contribute to inaccuracies in the calibrated relationships, and no matter how excellent the tree-ring records are, they cannot exceed the reliability of the climatic data used for calibration. Also, rain gauges, thermometers, and barometers all measure conditions at a point, and all contain some errors and local bias, and are therefore imperfect estimators of the macroclimate. As such, they may be poor recorders of those climatic variations that limit the plant and influence ring widths. Thus, many climatological, hydrological, and other environmental records now used for calibration can be averaged over space and time, or treated in some other fashion which facilitates and improves the climatic calibration.

New and improved statistical procedures will undoubtedly become available, as well as new climatological methods for attacking problems of climatic variation. Work has hardly begun on estimating future probabilities of climate or on attempts to forecast future climate. While leadership in many of these tasks will come directly from climatologists, the dendrochronologist should be an important part of this effort because of his excellent position to utilize tree-ring data to obtain the "long" view of past climatic variations.

The primary task for the dendroclimatologist in the future will be the extension of the tree-ring record backward in time, the utilization of new species, the sampling of trees in new regions of the world, and the development of new techniques of analysis. The International Tree-Ring Data Bank which is now operational at the Laboratory of Tree-Ring Research, University of Arizona, will hopefully provide long-term archival storage as well as quick retrieval of ring measurements and the chronologies derived from them. Although the present focus of the data bank is upon ring-width measurements usable for climatic reconstructions, the designed system is sufficiently flexible to allow for storage and retrieval of all types of tree-ring materials, including those usable only for dating or for study of environmental problems not necessarily involving climate.

It is hoped that the principles and procedures described in this book will have acquainted the reader with the variety of opportunities and techniques of the field and will stimulate his research in directions unimagined at present. Tree rings offer a rather remarkable record of those past conditions affecting tree growth. It is now up to the dendroclimatologist to skillfully use the many tools at his disposal, to extract that information, and to apply it to the many problems concerned with variations in the natural and man-affected environment.

Appendix
Scientific and Common Names of Trees

Scientific Names *Common Names*

Abies Mill. — fir
Abies amabilis (Dougl.) Forbes — Pacific silver fir
Abies concolor (Gord. and Glend.) Lindl. — white fir
Abies lasiocarpa (Hook.) Nutt. — subalpine fir
Abies magnifica A. Murr. — red fir
Abies procera Rehd. — noble fir
Acer L. — maple
Acer rubrum L. — red maple
Acer saccharum Marsh. — sugar maple
Agathis R. A. Salisbury — kauri
Alnus B. Ehrh. — alder
Amelanchier Med. — serviceberry
Araucaria A. L. Jussieu — araucaria, monkey puzzle
Artemisia L. — sagebrush
Artemisia tridentata Nutt. — big sagebrush
Austrocedrus chilensis D. Don Florin et Boutelje — ciprés, southern incense cedar (South America) [was *Libocedrus chilensis* (D. Don) Endl.]

Betula L. — birch
Carya Nutt. — hickory
Cedrus Trew — true cedar
Celtis L. — hackberry
Cercocarpus H. B. K. — mountain-mahogany
Cornus L. — dogwood
Cupressus L. — cypress
Cupressus arizona Greene — Arizona cypress
Ephedra L. — Mormon tea
Fagus L. — beech
Fagus grandifolia Ehrh. — American beech
Fagus silvatica L. — European beech
Fitzroya cupressoides (Mol.) Johnst. — alerce
Fraxinus L. — ash
Fraxinus americana L. — white ash
Juniperus L. — juniper, cedar

Juniperus communis L.	common juniper
Juniperus occidentalis Hook	Western juniper
Juniperus osteosperma (Torr.) Little	Utah juniper
Juniperus virginiana L.	Eastern red cedar
Larix Mill.	larch
Larix decidua Mill.	European larch
Larix occidentalis Nutt.	Western larch, tamarack
Libocedrus Endl.	incense cedar
Liquidambar styraciflua L.	sweetgum
Liriodendron L.	yellow poplar, tuliptree
Opuntia Mill.	prickly pear, cholla cactus
Picea A. Dietr.	spruce
Picea abies (L.) Karst.	Norway spruce
Picea excelsa Link	[same as *Picea abies*]
Picea glauca (Moench) Voss	white spruce
Picea rubens Sarg.	red spruce
Picea sitchensis (Bong.) Carr.	Sitka spruce
Pinus L.	pine
Pinus aristata Engelm.	Rocky Mountain bristlecone pine
Pinus banksiana Lamb.	Jack pine
Pinus cembra L.	Swiss stone pine
Pinus cembroides Zucc.	Mexican pinyon pine
Pinus contorta Dougl.	lodgepole pine
Pinus echinata Mill.	shortleaf pine
Pinus edulis Engelm.	pinyon
Pinus excelsa	[same as *Picea abies*]
Pinus flexilis James.	limber pine
Pinus halepensis Mill.	Aleppo pine
Pinus jeffreyi Grev. and Balf.	Jeffrey pine
Pinus longaeva D. K. Bailey	Great Basin bristlecone pine
Pinus monophylla Torr. and Frém.	singleleaf pinyon
Pinus monticola Dougl.	Western white pine
Pinus ponderosa Laws.	ponderosa pine, western yellow pine
Pinus radiata D. Don	Monterey pine
Pinus resinosa Ait.	red pine
Pinus strobus L.	Eastern white pine
Pinus sylvestris L.	Scotch pine, Scots pine
Pinus taeda L.	loblolly pine
Pistacia L.	pistachio
Podocarpus L'Héritier ex Persoon	podocarpus
Populus L.	poplar
Populus deltoides Bartr.	Eastern cottonwood
Populus tremuloides Michx.	quaking aspen
Pseudotsuga Carr.	Douglas-fir
Pseudotsuga menziesii (Mirb.) Franco	Douglas-fir
Purshia D. C.	antelope brush, bitterbrush
Quercus L.	oak
Quercus alba L.	white oak
Quercus coccinea Muenchh.	scarlet oak
Quercus ellipsoidalis E. J. Hill	Northern pine oak

Quercus prinus L.	chestnut oak
Quercus robur L.	English oak
Quercus rubra L.	red oak, eastern red oak, northern red oak
Sequoia Endl.	sequoia
Sequoia gigantea (Lindl.) Decne.	[same as *Sequoiadendron giganteum*]
Sequoiadendron giganteum (Lindl.) Büchholz	giant sequoia
Sequoia sempervirens Endl.	coast redwood
Taxodium L. C. Rich	bald cypress
Thuja L.	arborvitae
Tsuga Carr.	hemlock
Tsuga canadensis (L.) Carr.	Eastern hemlock
Tsuga heterophylla Sarg.	Western hemlock
Tsuga mertensiana (Bong.) Carr.	mountain hemlock
Ulmus L.	elm
Zygophyllum L.	bean caper

Bibliography

Agerter, S. R. and Glock, W. S. (1965). "An Annotated Bibliography of Tree Growth and Growth Rings 1950–1962." University of Arizona Press, Tucson.
Alestalo, J. (1971). Dendrochronological interpretation of geomorphic processes. *Soc. Geogr. Fenn.* **105**, 1–140.
Alvim, P. de T. (1964). Tree growth periodicity in tropical climates. *In* "The Formation of Wood in Forest Trees" (M. H. Zimmermann, ed.) pp. 479–495. Academic Press, New York, London and San Francisco.
Antevs, E. (1939). Precipitation and water supply in the Sierra Nevada, California. *Bull. Amer. Meteorol. Soc.* **20**, 89–91.
Antevs, E. (1952). Cenozoic climates of the Great Basin. *Geol. Rundsch.* **40**, 94–107.
Arakawa, H. (1954). Fujiwhara on five centuries of freezing data of Lake Suwa in the Central Japan. *Arch. Meteorol. Geophys. Bioklimatol. Ser. B* **6**, 152–166.
Arakawa, H. (1956). Climate change as revealed by the blooming dates of the cherry blossoms at Kyoto. *J. Meteorol.* **13**, 599–600.
Ashby, W. C. and Fritts, H. C. (1972). Tree growth, air pollution, and climate near LaPorte, Indiana. *Bull. Amer. Meteorol. Soc.* **53(3)**, 246–251.
Baggaley, A. R. (1964). "Intermediate Correlation Methods." Wiley Interscience, New York.
Bailey, D. K. (1970). Phytogeography and taxonomy of *Pinus* subsection of Balfourianae. *Ann. Missouri Bot. Gard.* **57**, 210–249.
Baillie, M. G. L. (1973). A dendrochronological study in Ireland with reference to the dating of medieval and post medieval timbers. Ph.D. Thesis, Queen's University, Belfast.
Baillie, M. G. L. and Pilcher, J. R. (1973). A simple crossdating program for tree-ring research. *Tree-Ring Bull.* **33**, 7–14.
Baker, F. S. (1950). "Principles of Silviculture." McGraw-Hill, New York.
Bakke, A. L. (1913). The effect of smoke and gases on vegetation. *Proc. Iowa Acad. Sci.* **20**, 169–188.
Bannan, M. W. (1955). The vascular cambium and radial growth in *Thuja occidentalis* L. *Can. J. Bot.* **33**, 113–138.
Bannister, B. (1969). Dendrochronology. *In* "Science in Archaeology" (Revised Edition) (D. Brothwell and E. Higgs, eds.), pp. 191–205. Thames and Hudson, London.
Barry, R. G. and Perry, A. H. (1973). "Synoptic Climatology Methods and Applications." Methuen, London.
Bendat, J. S. and Piersol, A. G. (1966). "Measurement and Analysis of Random Data." Wiley, New York.

Bergthorsson, P. (1962). Preliminary notes on past climate of Iceland. Data Presented to NCAR-USAF Camb. Res. Lab. Conference on the Climate of the 11th and 16th Centuries, June 16-24, 1962, Aspen, Colorado.

Billings, W. D. (1952). The environmental complex in relation to plant growth and distribution. *Quat. Rev. Biol.* **27(3)**, 251-265.

Billings, W. D. (1957). Physiological ecology. *Ann. Rev. Flant Physiol.* **8**, 375-392.

Billings, W. D. (1969). Vegetational pattern near alpine timberline as affected by fire-snowdrift interactions. *Veg. Acta Geobot.* **19**, 192-207.

Bitvinskas, T. T. (1974). "Dendroclimatic Investigations." Gidrometeoizdat, Leningrad.

Bjerknes, J. (1962). Synoptic survey of the interaction of sea and atmosphere in the North Atlantic. *Geofys. Norvegica* **24**, 115-157.

Bjerknes, J. (1966). A possible response of the atmospheric Hadley circulation to equatorial anomalies of ocean temperature. *Tellus* **18**, 320-829.

Blackman, R. B. and Tukey, J. W. (1958). "The Measurement of Power Spectra." Dover Publications, New York.

Blasing, T. J. (1974). Linear statistical transfer operators. Unpublished manuscript, Laboratory of Tree-Ring Res., Tucson.

Blasing, T. J. (1975). Methods for analyzing climatic variations in the North Pacific Sector and western North America for the last few centuries. Ph.D. Dissertation, Univ. of Wisconsin, Madison.

Blasing, T. J. and Fritts, H. C. (1975). Past climate of Alaska and Northwestern Canada as reconstructed from tree-rings. *In* "Climate of the Arctic" (G. Weller and S. A. Bowling, eds.), pp. 48-58. *Proc. 24th Alaska Sci. Conf.*, August 15-17, 1973, Fairbanks.

Blasing, T. J. and Fritts, H. C. (in press). Reconstruction of past climatic anomalies in the North Pacific and Western North America from tree-ring data. *Quaternary Res.*

Bormann, F. H. and Kozlowski, T. T. (1962). Measurements of tree ring growth with dial-gage dendrometers and vernier tree ring bands. *Ecology* **43(2)**, 289-294.

Box, G. E. P. and Jenkins, G. M. (1970). "Time Series Analysis, Forecasting, and Control." Holden-Day, San Francisco.

Bray, J. R. and Struik, G. J. (1963). Forest growth and glacial chronology in Eastern British Columbia, and their relation to recent climatic trends. *Can. J. Bot.* **41**, 1245-1271.

Brier, G. W. (1968). Long range prediction of the zonal westerlies and some problems in data analysis. *Rev. Geophys.* **6**, 525-551.

Brown, H. P. and Panshin, A. J. (1940). "Commercial Timbers of the United States." McGraw-Hill, New York.

Brown, J. M. (1968). The photosynthetic regime of some Southern Arizona ponderosa pine. Ph.D. Thesis, Univ. of Arizona, Tucson.

Bryson, R. A. (1974). A perspective on climatic change. *Science*, **184**, 753-760.

Bryson, R. A. and Dutton, J. A. (1961). Some aspects of the variance spectra of tree rings and varves. *Ann. New York Acad. Sci.* **95**, 580-604.

Buddemeier, R. W. (1974). Environmental controls over annual and lunar monthly cycles in hermatypic coral calcification. *Proc. Second Intern. Coral Reef Symp.* **2**, December, 1974, Great Barrier Reef Committee, Brisbane.

Budelsky, C. A. (1969). Variation in transpiration and its relationship with growth for *Pinus ponderosa* Lawson in Southern Arizona. Ph.D. Thesis, Univ. of Arizona, Tucson.

Charton, F. L. and Harman, J. R. (1973). Dendrochronology in northwestern Indiana. *Ann. Ass. Amer. Geogr.* **63(3)**, 302-311.

Chowdhury, K. A. (1940). The formation of growth rings in Indian trees. *Indian Forest Rec.* **2(1)**, 1939; **2(2)** and **2(3)**, 1940.

Clark, N., Blasing, T. J., and Fritts, H. C. (1975). Influence of interannual climatic fluctuations on biological systems. *Nature* **256(5515)**, 302–304.

Clayton, H. H. (1927). World weather records. *Smithsonian Misc. Coll.* **79**. Washington, D.C.

Cleaveland, M. K. (1975). Dendroclimatic relationships of shortleaf pine (*Pinus echinata* Mill.) in the South Carolina Piedmont. M.S. Thesis, Clemson Univ., Clemson, South Carolina.

Clevenger, J. F. (1913). The effect of the soot in smoke on vegetation. Univ. of Pittsburgh, Mellon Inst. of Industrial Res. and School of Specific Industries *Smoke Invest. Bull.* **7**, Pennsylvania.

Cooley, W. W. and Lohnes, P. R. (1971). "Multivariate Data Analysis." Wiley, New York.

Cooper, C. F., Blasing, T. J., Fritts, H. C., Oak Ridge Systems Ecology Group, Smith, F. M., Parton, W. J., Schreuder, G. F., Sollins, P., Zich, J., and Stoner, W. (1974) Simulation models of the effects of climatic change of natural ecosystems, pp. 550–562. *Proc. Third Climatic Impact Assessment Program* (CIAP) *Conf.*, Feb. 26–March 1, 1974, U.S. Dept of Transportation, Cambridge, Massachusetts.

Cooper, G. and Herrington, L. (1959). A remote reading band dendrograph. *Maine Farm Res.* **7(2)**, 10–11.

Craddock, J. M. (1957). An analysis of the slower temperature variations at Kew Observatory by means of mutually exclusive band pass filters. *J. Roy. Stat. Soc. Ser. A* **120**, 387–397.

Cunningham, G. L. and Fritts, H. C. (1970). Variation in water stress of one-year-old needles of ponderosa pine. *J. Ariz. Acad. Sci.* **6(2)**, 117–120.

Dahl, E. and Mork, E. (1959). On the relationships between temperature, respiration and growth in Norway spruce [*Picea abies* (L.) Karst.]. *Meddelelser fra Det norske skogforsøksvesen* **53**, 83–93.

Daniels, F. and Alberty, R. A. (1966). "Physical Chemistry" (3rd Edition) Wiley, New York.

Dansgaard, W., Johnsen, S. J., Clausen, H. B., and Langway, C. C., Jr. (1971). Climatic record revealed by the camp century ice core. *In* "The Late Cenozoic Glacial Ages" (K. Turekian, ed.), pp. 37–56. Yale Univ. Press, New Haven.

Daubenmire, R. F. (1945). An improved type of precision dendrometer. *Ecology* **26**, 97–98.

Daubenmire, R. F. (1959). "Plants and Environment: A Textbook of Plant Autecology" (2nd Edition). Wiley, New York.

Daubenmire, R. F. (1960). A seven-year study of cone production as related to xylem layers and temperature in *Pinus ponderosa*. *Amer. Midland Natur.* **64(1)**, 187–193.

Day, W. R. and Peace, T. R. (1934). The experimental production and the diagnosis of frost injury on forest trees. *Oxford Forest. Mem.* **16**.

Dean, J. S. (1969). Chronological analysis of Tsegi phase sites in Northeastern Arizona. *Pap. Lab. Tree-Ring Res.* **3**. Univ. of Arizona Press, Tucson.

Dean, J. S. (in press). Tree-ring dating in archeology. *Univ. of Utah Anthropol. Pap.*, **96**, *In* Univ. of Utah Press, Salt Lake City.

Decker, J. P. (1957). Further evidence of increased carbon dioxide production accompanying photosynthesis. *J. Solar Energy Sci. Eng.* **1**, 30–33.

de Martin, P. (1970). Les anneaux de croissance des arbres: dendroclimatologie et dendrochronologie. *Rev. Géogr. de L'est* **3–4**, 279–288.

Dempster, A. P. (1969). "Elements of Continuous Multivariate Analysis." Addison-Wesley, Reading, Massachusetts.

Denton, G. H. and Karlén, W. (1973). Holocene climatic variations—their pattern and possible cause. *Quaternary Res.* **3(2)**, 155–205.

Digby, J. and Wareing, P. F. (1966a). The effect of applied growth hormones on cambial division and the differentiation of the cambial derivatives. *Ann. Bot., N.S.* **30(119)**, 539–548.

Digby, J. and Wareing, P. F. (1966b). The relationship between endogenous hormone levels in the plant and seasonal aspects of cambial activity. *Ann. Bot., N.S.* **30(120)**, 607–622.

Dobbs, R. C. (1969). An electric device for recording small fluctuations and accumulated increment of tree stem circumference. *Forest. Chron.* **45**, 187–189.

Douglas, A. V. (1973). Past air-sea interactions off southern California as revealed by coastal tree-ring chronologies. M.S. Thesis, Univ. of Arizona, Tucson.

Douglas, C. L. and Erdman, J. A. (1967). Development of terminal buds in Pinyon pine and Douglas-fir trees. *Pearce-Sellards Texas Mem. Mus. Ser.* **8**, Austin, Texas.

Douglass, A. E. (1914). A method of estimating rainfall by the growth of trees. *In* "The Climatic Factor" (E. Huntington, ed.), pp. 101–122. *Carnegie Inst. Wash. Publ.* **192**.

Douglass, A. E. (1919). Climatic cycles and tree growth, Vol. I. *Carnegie Inst. Wash. Publ.* **289**.

Douglass, A. E. (1921). Dating our prehistoric ruins. *Natur. Hist.* **21(1)**, 27–30.

Douglass, A. E. (1929). The secret of the southwest solved by talkative tree rings. *Nat. Geogr. Mag.* **56(6)**, 736–770.

Douglass, A. E. (1935). Dating Pueblo Bonito and other ruins of the southwest. *Nat. Geogr. Soc. Contrib. Tech. Pap., Pueblo Bonito Ser.* **1**.

Douglass, A. E. (1936). Climatic cycles and tree growth, Vol. III. A study of cycles. *Carnegie Inst. Wash. Publ.* **289**.

Douglass, A. E. (1937). Tree rings and chronology. *Univ. Ariz. Bull.* **8(4)**, *Phys. Sci. Ser.* **1**.

Downs, R. J. (1962). Photocontrol of growth and dormancy in woody plants. *In* "Tree Growth" (T. T. Kozlowski, ed.), pp. 133–148. Ronald Press, New York.

Draper, N. and Smith, H. (1966). "Applied Regression Analysis." Wiley, New York.

Drew, A. P. (1967). Stomatal activity in semi-arid site Ponderosa pine. M.S. Thesis, Univ. of Arizona, Tucson.

Drew, A. P., Drew, L. G., and Fritts, H. C. (1972). Environmental control of stomatal activity in mature semiarid site ponderosa pine. *J. Ariz. Acad. Sci.* **7(2)**, 85–93.

Druce, A. P. (1966). Tree-ring dating of recent volcanic ash and lapilli, Mt. Egmont. *New Zealand J. Bot.* **4(1)**, 3–41.

Duff, G. H. and Nolan, N. J. (1953). Growth and morphogenesis in the Canadian forest species. I. The controls of cambial and apical activity in *Pinus resinosa* Ait. *Can. J. Bot.* **31**, 471–513.

Eardley, A. J. and Viavant, W. (1967). Rates of denudation as measured by bristlecone pines, Cedar Breaks, Utah. *Utah Geol. Min. Serv. Spec. Stud.* **21**, Salt Lake City.

Eckstein, D. and Bauch, J. (1969). Beitrag zur Rationalisierung eines dendrochronologischen Verfahrens und zur Analyse seiner Aussagesicherheit. *Forstwissenschaftliches Cent.* **88**, 230–250.

Eis, S. (1970). Natural root grafts in conifers and the effect of grafting on tree growth. *In* "Tree-Ring Analysis with Special Reference to Northwest America" (J. H. G. Smith and J. Worrall, eds.), pp. 25–29. Univ. of British Columbia *Fac. Forest. Bull.* **7**, Vancouver.

Eklund, B. (1957). The annual ring variations in spruce in the centre of northern Sweden and their relation to the climatic conditions. *Statens Skogsforskningsinstitut* **47(1)**, 2–63.

Elling, W. (1966). Untersuchungen über das Jahrringverhalten der Schwarzerle. *Flora Abt. B, Bd.* **156**, 155–201.

Erdman, J. A., Douglas, C. L., and Marr, J. W. (1969). The environment of Mesa Verde, Colorado, Wetherill Mesa studies. *Nat. Park Serv. Archaeol. Res. Ser.* **7B**. U.S. Dept. of the Interior, Washington, D. C.

Ermich, K. (1955). The dependence of diameter growth of trees from Tatra Mountains on the climatic fluctuations. *Acta Soc. Bot. Poloniae* **24**, 245–273.

Estes, E. T. (1970). The dendrochronology of black oak (*Quercus velutina* Lam.), white oak (*Quercus alba* L.), and shortleaf pine (*Pinus echinata* Mill.), in the Central Mississippi Valley. *Ecol. Monogr.* **40**, 295–316.

Ezekeil, M. and Fox, K. A. (1959). "Methods of Correlation and Regression Analysis" (3rd Edition). Wiley, New York.

Fahn, A., Wachs, N., and Ginzburg, C. (1963). Dendrochronological studies in the Negev. *Israel Explor. J.* **13(4)**, 291–299.

Farrar, J. L. (1961). Longitudinal variation in the thickness of the annual ring. *Forest. Chron.* **37(4)**, 323–331.

Fayle, D. C. F. (1968). Radial growth in tree roots. *Fac. Forest. Tech. Rep.* **9**, Univ. of Toronto, Ontario.

Ferguson, C. W. (1964). Annual rings in big sagebrush. *Pap. Lab. Tree-Ring. Res.* **1**, Univ. of Arizona Press, Tucson.

Ferguson, C. W. (1968). Bristlecone pine: science and esthetics. *Science* **159(3817)**, 839–846.

Ferguson, C. W. (1970a). Concepts and techniques of dendrochronology. *In* "Scientific Methods in Medieval Archaeology" (R. Berger, ed.). pp. 183–200. Univ. of California Press, Berkeley.

Ferguson, C. W. (1970b). Dendrochronology of bristlecone pine, *Pinus aristata*: establishment of a 7484-year chronology in the White Mountains of Eastern-Central California, U.S.A. *In* "Radiocarbon Variations and Absolute Chronology" (I.U. Olsson, ed.), pp. 237–259. *Nobel Symp.* **12**, Almqvist and Wiksell, Stockholm, and Wiley, New York.

Fons, W. L., Bruce, H. D., and McMasters, A. (1960). Tables for estimating direct beam solar radiation on slopes at 30° and 46° latitude. U.S. Dept. Agr., Pacific Southwest Forest and Rang Exp. Sta., Berkeley, California.

Fraser, D. A. (1962). Apical and radial growth of white spruce [*Picea glauca* (Moench) Voss] at Chalk River, Ontario, Canada. *Can. J. Bot.* **40**, 659–668.

Freeland, R. O. (1944). Apparent photosynthesis in some conifers during the winter. *Plant Physiol.* **19**, 179–185.

Freeland, R. O. (1952). Effect of age of leaves upon the rate of photosynthesis in some conifers. *Plant Physiol.* **27**, 685–690.

Freund, J. E. (1962). "Mathematical Statistics." Prentice-Hall, Englewood Cliffs, New Jersey.

Friesner, R. C. (1950). Growth-rainfall trend coefficients shown by six species of hardwoods in Brown County, Indiana. *Butler Univ. Bot. Stud.* **9**, 159–166.

Friesner, R. C. and Friesner, G. M. (1941). Relation of annual ring formation to rainfall. *Butler Univ. Bot. Stud.* **5(1-8)**, 95–111.

Fritts, H. C. (1958). An analysis of radial growth of beech in a central Ohio forest during 1954–1955. *Ecology* **39(4)**, 705–720.

Fritts, H. C. (1959). The relation of radial growth to maximum and minimum temperatures in three tree species. *Ecology* **40**, 261–265.

Fritts, H. C. (1960). Multiple regression analysis of radial growth in individual trees. *Forest Sci.* **6(4)**, 334–349.

Fritts, H. C. (1962a). An approach to dendroclimatology: screening by means of multiple regression techniques. *J. Geophys. Res.* **67(4)**, 1413–1420.

Fritts, H. C. (1962b). The relation of growth ring widths in American beech and white oak to variations in climate. *Tree-Ring Bull.* **25(1–2)**, 2–10.

Fritts, H. C. (1962c). The relevance of dendrographic studies to tree-ring research. *Tree-Ring Bull.* **24**, 9–11.

Fritts, H. C. (1965). Tree-ring evidence for climatic changes in Western North America. *Mon. Weather Rev.* **93**, 421–443.

Fritts, H. C. (1966). Growth-rings of trees: their correlation with climate. *Science* **154(3752)**, 973–979.

Fritts, H. C. (1969). Bristlecone pine in the White Mountains of California; growth and ring-width characteristics. *Pap. Lab. Tree-Ring Res.* **4**, Univ. of Arizona Press, Tucson.

Fritts, H. C. (1971). Dendroclimatology and dendroecology. *Quaternary Res.* **1(4)**, 419–449.

Fritts, H. C. (1974). Relationships of ring widths in arid-site conifers to variations in monthly temperature and precipitation. *Ecol. Monogr.* **44(4)**, 411–440.

Fritts, H. C. and Cathey, E. H. (1971). Dendroclimatic history of southwestern United States. Final Report Contract 1-35241 National Oceanic and Atmospheric Administration. U.S. Dept. of Commerce, Washington, D.C.

Fritts, H. C. and Fritts, E. C. (1955). A new dendrograph for recording radial changes of a tree. *Forest Sci.* **1(4)**, 271-276.

Fritts, H. C. and Holowaychuck, N. (1959). Some soil factors affecting the distribution of beech in a central Ohio forest. *Ohio J. Sci.* **59**, 167–186.

Fritts, H. C. and Shatz, D. J. (1975). Selecting and characterizing tree-ring chronologies for dendroclimatic analysis. *Tree-Ring Bulletin* **35**, 31–40.

Fritts, H. C., Smith, D. G., Budelsky, C. A., and Cardis, J. W. (1965a). The variability of ring characteristics within trees as shown by a reanalysis of four ponderosa pine. *Tree-Ring Bull.* **27(1–2)**, 3–18.

Fritts, H. C., Smith, D. G., Cardis, J. W., and Budelsky, C. A. (1965b). Tree-ring characteristics along a vegetation gradient in Northern Arizona. *Ecology* **46**, 393–401.

Fritts, H. C., Smith, D. G., and Stokes, M. A. (1965c). The biological model for paleoclimatic interpretation of Mesa Verde tree-ring series. *Amer. Antiq.* **31(2)**, (part 2), 101–121.

Fritts, H. C., Mosimann, J. E., and Bottorff, C. P. (1969). A revised computer program for standardizing tree-ring series. *Tree-Ring Bull.* **29**, 15–20.

Fritts, H. C., Blasing, T. J., Hayden, B. P., and Kutzbach, J. E. (1971). Multivariate techniques for specifying tree-growth and climate relationships and for reconstructing anomalies in paleoclimate. *J. Appl. Meteorol.* **10(5)**, 845–864.

Fry, K. E. and Walker, R. B. (1967). A pressure-infiltration method for estimating stomatal opening in conifers. *Ecology* **48(1)**, 155–157.

Gagnon, D. (1961). Rainfall and the width of annual rings in planted white spruce. *Forest. Chron.* **37(2)**, 96–101.

Gale, J. (1972). Availability of carbon dioxide for photosynthesis at high altitudes: theoretical considerations. *Ecology* **53**, 494–497.

Garrett, P. W. and Zahner, R. (1973). Fascicle density and needle growth response of red pine to water supply over two seasons. *Ecology* **54(6)**, 1328–1334.

Gates, D. M. (1962). "Energy Exchange in the Biosphere." Harper and Row Biol. Monogr., New York.

Gates, D. M. (1965). Energy, plants, and ecology. *Ecology* **46(1–2)**, 1–13.

Gates, D. M. (1967). Water balance in terrestrial ecosystems. *In* "Transport Phenomena in Atmospheric and Ecological Systems" (F. Kreith, ed.), pp. 21–36. Amer. Soc. Mech. Eng., New York.

Gates, D. M. (1968a). Energy exchange between organisms and environment. *Australian J. Sci.* **31(2),** 67–74.

Gates, D. M. (1968b). Transpiration and leaf temperature. *Ann. Rev. Plant Physiol.* **19,** 211–238.

Gates, W. L. and Mintz, Y. (1975). Understanding climatic change: a program for action. Report of the Panel on Climatic Variation of the U.S. Committee for GARP (Global Atmos. Res. Proj.), Nat. Res. Council, Nat. Acad. Sci., Washington, D.C.

Geiger, R. (1965). "The Climate Near the Ground." Harvard Univ. Press, Cambridge, Massachusetts.

Giddings, L. J., Jr. (1943). Some climatic aspects of tree growth in Alaska. *Tree-Ring Bull.* **9,** 26–32.

Giddings, L. J., Jr. (1953). Yukon river spruce growth. *Tree-Ring Bull.* **20,** 2–5.

Giertych, M. M. (1964). Endogenous growth regulators in trees. *Bot. Rev.* **30,** 292–311.

Glahn, H. R. (1968). Canonical correlation and its relationship to discriminant analysis and multiple regression. *J. Atmos. Sci.* **25,** 23–31.

Glerum, C. (1970). Drought ring formation in conifers. *Forest Sci.* **16(2),** 246–248.

Glerum, C. and Farrar, J. L. (1966). Frost ring formation in the stems of some coniferous species. *Can. J. Bot.* **44,** 879–886.

Glock, W. S. (1937). Principles and methods of tree-ring analysis. *Carnegie Inst. Wash. Publ.* **486.**

Glock, W. S. (1942). A rapid method of correlation for continuous time series. *Amer. J. Sci.* **240,** 437–442.

Glock, W. S. (1950). Tree growth and rainfall—A study of correlation and methods. *Smithsonian Misc. Coll.* **3(18).** Smithsonian Inst. Publ. **4016,** Washington, D.C.

Glock, W. S. (1951). Cambial frost injuries and multiple growth layers at Lubbock, Texas, *Ecology* **32(1),** 28–36.

Glock, W. S. (1955). Tree growth II, growth rings and climate. *Bot. Rev.* **21(1-3),** 73–188.

Glock, W. S. and Agerter, S. R. (1962). Rainfall and tree growth. *In* "Tree Growth" (T. T. Kozlowski, ed.), pp. 23–56. Ronald Press, New York.

Glock, W. S., Studhalter, R. A., and Agerter, S. R. (1960). Classification and multiplicity of growth layers in the branches of trees at the extreme lower forest border. *Smithsonian Misc. Coll.* **140.** Smithsonian Inst. Publ. **4421,** Washington, D.C.

Glock, W. S., Germann, P. J., and Agerter, S. R. (1963). Uniformity among growth layers in three ponderosa pine. *Smithsonian Misc. Coll.* **145.** Smithsonian Inst. Publ. **4508,** Washington, D.C.

Graham, B. F., Jr. and Bormann, F. H. (1966). Natural root grafts. *Bot. Rev.* **32 (3),** 255–292.

Grier, C. C. and Waring, R. H. (1974). Conifer foliage mass related to sapwood area. *Forest Sci.* **20(3),** 205–206.

Haase, E. F. (1970). Environmental fluctuations on south-facing slopes in the Santa Catalina Mountains of Arizona. *Ecology* **51(6),** 959–974.

Hall, R. C. (1944). A vernier tree-growth band. *J. Forest.* **42,** 742–743.

Hari, P. and Luukkanen, O. (1973). Effect of water stress, temperature, and light on photosynthesis in alder seedlings. *Physiol. Plant.* **29(1),** 45–53.

Hari, P. and Sirén, G. (1972). Influence of some ecological factors and the seasonal stage of development upon the annual ring width and radial growth index. Royal College of Forestry *Res. Notes* **40,** Stockholm.

Hari, P., Leikola, M., and Räsänen, P. (1970). A dynamic model of the daily height increment of plants. *Ann. Bot. Fenn.* **7,** 375–378.

Harman, J. R. and Elton, W. M. (1971). The LaPorte, Indiana, precipitation anomaly. *Ann. Ass. Amer. Geogr.* **61(3)**, 468–480.

Hart, J. H. and Wargo, P. M. (1965). Increment borer wounds—penetration points for *Ceratocystis fagacearum*. *J. Forest.* **63(1)**, 38–39.

Haury, E. W. (1962). HH-39: recollections of a dramatic moment in southwestern archaeology. *Tree-Ring Bull.* **24(3-4)**, 11–14.

Heger, L., Parker, M. L., and Kennedy, R. W. (1974). X-ray densitometry: a technique and an example of application. *Wood Sci.* **7**, 140–148.

Heinselman, M. L. (1969). Diary of the canoe country's landscape. *Naturalist* **20**, 2–13.

Heinselman, M. L. (1973). Fire in the virgin forests of the boundary waters canoe area, Minnesota. *Quaternary Res.* **3(3)**, 329–382.

Heizer, R, F. (1956). The first dendrochronologist. *Amer. Antiq.* **22**, 186–188.

Helley, E. J. and LeMarche, V. C., Jr. (1973). Historic flood information for northern California streams from geological and botanical evidence. *U.S. Geol. Surv. Prof. Pap.* **485-E**, 1–16.

Hollstein, E. (1975). Personal communication. Rheinisches Landesmuseum von Trier, D-55 Trier, Ostallee 44, West Germany.

Holmsgaard, E. (1955). Årringsanalyser af Danske Skovtraeer. (Tree-ring analyses of Danish forest trees.) *Det Forstlige Forsøgsvoesen i Danmark*, **22**, 1–246.

Holmsgaard, E. (1962). Influence of weather on growth and reproduction of beech. *Comm. Inst. Forest. Fenn.* **55**, 1–5.

Huber, B. (1941). Aufbau einer mitteleuropäischen Jahrring-Chronologie. *Mitteilung Akad. Dtsch. Forstwiss.* **1**, 110–125.

Huber, B., Schmidt, E., and Jahnel, H. (1937). Untersuchungen über den Assimilatstrom I. *Tharandt. forstl. Jahrb.* **88**, 1017–1050.

Hughes, J. F. (1965). Tension wood: a review of literature. *Forest. Abstr.* **26**, 1–16.

Huntington, E. (1914). The climatic factor as illustrated in arid America. *Carnegie Inst. Wash. Publ.* **192**.

Hustich, I. (1945). The radial growth of the pine at the forest limit and its dependence on the climate. *Soc. Sci. Fenn. Comment. Biol.* **9(2)**, 1–30.

Hustich, I. and Elfving, G. (1944). Die Radialzuwachsvariationen der Waldgrenzkiefer. (The radial growth variations of forest border pine.) *Soc. Sci. Fenn. Comment. Biol.* **9(8)**, 1–18.

Imbrie, J. and Kipp, N. G. (1971). A new micropaleontological method for quantitative paleoclimatology: application to a late Pleistocene Caribbean core. *In* "The Late Cenozoic Glacial Ages" (K. K. Turekian, ed.), pp. 71–181. Yale Univ. Press, New Haven.

Impens, I. I. and Schalck, J. M. (1965). A very sensitive electric dendrograph for recording radial changes of a tree. *Ecology* **46**, 183–184.

Jemison, G. M. (1944). The effect of basal wounding by forest fires on the diameter growth of some southern Appalachian hardwoods. *Duke Univ. Sch. Forest. Bull.* **9**, Durham.

Jenkins, G. M. and Watts, D. G. (1968). "Spectral Analysis and Its Applications." Holden-Day, San Francisco.

Jenny, H. (1941). "Factors of Soils Formation. A System of Quantitative Pedology." McGraw-Hill, New York.

Jenny, H. (1958). Role of the plant factor in the pedogenic functions. *Ecology* **39**, 5–16.

Jenny, H. (1961). Derivation of state factor equations of soil and ecosystems. *Soil Sci. Soc. Amer. Proc.* **25**, 385–388.

Jones, F. W. and Parker, M. L. (1970). G.S.C. tree-ring scanning densitometer and data acquisition system. *Tree-Ring Bull.* **30**, 23–31.

Jones, R. H. (1975). Estimating the variance of time averages. *J. Appl. Meteorol.* **14**, 159-163.
Jonsson, B. (1969). Studier över den av väderleken orsakade variationen i årsringsbredderna hos tall roch gran i Sverige. (Studies of variations in the widths of annual rings in Scots pine and Norway spruce due to weather conditions in Sweden.) Royal College of Forestry *Res. Notes* **16**, Stockholm.
Jonsson, B. and Sundberg, R. (1972). Has the acidification by atmospheric pollution caused a growth reduction in Swedish forests? Royal College of Forestry *Res. Notes* **20**, Stockholm.
Julian, P. R. (1970). An application of rank-order statistics to the joint spatial and temporal variations of meteorological elements. *Mon. Weather Rev.* **98(2)**, 142-153.
Julian, P. R. and Fritts, H. C. (1968). On the possibility of quantitatively extending climatic records by means of dendroclimatological analysis. *Proc. First Statist. Meteorol. Conf.*, Amer. Meteorol. Soc., pp. 76-82, Hartford, Connecticut.
Keen, F. P. (1937). Climatic cycles in eastern Oregon as indicated by tree rings. *Mon. Weather Rev.* **65(5)**, 175-188.
Kelsey, H. P. and Dayton, W. A. (1942). "Standardized Plant Names" (2nd edition). J. Horace McFarland, Harrisburg, Pennsylvania.
Kienholz, R. (1934). Leader, needle, cambial, and root growth of certain conifers and their interrelations. *Bot. Gaz.* **96**, 73-92.
Kilgore, B. (1970). Restoring fire to the Sequoias. *Nat. Parks and Conserv.* **44(277)**, 16-22.
Kisiel, C. C. (1969). Time series analysis of hydrologic data. *In* "Advances in Hydroscience Vol. 5" (V. T. Chow, ed.), pp. 1-120. Academic Press, New York.
Kochenderfer, J. N. (1973). Root distribution under some forest types native to West Virginia. *Ecology* **54**, 445-448.
Koerber, T. W. and Wickman, B. E. (1970). Use of tree-ring measurements to evaluate impact of insect defoliation. *In* "Tree-Ring Analysis with Special Reference to Northwest America" (J. H. G. Smith and J. Worrall, eds.), pp. 101-106. Univ. of British Columbia *Fac. Forest. Bull.* **7**, Vancouver.
Kohler, M. A. (1949). On the use of double-mass analysis for testing the consistency of meteorological records and for making required adjustments. *Bull. Amer. Meteorol. Soc.* **30**, 188-189.
Kozlowski, T. T. (1964). Shoot growth in woody plants. *Bot. Rev.* **30(3)**, 335-392.
Kozlowski, T. T. (1965). Expansion and contraction of plants. *In* "Advancing Frontiers of Plant Science", Vol. 10 (L. Chandra, ed.), pp. 63-74. Inst. for the Advance. of Sci. and Culture, New Delhi 16, India.
Kozlowski, T. T. (1968). Introduction. *In* "Water Deficits and Plant Growth, I" (T. T. Kozlowski, ed.), pp. 1-21. Academic Press, New York.
Kozlowski, T. T. (1971a). "Growth and Development of Trees, I. Seed Germination, Ontogeny and Shoot Growth." Academic Press, New York.
Kozlowski, T. T. (1971b). "Growth and Development of Trees, II. Cambial Growth, Root Growth, and Reproductive Growth." Academic Press, New York.
Kozlowski, T. T. and Peterson, T. A. (1962). Seasonal growth of dominant, intermediate and suppressed red pine trees. *Bot. Gaz.* **124(2)**, 146-154.
Kozlowski, T. T. and Winget, C. H. (1964). Diurnal and seasonal variation in radii of tree stems. *Ecology* **45**, 149-155.
Kozlowski, T. T., Winget, C. H., and Torrie, J. H. (1962). Daily radial growth of oak in relation to maximum and minimum temperature. *Bot. Gaz.* **124(1)**, 9-17.
Kramer, P. J. (1937). The relation between rate of transpiration and rate of absorption of water in plants. *Amer. J. Bot.* **24**, 10-15.

Kramer, P. J. (1969). "Plant and Soil Water Relationships." McGraw-Hill, New York.

Kramer, P. J. and Decker, J. P. (1944). Relation between light intensity and rate of photosynthesis of loblolly pine and certain hardwoods. *Plant Physiol.* **19,** 350–358.

✓Kramer, P. J. and Kozlowski, T. T. (1960). "Physiology of Trees." McGraw-Hill, New York.

Krueger, K. W. and Trappe, J. M. (1967). Food reserves and seasonal growth of Douglas-fir Seedlings. *Forest Sci.* **13,** 192–202.

Kukla, G. J. and Kukla, H. J. (1974). Increased surface albedo in the northern hemisphere. *Science* **183,** 709–714.

Kuo, M. and McGinnes, E. A., Jr. (1973). Variation of anatomical structure of false rings in eastern redcedar. *Wood Sci.* **5(3),** 205–210.

Kuroiwa, K. (1957). Daily growth curve of *Paulownia* [I] on the "Mirror Dendrometer." *J. Japanese Forest. Soc.* **39,** 89–91.

Kuroiwa, K. (1959). Measurement of radial change of stems by strain guage. *J. Japanese Forest. Soc.* **41(9),** 331–333.

Kutzbach, J. E. (1970). Large-scale features of monthly mean northern hemisphere anomaly maps of sea-level pressure. *Mon. Weather Rev.* **98(9),** 708–716.

Ladefoged, K. (1952). The periodicity of wood formation. *Danske Videnskabernes Selskab. Biol. Skrifter* **7,** 1–98.

Ladurie, E. LeR. (1971). "Times of Feast, Times of Famine: A History of Climate Since the Year 1000." Doubleday, Garden City, New York.

LaMarche, V. C., Jr. (1966). An 800-year history of stream erosion as indicated by botanical evidence. *U.S. Geol. Surv. Prof. Pap.* **550-D,** 83–86, Washington, D.C.

LaMarche, V. C. Jr. (1968). Rates of slope degradation as determined from botanical evidence, White Mountains, California. *U.S. Geol. Surv. Prof. Pap.* 352–I, Washington, D.C.

LaMarche, V. C. Jr. (1970). Frost-damage rings in subalpine conifers and their application to tree-ring dating problems. *In* "Tree-Ring Analysis with Special Reference to Northwest America" (J. H. G. Smith, and J. Worrall, eds.), pp. 99–100. Univ. of British Columbia *Fac. Forest. Bull.* **7,** Vancouver.

La Marche, V. C. Jr. (1973). Holocene climatic variations inferred from treeline fluctuations in the White Mountains, California. *Quaternary Res.* **3,** 632–660.

LaMarche, V. C. Jr. (1974a). Paleoclimatic inferences from long tree-ring records. *Science* **183,** 1043–1048.

La Marche, V. C. Jr. (1974b). Frequency-dependent relationships between tree-ring series along an ecological gradient and some dendroclimatic implications. *Tree-Ring Bull.* **34,** 1–20.

LaMarche, V. C., Jr. and Fritts, H. C. (1971a). Anomaly patterns of climate over the Western United States, 1700–1930, derived from principal component analysis of tree-ring data. *Mon. Weather Rev.* **99,** 138–142.

LaMarche, V. C., Jr. and Fritts, H. C. (1971b). Tree rings, glacial advance, and climate in the Alps. *Z. für Gletscherkunde und Glazialgeologie* **7** (**1–2**), 125–131.

LaMarche, V. C., Jr. and Fritts, H. C. (1972). Tree-rings and sunspot numbers. *Tree-Ring Bull.* **32,** 19–33.

LaMarche, V. C., Jr. and Harlan, T. P. (1973). Accuracy of tree-ring dating of bristlecone pine for calibration of the radiocarbon time scale. *J. Geophys. Res.* **78(36),** 8849–8858.

LaMarche, V. C., Jr. and Mooney, H. A. (1967). Altithermal timberline advance in Western United States. *Nature* **213,** 980–982.

LaMarche, V. C., Jr. and Mooney, H. A. (1972). Recent climatic change and development of the bristlecone pine (*P. longaeva* Bailey) krummholz zone, Mt. Washington, Nevada. *Arctic Alpine Res.* **4(1),** 61–72.

LaMarche, V. C., Jr. and Stockton, C. W. (1974). Chronologies from temperature-sensitive bristlecone pines at upper treeline in Western United States. *Tree-Ring Bull.* **34,** 21–45.

LaMarche, V. C., Jr. and Wallace, R. E. (1972). Evaluation of effects on trees of past movements on the San Andreas fault, northern California. *Geol. Soc. Amer. Bull.* **83,** 2665–2676.

Lamb, H. H. (1963). On the nature of certain climatic epochs which differed from the modern (1900–1939) normal. *In* "Changes of Climate." *UNESCO Arid Zone Res. Ser.* **20,** 125–150, Paris.

Lamb, H. H. (1966). Climate in the 1960's—changes in the world's wind circulation reflected in prevailing temperatures, rainfall patterns, and the levels of African lakes. *Geogr. J.* **132(2),** 183–212.

Lamb, H. H. (1969a). Climatic fluctuations. *In* "World Survey of Climatology, II. General Climatology" (H. Flohn, ed.), pp. 173–249. Elsevier, Amsterdam.

Lamb, H. H. (1969b). The New Look of Climatology. *Nature* **223,** 1209–1215.

Lamb, H. H. (1972). "Climate: Present, Past, and Future, I, Fundamentals and Climate Now" Methuen, London.

Lamb, H. H. and Johnson, A. I. (1959). Climate variation and observed changes in the general circulation, I and II. *Geogr. Ann.* **41,** 94–134.

Lamb, H. H., Johnson, M. A. and Johnson, A. I. (1966). Secular variations of the atmospheric circulation since 1750. Meteorological Offic **14(5),** *Geophys. Mem.* **110,** London.

Landsberg, H. E., Mitchell, J. M., Jr., Crutcher, H. L., and Quinlan, F. T. (1963). Surface signs of the biennial atmospheric pulse. *Mon. Weather Rev.* **91,** 549–556.

Lanner, R. M. (1970). Origin of the summer shoot of pinyon pines. *Can. J. Bot.* **48,** 1759–1765.

LaPoint, G. and VanCleve, K. (1971). A portable electronic multichannel dendrograph and environmental factor recording system. *Can. J. Forest Res.* **1,** 273–277.

Larson, P. R. (1957). Effect of environment on the percentage of summerwood and specific gravity of slash pine. *Yale Univ. Sch. Forest. Bull.* **63,** New Haven.

Larson, P. R. (1960). A physiological consideration of the springwood summerwood transition in red pine. *Forest Sci.* **6(2),** 110–122.

Larson, P. R. (1962). Auxin gradients and the regulation of cambial activity. *In* "Tree Growth" (T. T. Kozlowski, ed.), pp. 97–117. Ronald Press, New York.

Larson, P. R. (1964). Some indirect effects of environment on wood formation. *In* "The Formation of Wood in Forest Trees" (M. H. Zimmermann, ed.), pp. 345–365. Academic Press, New York.

Laufersweiler, J. D. (1955). Changes with age in the proportion of the dominants in a beech-maple forest in central Ohio. *Ohio J. Forest Sci.* **55(2),** 73–80.

Lawrence, D. B. (1950). Estimating dates of recent glacial advances and recession rates by studying tree growth layers. *Trans. Amer. Geophys. Union* **31,** 243–248.

Lawrence, D. B. (1952). Evidence of the age of beaver ponds. *J. Wildlife Manage.* **16,** 69–79.

Lawrence, D. B. and Lawrence, E. G. (1958a). Bridge of the Gods legend, its origin, history and dating. *Mazama* **60,** 1–9.

Lawrence, D. B. and Lawrence, E. G. (1958b). Historic landslides of the Gros Ventre Valley, Wyoming. *Mazama* **60,** 10–20.

Lawrence, D. B. and Lawrence, E. G. (1961). Response of enclosed lakes to current glaciopluvial climatic conditions in middle latitude Western North America. *Ann. New York Acad. Sci.* **95,** 341–350.

Lawson, M. P. (1972). The climate of the Great American Desert. Ph.D. thesis. Clark Univ., Worcester, Massachusetts.

Lawson, M. P. (1974). The climate of the Great American Desert, reconstruction of the climate of Western interior United States, 1800–1850. *Univ. Nebraska Stud. N.S.* **46,** Lincoln.

Leopold, A. C. (1964). "Plant Growth and Development." McGraw-Hill, New York.

Levitt, J. (1956). "The Hardiness of Plants." Academic Press, New York.

Li, J. C. R. (1964). "Statistical Inference, I." Edwards Brothers, Ann Arbor, Michigan.

Liming, F. G. (1957). Homemade dendrometers. *J. Forest.* **55,** 575–577.

Little, E. L., Jr. (1953). "Check List of Native and Naturalized Trees of the United States (Including Alaska)." *U.S. Forest Serv. Agr. Handbook* **41,** Washington, D.C.

Lorenz, E. N. (1956). Empirical orthogonal functions and statistical weather prediction. *M.I.T. Stat. Forecasting Proj. Sci. Rep.* **1,** Contract No. AF 19 (604)–1566.

Lowry, W. P. (1966). Apparent meteorological requirements for abundant cone crop in Douglas-fir. *Forest Sci.* **12(2),** 185–192.

Ludlum, D. M. (1966). "The History of American Weather: Early American Winters, I, 1604–1820." Amer. Meteorol. Soc., Boston.

Ludlum, D. M. (1968). "The History of American Weather: Early American Winters, II, 1821–1870." Amer. Meteorol. Soc., Boston.

Lund, I. A. (1963). Map-pattern classification by statistical methods. *J. Appl. Meteorol.* **2(1),** 56–65.

Lundegårdh, H. (1931). "Environment and Plant Development." (E. Ashby, Trans.) Edward Arnold, London.

Lyon, C. J. (1936). Tree-ring width as an index of physiological dryness in New England. *Ecology* **17(3),** 457–478.

Lyon, C. J. (1943). Water supply and the growth rates of conifers around Boston. *Ecology* **24(3),** 329–344.

MacDougal, D. T. (1924). Dendrographic measurements. *In* "Growth in Trees and Massive Organs of Plants." (D. T. MacDougal and F. Shreve, eds) pp. 3–88, *Carnegie Inst. Wash. Publ.* **350.**

Mariaux, A. (1967). Les cernes dans les bois tropicaux Africains, nature et périodicité. *Rev. Bois et Forêts des Tropiques* **113,** 3–14, and 23–37.

Mason, H. L. and Langenheim, J. H. (1957). Language analysis and the concept Environment. *Ecology* **38(2),** 325–340.

Matalas, N. C. (1962). Statistical properties of tree-ring data. *Intern. Ass. Sci. Hydrol. Publ.* **7,** 39–47.

McDonald, J. E. (1957). A note on the precision of estimation of missing precipitation data. *Trans. Amer. Geophys. Union* **38,** 657–661.

McGinnies, W. J. (1967). Correlation between annual rings of woody plants and range herbage production. *J. Range Manage.* **20(1),** 42–45.

McInteer, B. B. (1947). Tree ring study in Kentucky. *Castanea: J. Southern Appalachian Bot. Club* **12,** 38–50.

Meyer, B. S., Anderson, D. B., Böhning, R. H., and Fratianne, D. G. (1973). "Introduction to Plant Physiology" (2nd Edition), Van Nostrand, Princeton.

Mikola, P. (1952). The effect of recent climatic variations on forest growth in Finland. *Fenn.* **75,** 69–76.

Mikola, P. (1962). Temperature and tree growth near the northern timber line. *In* "Tree Growth" (T. T. Kozlowski, ed.), pp. 265–274. Ronald Press, New York.

Miller, C. W. (1950a). The effect of precipitation on annular-ring growth in three species of trees from Brown County, Indiana. *Butler Univ. Bot. Stud.* **9**, 167–175.
Miller, C. W. (1950b). Growth data from nine sections of *Acer saccharum* from Montgomery County, Indiana. *Butler Univ. Bot. Stud.* **10**, 12–19.
Mitchell, J. M., Jr., Dzerdzeevskii, B., Flohn, H., Hofmeyr, W. L., Lamb, H. H. Rao, K. N., and Wallen, C. C. (1966). Climatic change. *World Meteorol. Organ. Tech. Note* **79**, Geneva.
Mitchell, V. L. (1967). An investigation of certain aspects of tree growth rates in relation to climate in the Central Canadian Boreal Forest. Univ. of Wisconsin, *Dept. Meteorol. Tech. Rep.* **33**, Task NR 387–022, ONR Contract 1202(07), NSF GP-5572X, Madison.
Mitscherlich, G., Moll, W., Künstle, E., and Maurer, P. (1966). Ertragskundlich-ökologische Untersuchungen im Rein- und Mischbestand, VI. Zuwachsbeginn und-ende, Stärkenänderung und jährlicher Durchmesserzuwachs. *Allgemeine Forst und Jagdzeitung* **137**, 72–91.
Möller, C. M., Müller, M. D., and Nielsen, J. (1954). Graphic presentation of dry matter production of European beech. *Det. Forstlige Forsøgsvoesen i Danmark* **21**, 327–335.
Mooney, H. A. (1972). The carbon balance of plants. *Ann. Rev. Ecol. Syst.* **3**, 315–346.
Mooney, H. A., St. Andre, G., and Wright, R. D. (1962). Alpine and subalpine vegetation patterns in the White Mountains of California. *Amer. Midland Natur.* **68**, 257–273.
Mooney, H. A., West, M., and Brayton, R. (1966). Field measurements of the metabolic responses of bristlecone pine and big sagebrush in the White Mountains of California. *Bot. Gaz.* **127**, 105–113.
Mork, E. (1928). Die Qualität des Fichtenholzes unter besonderer Rücksichtnahme auf Schleif- und Papierholz. *Papier-Fabrik* **26(48)**, 741–747.
Morrison, D. F. (1967). "Multivariate Statistical Methods." McGraw-Hill, New York.
Munaut, A. V. (1966). Recherches dendrochronologiques sur *Pinus silvestris*, II. Première application des méthodes dendrochronologiques a l'étude de pins sylvestres sub-fossiles (Terneuzen, Pays-Bas). *Agricultura* **14**, Ser. 2, 361–389.
Namias, J. (1953). Thirty-day forecasting—A review of a ten-year experiment. *Meteorol. Monogr.* **2**, Amer. Meteorol. Soc., Boston.
Namias, J. (1960). Factors in the initiation, perpetuation, and termination of drought. *IUGG Intern. Ass. Sci. Hydrol. Publ.* **51**, 81–94, Assembly General of Helsinki.
Namias, J. (1969). Seasonal interactions between the North Pacific Ocean and the atmosphere during the 1960's. *Mon. Weather Rev.* **97**, 173–192.
Namias, J. (1971). The 1968–69 winter as an outgrowth of air and sea coupling during antecedent seasons. *J. Phys. Oceanogr.* **1**, 65–81.
Namias, J. (1973). Thermal communication between the sea surface and the lower troposphere. *J. Phys. Oceanogr.* **3**, 373–378.
Namias, J. (1974). Longevity of a coupled air-sea-continent system. *Mon. Weather Rev.* **102**, 638–648.
Nash, T. H., Fritts, H. C., and Stokes, M. A. (1975). A technique for examining nonclimatic variation in widths of annual tree rings with special reference to air pollution. *Tree-Ring Bull.* **35**, 15–24.
Neter, J. and Wassermann, W. (1974). "Applied Linear Statistical Models." Richard Irwin, Holmwood, Illinois.
Odin, H. and Openshaw, A. (1971). Electrical methods for measuring changes in shoot length and stem diameter. Royal College of Forestry *Res. Notes* **29**, Stockholm.
Olsson, I. U., ed. (1970). Radiocarbon variations and absolute chronology. *Nobel Symp.* **12**, Almqvist and Wiksell, Stockholm, and Wiley, New York.

O'Neil, L. C. (1962). Some effects of artificial defoliation on the growth of Jack Pine (*Pinus banksiana* Lamb.) *Can. J. Bot.* **40**, 273-280.

Page, R. (1970). Dating episodes of faulting from tree rings: effects of the 1958 rupture of the Fairweather fault on tree growth. *Geol. Soc. Amer. Bull.* **81**, 3085-3094.

Palmer, W. C. (1965). Meteorological Drought. *U.S. Weather Bur. Res. Pap.* **45**, U.S. Government Printing Office, Washington, D.C.

Panofsky, H. A. and Brier, G. W. (1968). "Some Applications of Statistics to Meteorology." Pennsylvania State Univ., University Park.

Parker, J. (1968). Drought-resistance mechanisms. *In* "Water Deficits and Plant Growth, I" (T. T. Kozlowski, ed.), pp. 195-234. Academic Press, New York.

Parker, J. (1971). Heat resistance and respiratory response in twigs of some common tree species. *Bot. Gaz.* **132(4)**, 268-273.

Parker, M. L. (1971). Dendrochronological techniques used by the Geological Survey of Canada. *Geol. Surv. Can. Pap.* **71-25**. Ottawa, Ontario.

Parker, M. L. and Henoch, W. E. S. (1971). The use of Engelmann spruce latewood density for dendrochronological purposes. *Can. J. Forest Res.* **1(2)**, 90-98.

Parker, M. L. and Kennedy, R. W. (1973). The status of radiation densitometry for measurement of wood specific gravity. *Proc. Division 5 Meetings Intern. Union Forest Res. Organizations*, September and October, 1973, Capetown and Pretoria, South Africa.

Parker, M. L. and Meleskie, K. R. (1970). Preparation of X-ray negatives of tree-ring specimens for dendrochronological analysis. *Tree-Ring Bull.* **30**, 11-22.

Parzen, E. (1967). "Stochastic Processes" (3rd Printing). Holden-Day, San Francisco.

Perry, T. O. (1971). Winter-season photosynthesis and respiration by twigs and seedlings of deciduous and evergreen trees. *Forest Sci.* **17**, 41-43.

Philip, J. R. (1957). The physical principles of soil moisture movement during the irrigation cycle. *Proc. 3rd Congr. Intern. Comm. Irrig. Drainage* **3**, Question 8, 125. San Francisco.

Phipps, R. L. (1961). Analysis of five years dendrometer data obtained within three deciduous forest communities of Neotoma. *Ohio Agr. Exp. Sta. Res. Circ.* **105**.

Phipps, R. L. (1967). Annual growth of suppressed Chestnut Oak and Red Maple, a basis for hydrologic inference. *U.S. Geol. Surv. Prof. Pap.* **485-C**.

Phipps, R. L. (1972). Tree rings, stream runoff, and precipitation in central New York. *U.S. Geol. Surv. Prof. Pap.* **800-B**, B259-B264.

Phipps, R. L. and Gilbert, G. E. (1960). An electric dendrograph. *Ecology* **41**, 389-390.

Phipps, R. L. and Yater, W. M., Jr. (1974). Three types of remote-reading dendrographs. *Ecology* **55(2)**, 454-457.

Plumb, R. C. and Bridgman, W. B. (1972). Ascent of sap in trees. *Science* **176**, 1129-1131.

Polge, H. (1966). Établissement des courbes de variation de la densité du bois par exploration densitométrique de radiographie d'échantillons prélevés à la tarière sur des arbres vivants, applications dans les domains technologique et physiologique. *Ann. Sci. Forest.* **23**, 1-206.

Polge, H. (1970). The use of X-ray densitometric methods in dendrochronology. *Tree-Ring Bull.* **30(1-4)**, 1-10.

Potter, N., Jr. (1969). Tree-ring dating of snow avalanche tracks and the geomorphic activity of avalanches, Northern Absaroka Mountains, Wyoming. U.S. Contrib. to Quaternary Res., *Geol. Soc. Amer. Spec. Pap.* **123**, 141-165.

Press, S. J. (1972). "Applied Multivariate Analysis." Holt, Rinehart, and Winston, New York.

Quenouille, M. H. (1952). "Associated Measurements." Butterworths Sci. Publ., London.

Rangenekar, P. V. and Forward, D. F. (1973). Foliar nutrition and wood growth in red pine: effects of darkening and defoliation on the distribution of ^{14}C-photosynthate in young trees. *Can. J. Bot.* **51,** 103–108.
Ray, P. M. (1972). "The Living Plant" (2nd Ed.) Holt, Rinehart, and Winston, New York.
Rees, L. W. (1929). Growth studies in forest trees. *Picea rubra* Link. *J. Forest.* **27,** 384–403.
Reineke, L. H. (1948). Dial gage dendrometers. *Ecology* **29,** 208.
Richardson, S. D. (1961). A biological basis for sampling in studies of wood properties. *Tech. Ass. Pulp Paper Ind.* **44,** 170–173.
Roberts, W. O. (1973). Relationships between solar activity and climate change. Unpublished manuscript, Univ. Corp. for Atmos. Res., Boulder, Colorado.
Romberger, J. A. (1963). Meristems, growth, and development in woody plants. U.S. Dept. of Agr., *Forest Serv. Tech. Bull.* **1293.**
Roughton, R. D. (1962). A review of literature on dendrochronology and age determination of woody plants. State of Colorado Dept. of Game and Fish *Tech. Bull.* **15.**
Samish, R. M. (1954). Dormancy in woody plants. *Amer. Rev. Plant Physiol.* **5,** 183–204.
Sampson, A. W. (1940). The dendrochronology enigma. *J. Forest.* **38,** 966–968.
Sampson, A. W. and Glock, W. S. (1942). Tree growth and the environmental complex: A critique of "ring" growth studies with suggestions for future research. *J. Forest.* **40,** 614–620.
Schulman, E. (1945). Root growth-rings and chronology. *Tree-Ring Bull.* **12,** 2–5.
Schulman, E. (1947). Tree-ring hydrology in Southern California. *Lab. of Tree-Ring Res. Bull.* **4,** *Univ. Ariz. Bull.* **18(3),** Tucson.
Schulman, E. (1951). Tree-ring indices of rainfall, temperature, and river flow. *In* "Compendium of Meteorology" (Amer. Meteorol. Soc., eds.), pp. 1024–1029, Boston.
Schulman, E. (1956). "Dendroclimatic Changes in Semiarid America." Univ. of Arizona Press, Tucson.
Schulman, E. (1958). Bristlecone pine, oldest known living thing. *Nat. Geogr.* **113(3),** 355–372.
Schulman, M. D. and Bryson, R. A. (1965). A statistical study of dendroclimatic relationships in south central Wisconsin. *J. Appl. Meteorol.* **4(1),** 107–111.
Schulze, E. D., Mooney, H. A., and Dunn, E. L. (1967). Wintertime photosynthesis of bristlecone pine *(Pinus aristata)* in the White Mountains of California. *Ecology* **48,** 1044–1047.
Scott, D. (1972). Correlation between tree-ring width and climate in two areas in New Zealand. *J. Roy. Soc. New Zeal.* **2(4),** 545–560.
Seal, H. L. (1968). "Multivariate Statistical Analysis for Biologists" (3rd Printing). Methuen, London.
Searles, S. R. (1966). "Matrix Algebra for the Biological Sciences." Wiley, New York.
Sellers, W. D. (1965). "Physical Climatology." Univ. of Chicago Press, Chicago.
Sellers, W. D. (1968). Climatology of monthly precipitation patterns in Western United States, 1931–1966. *Mon. Weather Rev.* **96,** 585–595.
Serre, F. (1973). Contribution à l'étude dendroclimatologique du pin d'alep *(Pinus halepensis* Mill.) Ph.D. thèse, l'université D'Aix-Marseille III.
Serre, F., Lück, H., and Pons, A. (1966). Premières recherches sur les relations entre les variations des anneaux ligneux chez *Pinus halepensis* Mill. et les variations annuelles du climat. *Oecol. Plant.* **1(1),** 117–135.
Sheppard, P. A. (1966). Preface. *In* "World Climate from 8000 to 0 B.C.," pp. 1–2. *Proc. Roy. Meteorol. Soc. Symp.* 18–19 April, 1966. London.

Siegel, S. (1956). "Nonparametric Statistics: For the Behavioral Sciences" (Intern. Student Edition). McGraw-Hill, New York.

Sigafoos, R. S. (1964). Botanical evidence of floods and flood-plain deposition. *U.S. Geol. Surv. Prof. Pap.* **485-A,** Washington, D.C.

Sigafoos, R. S. and Hendricks, E. L. (1961). Botanical evidence of the modern history of Nisqually glacier, Washington, *U.S. Geol. Surv. Prof. Pap* **387-A,** 1-20, Washington, D.C.

Sigafoos, R. S. and Hendricks, E. L. (1969). The time interval between stabilization of alpine glacial deposits and establishment of tree seedlings. *U.S. Geol. Surv. Prof. Pap.* **650-B,** 89-93. Washington, D.C.

Sigafoos, R. S. and Sigafoos, M. D. (1966). Flood history told by tree growth. *Natur. Hist.* **75(7),** 50-55.

Silver, G. T. (1962). The distribution of Douglas-fir foliage by age. *Forest. Chron.* **38,** 433-438.

Simpson, G. G., Roe, A., and Lewontin, R. C. (1960). "Quantitative Zoology." Harcourt, Brace, and World, New York.

Sirén, G. (1961). Skogsgränstallen som indikator för klimatfluktuationerna i norra fennoskandien under historisk tid. *Commun. Inst. Forest. Fenn.* **54(2),** 1-66.

Sirén, G. (1963). Tree rings and climate forecasts. *New Scientist* **346,** 18-20.

Sirén, G. and Hari, P. (1971). Coinciding periodicity in recent tree rings and glacial clay sediments. *Rep. KEVO Subarctic Res. Sta.* **8,** 155-157.

Skene, D. S. (1972). The kinetics of tracheid development in *Tsuga canadensis* Carr. and its relation to tree vigour. *Ann. Bot.* **36,** 179-187.

Slastad, T. (1957). Arringundersøkelser I gudbrandsdalen. (Tree-ring analyses in gudbrandsdalen.) Results of tree-ring invest. Supported by the Norwegian Res. Council for Sci. and the Humanities, **6,** *Meddelelser fra Det Norske Skogforsøksvesen* **48,** Oslo.

Slatyer, R. O. (1967). "Plant-Water Relationships." Academic Press, London & New York.

Small, J. A. and Monk, C. D. (1959). Winter changes in tree radii and temperature. *Forest Sci.* **5,** 229-233.

Smiley, T. L. (1958). The geology and dating of Sunset Crater, Flagstaff, Arizona. In "Guidebook of the Black Mesa Basin, Northeastern Arizona" (R. Y. Anderson and J. W. Harshberger, eds.), pp. 186-190. N. Mex. Geol. Soc., Socorro, New Mexico.

Smith, D. M. and Wilsie, M. C. (1961). Some anatomical responses of loblolly pine to soil-water deficiencies. *Tech. Ass. Pulp Paper Ind.* **44(3),** 179-185.

Snedecor, G. W. (1956). "Statistical Methods Applied to Experiments in Agriculture and Biology" (5th Edition). Iowa State College Press, Ames

Soil Survey Staff (1951). "Soil Survey Manual." *U.S. Dept. Agr. Handbook* 18, U.S. Government Printing Office, Washington, D.C.

Soil Survey Staff (1960). "Soil Classification, a Comprehensive System—7th Approximation." U.S. Dept. Agr., U.S. Government Printing Office, Washington, D.C.

Soil Survey Staff (1962). "Supplement to USDA Handbook 18, Soil Survey Manual (replacing pp. 173-188)." U.S. Dept. Agr., U.S. Government Printing Office, Washington, D.C.

Soil Survey Staff (1967). "Supplement to Soil Classification, a Comprehensive System—7th Approximation." U.S. Dept. Agr., U.S. Government Printing Office, Washington, D.C.

Spencer, D. A. (1964). Porcupine population fluctuations in past centuries revealed by dendrochronology. *J. Appl. Ecol.* **1(1),** 127-149.

Spiegel, M. R. (1961). "Theory and Problems of Statistics." Schaum, New York.
Squillace, A. E. and Silen, R. R. (1962). Racial variation in Ponderosa Pine. *Forest Sci. Monogr.* **2.**
Stallings, W. S., Jr. (1949). "Dating Prehistoric Ruins by Tree-Rings" (Revised Edition). Lab. of Tree-Ring Res., Tucson.
States, J. B. (1968). Growth of Ponderosa Pine on three geologic formations in Eastern Wyoming. M.S. Thesis. Univ. of Wyoming, Laramie.
Steel, R. G. D. and Torrie, J. H. (1960). "Principles and Procedures of Statistics, with Special Reference to Biological Sciences." McGraw-Hill, New York.
Stockman, W. B. (1904). Invariability of our winter climate. *Mon. Weather Rev.* **32,** 224–226.
Stockton, C. W. (1971). The feasibility of augmenting hydrologic records using tree-ring data. Ph.D. Thesis, Univ. of Arizona, Tucson.
Stockton, C. W. (1975). Long term streamflow records reconstructed from tree rings. *Pap. Lab. Tree-Ring Res.* **5,** Univ. Arizona Press, Tucson.
Stockton, C. W. (1976). Long-term streamflow reconstruction in the upper Colorado river basin using tree rings. *In* "Colorado River Basin Modeling Studies" (C. G. Clyde, D. H. Falkenborg, J. P. Riley, eds.), pp. 401–441. July 16–18, 1975, Utah State Univ., Logan.
Stockton, C. W. and Fritts, H. C. (1971a). Augmenting annual runoff records using tree-ring data. *Proc. Meetings Amer. Water Resources Ass.*, Arizona Section, April 22–23, 1971, pp. 1–12, Tempe.
Stockton, C. W. and Fritts, H. C. (1971b). Conditional probability of occurrence for variations in climate based on width of annual tree rings in Arizona. *Tree-Ring Bull.* **31,** 3–24.
Stockton, C. W. and Fritts, H. C. (1973). Long-term reconstruction of water level changes for Lake Athabasca by analysis of tree rings. *Water Resources Bull.* **9,** 1006–1027.
Stockton, C. W. and Jacoby, G. C., Jr. (in press). Long-term surface-water supply and streamflow trends in the upper Colorado river basin based on tree-ring analysis. Univ. of California (Los Angeles) Inst. of Geophys. and Planetary Phys., *Lake Powell Res. Proj. Bull.*
Stokes, M. A. (1965). The differentiation of tracheary elements from the cambium of *Pinus edulis* Engelm.: The correlation of differentiation with measured ring width and environmental factors. M.S. Thesis, Univ. of Arizona, Tucson.
Stokes, M. A. and Smiley, T. L. (1968). "An Introduction to Tree-Ring Dating." Univ. of Chicago Press, Chicago.
Stokes, M. A., Drew, L. G., and Stockton, C. W. (eds.) (1973). Tree-ring chronologies of Western America, I. Selected tree-ring stations. *Lab. Tree-Ring Res. Chronology Ser.* **1,** Univ. of Arizona, Tucson.
Stone, E. L. (1974). The communal root system of Red Pine: growth of girdled trees. *Forest Sci.* **20,** 294–305.
Studhalter, R. A. (1955). Tree growth: I. Some historical chapters. *Bot. Rev.* **21,** 1–72.
Studhalter, R. A. (1956). Early history of crossdating. *Tree-Ring Bull.* **21(1–4),** 31–35.
Studhalter, R. A., Glock, W. S., and Agerter, S. R. (1963). Tree growth—some historical chapters in the study of diameter growth. *Bot. Rev.* **29,** 245–365.
Swain, A. M. (1973). A history of fire and vegetation in Northeastern Minnesota as recorded in lake sediments. *Quaternary Res.* **3(3),** 383–396.
Sweet, G. B. and Wareing, P. F. (1966). Role of plant growth in regulating photosynthesis. *Nature* **210,** 77–79.
Tate, M. W. and Clelland, R. C. (1957). "Nonparametric and Shortcut Statistics." Interstate, Danville, Illinois.

Taylor, O. C. (1973). Oxidant air pollutant effects on a Western coniferous forest ecosystem. Statewide Air Pollution Res. Center, Univ. of California, Riverside.

Thornthwaite, C. W. and Mather, J. R. (1955). The water balance. *Drexel Inst. Tech. Publ. Climatol.* **8(1)**, 1-104, Centerton, New Jersey.

Thornthwaite, C. W. and Mather, J. R. (1957). Instructions and tables for computing potential evapotranspiration and the water balance. *Drexel Inst. Tech. Publ. Climatol.* **10(3)**, 184-311, Centerton, New Jersey.

Tranquillini, W. (1964a). The physiology of plants at high altitudes. *Ann. Rev. Plant Physiol.* **15**, 345-362.

Tranquillini, W. (1964b). Photosynthesis and dry matter production of trees at high altitudes. *In* "The Formation of Wood in Forest Trees" (M. H. Zimmermann, ed.), pp. 505-518. Academic Press, New York, London and San Francisco.

Tryon, E. H. and True, R. P. (1958). Recent reductions in annual radial increments in dying scarlet oaks related to rainfall deficiencies. *Forest Sci.* **4(3)**, 219-230.

Tryon, E. H., Cantrell, J. O., and Carvell, K. L. (1957). Effect of precipitation and temperature on increment of yellow-poplar. *Forest Sci.* **3(1)**, 32-44.

Tryon, R. C. and Bailey, D. E. (1970). "Cluster Analysis." McGraw-Hill, New York.

Tukey, J. W. (1950). The sampling theory of power spectrum estimates. *Symposium on Application of Autocorrelation Analysis to Physical Problems.* Woods Hole, 13 June 1949. U.S. Office of Naval Res., NAVEXOS-P-735, pp. 47-67, Washington, D.C.

Vinš, B. (1965). A method of smoke injury evaluation—determination of increment decrease. *Commun. Inst. Forest. Čech.*, pp. 235-245, Praha.

Vinš, B. (1970). Methods and use of tree ring analyses in Czechoslovakia. *In* "Tree-Ring Analysis with Special Reference to Northwest America" (J. H. G. Smith and J. Worrall, eds.), pp. 67-73. Univ. of British Columbia *Fac. Forest. Bull.* **7**, Vancouver.

Vinš, B. and Tesař, V. (1969). Increment loss due to smoke immissions in the region of Trutnov. Výzkumný ústav, lesního hospodářství a myslivosti, zbraslav-strnady, Czechoslovakia, *Práce Vúlhm* **38**, 141-158.

Wahl, E. W. and Lawson, T. L. (1970). The climate of the midnineteenth century United States compared to the current normals. *Mon. Weather Rev.* **98**, 259-265.

Walker, G. T. and Bliss, E. W. (1932). World weather V. *Mem. Roy. Meteorol. Soc.* **4**, 53-84. London.

Wallis, A. W. and Roberts, H. V. (1956). "Statistics: A New Approach." The Free Press, Glencoe, Illinois.

Wardlaw, I. F. (1968). The control and pattern of movement of carbohydrates in plants. *Bot. Rev.* **34**, 79-105.

Wardrop, A. B. (1964a). The structure and formation of the cell wall in xylem. *In* "The Formation of Wood in Forest Trees" (M. H. Zimmermann, ed.), pp. 87-134. Academic Press, New York, London and San Francisco.

Wardrop, A. B. (1964b). The reaction anatomy of arborescent angiosperms. *In* "The Formation of Wood in Forest Trees" (M. H. Zimmermann, ed.), pp. 405-456. Academic Press, New York, London and San Francisco.

Weakley, H. E. (1943). A tree-ring record of precipitation in western Nebraska. *J. Forest.* **41(11)**, 816-819.

Weaver, H. (1951). Fire as an ecological factor in southwestern Ponderosa pine forests. *J. Forest.* **49**, 93-98.

Webb, T., III and Bryson, R. A. (1972). Late- and postglacial climatic change in the northern Midwest, U.S.A.: quantitative estimates derived from fossil pollen spectra by multivariate statistical analysis. *Quaternary Res.* **2**, 70-115.

Westing, A. H. (1968). Formation and function of compression wood in gymnosperms, II. *Bot. Rev.* **34(1)**, 51–78.
Westing, A. H. and Schulz, H. (1965). Erection of a leaning eastern hemlock tree. *Forest Sci.* **11(3)**, 364–367.
Williams, C. B., Jr. (1968). Seasonal height growth of upper-slope conifers. U.S. Dept. of Agr., *Forest Serv. Res. Pap.* PNW-62.
Willis, J. C. (1973), "A Dictionary of the Flowering Plants and Ferns" (8th Edition). Cambridge Univ. Press, England.
Wilson, B. F. (1964). A model for cell production by the cambium of conifers. *In* "The Formation of Wood in Forest Trees" (M. H. Zimmermann, ed.), pp. 19–36. Academic Press, New York, London and San Francisco.
Wilson, B. F. and Howard, R. A. (1968). A computer model for cambial activity. *Forest Sci.* **14**, 77–90.
Wilson, B. F., Wodzicki, T. J., and Zahner, R. (1966). Differentiation of cambial derivatives: proposed terminology. *Forest Sci.* **12**, 438–440.
Winget, C. H. and Kozlowski, T. T. (1965). Seasonal basal growth area as an expression of competition in northern hardwoods. *Ecology* **46(6)**, 786–793.
Winstanley, D. (1973). Rainfall patterns and general atmospheric circulation. *Nature, Lond.* **245**, 190–194.
Wodzicki, T. J. (1964).Photoperiodic control of natural growth substances and wood formation in larch (*Larix decidua* D.C.). *J. Exp. Bot.* **15(45)**, 584–599.
Wodzicki, T. J. (1971). Mechanism of xylem differentiation in *Pinus silvestris* L. *J. Exp. Bot.* **22**, 670–687.
Wolter, K. E. (1968). A new method for marking xylem growth. *Forest Sci.* **14**, 102–104.
Worrall, J. (1966). A method of correcting dendrometer measures of tree diameter for variations induced by moisture stress change. *Forest Sci.* **12**, 427–429.
Wright, R. D. and Mooney, H. A. (1965). Substrate-oriented distribution of bristlecone pine in the white mountains of California. *Amer. Midland Natur.* **73(2)**, 257–284.
Young, H. E. (1952). Practical limitations of the dial gauge dendrometer. *Ecology* **33**, 568–570.
Young, H. E. and Kramer, P. J. (1952). The effect of pruning on the height and diameter growth of loblolly pine. *J. Forest.* **50(6)**, 474–479.
Zahner, R. (1968). Water deficits and growth of trees. *In* "Water Deficits and Plant Growth, II" (T. T. Kozlowski, ed.), pp. 191–254. Academic Press, New York, London and San Francisco.
Zahner, R. and Stage, A. R. (1966). A procedure for calculating daily moisture stress and its utility in regressions of tree growth on weather. *Ecology* **47**, 64–74.
Zasada, J. C. and Zahner, R. (1969). Vessel element development in the earlywood of red oak (*Quercus rubra*). *Can. J. Bot.* **47**, 1965–1971.
Zimmermann, M. H. (1969). Translocation velocity and specific mass transfer in the sieve tubes of *Fraxinus americana* L. *Planta Berlin* **84**, 272–278.
Žumer, M. (1969). Annual ring formation on Norway spruce in mountain forest. *Meddelelser fra Det Norske Skogforsøksvesen* **27**, 165–184.

Glossary

ABSCISE. Separate by formation of a specialized tissue which forms a line of weakness.
ABSORPTIVITY. With reference to the energy balance (q.v.), the capacity of a plant to absorb radiant energy.
ACROPETAL. Direction from the midportions of a plant toward the apex. In the case of a stem, from the ground level toward the stem tip. Opposite of basipetal. (q.v.).
ACTIVE ABSORPTION. See *active transport*.
ACTIVE TRANSPORT. The movement of minerals or ions through cell membranes against a free energy gradient (q.v.) by the expenditure of metabolic energy, resulting in the accumulation of salts such as occurs in roots. Cf. *diffusion*.
ADENOSINE TRIPHOSPHATE (ATP). In living cells, a high-energy coenzyme (q.v.) which acts as a carrier of chemical energy.
AEROBIC RESPIRATION. Respiration which utilizes free oxygen and food, such as glucose, and releases energy, carbon dioxide, and water. Cf. *anaerobic respiration*.
A HORIZON. In soils, the top mineral horizon in which eluviation (q.v.) occurs. Cf. *B horizon*.
AIR MASS. A large body of air which is relatively homogeneous in temperature and moisture content.
ALBEDO. Reflectivity of a surface when viewed from above.
ALDEHYDES. Highly reactive organic compounds characterized by a specific bonding configuration of hydrogen and oxygen units.
AMPLITUDE, ECOLOGICAL. See *ecological amplitude*.
AMPLITUDES OF EIGENVECTORS. The products of eigenvectors (q.v.) and the data from which the eigenvectors were derived expressing the importance of each eigenvector in each observation set. Also referred to as *factor scores*.
AMYLASE. A plant enzyme which catalyzes the condensation–hydrolysis reaction (q.v.) involving starch and glucose.
ANAEROBIC RESPIRATION. Respiration (q.v.) which occurs in living cells in the absence of free oxygen, involving changes in chemical bonding, the formation of intermediate organic molecules, and the release of carbon dioxide. Cf *aerobic respiration*.
ANALYSIS OF VARIANCE. A statistical procedure for measuring the variation attributable to different sources.
ANGIOSPERM. The flowering seed plants, as differentiated from gymnosperms, the nonflowering seed plants.
ANTIAUXIN. A growth regulator or hormone which inhibits the action of auxins (q.v.) and thereby inhibits cell enlargement.

ANTICYCLONE. A high pressure system in the atmosphere characterized by dry descending air and clear weather. Cf. *cyclone*.

APEX. The tip or pointed end of an organ or plant part, such as the tip of a root, shoot, leaf, or bud.

APICAL DOMINANCE. Inhibition of lateral bud growth by vigorously growing apices of a plant, usually attributed to hormones produced by the growing tissues.

APICAL MERISTEM. The tissue at the tips of stems and roots composed of small, thin-walled cells which are capable of dividing and differentiating into the various tissues of the stem and root.

ASSIMILATION. The utilization of carbohydrates, fats, and proteins to synthesize the protoplasm, cell walls, and numerous other substances making up the enzyme systems, pigments, and structures of an organism.

ATP. Adenosine triphosphate (q.v.).

AUTOCORRELATION COEFFICIENT. A statistic describing serial dependence or association in a time series with previous conditions or states. See *serial correlation*.

AUTOCORRELATION FUNCTION. The result of computing the autocorrelation coefficients (q.v.) from one to a prescribed number of lags and then plotting these coefficients as a function of lag.

AUTOREGRESSION. A regression (q.v.) in which one value in a time series is regressed upon one or more variables which precede it in time.

AUXIN. A plant growth regulator or hormone which is responsible for cell enlargement.

AVERAGE. Mean (q.v.).

BAR. Unit of atmospheric pressure equal to 1,000,000 dynes per square centimeter (about 14·50 pounds per square inch).

BASIPETAL. Direction from the apex toward the midportions. In the case of a stem, from stem tips toward the ground level. Opposite of acropetal. (q.v.).

B HORIZON. In soils, a mineral horizon below the A horizon (q.v.) in which illuviation (q.v.) occurs.

BOLE. The trunk or stem of a tree.

BORDERED PITS. In conifers, specialized thin areas in the walls of tracheids (q.v.) surrounded by a raised lip.

BUTTRESS ROOT. A root which has elongated vertically, been exposed to the atmosphere, and taken on some of the characteristics of a stem.

C-3 PLANTS. Plants which carry on respiration which involves light energy.

C-4 PLANTS. Plants which carry on respiration which is not affected by light.

CALIBRATION. In dendroclimatology, the process of obtaining a transfer function (q.v.) or response function (q.v.) which can be used to estimate one or more predictand (q.v.) variables from a set of predictors (q.v.).

CAMBIAL INITIALS. The nonvacuolated cells within the cambium from which all elements of the radial file (q.v.) are derived by cell division.

CAMBIUM. A thin layer of meristematic cells (q.v.) such as the vascular cambium (q.v.) and the cork cambium (q.v.) formed in stems and roots.

CANONICAL ANALYSIS. A multivariate statistical technique which includes both canonical correlation (q.v.) and canonical regression (q.v.).

CANONICAL CORRELATION. A statistical technique designed to specify orthogonal (q.v.) modes of intercorrelation between two sets of variables.

CANONICAL REGRESSION. Determination of a set of regression coefficients from a canonical correlation (q.v.) which are used to estimate values of one set of variables from values of another. Cf. *regression*.

CARBON ASSIMILATION. Photosynthesis (q.v.).

GLOSSARY

CASPARIAN STRIP. A layer of suberin (q.v.) deposited on the radial and horizontally oriented walls of the endodermis which is impervious to the movement of water along the cell wall.

CELL SAP. The solution of water and dissolved substances found in the vacuoles (q v.) of living cells.

CELLULOSE. The primary framework substance in a plant cell wall.

CHINOOK. See *foehn wind*.

CHLORENCHYMA. Green chlorophyll-containing tissues found in plants.

CHLOROPHYLL. The green pigment in plant cells which is necessary for the trapping of light energy used in the process of photosynthesis (q.v.).

CHLOROPLASTS. Chlorophyll-containing bodies within the cytoplasm (q.v.) of chlorenchyma (q.v.) cells.

C HORIZON. In soils, the lowermost horizon which includes undifferentiated parent material (q.v.).

CHRONOLOGY. Ring-width chronology (q.v.).

CHRONOLOGY BUILDING. The dating and processing of ring widths in many trees from a given region or site to produce long homogeneous ring-width chronologies used for crossdating and for deducing past climate.

CLIMATE. The aggregate of weather, including all of meteorological phenomena occurring over a relatively long period of time. Cf. *weather*.

CLIMATE SYSTEM. The properties and processes that are responsible for climate and its variations. The *properties* are classified as thermal, kinetic, aqueous, and static, while the *processes* include precipitation, evaporation, radiation, and the transfer of heat, mass, and momentum.

CLIMATIC STATE. The average, variability, and other statistics of climatic variables over a specified period of time in a specified domain of the earth–atmosphere system.

CLIMATOGRAPHY. The mapping of climatic patterns. Cf. *dendroclimatography* and *synoptic climatology*.

COEFFICIENT. In an equation, a constant number by which a variable is multiplied.

COENZYME. A substance whose presence is necessary for the activity of an enzyme.

COHERENCY. An estimate of association between two time series (analogous to the square of the correlation coefficient [q.v.]) expressed as a function of frequency. Also called *coherence*.

COLLOID. A suspension of molecule aggregates (such as cytoplasm [q.v.]) which is made up of particles that cannot readily diffuse through membranes (q.v.).

COMPANION CELL. A small specialized cell in the phloem associated with a large, thin-walled sieve tube and containing its nucleus.

COMPLACENT. A dendrochronological term referring to the lack of ring-width variability, which indicates that the growth of a particular tree is relatively unaffected by variations in climate.

COMPONENT VARIANCES. In analysis of variance (q.v.), the estimates of variances attributable to and partitioned into each source of variation.

COMPRESSION WOOD. The name given reaction wood (q.v.) which in coniferous species occurs on the lower side of the leaning stems and lateral branches. Cf. *tension wood*.

CONCENTRATION. The relative number of molecules of a given substance per unit of volume or weight.

CONDENSATION. The plant process in which molecules of soluble substances such as glucose combine with one another in the presence of an enzyme to form water and insoluble and chemically more complex substances such as starch; the reverse of hydrolysis (q.v.).

CONDENSATION–HYDROLYSIS REACTION. A class of reversible chemical reactions usually involving a change in solubility of the reactants but no change in their chemically bound energy. See also *condensation* and *hydrolysis*.

CONIFERS. The cone-bearing gymnosperms (q.v.)

CORK CAMBIUM. A thin cylinder of meristematic tissue in the outer bark which produces the cork. As the girth of the stem increases and the outer bark exfoliates, new cork cambia differentiate from beneath the younger tissue inside the bark.

CORRELATION. Association of two variables without implying the presence or the direction of dependence. Cf. *regression* and *coupling*.

CORRELATION COEFFICIENT. A statistic which expresses the amount of interdependence or association between two data sets. It ranges in values from $+1$, which indicates perfect and direct association, to -1, which indicates perfect and inverse association. A value of 0 indicates a complete lack of interdependence.

CORTEX. Parenchyma (q.v.) tissue found in young stems and roots which lies outside the vascular tissues and inside the epidermis.

COUPLING. The degree of cause-and-effect linkage between elements of a system, that is, the influence which one phenomenon has upon another. Cf. *correlation*.

CROSSDATING. The procedure of matching ring-width variations and other structural characteristics among trees that have grown in nearby areas, allowing the identification of the exact year in which each ring was formed.

CROSS POWER SPECTRUM ANALYSIS. A means of studying covariance between two time series expressed as a function of frequency. Cf. *power spectrum analysis* and *coherency*.

CUTICLE. A layer of wax formed on the external surface of the epidermis in many plants.

CUTIN. A water repellant waxy substance that forms the chief ingredient of the cuticle (q.v.). Cf. *suberin*.

CYCLONE. A storm or low pressure system carried in the general circulation of the atmosphere and characterized by counterclockwise motion (in the Northern Hemisphere) and by ascending air. Cf. *anticyclone*.

CYTOPLASM. In plants, the protoplasm of the cell exclusive of the nucleus, bounded by the cell wall on the outside and usually by a vacuole on the inside. It may include membranes (q.v.), plastids (q.v.), and a variety of other bodies.

DECIDUOUS. Foliage-shedding, as opposed to evergreen.

DEDUCTION. A conclusion drawn by reasoning from facts or premises, often involving a series of inferences (q.v.)

DEGREES OF FREEDOM. A statistical term referring to the number of items in a sample that can vary independently of one another.

DENDROCHRONOLOGY. The science that deals with the dating and study of annual growth layers in wood.

DENDROCLIMATOGRAPHY. A subfield of dendrochronology (q.v.) which utilizes dated tree rings to reconstruct and map the spatial variations in climate.

DENDROCLIMATOLOGY. A subfield of dendrochronology (q.v.) which utilizes dated tree rings to reconstruct and study past and present climate.

DENDROECOLOGY. A subfield of dendrochronology (q.v.) which utilizes dated tree rings to study ecological problems and the environment.

DENDROGEOMORPHOLOGY. A subfield of dendrochronology (q.v.) which utilizes dated tree rings to study land forms and geomorphic processes.

DENDROGRAPH. An instrument that continuously records size changes in stems of trees. Cf. *dendrometer*.

DENDROHYDROLOGY. A subfield of dendrochronology (q.v.) which utilizes dated tree rings to study hydrologic problems, such as river flow and flooding history.

DENDROMETER. An instrument that is used to measure the size of tree stems. Cf. *dendrograph*.

DEPENDENT DATA. Data used to obtain a calibration (q.v.).

DEWPOINT. Temperature of the air when water begins to condense.

DIAGONAL MATRIX. A square matrix (q.v.) with values along the main diagonal, and zeros on the off-diagonal elements (q.v.).

DIFFERENTIALLY PERMEABLE MEMBRANE. A membrane which varies in its permeability to various molecules and can therefore regulate the relative movement of substances within a plant.

DIFFERENTIATION. The process by which the young growing cells take on the characteristics of the tissue of which they will become a part. Also referred to as *maturation*.

DIFFUSE-POROUS WOODS. Certain angiosperms (q.v.), such as *Fagus, Acer,* and *Liriodendron,* in which vessels are of approximately the same diameter throughout the earlywood and latewood portions of the ring, as differentiated from ring-porous (q.v.) angiosperms.

DIFFUSION. The net movement of molecules (or ions) of a given substance from a region of high free energy (q.v.) of that substance to a region of low free energy of that substance. Cf. *mass movement*.

DIGESTION. Hydrolysis (q.v.).

DORMANCY. A condition within plants which prevents growth from occurring even though other physiological processes are active and environmental conditions are favorable. Dormancy may be classified as *temporary* (lasting a few days or a few weeks) or *permanent* (lasting for periods of many weeks or months).

DOUBLE-MASS ANALYSIS. A means of comparing several precipitation records to identify inhomogeneities due to station relocations and other causes.

EARLYWOOD. The wood produced in the annual ring during the early part of the growing season, characterized by large thin-walled cells. Earlywood is more porous than latewood (q.v.) and often lighter in color. Sometimes referred to as *springwood*.

ECOLOGICAL AMPLITUDE. The range of habitats over which a species may grow and reproduce.

ECOTYPE. A genetic variation within a species caused by isolation and environmental selection.

EFFECTIVE SAMPLE SIZE. The number of items in a time series which are independent of each other.

EIGENVALUES. A set of scalars or coefficients (q.v.) which are proportional to the amount of variance reduced by each eigenvector (principal component [q.v.]).

EIGENVECTORS. A set of orthogonal (q.v.) variables which are transformations representing the modes of uncorrelated behavior of a data set. There can be as many eigenvectors as there are variables in the original data set, but the transformations can be ranked from the most important ones which account for a large portion of the variance in the original data, and the least important ones which express only minor variations and orthogonality constraints in the original data. By discarding the smaller unimportant eigenvectors, it is possible to reduce the number of variables in a system. Also referred to as *principal components*. Cf. *amplitudes of eigenvectors*.

ELEMENT. In regard to a matrix (q.v.), the individual value of datum defined as being located in a particular row and column of a matrix.

ELUVIATION. In soils, the leaching or loss of minerals from a soil horizon (q.v.). Cf. *illuviation*.

EMISSIVITY. In plants, the ability to emit radiation.

ENDODERMIS. A sheath of cells most commonly occurring inside the cortex of roots but outside the vascular tissue of xylem and phloem, often with radially and horizontally oriented walls impregnated with suberin (q.v.) which form a layer referred to as the *casparian strip*.

ENERGY BALANCE. The disposition of energy within a given system such as a column of soil and the atmosphere above it, determined by rates of inflow and outflow, sources and sinks, and transformations of energy.

ENVIRONMENT. A general term which includes all phenomena affecting or impinging on an organism. Cf. *operational environment* and *potential environment*.

ENVIRONMENTAL FACTORS. The specific phenomena which are entering into an operational relationship with an organism at a given time. Cf. *operational environment* and *potential environment*.

EPIDERMIS. In plants, the single-cell tissue originating from the meristem which forms the outside layer covering leaves, young stems, young roots, flowers, and young fruits. As the young stem enlarges, the epidermis ruptures and is replaced by the cork produced in the bark.

EQUALLY PROBABLE CLASSES. Groups chosen so that each one has an equal likelihood of occurring.

ESTIMATES. The numerical solutions of a statistical regression equation in which the values of the predictors (q.v.) representing the statistical input of the system have been substituted into the equation to obtain values (the estimates) of the predictand (q.v.). These are compared to actual values of the predictand to evaluate the degree of calibration that has been obtained.

EVAPOTRANSPIRATION. The loss of water due to both evaporation from soil and transpiration (q.v.).

EXPLAINED VARIANCE. Variance reduced (q.v.).

EXPONENTIAL FUNCTION. A nonlinear mathematical relationship in which one variable changes as a function of a second variable which is an exponent term associated with a constant.

FACTOR SCORES. Amplitudes (q.v.).

FALSE RING. A change in cell structure within an annual growth layer which resembles the boundary of a true annual ring, making it appear to be two or more growth layers instead of one. Also referred to as *intra-annual growth band*.

FASCICLE. A small number or bundle such as the bundles of needles in many *Pinus* species.

FIBERS. In angiosperms, elongated and relatively narrow cells with thick walls.

FIELD CAPACITY. The amount of moisture held in a particular soil which has been thoroughly wetted and then allowed to drain for several days until the rate of drainage has become very slow.

FILTERS. As used in this text, a set of numerical weights which are applied to a time series to emphasize variations at certain frequencies. When only the long-term or low-frequency variations are retained the weights are referred to as a *low-pass filter*. When only the short-term or high-frequency variations are retained, the weights are referred to as a *high-pass filter*.

GLOSSARY

FIRST DIFFERENCES. The sequence of values from a time series obtained by subtracting the value of each item in a time series from its immediate successor.

FIRST-ORDER AUTOCORRELATION. The statistical association (correlation [q.v.]) of each value in a time series with the value of its immediate predecessor.

FOEHN WIND. A warm dry wind blowing down the side of a mountain. Same as *chinook*.

FOOD. As used in this text, food refers only to organic molecules classed as carbohydrates, fats, or proteins which contain chemically bound energy that can be released by oxidation (q.v.).

FREE ENERGY. As in diffusion (q.v.), the molecular energy capable of doing work involving concentration (the relative number of molecules for a substance) and the molecular activity (kinetic energy).

FREE ENERGY GRADIENT. The ratio of the difference in free energy of a given substance between two points and the distance between these points, governing the rate of diffusion. Cf. *diffusion*.

FRONT. A narrow zone of discontinuity between two air masses (q.v.) of different character.

FROST RING. Distorted xylem tissue damaged by freezing in the growing season during which the cells of the tissue were being formed.

FUSIFORM INITIALS. Elongated, spindle-shaped cells which make up approximately 90% of the cambium (q.v.) and which give rise to the longitudinally arranged cells in the xylem and phloem. Growth in the stem in a radial direction occurs by division of the fusiform initials throughout the entire length of the cell and by formation of the new wall along a tangential plan.

GENERATING PROCESS. The manner in which a time series (q.v.) such as ring width is generated by the growth-controlling factors; sometimes represented by a mathematical function.

GENOTYPE. Hereditary potential controlled by genes. Cf. *phenotype*.

GIBBERELLIN. A plant growth regulator or hormone which affects stem elongation.

GOODNESS OF FIT. The closeness of statistical estimates to the actual predictand data. Cf. *variance reduced*.

GRAND PERIOD OF GROWTH. The interval within the growing season when growth occurs most rapidly, excluding the period of increasing growth rate at the beginning of the season and the period of decreasing growth rate at the end of the season.

GRAVITATIONAL WATER. Water within a soil in excess of field capacity (q.v.) which percolates downward due to the pull of gravity.

GROSS PHOTOSYNTHESIS. The total amount of glucose formed or carbon dioxide utilized in photosynthesis without regard to the glucose utilized and carbon dioxide released by respiration. Cf. *net photosynthesis*.

GROWTH REGULATORS. Organic compounds produced in small amounts which promote, inhibit, or qualitatively modify growth of plants. Types of hormones (q.v.).

GUARD CELLS. Two cells which surround each stomate (q.v.), causing it to open or close. Walls of the guard cells are thicker near the pore opening than on the side away from the pore, and the two cells arch away from each other as the volume of the guard cells increases, opening the pore.

GUTTATION. The discharging or exudation of water droplets from turgid leaves.

GYMNOSPERM. The nonflowering seed plants, as differentiated from angiosperms, the flowering seed plants.

HEARTWOOD. The darker colored central portion of a stem in which the wood rays (q.v.) are dead, as differentiated from the lighter colored sapwood.

HEMICELLULOSE. An important carbohydrate constituent of xylem cell walls.

HIGH-FREQUENCY VARIANCE. In this text, all variations in ring width that are of shorter duration than eight years (representing cycles and other variations with wavelengths or durations of eight years or less). Cf. *low-frequency variance*.

HIGH-PASS FILTER. See *filters*.

HORMONES. A class of organic compounds produced in small amounts which promote, inhibit, or qualitatively modify growth, development, and life activities of plants. Cf. *growth regulators*.

HYDROLYSIS. The plant process in which relatively complex insoluble substances are converted to smaller soluble ones in the presence of enzymes, as in the conversion of starch to glucose; the reverse of the process of condensation (q.v.).

HYDROSTATIC. Relating to the pressure from overlying or confined liquids.

HYPOTHESIS. An inference (q.v.) to be tested.

ILLUVIATION. In soils, the accumulation of minerals in a soil horizon (q.v.). Cf. *eluviation*.

INCREMENT CORER. A field tool used to bore and extract a thin cylinder of wood from tree stems.

INDEPENDENT DATA. Data not used for calibration (q.v.) which can be used as verification of reconstructions. Verification is established when the estimates from the independent predictor (q.v.) set resemble the predictands (q.v.) of the independent data set.

INFERENCE. A generalization derived from given information usually including a number of facts.

INHIBITOR. A growth regulator or hormone which reduces or stops growth and can induce either temporary or permanent dormancy.

INTERACTION. The influence of a factor upon the effect of another which differs from the individual or mean effects of the two factors.

INTRA-ANNUAL GROWTH BAND. False ring (q.v.).

ION. An electrically charged atom or group of atoms.

JUVENILE WOOD. Relatively thin-walled lignified xylem tissue which is low in density and is formed in young trees or in tissues located near the stem apex.

KININ. A growth regulator or hormone which promotes cell division.

KRUMMHOLZ. Gnarled and stunted trees which grow in exposed sites usually near the upper elevational tree line.

LAPSE RATE. The rate of change of air temperature per unit increase in altitude.

LATENT HEAT. The heat absorbed when water evaporates.

LATEWOOD. Dense and often dark wood produced in the annual ring during the later part of the growing season, characterized by small, thick-walled cells. Also referred to as *summerwood*. Cf. *earlywood*.

LEAST SQUARES. A numerical method for fitting a function, such as a straight line, to a set of data in such a way that the sum of squares of the residuals (q.v.) is minimized.

LENTICELS. Ruptures in the bark of a stem through which substances such as water vapor, oxygen, and carbon dioxide can move.

LIGNIN. A material added to the thickening walls of living xylem cells after cell enlargement has ceased, which gives wood its hardness and increases its density.

LOCALLY ABSENT RING. A growth ring which is discontinuous around the stem so that it is absent along certain radii; also referred to as *partial* or *missing ring*.

LOW-FREQUENCY VARIANCE. In this text, all variations in ring width that last longer than eight years (represented by cycles and other variations with

GLOSSARY

wavelengths or durations greater than eight years). Cf. *trend* and *high-frequency variance*.

LOW-PASS FILTER. See *filters*.

MACROCLIMATE. The aggregate of climatic conditions for a region, often based upon 30 or more years of observation. Cf. *microclimate*.

MASS MOVEMENT. The transfer of molecules or particles by flow from one place to another driven by application of some external force, as opposed to diffusion (q.v.).

MATRIX. An array of data ordered in rows and columns. Cf. *element*.

MATRIX POTENTIAL (Ψm). The water potential forces attributed to surface phenomena; as opposed to osmotic potential (q.v.) and pressure potential (q.v.).

MATURATION. Differentiation (q.v.)

MEAN. The average; the value that is closest to all values in a data set.

MEAN SENSITIVITY. A statistic measuring the mean relative change between adjacent ring widths; that is, the average relative difference from one ring width to the next. Cf. *sensitivity*.

MEAN STANDARDIZED INDICES. Ring-width chronology (q.v.).

MEMBRANE. A thin layer lying along the outer limits of living protoplasm (inside the cell walls) and along various other boundaries, such as between protoplasm and vacuole. It may vary in its permeability to different molecules. Cf. *differentially permeable membrane*.

MERISTEM. A plant tissue composed of small thin-walled cells that are capable of dividing. Meristematic tissue is found in all buds, root tips, and growing regions including the cambia. Cf. *apical meristem*.

MESOPHYLL. In plants, all cells located between the upper and lower epidermis of a leaf.

MICROCLIMATE. The aggregate of meteorological conditions within a localized site occurring over a period of time. Cf. *macroclimate*.

MICROENVIRONMENT. A localized environment such as that surrounding a leaf, bud, or root.

MICROMETEOROLOGY. The meteorological conditions of a localized site or the study thereof. Cf. *microclimate*.

MIDDLE LAMELLA. A cementing layer between cell walls in plants.

MISSING RING. Locally absent ring (q.v.) or partial ring.

MODEL. A statement, equation, or diagram which represents a basic set of facts and their interrelationships. They can range from extremely simple preconceived notions to highly complex systems.

MOTHER CELL. A derivative of the initial cell of a cambium that can still divide.

MOVING AVERAGES. Averages of overlapping segments of a time series usually calculated at regular intervals in the time sequence. Also referred to as *running means*.

MULTINODAL. Having the capacity to produce several series of leaves during one season's growth, as occurs in *Pinus halepensis*, *P. echinata*, and *P. taeda*.

MULTIPLE CORRELATION. The correlation coefficient calculated from a multiple regression (q.v.) relationship.

MULTIPLE REGRESSION. A regression (q.v.) involving more than one predictor variable. The multiple regression coefficients describe the relative effect of each predictor (q.v.) variable upon the predictand (q.v.), adjusting for the intercorrelations between the predictor variables.

NET PHOTOSYNTHESIS. Gross photosynthesis (q.v.) minus respiration (q.v.),

T

expressing the net increase in glucose and the net utilization of carbon dioxide.

NOISE. In this text, the random, residual (q.v.), or background variation in a time series that cannot be attributed to a detectable quality, pattern, or variation, as opposed to signal (q.v.), which represents meaningful information such as the ring-width variations attributed to climate, environment, or biological conditions.

NORMAL DISTRIBUTION. A bell-shaped distribution of data which closely resembles the normal density function with the mean value corresponding to the point at which the distribution function is a maximum and with approximately two-thirds of the data within a distance of one standard deviation (q.v.) on each side of the mean (q.v.).

NORMALIZED DATA. Standard normal variates (q.v.).

NUTRIENT. In a botanical context, mineral salts which are essential in small amounts for the normal growth and survival of the plant. They are not considered foods in this text, because they are not a significant source of energy.

NUCLEUS. The dense portion of the protoplasm which contains the hereditary material.

OPERATIONAL ENVIRONMENT. The environmental phenomena which are actually entering into a relationship with an organism at a particular time. The specific phenomena which enter such operational relationships are considered environmental factors (q.v.).

OROGRAPHIC. Relating to mountains or elevational relief.

ORTHOGONAL. Uncorrelated, independent of one another.

OSMOSIS. Diffusion (q.v.) of water through a differentially peremable membrane (q.v.). The process by which water moves into and out the living protoplasm of cells.

OSMOTIC POTENTIAL (Ψs). The water potential (q.v.) forces which are due only to substances dissolved in the water.

OXIDATION. A class of chemical reactions which releases energy stored in food molecules; the reverse of reduction (q.v.) in which molecules increase in energy.

OXIDATION-REDUCTION REACTION. A general class of chemical reactions in which there is a change in energy.

PARAMETER. A particular variable or constant.

PARENCHYMA. A plant tissue made up of living thin-walled, and vacuolated cells.

PARENT MATERIAL. In soils, the undifferentiated weathered rock material from which a soil develops.

PARTIAL RING. Locally absent (q.v.) or missing ring.

PERMANENT WILTING POINT. The moisture content of a drying soil at the time plants first become permanently wilted, including moisture that is generally unavailable to the plant.

PHELLEM. The technical term for cork cells which are differentiated from the cork cambium and are impregnated with suberin (q.v.).

PHENOLOGY. The changing appearance or condition of an organism, often associated with its yearly cycle, such as the breaking of dormancy, beginning of growth, enlargement, flowering, fruiting, and onset of dormancy.

PHENOTYPE. Observable characteristics of an organism that result from its heredity. Cf. *genotype*.

PHLOEM. In trees, the region containing food-conducting tissue in the bark.

PHOTOPERIOD. The length of day or period of illumination, which can influence certain phenological conditions such as flowering and dormancy in plants.

PHOTO RESPIRATION. Respiration pathway involving light energy. Occurs in C-3 plants.

PHOTOSYNTHATE. Glucose, the food product of photosynthesis, sometimes including certain chemically altered fragments of glucose.
PHOTOSYNTHESIS. The process by which sugar (glucose) is manufactured from carbon dioxide and water by the chlorophyll-containing tissues (chlorenchyma) of plants in the presence of light, and oxygen is released as a by-product. Also referred to as *carbon assimilation*.
PHYSIOLOGICAL PRECONDITIONING. The induction of internal metabolic conditions which can influence the biochemical reactions and other processes of a plant during later stages in its life cycle.
PHYSIOLOGICAL SEASONS. In plants, periods of varying physiological activity during the year, usually determined by the climatic regime which preconditions and governs the processes, conditions, and structures of the plant.
PITH. In plants, the soft, thin-walled tissue at the center of the stem made up of parenchyma with unlignified cell walls.
PLASTIDS. Specialized bodies such as the chloroplasts (q.v.) contained in the cytoplasm (q.v.) of a cell.
POLYSACCHARIDES. A carbohydrate made up of several sugar molecules, a substance produced early in the development of the cell wall matrix of plants.
POTENTIAL ENVIRONMENT. Environmental phenomena which are capable of entering into a reaction with an organism but have not yet done so. Cf. *operational environment*.
POWER SPECTRUM ANALYSIS. A technique for studying the distribution of variance in a time series as a function of frequency.
PREDICTAND. In stastistics, the variable or variables of a system which are predicted (estimated, reconstructed). Cf. *predictor*.
PREDICTION. Estimates of conditions or data, not necessarily made for the future or forward in time. Cf. *reconstruction* and *retrodiction*.
PREDICTOR. In statistics, the input variable or variables of a system, values of which are used to obtain prediction (estimates, reconstructions) of the predictands (q.v.).
PRESSURE POTENTIAL (Ψp). The positive water potential forces attributed to pressure exerted on a cell wall which results in cell turgor.
PRINCIPAL COMPONENTS. Eigenvectors (q.v.).
PRINCIPLE. A general truth or essential nature of a phenomenon ascertained from experience and usually including one or more inferences (q.v.).
PRODUCT MEAN. A statistic useful for verification with independent data which takes into account both the sign and magnitude of the departure from the calibration mean; computed from the product of the departures of actual and estimated values from the mean value, with positive and negative products averaged separately.
PROTOPLASM. The viscous colloidal substance making up the living portion of the cell including the cytoplasm (q.v.) and nucleus (q.v.). The protoplasm is surrounded by a differentially permeable plasma membrane and, in plants, by a more or less rigid cell wall.
PROXY. In this text, a substitute record.
RADIAL FILE. A row or tier of cells such as the tracheids which arise from the division of a single cambial initial.
RAYS. Wood rays (q.v.).
RE Reduction of error (q.v.).
REACTION WOOD. Anomalous cells formed in the xylem of leaning stems and lateral

branches, often associated with an eccentric growth ring. In coniferous species, the reaction wood occurs on the lower side of the lean, and is referred to as *compression wood*; in angiosperm species it occurs on the upper side of the lean and is referred to as *tension wood*.

RECIPROCAL FILTERS. A pair of filters (q.v.) which have been designed to pass variance at opposite ends of the frequency range.

RECONSTRUCTION. As used in this text, statistical estimates or deductions of past conditions or events derived from various types of evidence, especially those from well-dated and replicated ring-width chronologies, Cf. *retrodiction* and *prediction*.

REDUCTION. A class of chemical reactions in which molecules gain chemical energy generally derived from ATP (q.v.); the reverse of oxidation (q.v.) in which molecules lose energy.

REDUCTION of ERROR (RE). A statistic useful for verification with independent data, having possible values ranging from $+1 \cdot 0$ to minus infinity. If it is applied to the dependent data set, the result is equivalent to the square of the correlation coefficient (q.v.) which measures the percent variance calibrated (q.v.) by the relationship.

REGRESSION. A general statistical term often represented by an equation which describes relationships where the values of one or more variables are expressed as a function of other variables. Cf. *correlation*.

RELEASE. In trees, a marked increase in growth due to removal of competitors. Cf. *suppression*.

REPRODUCTIVE BUDS. Buds that can produce stems with flowers, fruits, and seeds as well as leaves. Cf. *vegetative buds*.

REPRODUCTIVE ORGANS. In plants, the flowers, fruits, and seeds, as differentiated from the vegetative organs (q.v.).

RESIDUAL. The difference between an observation of actual data and an estimate (q.v.) obtained by applying a given function to the data set. The residuals represent the noise (q.v.) of the system and any signal not accounted for by the functional relationship. Cf. *least squares*.

RESIN DUCT. Long, narrow, intercellular channels in conifers surrounded by parenchyma and filled with resin.

RESPIRATION. A series of biochemical reactions which release the chemical energy stored in food such as that in the glucose molecule. Cf. *aerobic respiration* and *anaerobic respiration*.

RESPONSE FUNCTION. In this text, a statistical calibration equation which expresses the separate relative effects of several climatic factors on ring width. Climatic factors often are average temperature and total precipitation during each month of a 14-month period prior to and concurrent with the period of growth. A weight is associated with each factor which describes how the tree "responds" to variations in that climatic factor. Cf. *transfer function*.

RETRODICTION. A term proposed for statistical estimations made backward in time, as opposed to the term *prediction* (q.v.) implying estimates made forward in time.

RIDGE. In the atmosphere, an elongated high pressure area which is commonly formed at altitudes of several kilometers above the earth's surface. Cf. *trough*.

RING-POROUS WOODS. Certain angiosperms (q.v.), such as *Ulmus*, *Fraxinus*, and *Quercus*, in which the vessels in the earlywood are markedly wider than those in the latewood, as differentiated from diffuse-porous (q.v.) angiosperms.

RING-WIDTH CHRONOLOGY. The averaged standardized ring-width indices from a number of trees sampled from a particular site which can be used for crossdating and deducing past climate. Also called *mean standardized indices* and *chronology*.

GLOSSARY

RING-WIDTH INDICES. The transformed values of ring widths after standardization (q.v.) which have a mean of 1·0 (or 100) and a variance that is relatively homogeneous through time. Also referred to as *standardized ring-width chronology*.

ROOT CAP. A structure which covers the apical meristem (q.v.) of a root.

ROOT/SHOOT RATIO. The amount of roots relative to the amount of shoots.

RUNNING MEANS. Moving averages (q.v.)

SAPWOOD. The lighter colored outer water-conducting portion of the tree stem with living ray tissue, as differentiated from the darker dead heartwood portion in the center of the stem.

SENSIBLE HEAT. Heat added or lost when there is a temperature change.

SENSITIVITY. A dendrochronological term referring to the presence of ring-width variability in the radial direction within a tree which indicates that the growth response of the particular tree is "sensitive" to variations in climate. Opposite of complacency (q.v.). Cf. *mean sensitivity*.

SERIAL CORRELATION. The correlation between successive values in a time series (autocorrelation) or lagged correlation between two time series.

SIEVE TUBES. Large, nonwoody, thin-walled living phloem cells which are involved in translocation of organic substances throughout a tree. Cf. *companion cells*.

SIGMOID CURVE. A cumulative curve of growth in the shape of a prostate S.

SIGNAL. In this text, a detectable quality, pattern, or variation in a time series such as ring widths which can be attributed to meaningful information on climate, environment, or biological conditions of the trees, as opposed to the background variations or noise (q.v.) which provide no useful information about the trees and their environment.

SIGN TEST. A count of agreement or disagreement between signs, plus or minus, for two series of numbers and the statistical test used to determine whether the number of agreements is significantly different from the number of disagreements.

SITE SELECTION. Selection of tree-ring samples from limiting sites to maximize the information or signal contained in the rings and to minimize the undesirable variations representing noise.

SOIL HORIZON. A more or less distinct layer in a soil profile (q.v.) characterized by its morphological, physical, chemical, and mineralogical properties.

SOIL PROFILE. The vertical assemblage of soil horizons (q.v.) from the soil surface down to the lowermost layer of undifferentiated unweathered material

SOLUM, The portion of the soil represented by the A and B horizons (q.v.).

SPIKE TOP. The dead terminal branch of a tree, which may remain intact for many years. A feature characteristic of some extremely old, stress-site trees.

SPRINGWOOD. Earlywood (q.v.).

STANDARD DEVIATION. The mean scatter of a population of numbers from the population mean. The square root of the variance (q.v.). Cf. *normal distribution*.

STANDARD ERROR. An estimate of how much a particular statistic of a sampled population, such as the mean, can vary from its "true" theoretical value based on the entire population as represented by the statistical universe (q.v.).

STANDARD NORMAL VARIATES. A transformation converting a set of data to a population with a mean of zero and a standard deviation of 1·0 obtained by subtracting the mean and dividing by the standard deviation of the original data. Also referred to as *normalized data*.

STANDARDIZATION. The removal of the effects of increasing tree age on ring widths and their conversion to a time series of ring-width indices (q.v.) with a mean of 1·0 (or 100) and with a variance relatively homogeneous through time.

STANDARDIZED RING-WIDTH CHRONOLOGY. Ring-width indices (q.v.).

STATISTICAL UNIVERSE. A theoretical term referring to an entire population including all data, as opposed to the subset of that population represented by a sample. Since entire populations can rarely be measured, statistics derived from samples are used to estimate the theoretical values of entire populations. Cf. *standard error*.

STATISTICS. Any kind of numerical data, but most often applied to values used to describe characteristics of populations.

STOMATES. Microscopic holes in the epidermis of plants, through which diffusion (q.v.) of gasses such as water vapor, carbon dioxide, and oxygen occurs. Two cells called guard cells (q.v.) surround each stomate and govern its size.

SUBERIN. A fatty or waxy water-repellant substance which is usually found in the endodermis and cork. Cf. *cutin*.

SUMMERWOOD. Latewood (q.v.).

SUPPRESSION. In trees, a marked reduction in growth due to crowding or shading from neighboring trees. Cf. *release*.

SURROGATE. A substitute.

SYNOPTIC CLIMATOLOGY. A term applied to studies of climate in which the general circulation of the atmosphere must be considered over both space and time. Cf. *climatography*.

SYSTEM (NATURAL SYSTEM). A portion of the universe separated from its surroundings by recognizable boundaries and including one or more substances, any chemical reactions and interactions, as well as other processes affecting them.

TALUS SLOPE. A slope of disintegrated rock and earth material that accumulates at the base of cliffs and exposed rock surfaces.

TENSION WOOD. The name given to reaction wood (q.v.) which in angiosperm species occurs on the upper side of leaning stems and lateral branches. Cf. *compression wood*.

TEXTURE. In soils, the relative amounts of sand, silt, and clay.

TIME SERIES. A set of data representing a regular sequence of occurrences or events which are indexed as a function of time.

TRACHEIDS. In gymnosperms, dead, vertically oriented xylem cells with relatively thick lignified (q.v.) walls which may be perforated by bordered pits. (q.v.).

TRANSFER FUNCTION. A set of weights applied to ring-width index values to obtain estimates of climate. Growth records are "transferred" into reconstructions of climate. Cf. *response function*.

TRANSLOCATION. The long distance transport by mass movement (q.v.) of organic and inorganic substances in the phloem and xylem from one part of the plant to another, as contrasted with molecular diffusion (q.v.) of solutes and water from cell to cell.

TRANSMISSIVITY. The ability to transmit energy through a structure.

TRANSPIRATION. The loss of water by evaporation from the surfaces of all parts of the exposed plant, including leaves, stems, flowers, and fruits.

TREND. Extremely low-frequency variance (q.v.) with wavelengths which are greater than the length of the time series. Trend in a ring-width series results from such things as the changing growth potential of the tree due to increasing age, successional development in the forest community, geological changes, or very gradual changes in climate.

TROUGH. An elongated low pressure area which is commonly formed at altitudes several kilometers above the earth's surface. Cf. *ridge*.

TURGOR. The outward pressure exerted on the walls of living cells resulting from the pressure potential developed within them.

UNINODAL. Producing only one series of leaves each year as occurs in many *Pinus* species. Cf. *multinodal*.

VACUOLE. The central portion of a living cell containing water and dissolved substances referred to as cell sap (q.v.).

VAPOR PRESSURE DEFICIT. The difference between the actual vapor pressure of the air and the vapor pressure of an atmosphere at the same temperature that is saturated with water.

VARIABLE. In this text, a characteristic, feature, or factor which can assume different values in successive individual cases.

VARIANCE. The square of the mean scatter of a statistical population of numbers from the population mean. The square of the standard deviation (q.v.).

VARIANCE CALIBRATED. Variance reduced (q.v.).

VARIANCE REDUCED. The difference between the mean scatter or variance (q.v.) of a data set and the residual (q.v.) variance, representing the departure of the data from a statistical relationship or representation of the data; for example, variance reduced by regression or by eigenvectors (principal components) of the data set. Sometimes called *explained variance*. Cf. *goodness of fit*.

VASCULAR CAMBIUM. The thin cylinder of meristematic tissue (q.v.) between the wood and the bark which produces xylem (q.v.) on the inside and phloem (q.v.) on the outside.

VASCULAR TISSUE. The xylem (q.v.) and phloem (q.v.).

VEGETATIVE BUDS. Buds that can produce only stems and leaves. Cf. *reproductive buds*.

VEGETATIVE ORGANS. In plants, the roots, stems, and leaves, as differentiated from the reproductive organs (q.v.).

VERIFICATION. The proof of validity of results by comparing actual independent climatic data to reconstructions of climate.

VESSEL Large tube-like water-conducting element in the wood of angiosperms (hardwoods) in contrast to the smaller tracheid (q.v.) found in gymnosperms (q.v.).

VISCOSITY. The property of a fluid that resists internal flow because of strong cohesive forces among the molecules.

WATER BALANCE. The rates of movement and pathways governing the inflow, outflow, and water status in a system such as a plant.

WATER POTENTIAL (Ψ). The difference in free energy (q.v.) of water in the plant, soil, or atmospheric system under consideration and that of pure water at atmospheric pressure. Cf. *osmotic potential*.

WEATHER. All observable meteorological phenomena which vary over time scales of minutes to days, including precipitation, heat, temperature, light, wind, clouds, and related factors that influence the environments of the biosphere. Cf. *climate*.

WOOD RAYS. Thin-walled cells in the xylem (q.v.) forming lines or ribbons radiating from the pith (q.v.).

XYLEM. The water-conducting tissue of plants which makes up the wood cylinder of a tree inside the bark and surrounding the pith.

XYLEM RAYS. Wood rays (q.v.).

Author Index

A

Agerter, S. R., 9, 14, 62, 86, 87, 91, 95, 108, 119, 329, *511*, *517*, *527*
Alberty, R. A., *513*
Alestalo, J., 10, 53, 113, 219, 220, 221, 420, *511*
Alvim P de T., 12, 71, *511*
Anderson, D. B., 119, 121, 128, 143, 153, 157, 158, 163, 167, 168, 176, 198, *522*
Antevs, E., 222, 499, *511*
Arakawa, H., 499, *511*
Ashby, W. C., 43, 321, 381, *511*

B

Baggaley, A. R., 333, *511*
Bailey, D. E., 383, *528*
Bailey, D. K., 9, 22, *511*
Bailley, M. G. L., 31, *511*
Baker, F. S., 73, 185, 277, *511*
Bakke, A. L., 221, *511*
Bannan, M. W., 91, *511*
Bannister, B., 22, *511*
Barry, R. G., 436, *511*
Bauch, J., 31, 331, *514*
Bendat, J. S., 342, *511*
Bergthorsson, P., 499, *512*
Billings, W. D., 74, 221, 224, *512*
Bitvinskas, T. T., 10, 31, 266, *512*
Bjerknes, J., 436, *512*
Blackman, R. B., 295, *512*
Blasing, T. J., 11, 365, 400, 402, 403, 425, 426, 427, 428, 436, 444, 445, 449, 450, 455, 456, 457, 458, 459, 460, 461, 462, 463, 464, 465, 467, 468, 469, 474, 475, 476, 479, 480, 481, 482, 487, 488, 491, 492, 493, 494, 497, 499, *512*, *513*, *516*
Bliss, E. W., 436, *528*

Böhning, R. H., 119, 121, 128, 143, 153, 155, 157, 158, 163, 167, 168, 176, 198, *522*
Bormann, F. H., 75, 77, 153, 155, *512*, *517*
Bottorff, C. P., 264, *516*
Box, G. E. P., 296, *512*
Bray, J. R., 323, 332, *512*
Brayton, R., 169, 170, 176, 177, 182, *523*
Bridgman, W. B., 143, *524*
Brier, G. W., 254, 299, *512*, *524*
Brown, H. P., 60, 112, *512*
Brown, J. M., 129, 132, 143, 165, 167, 168, 172, 175, 179, 180, *512*
Bruce, H. D., 214, 215. *515*
Bryson, R. A., 300, 328, 341, 343, 344, 501, *512*, *525*, *528*
Buddemeier, R. W., 504, *512*
Budelsky, C. A., 86, 95, 100, 108, 109, 143, 148, 183, 193, 279, 301, 304, 306, 308, *512*, *516*

C

Cantrell, J. O., 332, 341, *528*
Cardis, J. W., 86, 95, 108, 109, 193, 279, 301, 304, 306, 308, *516*
Carvell, K. L., 332, 341, *528*
Cathey, E. H., 499, *516*
Charton, F. L., 295, *512*
Chowdhury, K. A., 95, *512*
Clark, N., 425, 426, 427, 428, *513*
Clausen, H. B., 504, *513*
Clayton, H. H., 479, *513*
Cleaveland, M. K., 44, *513*
Clelland, R. C., 330, 331, *527*
Clevenger, J. F., 221, *513*
Cooley, W. W., 353, *513*

Cooper, C. F., 403, *513*
Cooper, G., 77, *513*
Craddock, J. M., 270, *513*
Crutcher, H. L., 299, *520*
Cunningham, G. L., 126, *513*

D

Dahl, E., 34, *513*
Daniels, F., *513*
Dansgaard, W., 504, *513*
Daubenmire, R. F., 75, 119, 165, 332, *513*
Day, W. R., 199. *513*
Dayton, W. A., *519*
Dean, J. S., 4, 6, 7, 8, 19, 21, 22, 23, *513*
Decker, J. P., 162, 164, *513, 520*
Dempster, A. P., 353, *513*
Denton, G. H. 410, *514*
Digby, J,, 191, 192, *514*
Dobbs, R. C., 77, *514*
Douglas, A. V., 212, 421, 422, 423, 425, *514*
Douglas, C. L., 107, 203, 216, 217, 221, *514, 515*
Douglass, A. E., 5, 6, 7, 8, 19, 258, 266, 269, *514*
Downs, R. J., 195, *514*
Draper, N. R., 342, 344, 365, *514*
Drew, A. P., 131, 132, 143, 174, 175, 181, *514*
Drew, L. G., 131, 132, 143, 174, 175, 181, 439, *514, 527*
Druce, A. P., 221; *514*
Duff, G. H., 277, 278, 305, *514*
Dunn, E. L., 166, 169, 170, 182, *525*
Dutton, J. A., 300. *512*
Dzerdzeevskii, B., 254, 258, 269, 270, 295, 296, 322, 324, 436, *523*

E

Eardley, A. J., 115, *514*
Eckstein, D., 31, 331, *514*
Eis, S., 153, *514*
Eklund, B., 228, 327, 332, 341, 342, *514*
Elfving, G., 332, 341, *518*
Elling, W., 341, *514*
Elton, W. M., *517*
Erdman, J. A., 107, 203, 216, 217, 221, *514, 515*
Ermich, K., 331, *515*
Estes, E. T., 327, 332, *515*
Ezekiel, M., 256, 333, 342, 344, 350, *515*

F

Fahn, A., 14, 328, *515*
Farrar, J. L., 109, 110, 112, 199, *515, 517*
Fayle, D. C. F., 113, 114, *515*
Ferguson, C. W., 22, 31, 153, 187, *515*
Flohn, H., 254, 258, 269, 270, 295, 296, 322, 324, 436, *523*
Fons, W. L., 214, 215, *515*
Forward, D. F., 188, *525*
Fox, K. A., 256, 333, 342, 344, 350, *515*
Fraser, D. A., 86. *515*
Fratianne, D. G., 119, 121, 128, 143, 153, 155, 157, 158, 163, 167, 168, 176, 198, *522*
Freeland, R. O., 165, 176, *515*
Freund, J. E., 332, 333, *515*
Friesner, G. M., 331, *515*
Friesner, R. C., 331, *515*
Fritts, E. C., 76, 77, *516*
Fritts, H. C. 1, 29, 34, 35, 38, 39, 41, 42, 43, 53, 76, 77, 78, 83, 84, 86, 87, 89, 91, 93, 95, 105, 106, 107, 108, 109, 126, 131, 132, 136, 139, 143, 148, 153, 159, 174, 175, 181, 187, 188, 189, 190, 193, 200, 203, 210, 212, 215, 216, 217, 218, 228, 233, 238, 241, 264, 273, 274, 275, 276, 279, 292, 300, 301, 303, 304, 306, 308, 320, 321, 327, 328, 333, 334, 335, 336, 337, 339, 340, 341, 344, 345, 346, 348, 351, 352, 365, 371, 372, 373, 374, 377, 381, 382, 383, 384, 385, 388, 391, 393, 400, 402, 403, 404, 405, 406, 407, 408, 412, 415, 416, 418, 419, 425, 426, 427, 428, 437, 439, 440, 444, 445, 449, 461, 462, 474, 475, 476, 479, 482, 487, 491, 492, 493, 494, 497, 499, *511, 512, 513, 514, 515, 516, 519, 520, 523, 527*
Fry, K. E., 131, *516*

G

Gagnon, D., 332, *516*
Gale, J., 176, *516*
Garrett, P. W., 188, *516*
Gates, D. M., 128, 130, 164, 210, 211, 212, *516, 517*
Gates, W. L., 10, 11, 46, 47, 410, 436, 501, *517*
Geiger, R. 218, *517*
Germann, P. J., 108, 109, *517*
Giddings, L. J., Jr., 86, 228, *517*
Giertych, M. M., 191, 192, 193, *517*

AUTHOR INDEX

Gilbert, G. E., 77, *524*
Ginzburg, C., 14, 328, *515*
Glahn, H. R., 418, 438, *517*
Glerum, C., 199, 229, *517*
Glock, W. S., 9, 14, 17, 62, 86, 87, 91, 95, 108, 109, 119, 199, 200, 304, 326, 328, 329, 332, *511, 517, 525, 527*
Graham, B. F., Jr. 153, 155, *517*
Grier, C. C. 62, *517*

H

Haase, E. F., 215, *517*
Hall, R. C., 75, *517*
Hari, P., 67, 70, 184, 300, *517 526*
Harlan, T. P., 23, 39, 408, *520*
Harman, J. R., 295, *512, 518*
Hart, J. H., 29, *518*
Haury, E. W., 8, *518*
Hayden, B. P., 11, 365, 400, 402, 444, 445, 449, 482, *516*
Heger, L., 44, *518*
Heinselman, M. L., 221, *518*
Heizer, R. F., 6, *518*
Helley, E. J., 220, 221, *518*
Hendricks, E. L., 219, *526*
Henoch, W. E. S., 44, 332, *524*
Herrington, L., 77, *513*
Hofmeyer, W. L., 254, 256, 269, 270, 295, 296, 322, 324, 436, *523*
Hollstein, E., 281, *518*
Holmsgaard, E., 188, 332, 341, *518*
Holowaychuck, N., 136, 139, 217, *516*
Howard, R. A., 226, *529*
Huber, B., 9, 155, *518*
Hughes, J. F., 113, *518*
Huntington, E., 139, *518*
Hustich, I., 228, 327, 332, 341 *518*

I

Imbrie, J., *518*
Impens, I. I., 77, *518*

J

Jacoby, G. C., Jr. 414, 415, *527*
Jahnel, H., 155, *518*
Jemison, G. M., 153, 200, *518*
Jenkins, G. M., 259, 296, 322, *512, 518*
Jenny, H., 136, *518*
Johnsen, S. J., 504, *513*
Johnson, A. I., 11, 436, *521*
Johnson, M. A., 11, *521*

Jones, F. W., 43, *518*
Jones, R. H., 436, *519*
Jonsson, B., 53, 152, 190, 280, 328, 401, *519*
Julian, P. R. 29, 210, 233, 273, 274, 275, 341, 365, 455, *519*

K

Karlen, W., 410, *514*
Keen, F. P., 328, 332, *519*
Kelsey, H. P., *519*
Kennedy, R. W., 43, 44, *518, 524*
Kienholz, R., 70, *519*
Kilgore, B., 221, *519*
Kisiel, C. C., 296, *519*
Kochenderfer, J. N., 139, *519*
Kipp, N. G., *518*
Koerber, T. W., 223, *519*
Kohler, M. A., 252, 254, *519*
Kozlowski, T. T., 58, 60, 63, 65, 66, 68, 69, 70, 72, 75, 77, 80, 86, 90, 91, 111, 113, 123, 141, 142, 143, 147, 151, 152, 153, 154, 155, 158, 163, 164, 176, 177, 195, *512, 519, 520, 529*
Kramer, P. J., 60, 63, 69, 87, 88, 111, 113, 123, 124, 134, 139, 140, 141, 142, 143, 147, 151, 152, 153, 154, 155, 158, 163, 164, 176, 177, 195, *519, 520, 529*
Krueger, K. W., 69, *520*
Kukla, G. J., 436, *520*
Kukla, H. J., 436, *520*
Kunstle, E., 77, 86, *523*
Kuo, M., 20, *520*
Kuroiwa, K., 76, 77, *520*
Kutzbach, J. E., 11, 365, 400, 402, 436, 444, 445, 449, 482, *516, 520*

L

Ladefoged, K., 64, 84, *520*
Ladurie, E. LeR., 11, 499, *520*
La Marche, V. C., Jr. 11, 23, 39, 41, 114 115, 154, 188, 198, 200, 203, 220, 221, 222, 268, 275, 298, 299, 300, 325, 328, 373, 374, 391, 408, 409, 411, 437, 439, 440, 492, *518, 520, 521*
Lamb, H. H., 11, 254, 256, 269, 270, 295, 296, 322, 324, 410, 436, 499, 501, *521, 523*
Landsberg, H. E., 299, *521*
Langenheim, J. H., 48, 49, 50, *522*
Langway, C. C., Jr. 504, *513*
Lanner, R. M., 72, *521*

La Point, G., 77, *521*
Larson, P. R., 62, 93, 111, 112, 191, 192, 193, 230, *521*
Laufersweiler, J. D., 154, *521*
Lawrence, D. B., 219, 220, 221, 222, *521*
Lawrence, E. G., 219, 220, 222, *521*
Lawson, M. P., 317, 499, *522*
Lawson, T. L., 499, *528*
Leikola, M., 70, *517*
Leopold, A. C., 191, *522*
Levitt, J., 195, 197, 198, *522*
Lewontin, R. C., 282, 342, *526*
Li, J. C. R., 340, *522*
Liming, F. G., 75, *522*
Little, E. L., Jr. *522*
Lohnes, P. R., 353, *513*
Lorenz, E. N., 332, *522*
Lowry, W. P., 194, *522*
Lück, H., 327, 328, 332, 341, 343, 344, *525*
Ludlum, D. M., 470, 471, 472, *522*
Lund, I. A., 457, *522*
Lundegardh, H., 119, *522*
Luukkanen, O., 184, *517*
Lyon, C. J., 331, *522*

M

McDonald, J. E., 252, *522*
MacDougal, D. T., 142, 143, *522*
McGinnes, E. A., Jr. 20, *520*
McGinnies, W. J., 332, *522*
McInter, B. B., 331, *522*
McMasters, A., 214, 215, *515*
Mariaux, A., 95, *522*
Marr, J. W., 203, 216, 217, 221, *515*
Martin, P., de, 10, 269, *513*
Mason, H. L., 48, 49, 50, *522*
Matalas, N. C., 260, *522*
Mather, J. R., 209, 345, *527, 528*
Maurer, P., 77, 86, *523*
Meleskie, K. R., 43, *524*
Meyer, B. S., 119, 121, 128, 143, 153, 155, 157, 158, 163, 167, 168, 176, 198, *522*
Mikola, P., 228, *522*
Miller, C. B., 331, *523*
Miller, C. W., 331, *523*
Mintz, Y., 10, 11, 46, 47, 410, 436, 501, *517*
Mitchell, J. M., Jr 254, 258, 269, 270, 295, 296, 299, 322, 324, 436, *521, 523*
Mitchell, V. L., 279, 329, *523*
Mitscherlich, G., 77, 86, *523*
Moll, W., 77, 86, *523*

Möller, C. M., 177, 178, *523*
Monk, C. D., 77, *526*
Mooney, H. A., 153, 166, 169, 170, 171, 172, 176, 177, 178, 182, 190, 198, 203, 217, 218, *520, 523, 525, 529*
Mork, E., 34, 62, *513, 523*
Morrison, D. F., 353, *523*
Mosimann, J. E., 264, *516*
Müller, M. D., 177, 178, *523*
Munaut, A. V., 222, *523*

N

Namias, J., 421, 435, 436, 465, *523*
Nash, T. H., 53, *523*
Neter, J., 282, *523*
Nielsen, J., 177, 178, *523*
Nolan, N. J., 277, 278, 305, *514*

O

Oak Ridge Systems Ecology Group, 403, *513*
Odin, H., 77, *523*
Olsson, I. U., 152, *523*
O'Neal, L. C., 177, *524*
Openshaw, A., 77, *523*

P

Page, R., 220, 222, *524*
Palmer, W. C., 34, 210, *524*
Panofsky, H. A., 254, *524*
Panshin, A. J., 60, 112, *512*
Parker, J., 196, 198, *524*
Parker, M. L., 43, 44, 45, 200, 332, *518, 524*
Parton, W. J., 403, *513*
Parzen, E., 332, *524*
Pearce, T. R., 199, *513*
Perry, A. H, 323, 436, *511*
Perry, T. O, 166, *524*
Peterson, T. A., 86, *519*
Philip, J. R., 124, *524*
Phipps, R. L., 65, 77, 88, 89, 90, 91, 109, 112, 192, 277, *524*
Piersol, A. G., 342, *511*
Pilcher, J. R., 31, *511*
Plumb, R. C., 143, *524*
Polge, H., 43, 176, *524*
Pons, A., 327, 328, 332, 341, 343, 344, *525*
Potter, N. Jr. 220, 221, *524*
Press, S. J., 353, *524*

AUTHOR INDEX

Q
Quenouille, M. H., 344, *524*
Quinlan, F. T., 299, *521*

R
Rangenekar, P. V., 188, *525*
Rao, K. N., 254, 258, 269, 270, 295, 296, 322, 324, 436, *523*
Räsänen, P., 70, *517*
Ray, P. M., 162, 165, *525*
Rees, L. W., 86, *525*
Reineke, L. H., 75, *525*
Richardson, S. D., 278, *525*
Roberts, H. V., 321, *528*
Roberts, W. D., 322, *525*
Roe, A., 282, 342, *526*
Romberger, J. A., 58, *525*
Roughton, R. D., 219, *525*

S
Samish, R. M., 195, *525*
Sampson, A. W., *525*
Schalck, J. M., 77, *518*
Schmidt, E., 155, *518*
Schreuder, G. F., 403, *513*
Schulman, E., 9, 14, 30, 114, 327, 328, 332, 388, 405, *525*
Schulman, M. D., 328, 341, 343, 344, *525*
Schulz, H., 113, *528*
Schulze, E. D., 166, 169, 170, 182, *525*
Scott, D., 328, 332, 341, *525*
Seal, H. L., 353, *525*
Searles, S. R., 356, *525*
Sellers, W. D., 208, 209, 373, 374, 469, *525*
Serre, F., 72, 75, 107, 327, 328, 332, 341, 343, 344, *525*
Sheppard, P. A., 11, 501, *525*
Siegel, S., 331, *526*
Sigafoos, M. D., 220, *526*
Sigafoos, R. S., 114, 219, 220, 221, *526*
Silen, R. R., 74, *527*
Silver, G. T., 177, *526*
Simpson, G. G., 282, 342, *526*
Siren, G., 67, 300, 332, 341, 405, *517*, *526*
Skene, D. S., 63, 64, 86, *526*
Slàstad, T., 327, 332, *526*
Slatyer, R. O., 170, 171, 175, *526*
Small, J. A., 77, *526*
Smiley, T. L., 22, 31, 32, 221, 249, 264, *526*, *527*
Smith, D. G., 34, 78, 83, 84, 86, 87, 95, 108, 109, 148, 188, 189, 193, 203, 216, 233, 279, 301, 304, 306, 308, 351, 352, 405, 406, 407, *516*
Smith, D. M., 112, 277, 278, 328, *526*
Smith, F. M., 403, *513*
Smith, H., 342, 344, 365, *514*
Snedecor, G. W., 332, *526*
Soil Survey Staff., 138, *526*
Sollins, P., 403, *513*
Spencer, D. A., 222, *526*
Spiegel, M. R., 340, *527*
Squillace, A. E., 74, *527*
Stage, A. R., 34, 210, *529*
Stallings, W. S., Jr. 3, *527*
St. Andre, G., 217, *523*
States, J. B., 217, *527*
Steel, R. G. D., 282, *527*
Stockman, W. B., 472, *527*
Stockton, C. W., 40, 41, 42, 212, 217, 275, 276, 296, 297, 298, 318, 325, 328, 333, 334, 335, 336, 337, 339, 340, 391, 412, 413, 414, 415, 416, 418, 419, 439, 492, *521*, *527*
Stokes, M. A., 22, 31, 32, 34, 53, 78, 83, 84, 87, 107, 148, 188, 189, 203, 216, 233, 264, 351, 352, 405, 406, 407, 439, *516*, *523*, *527*
Stone, E. L., 153, *527*
Stoner, W., 403, *513*
Struik, G. J., 332, *512*
Studhalter, R. A., 6, 14, 62, 86, 87, 91, 95, *517*, *527*
Sundberg, R., 152, 190, *519*
Swain, A. M., 504, *527*
Sweet, G. B., 177, *527*

T
Tate, M. W., 330, 331, *527*
Taylor, O. C., 43, *527*
Tesai, V., 43, *528*
Thornthwaite, C. W., 209, 345, *527*, *528*
Torrie, J. H., 90, 282, *519*, *527*
Tranquillini, W., 163, 165, 166, 167, 169, 182, *528*
Trappe, J. M., 69, *520*
True, R. P., 327, 332, *528*
Tryon, E. H., 327, 332, 341, *528*
Tryon, R. C., 383, *528*
Tukey, J. W., 295, *512*, *528*

V

Van Cleve, K., 77, *521*
Viavant, W., 115, *514*
Vinš, B., 43, 176, 190, 217, 220, 221, 222, *528*

W

Wachs, N., 14, 328, *515*
Wahl, E. W., 499, *528*
Walker, G. T., 436, *528*
Walker, R. B., 131, *516*
Wallace, R. E., 154, 220, 222, *521*
Wallén, C. C., 254, 258, 269, 270, 295, 296, 322, 324, 436, *523*
Wallis, A. W., 321, *528*
Wardlaw, I. F., 155, *528*
Wardrop, A. B., 67, 113, *528*
Wareing, P. F., 177, 191, 192, *514, 527*
Wargo, P. M., 29, *518*
Waring, R. H., 62, *517*
Wassermann, W., 282, *523*
Watts, D. G., 259, 296, 322, *518*
Weakley, H. E., 332, *528*
Weaver, H., 221, *528*
Webb, T., *528*
West, M., 169, 170, 176, 177, 182, *523*
Westing, A. H., 113, 192, 193, *528*
Wickman, B. E., 223, *519*
Williams, C. B., Jr. 71, *529*
Willis, J. C., *529*
Wilsie, M. C., 112, 277, 278, 328, *526*
Wilson, B. F., 63, 64, 65, 67, 226, *529*
Winget, C. H., 77, 86, 90, *519, 529*
Winstanley, D., 435, *529*
Wodzicki, T. J., 64, 65, 192, *529*
Wolter, K. E., 75, *529*
Worrall, J., 78, *529*
Wright, R. D., 153, 171, 172, 177, 190, 217, 218, *523, 529*

Y

Yater, W. M., Jr, 77, *524*
Young, H. E., 77, 87, 88, *529*

Z

Zahner, R., 34, 64, 65, 90, 93, 188, 210, 229, 230, 233, *516, 529*
Zasada, J. C., 90, *529*
Zich, J., 403, *513*
Zimmermann, M. H., 155, *529*
Zumer, M., 86, *529*

Subject Index

Abcission, 194, 531
Abies, 14, 196
  amabilis, 71
  concolor, 73, 391
  lasiocarpa, 71
  magnifica, 73
  procera, 71
Absorptivity, 212, 531
Acer, 63
  rubrum, 90
  saccharum, 82, 103, 138
Acid rain, 152
Active absorption, 123, 151, 531
Active transport, *see* Active absorption
Adaptation, 196
Adenosine triphosphate, 158, 160, 531
Adhesion of molecules, 120, 142–143
Adventitious shoots, 114
Agathis, 14
Age, *see also* Longevity; Ring-width, as a function of age
  of cambium, 107–113, 277–278, 305–310
  of tissues in tree, 88, 114, 176, 306–310
  of tree, 24–25, 33, 72, 84, 86, 107–113, 115, 148, 177, 185, 219, 282, 305–310
Agriculture, 407
Air mass, 435, 531
Air pollution, *see* Pollution, environmental
Albedo, 209, 217, 236, 436, 531
Alcohol, 158
Aldehydes, 158, 531
Aleutian Low, 443–444, 458, 473–474
Alnus, 14
Altitude, *see* Elevation
Amelanchier, 14

Amino acid, 161–162, 190
Amplitude, ecological, 16–17, 531
Amplitude, of eigenvector, *see* Eigenvector, amplitude of
Amylase, 160, 531
Antiauxin, 191, 531
Anticyclone, 434–435, 453–454, 465, 532
Apical dominance, 72, 532
Apical growth, *see* Growth, of shoots
Apical meristem, *see* Meristem
a posteriori reasoning, 34–35
a priori reasoning, 33–34, 314, 325–326
Araucaria, 14
Archaeology, 2–3, 7–8, 405–407
Artemisia, 14
  tridentata, 153
Assimilation, 159–162, 227–228, 396–397, 532
Association, testing for, 329–340
Athabasca Lake study, 415–420
Atlantic Subtropical High, *see* Bermuda High
Atmosphere
  circulation of, 410–411, 434–437, *see also* Climate
  components of, 175–176
  tides in, 501
ATP, *see* Adenosine triphosphate
Austrocedrus, 14
  chilensis, 14, 30
Autocorrelation, 26–28, 150, 188, 257–259, 261–262, 266, 268, 289, 300, 302–310, 324–325, 343, 347, 365, 373, 379, 382, 390, 408, 432, 436, 492, 543
  correcting a time series for, 405–406

554

TREE RINGS AND CLIMATE

degrees of freedom adjustment, 323–325
Autocorrelation function, 259, 262, 295–297, 532
Autoregression, 7, 258, 262, 324, 532, *see also* Autocorrelation
Auxin, 190, 230, 532
Average, *see* Mean

B

Bag, polyethylene, *see* Tent, polyethylene
Bark, 58–59, 68, 77, 81, 153, 200
Bermuda High, 410, 465, 467
Betula, 14
Bole, 86, 102, 109–110, 184–185, 532 *see also* Stem
Bordered pit, 61, 68, 532
Boron, 190
Boundary layer of air, 124, 129, 212
Branch, 58, 72–73, 80, 86, 95, 105, 110, 113, 184–185, 187, 310
Breast height, sampling at, 310
Bud, 57–58, 72, 103, 176–177, 194, 399
Buttress root, 115, 532
Buttress trunk, 30, 154

C

C-3 plants, 165, 178, 532
C-4 plants, 165, 178, 532
Calcium, 189–190
Calibration, 27–28, 33–34, 53, 152, 313, 321, 405, 413–414, 417–418, 421–423, 425–426, 428–431, 450–453, 532
  procedure of, 313–320, 363–365, 371, 438–439
Calibration equation, 313, 321, 342–344, 364, 377, 413, 418–419, 421, 425, 430, 439, 451
Cambial dieback, 149, 153–154
Cambial initials, 65–67, 75, 532
Cambium, cork, 56–58, 68, 141
Cambium, vascular, 56–57, 63–69, 81, 277, 396–397, 532
  activity of the, 69, 80, 84–91, 96–101, 103–107, 114, 191, 194, 226–233, 389, 396–397, 532
  circumferential expansion of, 66, 68
  dormancy of, 64, 69, 72, 192, 194–198, 205, 227
  initiating activity of, 65–66, 69, 114, 191–192, 227–228, 310
  sampling of, 75
Canonical analysis, 438–439, 450–451, 532
  significance of, 439
Canonical correlation, 438, 532
Canonical regression, 418, 439, 532
Canopy, 111, 130, 165
Capillary movement, 120, 135–136, 142
Carbohydrate, 159–162
Carbon-14, 152
Carbon assimilation, *see* Photosynthesis
Carbon dioxide, 143–146, 157–158, 162–163, 173–176, 180, 211
Carya, 14
Casparian strip, 141, 532
Cedrus, 14
Cell
  diameter of, 93–96, 99–100, 105, 191–192
  differentiation of, 63–68, 93–96, 191–193, 226
  division of, 63, 66–67, 81, 91, 100, 105, 114, 191, 226–231, 396–400
  enlargement of, 63–65, 67, 93–96, 191–192, 229–230, 396–400
  maturation of, *see* Cell, differentiation of
Cell sap, 123, 533
Cell studies, 75, 96–107
Cell vacuole, 176
Cell wall, 62–64, 67–68, 74, 93, 100, 103, 122–123, 176, 192
Cell water status, 123–127
Cellulose, 67, 159, 533
Celtis, 14
Cercocarpus, 14
Chance variations, 256, 314, 321–326, 336, 344, 451, 485
Chinook, *see* Foehn wind
Chi square (χ^2), 333, 336–337
Chlorenchyma, 56, 127, 157, 533
Chlorine, 190
Chlorophyll, 56, 157–158, 190, 533
  wintertime loss of (deactivation), 165–167, 203–204, 399
Chloroplasts, 157, 533
Chronicles, 499
Chronology, 2, 6, 25, 266–268, 282, 285, 289–290, 314, 533, *see also* Ring-width indices
  error in, 23, 280, 290–293, 322–323

archaeological, 8
floating, 7, 8
reliability of, see Chronology, error in
Chronology building, 3, 6, 22–23, 277, 533
Climate, 46–48, 52–54, 434–437, 533
  macro-, 46–47, 53, 115, 205, 282, 285, 300, 539
  micro-, 47–50, 53, 205, 228, 394, 539
  nongrowing season, importance to growth, 239–241
Climate system, 15, 46, 209–210, 434–437, 533
Climatic anomaly, 35–38, 373, 402–405, 421, 473–474, see also Mapped anomalies of climate
Climatic change, 11, 29, 35, 267, 436, 501
  inadvertent, 11, 43, 402–405, 501
Climatic data, see Meteorological data
Climatic factors, see Factors, climatic
Climatic seasons, see Seasons, climatic
Climatic state, 47, 533
Climatic variations, 6, 11, 17, 35, 436, 488, 501–503
Climatic window, 238–240
Climatography, 436, 533, see also Dendroclimatography
Climatological studies, 455–468, 501
Climatology, synoptic, 436
Clouds, 91, 209–211, 228, 327, 396, 398–399, 426–427, 493
Cluster analysis, 383–391
Coefficient, 533
  of contingency, 333, 336–337
  of eigenvector, see Eigenvector, amplitude of
  general use of term, 28, 34, 341, 353
  partial regression, see Regression, coefficient of
Coherency, 298–299, 533
Cohesion of water molecules, 120, 143
Collecting ring samples, 29–31, 91, 247–249, 323, 402
Colloid, 124, 533
Companion cell, 68, 533
Competition,
  among trees, 51, 110–112, 116, 200, 244–245, 289
  within the tree, 80, 147–148, 187, 229
Complacency, 18–19, 300–303, 533
Component variances, see Variance, components

Compression wood, 113, 193, 221, 533
Computer processing, see Processing by computer
Concentration, 121, 160, 533
Condensation, 159, 210–211, 533
Condensation-hydrolysis reaction, 159–161, 534
Conditional probability, 36, 333–340
Confidence limits, of response function, 365–371, 377–379
Conservation term of Douglass, 7
Contingency analysis, 333–340, 477–479
Copper, 190
Corals, annual layers in, 504
Corer, increment, 5, 29
Cork, 58, 68
Cornus, 164
Correlated relationships, usability in reconstruction, 27, 52, 394
Correlating systems, 193–198
Correlation, 534
  analysis of, 293–295, 302–310
  as a function of distance, 273–275
  auto-, see Autocorrelation
  between error and mean index, 292–293
  coefficient of, 257, 332–333, 341, 345, 357, 473–474, 477, 479, 488, 534
  coefficient of, used with filters, 272–277
  coefficient of, used to study site variation, 300–304
  coefficient of, used to study within-tree variation, 305–310
  compared to analysis of variance, 293–295, 301–304
  multiple, 34–35, 341–353, 470, 539
  the phenomenon of, 52–53, 118, 197, 213, 293–295, 300–301, 345, 393, 435, 457
  of random numbers, 451
  serial, see Autocorrelation
  trend method, 332
Cortex, 58, 68, 141–142, 192, 534
Coupling, 52, 212, 400, 534
Covariance, 296, 332, 450
Crevice site, 149–151
Crossdating, 534
  accuracy of, 14, 21, 31, 250
  aid by computer analysis, 31, 250
  method of, 2–6, 15–16, 21–23, 32, 95, 198, 200, 249–250, 277, 295, 300–302

the necessity for, 13–14, 20–21, 95, 249–250
principle of, 20–23
in roots, 114–115
time required for, 31–32
verification of, 23, 250
Cross-power spectrum analysis, 36, 296–300, 401, 534
Crown
  shape or size variations, 72–74, 86–87, 102, 109–112, 142, 148, 153, 164, 191, 289
  vigor of, 193
Cupressus, 14
  arizonica, 12, 200
Curve-fitting, 33, 116, 263–270
Curvilinearity, 318, 320, 400–401
Cuticle, *see* Cutin
Cutin, 122, 127, 159, 197, 534
Cycles, 41, 258–260, 295–300
  fruiting, 260
  sunspot, *see* Sunspots
Cyclone, 435, 426–427, 434–435, 453–468, 534
Cytoplasm, 121–123, 534

D

Data Bank, International Tree-Ring, 247–248, 256, 505
Dating,
  archaeological, 2–3, 7–8, 23
  carbon-14, 222, *see also* Carbon-14
  necessity for accuracy of, 437, 500
  tree-ring, *see* Crossdating
Daylength, 70, 74, 163, 193–195, 228
Death, 49–50, 168, 186
Deduction, 35, 405, 534, *see also* Inference
Defoliation, 91, 188, 199–200, 223
Degrees of freedom, 323–325, 342–344, 350, 353, 365, 401, 440, 451, 534
Dendrochronology, 534
  historical background, 4–10
  principles and concepts, 14–28
  subfields, 10
Dendroclimatography, 10, 436–438, 534
Dendroclimatology, 2, 9, 10, 38–46, 52, 190, 313, 405–505, 534
  biological constraints, 11–14
  procedures of, 28–38, 246–375, 437–439, 455–457, 503

prospects for future, 503–505
value of, 10–11, 500–505
Dendroecology, 10, 46, 219–223, 535
Dendrogeomorphology, 10, 114–115, 420, 535
Dendrograph, 77–80, 83–84, 87, 96, 104–105, 106, 535
Dendrohydrology, 10, 41–43, 319, 412–420, 535
Dendrometer, 75–76, 84, 91, 96, 535, *see also* Dendograph
Denormalization, 402, 406
Densitometry, 43–45, 62, 96, 117
Density,
  of forest, 110, 116, 216, 244–245, 300–301
  of wood, *see* Wood density
Dependent data, 320, 477–479, 535
Dew, 209
Dewpoint, 146, 535
Diffuse-porous woods, 63, 66, 88, 112, 114, 191, 535
Diffusion, 120–123, 135, 141, 170, 176, 396, 535
  resistance to, 128, 131, 140–141, 171, 212
Digestion, 160, 535
Disease of tree, 29, 50, 63, 153, 223
Distance, separation between sites, 273–277, 430–432
Disturbance, 53, 273–275, *see also* Site factors; History of forest stand
Diurnal expansion and contraction of plant parts, 70, 76–84, 91, 106, 127, 142–146, 171, 229
Dormancy, 51, 64–66, 69, 72, 95, 113, 165, 176, 195–198, 398–400, 535
  breaking of, 192, 194, 227, 379, 398
Double-mass analysis, 252–254, 535
Douglass, Andrew E., 4–8, 266
Drainage, cold air, 48
Drought, 80, 86, 98, 136, 181–182, 289, 399, 465–468, 471, 490, 502
  an adjustment to prolonged, 116, 244–245
  The Great, 407
  index to severity of, 210
  the sub-Saharan, 435
Drought resistance, 196–197
Dry-matter production, 163, 177–178, 184–185

SUBJECT INDEX

Duff and Nolan classification, 277–279, 304–310
Dust, atmospheric, 209–211, 470, 501
Dust Bowl, 465

E

Earlywood, 60–63, 65, 89, 95, 100, 102, 111, 191–193, 230, 345–349, 535
Eccentricity in ring width, 113–114, 220–221
Ecotype, 74, 535
Effective sample size, 258, 323–325, 535
Efficiency of sampling, 292
Eigenvalue, 358, 535
Eigenvector, 34–35, 352–353, 356–363, 371–375, 417–418, 427, 437, 440–445, 450, 455, 483, 535
   amplitude of, 359–369, 372–374, 425–426, 442–445, 450, 468–469, 482, 531
   element of, 358–362
   negative and positive representation of, 359, 361, 455
   orthogonality 356, 358, 362, 371–372, 455, 540
   reflection property of, 210, 217
   value and significance of, 362–363, 371–372, 425
Elevation,
   conditions at high, 27, 39, 41, 84, 166–167, 169, 235–236, 298–299, 390–391, 396–399, 408
   as an ecological factor, 74, 107, 166–167, 176, 182–184, 188, 202–206, 218–219, 289, 302–304, 408, 540
Eluviation in soil, 138, 536
Emissivity, 212, 536
Endodermis, 127, 141, 536
Energy
   chemical, 156–159
   conservation of, 208
   forms of, in the climate system, 46–47
   free, 121, 123, 130, 142, 537
   kinetic, 120–121, 124
   light, 157
   transfer of, *see* Energy balance
Energy balance, 128, 179, 208–219, 228, 394, 435–436, 536
Energy budget, *see* Energy balance
Environment, 536
   history of, 50, 220
   operational, 48–50
   potential, 50
Environmental factors, 49–51, 127, 219–223, 536
Environmental gradient, 300–304
Environmental impact, 41–43, 219–223, 402–405, 415–420
Enzymes, 119, 157–162
Ephedra, 14
Epidermis, 56, 68, 127, 141, 212, 536
Equally probable classes, 328, 333–335, 536
Equilibrium, 121
Erethizon epixanthum, 222
Error
   in ring-width measurement, 250–252
   statistical, 91, 255–256, 353, 365, 368, 413–414, 433
Estimate, 35, 256, 263, 318–319, 536
   accuracy of, 27, 256
Evaporation, 50, 122, 139, 209–213, 426, *see also* Evapotranspiration
Evaporative stress, 343
Evapotranspiration, 104, 209–213, 301, 328, 346–349, 396, 412, 536
Experimental control, crossdating as an, 20
Explorer's records, 499
Exponential function, 263–265, 267–268, 310, 536

F

Factor scores, *see* Eigenvector, amplitude of
Factors
   biotic, 219–223
   climatic, 23, 46–54, 116, 194, 231–238.
   hereditary, 16, 51, 74, 88, 194, 196, 238–241
   nonclimatic, 17, 23, 53, 116, 219–224, 277, 285, 289, 433
Fagus, 14, 63, 136–137
   grandifolia, 90, 138–139, 345–347
   silvatica, 177, 178, 188
Fat, 159–162
Fatty acid, 161
Feedback relationship, negative, 390
Fibers, 56, 61, 63, 68, 93, 114, 536
Field capacity, 133–136, 536
Field data, 247

558   TREE RINGS AND CLIMATE

Field examination, 248–249
Filter, digital, 36, 268–277, 298, 325, 401, 536
　reciprocal, 269–277
　weights of, 270–271, 325
Fire, 68, 91, 153, 200–201, 221, 229, 400
Fire scar, see Injury
First differences, 254, 269, 329, 332, 537
Fitzroya cupressoides, 14
Flooding, 41–43, 114, 170, 328, 397–398, 400, 415–420
Flowering, 103, 194, 398, 499
Foehn wind, 218, 533, 537
Foliage, relationships involving, 73, 87, 110–111, 177, 188, 198
Food, 142, 151, 156–162, 537
　distribution of, 184–189
　reserves of, 183, 186, 204, 234–235, 396–397, 399, 403
　synthesis of, 156–162
Forecast improvement using reconstructions, 415, 490–491, 500–502
Forest interior sites, 108, 300–302
Forest limits
　lower, 16, 108, 202–203, 298, 300–302, 396
　upper, 16, 202–203, 298
Fourier analysis, 295–300
Fraxinus, 14, 63, 140
　americana, 155
Freezing of the stem, see Stem, freezing of
Freezing and thawing dates of lakes, 499
Frequency, 115–116, 259–260, 268–270, 295–300, 413–414, 436, see also Variance, frequency of
Front in atmosphere, 435, 454, 465, 537
Frost, 64, 68, 91, 167, 328, 343, 397, 471, 492
Frost hardiness, 194, 197–198, 204, 237, 397–399
Frost resistance, see Frost hardiness
Frost ring, see Ring, frost
Fruiting, 103, 194, 260, 398, 499
F statistic, 365, 368, 422–424
Fungi, associated with roots, 141
Fusiform initials, 66, 537

G

Generating function, see Generating process

Generating process, 255, 258, 273, 275, 413–414, 537
Genetic factors, see Factors, hereditary
Geological applications, 114–115, 219–222, 499, see also Dendrogeomorphology
Germination, 194
Gibberellin, 191, 537
Glaciers, 39, 41, 220, 328, 409–410
Glock, Waldo S., 8–9, 332
Glucose, see Sugar
Glycerol, 161
Goodness of fit, 322, 537
Growing season, 68–72, 84–91, 103–107, 203, 205, 232–233, 396
Growth, see also Diurnal expansion and contraction of plant parts
　abrupt change in, 244
　age as a factor in, 73–74, 84, 107–113
　anomaly in, 35–38, 91, 371–374, 421, 427, 491–499, see also Mapped anomalies of growth
　apical, see Growth of shoots
　asymmetric, 115, 153
　of buds, see Growth of shoots
　cessation in, 70–71, 86, 88, 95, 114, 203–205
　cumulative curve of, see Sigmoid curve
　flush in, 12, 70, 72, 192, 195
　grand period of, 70, 186, 537
　initiation of, 58, 63, 65, 69–70, 84–86, 91, 114, 118, 203–205
　interactions involving, see Interaction
　intermittent, 71
　multiple peaks in rate of, 89–91
　radial, 74–107
　rate of, 86–90, 105, 313
　of roots, 69, 93, 113–115, 139, 141–142, 398–399
　of shoots, 70–74, 93, 103–109, 191–194, 230, 398–399
Growth anomaly type, 492–499
Growth function, 112–113, 263–264, 266, 277–281
Growth regulators, 70, 142, 151, 190–193, 195, 230, 397–398, 537
Growth response, diversity in, 401–402, 407–412
Guard cells, 127–128, 537
Guttation, 120, 538

SUBJECT INDEX

H

Harvest logbooks, 499
Heartwood, 62, 538
Heat
 conduction of, 209–213, *see also* Insulation
 convection of, 210–213
 latent, 209–213
 sensible, 209–213
 storage in oceans, 211, 421, 426–427
Hemicellulose, 67, 538
Heredity, *see* Factors, hereditary
Historical records, 27, 499–500
History of forest stand, 110–113, 219–233, 263, 400
Homogeneity, lack of
 in variance of ring-width measurements, 260–266
 test, for climatic data, 252–254
Hormones, 190–193, 396, 538, *see also* Growth regulators
Humidity of the atmosphere, 124, 126, 128–130, 343, 347–349, 399
Hurricane, 435, 465
Hydrolysis, 160, 538
Hydrology, *see* Dendrohydrology
Hydrostatic forces, 76, 81, 124, 142–147, 538
Hypothesis, 25, 224–225, 248, 314, 335–336, 394–395, 403, 425, 538
 testing of, 315, 321, 324, 403, 405–407, 429

I

Illuviation, 138, 538
Independent data, 15, 33, 320–321, 373–374, 428–432, 479–488, 491–500, 538
Index, ring-width, *see* Ring-width indices
Inference, 12, 15, 25, 35, 225, 285, 311, 394–395, 405, 408, 410, 451, 468, 538
Infiltration technique, measuring stomatal opening, 131
Inhibitor, 191, 195, 538
Inhomogeneity of climatic data, *see* Homogeneity, lack of
Injury, 29, 75, 91, 114, 153, 197–201, 219–222, 289, 397–398
Insect, 50, 223, 400

Insulation
 by bark, 68, 200
 by snow, 236, 398
Integration by the tree, 52, 325, 429–430, 488
Interaction, 12, 69, 74, 91, 184–189, 193, 226, 232–235, 242–245, 282, 291, 340–341, 351–352, 382, 392, 397–400, 403, 538
 cancellation by, 329, 345, 393–394, 433
Intercorrelation, 345, 347, 352, 369, 382, 436, *see also* Interaction
Intra-annual growth band, *see* Ring, false; Ring, multiple
Ion, 120, 190, 538
 exchange of, 151
Iron, 158, 190
Isotopes, 151–152

J

Jet stream, 466–467
Juniperus, 12, 14, 72, 196
 communis, 196
 occidentalis, 71
 osteosperma, 153–154, 203, 351–352
 virginiana, 20, 196, 200
Juvenile wood, 191, 538

K

Kinin, 191, 538
Krummholz, 198, 538

L

Laboratory of Tree-Ring Research, 8, 31–33, 247, 249–252, 281, 304, 505
Lagging relationship, 26–27, 39, 183–184, 188, 193–198, 299, 364, 373, 379, 399, 408, 413–414, 419, 437, 450–455, 477–479, 492–493
Lake level, 41–43, 212, 328, 415–420
Landslide, 221, 400
Lapse rate, 218, 538
Larix, 14
 decidua, 192, 199
 occidentalis, 73
Latewood, 60–63, 89, 95, 100, 102, 111–112, 191–193, 230, 345–349, 538, *see also* Mork's definition of latewood
Latitude
 as an ecological factor, 18, 74, 84, 86, 163, 166, 214–215

studies at high, 34, 84, 86, 163, 184, 195, 235–236, 396
Law of Large Numbers, 322–323
Law of Limiting Factors, 15–16, 119, 292
Law of Relative Effects, 119
Leaf, *see* Leaves
Leaf litter, 138, 217
Least squares, 341–342, 538
Leaves, 56, 58, 80, 126–127, 131, 147, 164, 184–185, 298
  age of, 176, 187–189, 299
  efficiency of, 187
  number of, 188
  optical density of, 164
  size of, 188, 299
  temperature of, *see* Temperature
Lenticels, 128, 538
Libocedrus, 14
Liebig's Law of the Minimum, 119
Light, 48, 128, 131–132, 157–158, 162–165, 194–195, 328, 396
  saturation of, 163
Light intensity, minimum required, 165
Lightning, 91, 398
Light tolerance, *see* Light intensity, minimum required
Lignification, 67, 93, 107, 113, 192
Lignin, 67, 113, 159, 538
Limiting factors, 11, 13, 15–16, 51, 183, 193, 225, 300–302, 313, 340, 393, *see also* Law of Limiting Factors
Linearity, 316, 318–319, 329, 338, 351, 400–401, 438
Linkage, strength of climate-growth, 377, 379, 394, 400, 485, *see also* Coupling
Liquidambar styraciflua, 73
Liriodendron, 63
Little Ice Age, 39, 407
Longevity, 29–30, 147–151, 217

M

Magnesium, 158, 190
Manganese, 190
Mapped anomalies
  of climate, 35, 436–438, 446–449, *see also* Climatic anomaly
  of growth, 35–38, 436–438, 441, 444, 446, *see also* Growth, anomaly in
Mass movement, 120, 142–143, 539
Master horizon of a soil, *see* Soil horizon

Matrix, 353, 438, 539
  diagonal, 353, 365, 438, 535
  element of, 353–356
  multiplication of, 354–355
  notation using, 353–356
  scalar used with, 358
  transposition of, 356, 358
Matrix potential, 125, 539
Mean, 23, 29, 255, 271, 322–323, 539
Mean sensitivity, 18, 257–258, 261, 266, 269, 300–310, 432, 539
Membrane, differentially permeable, 121–123, 151, 190, 539
Meridional circulation, 465
Meristem, 57–58, 69, 91, 115, 176, 532, 539
Mesophyll, 56, 539
Metabolism
  energy released by, 160–162, 212–213
  wastes of, 62
Meteorological data, 2, 29, 33, 202–206, 210, 316, 325, 327–329, 434–437, 439–440, 479, 482–483, 499–500
  estimating missing values, 252
  length of record, 401
  local variations in, 48, 252–254, 430, 432
  modern record bias, 501–502
Meteorological factors, *see* Factors, climatic
Microenvironment, *see* Climate, micro-
Micrometeorology, *see* Climate, micro-
Middle lamella, 122, 539
Military fort records, 471–472, 499
Mineral salts, *see* Salts; Nutrient
Missing ring, *see* Ring, partial
Model, 539, *see also* Modeling
  of cambial activity, 66–77, 226–231
  insect infestation effects on growth, 223
  minimum-sized pressure, 468–470, 492
  network, 223–224, 227, 232–236, 242
  ring-width—climate system, 25–27, 182–184, 186, 193, 231–238, 242–245, 437–438
  simulation of biological processes, 224–225, 403–405
  statistical (mathematical), 2, 28, 224, 315–320, 325–326, 341, 413, 418, 421, 430, 438–439, 451
  of streamflow, 413–414
  suitability and limitations of growth model, 242–245
  testing of, 224

SUBJECT INDEX

Modeling
　atmospheric circulation, 11, 29, 501
　biological system, 314–315
　ring-width—climate system, 25–27, 33–35, 223–231, see also Model, ring-width—climate system
　spatial relationships in growth-climate system, 438–455, 468–470, 488
Moisture, see Soil moisture
Molecular motion, 121
Molybdenum, 190
Mork's definition of latewood, 62
Mother cell, 67, 539
Moving average, 266, 268–269, 325, 539
Multinodal growth, 72, 539
Multiple regression, see Regression, multiple
Multiple stems, 72–73
Multiplicative effects, 349, 351, 401
Multivariate statistics, see Statistics, multivariate

N

Needles, see Leaves
Nitrogen, 158, 161–162, 190
Noise, 313, 318, 322–323, 356, 432, 540
Nonhomogeneity, see Homogeneity, lack of
Nonrandomness, see specific types; Randomness in time series
Normal distribution, 255, 333–334, 342, 540
Normalized data, 333–335, 338, 359, 540
North Pacific High, 425, 463, 493
Nucleus, 122–123
Number matching, see Chance variations
Nutrient, 189–190, 217, 231, 540

O

Old age, see Longevity; Age of tree
Opuntia, 140
Orographic factor, see Elevation
Osmosis, 123, 141, 540
Osmotic potential, 124–125, 141, 540
Oxidation, 160–161, 540
Oxidation-reduction reaction, 160–161, 540
Oxygen
　atmospheric, 157–158, 161, 176
　of soil, 139, 142, 151, 155, 170, 176, 397–398
Ozone, 176

P

Paraboloid shape of growth layer, 57, 59
Parenchyma, 56, 58, 63, 68, 540
Parent material, 136, 138, 540
Partial regression coefficient, see Regression, coefficient of
Permanent wilting point, 134–135, 137, 149, 171, 175, 540
Permeability, 122, 142, 190
Persistence in time series, 258–260, 296, 324, 329, 492
Phellem, 68, 540
Phenology, 82–83, 85, 89, 96–101, 103–107, 195, 540
Phenotype, 16, 541
Phloem, 56–60, 66, 68, 80–81, 120, 141, 152–155, 192, 541
Phosphorus, 190
Photoperiod, 194–195, 541
Photorespiration, 162, 165, 541
Photosynthesis, 50, 103, 131–132, 157, 162–182, 203–206, 213, 223, 232–238, 298–299, 396–397, 541
　annual regime, 178–182
　efficiency of, 103, 176–177
　gross, 162, 168–169, 537
　net, 162, 168–169, 172–175, 179–182, 540
　in winter, 166, 179–181, 202–206, 215, 234, 380, 399
Physiological processes, 51, 119–120, 182–184, 202–206, 223–238, 393–400
Physiological adjustment, 171
Physiological seasons, see Seasons, physiological
Picea, 14, 62, 86, 196
　excelsa, 342
　glauca, 41, 199, 279, 328, 415–420
　rubens, 73
　sitchensis, 73–74
Pinus, 14, 72, 103, 112, 127, 196
　aristata, 22, 238–241, 391
　banksiana, 73, 177, 279
　cembra, 13, 41, 166–167, 328
　contorta, 71
　echinata, 72–73
　edulis, 72, 83, 86–87, 150, 189, 197, 203, 216, 222, 238–240, 274, 292–293, 351–352, 388–390, 404, 431
　flexilis, 153, 186, 203, 238–240, 303–304, 389, 431

halepensis, 72, 196, 327, 343, 380–381
jeffreyi, 44, 389
longaeva, 9, 22, 39, 72, 85–86, 104–107, 115–116, 147, 153, 166, 169–172, 176–177, 186–188, 190, 198, 203, 205, 216–218, 231, 238–240, 275, 282, 284, 286–288, 290–292, 298–299, 302–304, 320, 327, 378–380, 390–391, 399, 408–410
monophylla, 238–240, 303
monticola, 73
ponderosa, 7, 62, 73–74, 78–79, 84–86, 92, 94–103, 108–109, 126, 129, 131–132, 143–147, 165, 167–168, 172–175, 178–182, 197, 200, 203, 230, 233, 238–240, 273–275, 278–279, 304–310, 365–370, 377–379, 388–390, 395, 404, 431
radiata, 16, 177
resinosa, 112, 192
strobus, 63, 73
sylvestris, 13, 64, 116, 200–201, 381–382
taeda, 72–73, 88, 140, 164
Pistacia, 14
Pith, 58–59, 306, 541
Pits in xylem cells, 61, 68
Plastid, 122–123, 541
Podocarpus, 14
Pollen
 formation of, 194
 shedding of, 83, 103–107
Pollution, environmental, 43, 53, 151, 176, 190, 211, 221–222, 295, 402–405
Polyethylene enclosure, *see* Tent, polyethylene
Polynomial function, 263–265, 268, 280–281
Populus, 14, 63
 deltoides, 56
 tremuloides, 73
Porcupine, *see* Erethizon epixanthum
Potassium, 190
Power spectrum analysis, 36, 295–300, 401, 413–414, 541
Precipitation, 7, 34, 41, 48, 81, 91, 97, 104–106, 131, 134–135, 149, 181–183, 209, 218, 231–238, 295, 327–328, 338, 340, 342–343, 346–352, 367–370, 373, 377–391, 396–400, 402–405, 412, 426–427, 457–468

anomaly in, 35, 43, 373–374, 407–410, 453–454, 459–460, 464, 467, 479–482
correlations with, 19, 90, 232–237, 273, 275, 298, 408
data, 252–254, 314, 333–335, 356–357, 363, 428, 430–433
modeling effects of, *see* Model, ring-width–climate system
reconstruction of, 428–433
as snow, 216, 218, 228, 236–237, 302–304, 396, 398, 436
used in verification, 479–482
Preconditioning, 115, 184, 193–198, 233, 243–244, 380, 399
Predictand, 317–318, 345, 373, 428, 541
Prediction, 317, 541, *see also* Forecast improvement using reconstructions
Predictor, 317–318, 320, 345, 353, 363, 373, 428, 541
Preparation of tree-ring specimens, 31–32, 249
Pressure anomaly types, 455–468
 deriving, 457
 reconstruction and analysis of, 468–476, 488–491, 501–502
 verification, 477–488, 491–500
Pressure, atmospheric, 35, 425, 434–437, 439–440
 anomalies in, 35, 410–411, 443, 446–449, 456, 459–460, 462, 464, 467, 471–472
 reconstructions, 446–454, 471–472
 ridge, 410–411, 425, 458, 461, 466
 700 mb surface, 410–411
 trough, 410, 425, 461, 466–467
Pressure potential, 124, 541
Principal component, *see* Eigenvector
Principle, 14, 225, 541
Probability, 335, 338–340
Processing by computer, 32–33, 252, 264, 267, 281, 311
Product mean, 331–332, 431, 483, 485–486, 541
Protein, 159–162
Protoplasm, 122–123, 541
Proxy records, 4, 6, 27, 29, 499–500, 541
Pruning of branches, 91, 110
Pseudotsuga, 14
 menziesii, 69, 71–73, 83, 103, 177, 194, 199, 203, 216, 238–240, 260–262, 267,

SUBJECT INDEX

271–275, 297, 351–352, 388–390, 404–406, 412–413, 431
Purshia, 14

Q

Quasi-biennial oscillation, 299
Quercus, 14, 63, 88, 90, 103, 327
 alba, 82, 89, 164, 250–251, 345, 347–349, 380–381, 515
 ellipsoidalis, 90
 prinus, 90
 robur, 192
 rubra, 82, 89, 164, 295, 343

R

Radiation
 diffuse, 209–211
 down scatter of, 209–211
 longwave, 209–211, 228
 solar, 50, 104, 121, 128, 143–146, 173–175, 179, 209–215, 218, 228, 328, 349, 398, 501
Randomness in time series, 258–259, 323–325, 435–436
Rays in wood, 61–63, 66, 67, 546
RE, *see* Reduction of error
Reaction wood, *see* Compression wood; Tension wood
Reconstruction, 27–28, 33, 53, 318–320, 338–340, 405–433, 445–450, 470–477, 483–484, 488–491, 542
 by averaging arbitrary intervals of time, 473
 difficulties with summer climate, 491
 improvement using verification data, 491–500
 overestimation in, 489
 temporal accuracy of, 500–501
 of transition periods, 492–496
 use of, 29, 414–415, 420, 428, 500–505
Reduction, 160, 542
Reduction of error (RE), 332–333, 429, 432, 483–485, 542
Redundancy in variables, 372, 438
Regression, 34, 78, 258, 281, 294, 542
 canonical, *see* Canonical analysis
 coefficient of, 324, 341–349, 355, 364, 366, 368–370, 419, 425
 multiple, 324, 341–353, 363–371, 373, 418, 422–423, 540, *see also* Correlation, multiple
 stepwise multiple, 34, 346–350, 363–371, 425, 428–430
 variance loss in, 414
Repetition, *see* Replication
Replication, 23, 31–32, 248–249, 256–257, 289, 293, 402
Reradiation, *see* Radiation, longwave
Residual, in statistics, 318, 322, 333, 344–345, 356, 365, 542
Resin duct, 13, 59, 61–63, 127, 542
Respiration, 50, 158, 162–178, 213, 227–228, 380, 396, 399, 403, 542
 aerobic, 158, 531
 anaerobic, 158, 531
 measurement of, 162–163
 in winter, 166, 168, 396
Response function, 318, 340–341, 363–370, 428, 542
 compared to regression, 368–370
 direct precipitation relationships in, 397–398
 direct temperature relationships in, 395–396
 element of, 364–369, 377–383, 391–393
 examples of, 377–391
 frequency of positive and negative relationships, 391–393
 interpretation of, 393–400
 inverse precipitation relationships in, 398
 inverse temperature relationships, in, 396–397
 limitations of, 391–393, 400–401
 mean for samples of same species, 241
 seasonal difference in, 391–402
 the significance of, 400–402, 405
 of streamflow, 412
 variance reduced by, 400
 variations in, 382–393, 401–402, 408
Restitution, 197
Retrodiction, 317, 542, *see also* Reconstruction
Ring
 absent, *see* Ring, partial
 boundary of, 12–13, 20, 62–63, 93
 counting, as opposed to dating, 1, 219, 249–250
 double, *see* Ring, multiple

false, 20–21, 59, 62, 93–95, 98–103, 192–193, 250, 305–310, 536
frost, 64, 198–200, 243, 537, *see also* Injury
intra-annual, *see* Ring, multiple; Ring, false
marking and preparation of, 31–32
missing, *see* Ring, partial
multiple, 13, 20, 62, 72, 100, 305–310
partial, 2, 13, 21, 31, 249–250, 300–310
in roots, 113–115
Ring characteristics, synchronous variations in, *see* Crossdating
Ring-porous woods, 61, 63, 65, 88, 93, 103, 112, 114, 191, 543
Ring width
as a function of age, 24–25, 33, 107–113, 116, 149, 188, 259–261, 263–267, 277–281, 304–311
measurement of, 32, 250, 289
variations in, 2, 6, 17, 19, 23, 25, 30, 35, 53, 74–75, 80, 139, 190, 200–201, 216, 239, 261–270, 300–302
Ring-width chronology, *see* Chronology
Ring-width indices, 24–25, 33, 35, 266–268, 281, 543
Root, 58, 69, 80, 93, 113–115, 134–143, 151–152, 170
distribution of, 137–139, 141, 148, 196, 396
flare of, 109
grafts in, 153–155
Root cap, 58, 543
Root hair, 141
Root/shoot ratio, 142, 148–149, 186, 543
Rotation of axis, 358
Running mean, *see* Moving average
Runoff, 149, 210

S

Salts, *see also* Nutrient
in soil, 142, 198
uptake of, 151–152
Sample size
for ring collections, 23, 31–32, 256–257, 290–293, 322–323
statistical, 321–325
Sap tension in xylem, 78
Sapwood, 62, 543
Schulman, Edmund, 8–9, 408

Sea ice, 499
Seasons
climatic, 214–215, 393–400, 410, 421, 440, 450
physiological, 202–206, 437
Seedling, 2, 12, 57–58
Selection of sample, 323
Sensitivity, 18–19, 29, 148–151, 216, 258, 295, 300–303, 543
Sensitivity, mean, *see* Mean sensitivity
Sequoia, 14
  sempervirens, 154
Sequoiadendron giganteum, 14, 16
Serial correlation, *see* Autocorrelation
Shade tolerance, 165, 197
Ships' logs, 499
Shrinkage, *see* Diurnal expansion and contraction of plant parts
Sieve tubes, 68, 543
Sigmoid curve, 70–73, 87–88, 543
Sign test, 329–331, 431, 483, 485–486, 543
table of significance levels, 330
Signal, 313, 322, 543
climatic, 322
Signal-to-noise ratio, 313, 322
Significance, statistical, 258, 324–325, 342–345, 365–366, 368, 387–388
Site factors, 213–223, 300–304
Site selection, 17–19, 29–31, 402, 543
Site variation as an ecological factor, 112, 116, 213–223, 289, 300–304
Size, *see* Sample size
Slope
as a factor, 107, 115, 139, 183, 214–216, 289, 343
degradation, measurement of, 115
exposure of, 107, 214–216, 289
Smoothing, 269–272, 325, 328
Snow, *see* Precipitation, as snow
Snow slides, 221
Soil, *see also* Substrate
air in, 136–139, 237, 417, 419
available water in, 133–135
carbon dioxide in, 139
clay in, 134–139
dolomitic, 217, 231
drainage, 136–139, 398, 417, 419
effects upon ring width, 289
granite, 217
gravel, 135
loam in, 134–135

oxygen in, *see* Oxygen, of soil
percolation of water into, 120, 135, 139
sand in, 134–139
sandstone, 217
silt in, 134–138
unavailable water in, 133–134
Soil-forming factors, 136, 138–139
Soil horizon, 136–139, 543
Soil moisture, 7, 50, 97–99, 101, 104–105, 133–137, 143–149, 170–175, 178–182, 208–209, 217, 300–302, 328, 346–349, 380, 397–398, 400
  constants, 133–138
  gravitational water, 133–135
  movement of, 135–136, 396
  saturation of, 134, 170
Soil moisture accounting system, 209–210
Soil porosity, 133, 135–139
Soil profile, 136–139, 217, 543
  Soil system, 136–139
  Soil temperature, *see* Temperature of soil
Soil texture, 134–139, 217
Solar constant, 210, 214
Solubility, 160
Solum, 138, 543
Spatial coherence in climate, 435–436
Spatial patterns
  in climate, 35, 433, 435–437, 446–449, 455–503
  in growth, 35, 428–430, 433, 437, 441, 444, 446, 459–460, 464, 467, 491–499
Specific gravity, 111–112, 114
Specimen, preparation of, *see* Preparation of tree-ring specimens
Spike top tree, 30, 148, 543
Spiral-grained wood, 153
Springwood, *see* Earlywood
Standard deviation, 35, 38, 255, 268, 300–301, 324, 333–334, 543
Standard error, 256, 290–293, 342, 365, 544
Standardization, 24–25, 33, 116, 219, 260–268, 283
  significance of, 266–267, 544
Standardized ring-width chronology, *see* Ring-width indices
Standard normal variates, *see* Normalized data
Starch, 159–161
Starvation, 168

Statistical significance, *see* Significance, statistical
Statistical analysis, examples of, 300–310
Statistical universe, 256, 290–291, 323, 544
Statistics, 544
  excessive reliance upon, 250, 321–322
  general use of, 33–36, 246, 254–261, 268, 277, 300–302, 310–311, 321–325
  improvement of by lengthening the record through reconstruction, 29, 414–415, 420, 500–502
  misuse of, *see* Statistics, role of
  multivariate, 34–36, 224, 326, 329, 340–375, 382–384, 402, 438–455
  nonparametric procedures, 329–331
  power of, 35–36, 311, 321–322, 325–326, 439
  role of, 321–323
Stem
  circumferencial variation, 72–73, 109, 111, 114, 201, 305
  directions in, 60–61
  freezing of, 167, 197–200, 228
  girdle of, 153
  height, growth variations as a function of, 25, 87, 102, 109–113, 192–193, 277–279, 305–310
  lean of, 113, 220–221, 289–290, 420
  scar in, *see* Injury
  strip-bark, 30, 115, 154, 186–187
Stem form, slab-shaped, 153
Stomatal movement, 128, 131–133, 160, 165, 170–171, 173–175, 181
  mechanism affecting, 170–171
Stomate, 127–129, 544
  size measurement, 131
Storms, 421, 426–427, 434–436, 491–499
Strategy of sampling, 247–249, 292
Stratification and testing of samples, 225
Streamflow, 40–41, 319, 412–415
Structure, *see* Wood anatomy
Suberin, 68, 122, 141–142, 159, 196, 544
Substomatal cavity, 127
Substrate, 216–217, 302–304, *see also* Soil
Subtropics, *see* Tropics
Sugar, 157–162, 177, 184–185
Sulfur, 161–162, 190
Sulfur dioxide, 176
Summerwood, *see* Latewood
Sunshine, *see* Radiation, solar
Sunspots, 4, 41, 300, 501

Superabundance of a factor, 119–120
Suppression, 86, 111, 544
System
  atmospheric, 46–48, 208–213, 434–437
  biological, 25–27, 46–54, 224–241, 425–428

T

Talus slope, 114–115, 148, 544
Taxodium, 14
Temperature, 18, 39, 41, 46, 74, 76, 180, 327, 333–335, 338, 343, 345–352, 367–370, 377–391, 395–400, 402–405, 426–433, 457–468
  above optimum, 169, 205, 396
  absolute, 121
  as affecting cambial activity, 197, 227–229
  anomaly in, 408–410, 479–482, 485–488
  as data, 252–254, 314, 356–357, 363, 412, 428
  day and night averages of, 97, 104–107
  departure from that of air, 128–133, 171, 179–180, 212–213, 218, 228, 396, 399
  direct correlation with, 204, 216, 228, 235–237, 382, 391, 393–396, 408, 491–492
  effects on biological processes, 121, 129–133, 144–146, 155, 157–158, 165–170, 182–184, 195
  in energy balance, 208–213
  freezing effects, 166–167, 197–200
  inverse correlation with, 204, 232–235, 396, 408
  maximum, 180, 202–206, 342
  minimum, 70, 180, 197, 202–206
  modeling effects of, *see* Model, ring-width–climate system
  reconstruction of, 428–433, 482–488
  sea surface, 211–212, 421–427, 436
  of soil, 134, 142, 151, 217–218, 228, 396
  thresholds of, 182, 202–206, 327, 342
  upper limits of, 168, 200–201, 229
  used for verification, 470–472, 479–488, 500
Tension in water columns, 142–143
Tension wood, 113, 220–221, 544
Tent, polyethylene, 85, 131–132, 143, 172–175, 179
Thinning, 86, 112

Thuja, 14
Thunderstorms, 435
Thunnus alalunga, 425–428
Tilting of stem, *see* Stem, lean of
Time, as a statical variable, 344
Tissues, 56–58
Topography as an ecological factor, 136, 214–216, 302–304
Tornado, 435
Toxic substances, 158
Tracheids, 60, 67, 92–95, 191, 544
Transfer function, 35, 318, 371–375, 402, 408, 412–433, 438–439, 451, 544
  elements of, 418–420, 426–427
  interpretation of, 418–420
Transformation, mathematical or statistical
  by calibration, 314, 316
  logarithmic, 263, 280–281, 318, 328, 331, 343, 351
  power function, 318
  reciprocal, 318
  square root, 318
Transient factors, 80, 244
Translocation, 68, 143, 152–155, 170, 398, 544
Transmissivity, 212, 545
Transpiration, 80–81, 127–133, 140, 143–147, 173–175, 545, *see also* Evapotranspiration; Water, loss of
Tree form, 29–30, 72, 147–148
Tree line
  arid, *see* Forest limits, lower
  high elevation, 39, 41, 169, 198, 280, 298, 381
  polar, 41, 86, 280, 381, 491
Trend, 258–260, 262–266, 295–297, 329, 545
Tropics, 12, 71, 84, 95–96, 168, 195, 504
t statistic, 331–332, 344
Tsuga, 14
  canadensis, 63–64
  heterophylla, 71
  mertensiana, 71
Tuna, albacore, *see Thunnus alalunga*
Turgidity, relative, 181
Turgor, 124–126, 141, 545

U

Ulmus, 14, 63, 72

SUBJECT INDEX

Uniformitarian principle, 14–15, 410
Uninodal stem, 72, 545
United States Geological Survey, 250, 252

V

Vacuole, 122–123, 125, 545
Vapor pressure, 349
Vapor pressure deficit, 126–127, 545
Variance, 23, 46, 255–256, 259–260, 266–268, 271–272, 277, 295–300, 356, 439–440, 450, 545
   accounted for, see Variance, reduced
   analysis of, 249, 282–293, 302–304, 323
   calibrated, see Variance, reduced
   frequency of, 295–299
   high-frequency, 260, 268–277, 298, 538
   low-frequency, 259–260, 262, 268–270, 295–300, 408, 413
   ratio of high-pass to low-pass, 277
   reduced, 314, 322, 342, 345, 358, 360, 362, 365, 371–372, 422, 431–432, 451–454, 469, 545
   sources of, 282–290
   within tree stem, 304–310
Variance components, 283–295, 302–304, 533
Vascular tissue, see Xylem; Phloem
Verification, 545
   importance of, 321
   of reconstructions, 27–28, 33, 320–321, 407, 410, 412, 414–415, 420, 424, 428–433, 477–488, 491–500
   of response functions, 395
   statistics used for, 329–333
Vessel, 61, 63, 88, 93, 114, 545
Vintage dates, 499
Viscosity, 64, 142, 545
Volcanic eruption, 221, 470, 501
V^2 stastistic, 469

W

Water
   absorption of, 80, 90, 123–127, 134–136, 140–143, 171, 398
   affecting cambial activity directly, 229–230
   loss of, 50, 80, 126–128, 140, 144–145, 173–175, 213, see also Transpiration
   as a physiological factor, 170–175, 229–230
Water balance, 131, 136, 208–213, 379–380, 394, 399, 545
Water columns, breakage of, see Tension in water columns
Water deficit, 70, 80–81, 170–171, 175, 229–230, see also Water stress
Water level, see Lake level
Water potential, 123–128, 133–136, 140–142, 170, 175, 197, 229, 545
Water status, see Water balance
Water stress, 34, 77, 93, 147–148, 171, 175, 196–198, 202–206, 213, 229–230, 232–235, 378–379, 396–397, 403, see also Water deficit
Water table, 139
Weather, 46–48, 434–437, 545
Weather records, see Meteorological data
Weight, see Coefficient
Wilting of plant, 134, 171
Wilting point, see Permanent wilting point
Wind, 48, 128–130, 179, 198, 212–213, 216, 228, 302–303, 396, 398, see also Foehn wind
   direction of, 435, 491–500
Winter hardening, see Frost hardiness
Wood anatomy, 55–63
Wood density, 43–46, 74, 103, 111–113, 117, 191, 230, 278
Wood rays, see Rays in wood
Wound, see Injury

X

X-ray densitometry, see Densitometry
Xylem, 56–63, 65–66, 67, 75, 80–81, 87, 105, 140–143, 152, 192, 546
Xylem sheath, 12, 57–58, 110–113

Y

Year without a summer, the, 470–471

Z

Zinc, 190
Zygophyllum, 14